Rethinking Cancer

Vienna Series in Theoretical Biology
Gerd B. Müller, editor-in-chief
Thomas Pradeu and Katrin Schäfer, associate editors

The Evolution of Cognition, edited by Cecilia Heyes and Ludwig Huber, 2000

Origination of Organismal Form, edited by Gerd B. Müller and Stuart A. Newman, 2003

Environment, Development, and Evolution, edited by Brian K. Hall, Roy D. Pearson, and Gerd B. Müller, 2004

Evolution of Communication Systems, edited by D. Kimbrough Oller and Ulrike Griebel, 2004

Modularity: Understanding the Development and Evolution of Natural Complex Systems, edited by Werner Callebaut and Diego Rasskin-Gutman, 2005

Compositional Evolution: The Impact of Sex, Symbiosis, and Modularity on the Gradualist Framework of Evolution, by Richard A. Watson, 2006

Biological Emergences: Evolution by Natural Experiment, by Robert G. B. Reid, 2007

Modeling Biology: Structure, Behaviors, Evolution, edited by Manfred D. Laubichler and Gerd B. Müller, 2007

Evolution of Communicative Flexibility, edited by Kimbrough D. Oller and Ulrike Griebel, 2008

Functions in Biological and Artificial Worlds, edited by Ulrich Krohs and Peter Kroes, 2009

Cognitive Biology, edited by Luca Tommasi, Mary A. Peterson and Lynn Nadel, 2009

Innovation in Cultural Systems, edited by Michael J. O'Brien and Stephen J. Shennan, 2009

The Major Transitions in Evolution Revisited, edited by Brett Calcott and Kim Sterelny, 2011

Transformations of Lamarckism, edited by Snait B. Gissis and Eva Jablonka, 2011

Convergent Evolution: Limited Forms Most Beautiful, by George McGhee, 2011

From Groups to Individuals, edited by Frédéric Bouchard and Philippe Huneman, 2013

Developing Scaffolds in Evolution, Culture, and Cognition, edited by Linnda R. Caporael, James Griesemer, and William C. Wimsatt, 2013

Multicellularity: Origins and Evolution, edited by Karl J. Niklas and Stuart A. Newman, 2016

Vivarium: Experimental, Quantitative, and Theoretical Biology at Vienna's Biologische Versuchsanstalt, edited by Gerd B. Müller, 2017

Landscapes of Collectivity in the Life Sciences, edited by Snait B. Gissis, Ehud Lamm, and Ayelet Shavit, 2017

Rethinking Human Evolution, edited by Jeffrey H. Schwartz, 2018

Rethinking Cancer

A New Paradigm for the Postgenomics Era

Edited by Bernhard Strauss, Marta Bertolaso, Ingemar Ernberg, and Mina J. Bissell

The MIT Press
Cambridge, Massachusetts
London, England

This book was set in Times New Roman by Westchester Publishing Services. Printed and bound in the United States of America.

Library of Congress Cataloging-in-Publication Data

Names: Strauss, Bernhard, editor. | Bertolaso, Marta, editor. | Ernberg, Ingemar, 1948- editor. | Bissell, Mina, editor.
Title: Rethinking cancer : a new paradigm for the postgenomics era / edited by Bernhard Strauss, Marta Bertolaso, Ingemar Ernberg, and Mina J. Bissell.
Other titles: Vienna series in theoretical biology.
Description: Cambridge, Massachusetts : The MIT Press, [2021] | Series: Vienna series in theoretical biology | Includes bibliographical references and index.
Identifiers: LCCN 2020029807 | ISBN 9780262045216 (hardcover)
Subjects: MESH: Neoplasms--genetics | Neoplasms--therapy
Classification: LCC RC268.4 | NLM QZ 210 | DDC 616.99/4042--dc23
LC record available at https://lccn.loc.gov/2020029807

10 9 8 7 6 5 4 3 2 1

Contents

Series Foreword

Biology is a leading science in this century. As in all other sciences, progress in biology depends on the interrelations between empirical research, theory building, modeling, and societal context. But whereas molecular and experimental biology have evolved dramatically in recent years, generating a flood of highly detailed data, the integration of these results into useful theoretical frameworks has lagged behind. Driven largely by pragmatic and technical considerations, research in biology continues to be less guided by theory than seems indicated. By promoting the formulation and discussion of new theoretical concepts in the biosciences, this series intends to help fill important gaps in our understanding of some of the major open questions of biology, such as the origin and organization of organismal form, the relationship between development and evolution, and the biological bases of cognition and mind. Theoretical biology has important roots in the experimental tradition of early twentieth-century Vienna. Paul Weiss and Ludwig von Bertalanffy were among the first to use the term *theoretical biology* in its modern sense. In their understanding, the subject was not limited to mathematical formalization, as is often the case today, but extended to the conceptual foundations of biology. It is this commitment to a comprehensive and cross-disciplinary integration of theoretical concepts that the Vienna Series intends to emphasize. Today, theoretical biology has genetic, developmental, and evolutionary components, the central connective themes in modern biology, but it also includes relevant aspects of computational or systems biology and extends to the naturalistic philosophy of sciences. The Vienna Series grew out of theory-oriented workshops organized by the KLI, an international institute for the advanced study of natural complex systems. The KLI fosters research projects, workshops, book projects, and the journal *Biological Theory*, all devoted to aspects of theoretical biology, with an emphasis on—but not restriction to—integrating the developmental, evolutionary, and cognitive sciences. The series editors welcome suggestions for book projects in these domains.

Gerd B. Müller, Thomas Pradeu, and Katrin Schäfer

Preface

This book is published at an important juncture in the history of cancer research. Never before have we known so much about the individual cancer cell, yet never before has it been so unclear how to translate this knowledge into treatment success. This book is also published over a year into the global COVID-19 pandemic. Apart from its many other devastating consequences, the pandemic has caused many millions of cancer patients to have not been treated or diagnosed. Moreover, cancer research spending has dropped significantly. In October 2020 the UK's National Cancer Research Institute released figures projecting a 24 percent drop in the UK's overall cancer research spending, driven by a 46 percent fall in charity sector funding. The impact of the pandemic on cancer patients and cancer research will be felt for years to come, and will make it all the more important to determine what to focus on with the funds available.

Great technological progress over the past four decades has enabled earlier diagnosis, better surgery, disease monitoring, and follow-up, and it just begins to show in the cancer survival statistics. What is still hard to show is any significant extension of life span after treatment of late-stage disease, the real measure of our ability to effectively cure cancer. This is urgently needed, however, in the face of a worldwide, rapidly increasing cancer incidence. It seems that we are still waiting for the progress that was promised at the time of the "genomics revolution" by the sequencing of the first human genome twenty years ago.

The early 2000s were a time of great optimism in biomedical research, as it was generally assumed that once we know every single human gene, applications would be easy to engineer, and tangible benefits for human health would inevitably follow. But cures based on this "complete" knowledge of our genetic blueprint have remained largely elusive. Targeted therapy, as in precision cancer medicine (PCM), is still only applicable to small subsets of patients, and treatment outcomes are often not as expected. Cancer immunotherapy, after fifty years of basic research, could finally be translated into clinical practice during the last decade, but is so far successfully applicable to only a few types of cancer. Over the same time span, precise manipulation of the genome has become even easier than anybody would have thought back then. In addition, novel computational methods have enabled in-depth analysis of vast amounts of genomics data, be it at the single cell or tumor level, or in large cohorts of cancer patients, with the aim to uncover the genes and molecular pathways that cause cancer. The genomics era was characterized by a sense of relief, as one was under the impression that finally the protocol for understanding and manipulating life had truly arrived. It appeared that with a few minor technical optimizations, any problem in biology, including in humans, would become solvable, at least "in principle."

In the second decade of the twenty-first century, however, it has become clear that simple correlations between genes, mutations, and cancer that have diagnostic or therapeutic value are not to be found exclusively at the DNA sequence level. It seemed that we had reached "peak genomics." Even among cancer systems biologists, consensus started to build that understanding cancer as a perturbation in a complex multimodal, molecular network will not lead to straightforward actionable treatments, despite impressive recent advancements in computational powers and single-cell analysis methods. What these approaches have uncovered instead is enormous heterogeneity at the genomic level, often presented as "complexity": not only between different cancers but also of the same cancer type in different patients, and even between the individual cancer cells within a single tumor in one patient. This ubiquitous observation has led to the declaration of a "complexity crisis" in the cancer genomics field. On the one hand, this admission has relativized the significance of the large amounts of cancer data that have been accumulated and was often used to explain the failures of new cancer drugs in the clinic; on the other hand, the implicit understanding was that doubling down on the acquisition of DNA sequence data and on throughput of analysis (using artificial intelligence) will lead to important breakthroughs in the foreseeable future.

Despite the persistence of the central causal narrative in cancer biology, which holds that cancer is caused by mutations in certain genes, many researchers have begun to doubt that DNA-level information is sufficient for understanding the cancer phenotype and have moved on to cancer epigenomics and other kinds of -omics. Hence, there is now widespread commentary about the arrival of the "postgenomics era" in cancer research. Originally, this term had an optimistic connotation, supposed to mean that things will be easier from then on, say with the ability to personalize treatments, and to tailor therapies to the exact specifications of a patient's disease. Although indeed technically achievable, it now appears very unlikely that these approaches can ever become standard of care due to their enormous complexities and thus inevitable diagnostic and therapeutic uncertainties, not to mention the high costs of the methodologies involved. Meanwhile, data on the success rates of "targeted" drugs based on genomic information show that these drugs have overall been far less successful in the clinic than expected.

Perhaps, "postgenomics" really is meant to announce a reboot: we have tried genomics, with only modest success in curing cancer, and now need to move on to something else—but where to? This is what this book is about.

Applying loosely a historical framework as presented by Thomas S. Kuhn in his 1962 book *The Structure of Scientific Revolutions*, it appears that an increasing number of scientists would agree today that the current consensus that defines what is considered "normal science" in cancer research, or its contemporary scientific "paradigm" in Kuhn's terminology, has been insufficient to answer basic questions about carcinogenesis.

Emerging from this wider historic perspective and from the very concrete results of our own scientific work and that of others, the following two premises motivated the creation of this volume:

1) The current paradigm, namely that a number of specific genes, when mutated or mis-regulated, cause cancer, has not by itself led to a cure for cancer—a failure clearly not due to lack of financial investments or intellectual effort. Therefore, a new theoretical frame-work for causally understanding and treating cancer is required. (We are not criticizing the general understanding of how genes function and their causal role in biology.)

What went wrong? We believe that first and foremost, we have applied an incomplete or incorrect theoretical framework in our attempts to explain carcinogenesis. This concerns specifically how a simplistic understanding of the causal role of individual genes has been applied to cancer.

2) Several lines of evidence, supported by comprehensive data over the past two decades, can be identified that challenge the current paradigm. These are now converging toward a more widely accepted systems view of cancer and are presented in this volume along different "dimensions" of cancer. This view has, however, not yet led to a change in research practices or to fundamentally new experimental approaches in mainstream cancer research.

At the core of this premise is the understanding that models of linear causation, based on single (mutated) genes or networks thereof, "in principle" cannot explain the cancer phenotype and therefore cannot be used to formulate a cure. This view is now increasingly supported by the very data that were initially collected with the aim to find simple answers. However, the logical structure of most current cancer research efforts still appears to follow a mind-set that wants to find *the* few most relevant cancer genes for a given tumor or *the* corresponding drug that targets such genes in a precise manner—while it is becoming more and more obvious that cancer is certainly more complex than that. To change this mind-set, we believe, requires active concerted efforts in order to translate alternative concepts into scientific practice—instead of waiting for "linear" science to take its course, while hoping that the relevant breakthroughs will emerge eventually "anyway." From the patients' perspective, that is certainly not good enough value for research money.

This volume aims to reemphasize the point that it is primarily novel conceptual or theo-retical thinking that is required to drive progress in cancer research. Ultimately, only a clearly detectable change in research practices and funding policies will tell whether new thinking has arrived. New thinking becomes particularly important as currently, an increasing number of formerly "solid" fundamental concepts that are still constituting elements of the current paradigm, such as "oncogene," "clonal expansion," "tumor suppressor gene," and "driver mutation," to name a few, are becoming increasingly "softened" and attached with various disclaimers regarding their explanatory power, often by the very scientists who introduced them earlier. It seems therefore rather surprising that this has not led to a frantic search for additional conceptual building blocks, if not new frameworks.

This highlights also one central issue with theoretical thinking in cancer research: the fact that novel concepts emerge mainly in the basic sciences where they can be dynamic and evolving, but cancer clinicians remain suspended in the tension between their own empirical insights based on clinical practice and the concepts from the basic sciences they apply (usually with some delay) to the cancer context in humans. When clinical outcomes are not in agreement with the prevailing paradigm, for instance, when no plausible "cancer mutations" are found in a tumor, or rationally designed, target-selective drugs do not work as expected or even make the tumor more aggressive, then clinicians usually assume that the principles established in cell culture, animal models, or small cohorts may not apply, because humans might be just too complex and too diverse. They would certainly not question the scientific foundations that guide clinical practice, let alone assume flaws therein that needed addressing. Despite the fact that the number of clinician-scientists (MD-PhDs) is increasing, and integration between clinical and basic sciences is steadily improving, it seems by now quite clear that beyond such logistical advancements, theory-driven cancer biology needs to lead the way with conceptual innovation.

Over the past four decades, the dedication to explain cancer exclusively from the gene level up has been so all-encompassing that even trying to conceptualize alternative ideas has become difficult for at least two generations of cancer scientists. It has also discouraged research into other causally relevant processes beyond the single cancer cell level, depriving us of not only the theoretical but also the experimental tools to study them. But where do we start if we are to create a new theory framework, and what would its conceptual building blocks be?

Here we have gathered contributions from a number of theory-minded cancer scientists who present ideas and results that are covering many different aspects of cancer but are united by the view that it is paramount to revise the current somatic mutation paradigm if we are to make more progress in finding a cure for cancer. Their thinking comes from different areas of basic cancer cell biology, clinical research, and theoretical investigation, but they share a systemic and dynamic understanding of cancer that goes well beyond the idea that specific mutations in certain genes cause cancer and that assembling them into linear schemes of causation would be sufficient to explain carcinogenesis.

We are aware that this volume had to remain incomplete, as many colleagues who have contributed over the years with their work and theoretical thinking to a possibly emerging new paradigm could not be included here because of time and space constraints. In particular, the areas of inflammation and cancer immunology, two of the most dynamic fields in recent cancer research, are not explicitly represented, although references to relevant findings in these areas are made repeatedly by contributing authors throughout the book.

We believe that not only has sufficient solid evidence accumulated to warrant a change of the current paradigm based on scientific reasoning, but also that over the past decade the readiness for change has increased within the scientific community—despite the fact that funding agencies and mainstream research efforts still largely adhere to outdated concepts.

In this volume, we do not argue yet another critique of current research practices as this has been done by some of the authors and others in the past. Instead, we wish to present a number of conceptual stepping-stones that should lead the reader to a new vantage point from where a coherent new theory framework for cancer research might become visible.

The editors

1 Introduction and Overview

Bernhard Strauss

1.1 The Theory Dimension of Cancer

Cancer research is viewed by society as an applied science that has to be assessed ultimately with regards to progress in achieving its constituting goal, namely, the eradication of cancer. For some, sufficient progress has been made over the past decades to justify continuing with current scientific practices. For others, there appears to be a considerable gap between investments made and real benefits for patients, and they therefore call for a change of scientific paradigm to close that gap. The fact that such differing opinions exist at all warrants a closer look at how we can actually measure progress in cancer research. In addition, when proposing a change of scientific paradigm, one needs to first establish what the theoretical foundations of current cancer research are and whether they in fact constitute a paradigm that might be changing or needs to change, not an obvious question for a science that is perceived as solving an urgent practical problem and hence is presumed to be resting on solid foundations.

The preconditions for a fundamental paradigm shift in cancer research may have recently emerged, as **Marta Bertolaso** and **Bernhard Strauss** suggest in their philosophy and history of science analysis. They argue that several theoretical concepts that underpin the current paradigm (namely, that certain genes or mutations cause cancer) have undergone epistemological shifts over the past two decades, converging on a systemic view of cancer that embraces elements of nonlinear and reciprocal (multi)causality and a cellular "ecosystem" view of cancer. In addition, several concepts that have evolved over the same period of time, based on many decades of empirical evidence, are finding recently wider acceptance in the cancer research community. These conceptual expansions concern the tissue/tumor microenvironment, the role of systemic immune response and inflammation, and the temporal dimension of cancer progression analyzed with models from evolutionary theory, as well as concepts from dynamical complex systems theory. Such conceptual inputs have introduced a theory background to cancer research that in many ways contradicts the tenets of the prevailing somatic mutation paradigm that has dominated the field for the past fifty years. Currently, these contradictions are not acknowledged by most of mainstream cancer research busy doubling down on efforts dedicated to cement and uphold the old paradigm. What is still unclear at the moment is how novel conceptual frameworks that propagate a relational epistemology of theory-based reasoning can be integrated such that it motivates and guides a new scientific practice (which then also needs to be funded accordingly). Is the perception of increasing causal "complexity" in cancer research in fact a sign that discontent with current thinking is mounting, causing further questioning of existing concepts, followed by a quick "phase transition" away from the current paradigm, which then enables a change into new directions—as one would expect from any complex dynamical system, such as a scientific community?

Complex problems have always tested the foundations of any science, and cancer is no exception, as **Thea Newman** elegantly demonstrates in her section with examples from physics. Clearly, other areas of science have been there before, and there are lessons to be learned from physics on how deliberate selection of a different "granularity" in our theoretical thinking can make difficult problems suddenly accessible, leading sometimes to unexpected simplicity, greater explanatory power, and new ways forward, as Newman also shows for two longstanding observations in the cancer field. Thus, the right choice of "entity identity" to which we assign causal explanatory function within a phenomenon matters profoundly. As we know now from other sciences, it is not always the smallest known subentity of a given phenomenon that explains its overall behavior. Many would agree that complete understanding of a single ant's genome or of its nervous system might still not explain the existence of anthills. But to suggest, as Newman does, that to really understand cancer, we might need to look for different functional entities above the single gene or cell level will be a difficult leap of faith for many cancer scientists to make. After all, "the cancer cell" has been for over a century the causal explanatory entity of the complex biological phenomenon that cancer is. Such a change of conceptual level requires courage, as it is clearly very hard to start building a new theory from different first principles by disregarding decades' worth of highly detailed information and the corresponding, carefully erected edifice of interpretation that we believe are the ground "facts" about the phenomenon to be explained. It has been done in science before, but it is fair to say that cancer research in particular has not tried recently.

1.2   The Systems Dimension of Cancer

What we have learned from complex systems, however, not only in biology, is the fact that their inherent nonlinearity often produces outcomes that seem counterintuitive to human "rational" thinking. Thus, when human intervention interferes with such systems, the results can appear often "paradoxical," as for example when perturbing cancer with different treatment modalities. As **Sui Huang** explains clearly from first principles in his section, we can indeed understand why after therapeutic intervention cancer more often than not recurs and almost always more aggressive and resistant to further treatment. Once a solid theory framework for the "cancer system" is established, then also its inherent logic of interactions emerges. It is this logical structure that one needs to understand first to be able to interfere with a complex system successfully. As Huang demonstrates by applying a complex dynamical systems theory framework to empirical experimental and clinical examples, this has profound consequences for the design of therapeutic interventions. Taking a complex dynamical systems logic into account from the start, seemingly paradoxical actions, such as *not* attempting to kill all cancer cells or choosing lower than the maximum kill dose of a given drug, make perfect sense and can deliver desired results in cancer treatments.

Theories, however, are only as good as their fundamental principles, or rather, as good as these principles can accurately explain a phenomenon in the real world by consistently applying the theory. Thus, careful (and experimental) observation of real phenomena has always been the way to validate the importance of more abstract concepts. One such fundamental concept of complex systems theory as applied to biology is the "cellular attractor" (introduced first in the chapter by Sui Huang). Although intuitively plausible and conceived long before the molecular tools were available to test its explanatory power in real cells, recently developed single-cell analysis methods now allow direct observation of the attractor concept in populations of cells. As **Ingemar Ernberg** demonstrates in his chapter, using experimental data obtained by these methods, cellular phenotypic traits, such as absolute amounts of expressed proteins, vary stochastically yet stay within a certain range so as to be considered of the "same" phenotype—or as occupying the same attractor. This confirmation of a theoretical concept in real cells, however, poses new important questions that need to be answered, such as how dynamic can a cellular phenotype be, including transitions to another phenotype state, that we perceived traditionally as a static set of values, and how is "decision making" executed in cells when switching phenotypes. Given that such decisions are the result of intracellular molecular interactions, the question arises to what extent the application of standard physics and chemistry/biochemistry theory frameworks is sufficient to describe the ultra-high molecular density environment of millions of simultaneously occurring chemical reactions within a cell. Applying the cellular attractor concept to the cancer context can now be explored and tested empirically, as well as inspire novel approaches to therapy and new ways of thinking, as any sound theoretical concept should.

Well-characterized examples of cancer-relevant complex dynamical systems are now available to apply and test the kind of thinking that they demand. The intracellular gene regulatory network that brings about cellular phenotypes, normal and diseased, has become conceptually accessible over the past two decades due to enormous advances in network theory and modeling. They have made possible a dynamically networked view of gene interactions and delivered the insight that enormous phenotypic plasticity and adaptability are built in within the same genome, to be able to respond to changes and adapt, without the requirement for mutations. As **Peter Csermely** demonstrates in his chapter, such models have revealed that the interactions between genes and the adaptive responses of a gene regulatory network are strongly influenced and channeled by structural properties of the network architecture itself. This can lead to often nonobvious, counterintuitive network behaviors when highly adaptive responses of such networks emerge upon perturbation of the system, for example, through stress factors in the tumor microenvironment or through drugs intended to only kill cancer cells. A deeper understanding of these dynamic structural units of molecular interaction networks can now guide drug design strategies. For example, there is now good evidence from molecular network analyses in cancer cells that helps us understand why it is *not* necessarily the gene that is most prominent (or prominently disrupted) within a gene regulatory network neighborhood but some easy to miss neighbors that might in fact need to be targeted to affect a nominal "target gene" in the intended manner.

What a systems perspective of cancer also teaches us is that we need to go off-piste a lot more than we normally do and explore many more areas outside the well-funded highways to professorship. One example of such an area is what constitutes 99 percent of the genome, namely, the noncoding DNA, formerly discredited as "junk." That such areas can be productive grounds for new discoveries is explained by **Kahn Rhrissorrakrai** and **Laxmi Parida**, who present insights into the so-called dark matter of our genome that are highly relevant to our understanding of cancer genomics. To elucidate this vast, little-explored part of our genome requires powerful algorithms to uncover patterns and regularities where we have so far failed to find much "meaningful" information with the tools that have worked well with the 1 percent of the genome that encodes proteins. The analysis methods to search for correlations between cancer and genomic dark matter have only just arrived, and some are briefly introduced in this chapter, but first results indicate that dark matter might soon enlighten our understanding of cancer at the genomic level.

1.3   The Time Dimension of Cancer

Single-cell analysis technologies have shown over the past decade beyond doubt that tumor tissue is extremely heterogenous (often called "complex") at the cellular and genomic level. This has led to the wide acceptance of the view that the temporal dimension of cancer progression is of utmost importance for understanding the disease and for finding suitable treatment options. The progression from a few abnormal cells to a detectable tumor cannot be observed directly in patients, but it can be analyzed retrospectively at the genome level of individual cancer cells. This involves applying models from evolutionary theory that help explain how the heterogeneity of late-stage tumors might be the result of evolutionary forces acting on the various cell types that seemingly compete for survival within a stressful, abnormal cancerous tissue environment. The cellular evolutionary dynamics under these conditions resemble very much conditions in complex ecological systems, and concepts from ecological modeling have been successfully applied to the cancer context in the past decade. The fact that diversity of species emerges in complex ecosystems in response to external stress factors can only be explained by evolutionary theory when also the selective forces that act on such a system are factored in. Therefore, when using evolutionary models in cancer research, it matters what kind of evolutionary thinking exactly is applied, as **Jacob Scott**, **David Basanta**, and **Andriy Marusyk** argue in their chapter. To explain intratumor cell heterogeneity in evolutionary terms, it is clearly not sufficient to look at mutational changes alone, as is currently the case for most available cancer evolution data. The idea that only the "driver mutations" would be the driving forces of evolutionary processes overlooks the fact that what makes any mutation relevant in the cancer context are the selective pressures that act on it in the first place. Most of the mutations found in a tumor are usually recovered at a late stage of cancer progression and therefore might be the *consequence* of selective pressures within the cancer tissue and *not* necessarily the *cause* for the tissue conditions that have triggered and promoted malignancy in the first place. In their chapter, they call for a return to truly Darwinian thinking and for including models of currently ill-defined selective forces acting on cancer cells as a necessity to understand how the evolutionary dynamics within tumors shape cancer initiation and progression. Only then can we successfully interfere with the very evolutionary forces that bring about observed cancer cell behaviors, such as high adaptability and therapy resistance. Rather than aiming for short-term maximal cancer cell killing, understanding the evolutionary dynamics over time in cancers should help with the design of therapeutic interventions that make purposeful use of them to the long-term benefit of the patient.

Whichever selective forces or microenvironmental changes act upon cells within a tumor, the cellular response to them will not be random (one popular misunderstanding about evolution) but rather "channeled" and shaped by the evolutionary history of the genome itself and thus by the structural constraints embodied by the gene regulatory network. Thus, the current structure of the genome will allow only certain adaptive responses and not others, and some regions of the genome will have higher propensities for acquiring mutations than others, simply due to its evolved molecular "anatomy." Like any other evolved complex structure in biology, some genes represent highly conserved, "backbone" entities that underpin fundamental functions of all cellular life since its unicellular origins, and others are later evolved, often redundant additions and optimizations for specific functions within more complex, multicellular organisms and tissues. As **Kimberly Bussey** and **Paul Davies** propose in their chapter, the cancer state in multicellular organisms might represent the reversion of cellular behaviors to a unicellular functional phenotype, representing a rewiring of gene interactions into an "ancient" (premetazoan) functional configuration, involving often genes that are also old in evolutionary terms—reminding of an atavism. Such a functional switch makes sense for individual cancer cells in the context of disorganized cancer tissue, with altered mechanical, metabolic, and cell communication contexts that act as stress factors. Under these circumstances, cells do not maintain and receive the stabilizing signals of a normal tissue environment and therefore rather need to compete with each other for survival (instead of cooperating within a normal tissue to the benefit of the organism). To optimize survival, the fallback to a single-cell phenotype becomes then the evolutionary "sensible" strategy to maximize survival of the individual cancer cell and to evolve the greatest number of possible adaptive phenotypes to cope with the continuously changing stresses within progressing tumor tissue. This view also implies that it is not necessarily certain mutations that elicit "cancer hallmark" cell behavior but rather genome intrinsic rewiring options that are still available within the structure of the current genome. The fact that such regulatory wiring options are still readily accessible in response to a variety of stresses is an expected manifestation of the existence of attractor states in complex systems, as presented in the chapters by Huang, Ernberg, and Csermely. Such a perspective of cancer has implications for the design of therapeutic strategies that might aim to induce cancer cells to switch their rewired cancer genome "back to normal." Clearly, such a systemic switch cannot be achieved by targeting single mutations in single genes.

While it has become the norm to think of cancer in terms of disrupted cellular functions at the molecular or gene level, the clinical oncologist is still dealing on a daily basis with the "whole beast" that cancer is in real human patients. Although debates on the success rates of cancer treatments often lament the gap between results in the molecular biology laboratories and outcomes in the clinic, it is rarely discussed whether basic cancer biology could learn valuable lessons from the clinic. As **Larry Norton** illuminates in his section, the overall response of tumor tissue to treatment allows important insights into the fundamental biology of cancers. As historically the effects of drugs in cancer cells have been inferred from highly abnormal, exponentially growing cell lines in a dish, it has become quite clear by now that real tumors grow not only with different kinetics but also respond to drugs in a highly tissue type, disease stage, and tumor geometry dependent manner that affects their overall growth rate. Thus, drug scheduling, which is currently only considered *after* drug design and clinical testing, might be as important for treatment efficacy as the exact mechanistic effect the drug is supposed to have in individual tumor cells. As a drug has different effects with different outcomes at different stages of cancer progression, clinical trials need to be designed in a new way, as currently most candidate drugs are only tested in late-stage patients. The empirical fact, known to many clinicians, that overall tumor shape affects proliferation rates and therefore the effect that an antiproliferative drug can have also means that we need to understand the basic cell biology of tumor growth in three dimensions much better than we currently do. As Norton demonstrates in his chapter using mathematical arguments, we can model the overall growth kinetics of tumors, but what is still missing is an understanding of the exact cell biological mechanisms in the three-dimensional tissue context that drive these dynamics. Understanding the "units of tumor growth" at the basic science level will have wide-ranging implications for treatment design.

1.4   The Micro-/Environment Dimension of Cancer

Few other concepts in cell biology have changed our understanding of cellular function to such a degree than the idea of cell phenotype being defined by constant interaction with the microenvironment. Its original meaning, however, has become recently blurred by overuse in a number of very specific contexts, particularly in cancer biology, such as when the term is sometimes used exclusively for invading immune cells or only for stromal cells. The notion of a complex microenvironment with noncellular *and* cellular components that plays an *active* role has shifted our understanding from a cell-autonomous, single-cell view of cellular function toward a contextual tissue perspective. In such a view, cellular fate and differentiation as well as specific cell functions are the result of intricate positive and negative feedback interactions between the cell and its microenvironment. Although postulated earlier on theoretical grounds by a number of biologists, such as C. H. Waddington, or in the cancer context by S. Paget as part of his "seed and soil" hypothesis of metastasis, empirical experimental evidence had to build for many decades until the scientific community fully accepted its importance only recently. Fundamental discoveries in the past four decades have paved the way for conceptualizing the tissue/tumor microenvironment based on an understanding of the molecular mechanisms involved. Owing to technological advances, it also became clear that not only the molecular *composition* of the microenvironment but also its *three-dimensional topology* and *physicochemical properties*, such as mechanical forces, pH, or oxygen concentration, are perceived by the cell as essential "instructive" signals determining its phenotype. This is of particular importance for the cancer phenotype, as cancers are characterized only in part by abnormal proliferation but even more so by disrupted tissue architecture and associated changes in composition and physical properties of the surrounding extracellular matrix (ECM). It is these changes that elicit, and feed back into, the "cancer hallmark" cell behaviors that are characteristic for most cancers. Only over the past two decades has convincing evidence been found for how signals from these microenvironmental factors interact with each other in healthy and disease contexts.

As **Roger Oria**, **Dhruv Thakar**, and **Valerie Weaver** show in their chapter reviewing a large body of literature, the properties of the microenvironment, such as mechanical tissue stiffness and metabolic regulation, influence each other directly. Both being altered in carcinogenesis triggers major phenotypic changes in cells because mechanical and metabolic inputs can cause major, genome-wide changes in epigenetic regulatory states. In particular, the transition from nonmalignant to malignant cell phenotypes appears to be driven by long-term perturbation of the normal microenvironment parameters, for instance, as a result of chronic inflammation. Importantly, a stiffened ECM, clinically known as fibrosis, can directly influence chromatin organization and gene transcription, even in the absence of any DNA mutations. Such changes, however, are indirectly linked to a rewiring of metabolic pathways that drives cells further toward malignancy and tumor aggressiveness. Thus, a deeper understanding of the microenvironment has made us aware that we are still facing a chicken and egg problem in most cases when we want to evaluate the *causal* role of mutations in carcinogenesis. It appears we have been focusing for far too long on the egg—or was it the chicken?

Now that we better understand how different properties of the microenvironment can affect cell metabolism, we can attempt in turn to look in cancer cells for possibilities to change the malignancy-promoting effects that altered metabolism itself has through its abnormal metabolites that change the microenvironment. Over the years, different cancer-specific, modified metabolic pathways have been found, and how they are regulated by various signaling inputs is increasingly well understood. This detailed knowledge has made it clear that this metabolic adaptability in response to tissue stress factors is a very common feature of metazoan cells that does not necessarily require any specific mutations in metabolic pathway genes. **Maša Ždralević** and **Jacques Pouysségur** elucidate in their chapter why targeting altered metabolic pathways can be an effective treatment approach as this will break the positive feedback loops that promote more abnormal adaptive cell behavior induced by more abnormal metabolites causing further tumor progression and malignancy. Such attempts to target abnormal, cancer-specific metabolic pathways that will also affect microenvironmental properties of the tumor can, for example, induce programmed cell death or support the local immune response against cancer cells.

Disrupted interactions with the microenvironment are not only causally relevant for solid tumors, as one might think, but even for cancers of the blood, such as lymphoma and different types of leukemia, which historically have been viewed as *the* prototypical cancers caused by somatic DNA alterations. As **Luca Vincenzo Cappelli**, **Liron Yoffe**, and **Giorgio Inghirami** elucidate in their chapter, summarizing a wealth of recent, pioneering experimental data, leukemias that are not displaying any "classic" mutations are the result of "misinstruction" of blood progenitor cells by the microenvironment of a disrupted hematopoietic niche. The key interactors in translating these inappropriate signals and promoting a malignant phenotype are endothelial cells that form the inner lining of blood vessels. This also emphasizes the importance of these communications more generally in tumor vascularization, when the same interactions between tissue-specific microenvironments and tumor cells instruct endothelial cells to vascularize tumors that run out of nutrients. The authors discuss how a better understanding of these interactions between endothelial cells and the tumor microenvironment can open up novel treatment approaches. It seems that even the most widely traveling cells in the body, such as blood cells, can be shaped by bad influence of the neighborhood in which they spent their formative early stages: not a new phenomenon, given what we know about human nature.

Once a patient is diagnosed with cancer, treatment efforts usually focus on the local tumor, although it is a long-known fact that most cancer patients die of metastatic spread of the disease and not of the primary tumor. It is the process of cancer cell spreading that makes cancer a truly systemic disease, as the "success" of disseminating cancer cells in forming a malignant tumor at a distant site depends not only on their intrinsic abnormal, autonomous capabilities that they have acquired in the course of the cancer's evolution but also on the microenvironmental susceptibility in the tissues they "attempt" to colonize. This also means that at the time when metastatic growth becomes clinically apparent, most of the cells in a metastatic tumor may have been kept under control by local and systemic suppression mechanisms for a long time, often decades. Therefore, knowing the exact time when cancer cells begin to disseminate is very important for interventions that seek to prevent metastasis as early as possible and therefore potentially a large proportion of cancer deaths, as **Courtney König** and **Christoph Klein** propose in their chapter. Although it was traditionally believed that metastasis would be a characteristic feature only of late-stage tumors, single-cell analysis tools have recently revealed that tumor cells begin to disseminate at a very early stage, even before diagnosis of the primary tumor. During this "invisible phase of cancer spread" (as we currently do not have the technology to detect very small emerging tumors in a clinical setting), disseminating cancer cells are battling against various systemic defense mechanisms of the body as well as against unfavorable (for them) microenvironments at different distant locations. Why they ultimately win in many patients, in some earlier and in others later, is one of the unresolved questions discussed in this chapter. A much better understanding of the specific local control mechanisms within different tissue microenvironments that define healthy tissue homeostasis and fend off invading neoplastic cells is, however, required. It will enable treatment strategies that aim at cutting short the journey of metastatic cells, instead of aiming to directly kill individual cancer cells in tumors. This should involve approaches that make metastatic cells lose against the natural tissue maintenance and defense mechanisms that are already in place. Technologies to study early metastatic spread at the single-cell level in patients have just arrived, and thus the possibility to intervene in the tug of war between rouge invaders and health-maintaining mechanisms early enough to prevent a hostile takeover at a later stage when the patient's defenses might be already weakened by other modes of cancer treatments.

With the microenvironment concept, also the description of its phenotype-defining properties, including its topology, at tissue-specific locations as a "niche" that forms the normal "habitat" of any given cell type has become commonplace (e.g., the stem cell niche). This intuitively plausible idea of a phenotype-defining niche that is shaped by the organism that inhabits it was originally developed over one hundred years ago in ecological studies of animal species on how their interactions with their environment change its properties and thus change the evolutionary forces that act on the organisms' own evolution in a nonrandom manner. Applying this concept now to the context of cancer cells again seems apt as we have now enough empirical data suggesting that their own evolution appears to be strongly influenced by the changes in the microenvironment that they have introduced through abnormal cell–microenvironment interactions. However, once the theory background concerning niches from evolutionary theory is applied to cancer, far-reaching implications for our causal understanding of the disease become apparent, as **Emmy Verschuren** explains in her chapter. One of the consequences she discusses is that tissue origin or tissue niche factors are more defining for a specific cancer type than its mutational signature, or genotype, as Verschuren demonstrates with her own data from lung cancers. Another conceptual innovation that can be derived from consequent application of evolutionary thinking to the cancer niche is the possibility that phenotypic similarities between different cancer types (cancer hallmarks) might represent "units of selection," as similar selective pressures seem to arise in different diseased tissues. A better understanding of these units and their interactions with their respective microenvironmental parameters should open up novel ways of effectively targeting a broad range of tumors, without the need for tailoring treatments to the genetic signature of the individual tumor tissue.

Finally, Verschuren proposes that the lessons learned from niche construction and its effect on the evolution of animal and plant species, as well as whole communities of organisms, enable us now to speculate about ways of reconstructing niches that got perturbed causing disease. In the case of the human body, this means reestablishing health-promoting niche properties. This is, however, only possible when we understand one fundamental aspect of the niche concept, namely, the interconnectedness of different niches across levels of organization, from the global to the behavioral down to the molecular, and how they in a reciprocal manner define and shape the coevolution of each other's properties. No niche can exist in isolation; it is always embedded in an encompassing niche that enables its existence. Although at first a seemingly sweeping conceptual proposition for the cancer field, so far it appears that no cell or organism on this planet has ever survived complete destruction of its surrounding niches or has been able to escape from them alive. Thus, tackling the very tissue-specific question of cancer prevention and treatment might require a much greater awareness of the bigger picture into which human health is embedded. The better we understand the causal chains that connect the different niches that surround us and connect our body with our environment (physical and social), the greater the chances that we can find therapeutic approaches that use this causal connectivity to help with the reconstruction/healing of diseased cellular niches. Early attempts in this direction look indeed very promising.

I REDEFINING THE PROBLEM

The Theory Dimension of Cancer

2 The Search for Progress and a New Theory Framework in Cancer Research

Marta Bertolaso and Bernhard Strauss

Overview

The question of whether we have made real progress in curing cancer over the past fifty years is still a matter of debate, despite the enormous research efforts and financial investment dedicated to the cause. In particular, it is now recognized that the genomics "revolution" after the sequencing of the human genome twenty years ago and the resulting "targeted" therapies have clearly not delivered the progress that was hoped for at the time. This lack of progress has been analyzed from various angles, but whether the scientific, theoretical foundations of cancer research—namely, a gene/DNA/mutation-centric paradigm in explaining cancer—might be the reason for this lack of progress has not received much attention in mainstream cancer research. The still persistent paradigm that certain "cancer genes" and "cancer mutations" cause cancer is, however, under attack from various areas of cancer research based on empirical evidence. We propose here that it has been the cumulative results of scientific practice, not deliberate conceptual innovation aiming to develop a new causal paradigm, that are currently preparing the ground for such an urgently needed paradigm shift. Evidence for a convergence toward a new kind of theory framework can be detected in the following fundamentals of cancer research and specifically in how causal explanations of carcinogenesis have changed (or advanced/progressed) over the past decades at the level of scientific practice. We distinguish four areas of conceptual advancement that concern (1) the definition of cancer, (2) the biological processes that are considered causally relevant to cancer, (3) the levels of biological organization that are considered causally relevant to cancer, and (4) the explanatory relevance of the tissue microenvironment. A fifth area, very much related to the first one, that has been fully conceptualized only recently is tumor heterogeneity, which appears currently as *the* defining characteristic of malignancies that needs to be understood and dealt with in any therapeutic approach to be successful. We propose that real advancement/progress—a concerted change in epistemological assumptions and research practices—is indeed taking place, even though most of current research practice is still firmly adhering to an outdated paradigm that has clearly failed in the clinic. Explicit formulation of the emerging conceptual change in terms of a new integrative theory framework (a true paradigm shift) is still lacking. The new epistemological perspective is shifting our understanding of carcinogenesis toward the notion that cancer is ultimately a disease of the life history of the organism. This reconceptualization requires an explicit, new epistemological practice, which we define in terms of a *relational ontology*. It is based on the reciprocal, *integrative* functionality of cells, which results from their relations that act *across* levels of organismic organization. As cancer is a dynamical process, the epistemic focus needs to shift from a "thing" or "parts" view (cells, genes, molecules, etc.) to the *relational processes* that define and maintain the parts.[1] This conceptual framework should also inspire novel experimental approaches in the scientific practice of cancer research and help to make further progress toward its constituting goal, the eradication of cancer.

2.1 The Historical and Philosophical Contexts of Current Cancer Research

2.1.1 The Historical Context: Who Is Winning "The War on Cancer"?

In 1971, the U.S. Congress under President Nixon signed into law the "National Cancer Act," commonly referred to as the beginning of the "war on cancer." The act presented cancer as a serious threat to society that required raising public awareness on a national and international scale in order to be able to assemble diverse forces and resources to counterattack against the threat and "win the war." Importantly, this vision template assumed that the basic science was "already there" and that a quantitative boost in resources was all that was needed to inevitably bring about victory. However, explicitly and implicitly, this assumption has fundamentally shaped the epistemological framework of modern cancer research. It implied from the outset that any failure in achieving the ultimate aim of curing and preventing cancer may not have anything to do with the science (as it is already there) but must therefore be a consequence of some unanticipated aspect of the enemy (cancer). Once the gene/DNA level was established as the causally relevant level of analysis, as originally proposed by the scientists advising the U.S. government at the time who assumed that a limited number of viral genes might be the causal agent of carcinogenesis, it has remained at the core of the explanatory paradigm ever since. Interestingly, the linear, DNA-centered causal narrative has become ever more successful, despite the fact that our understanding of the exact identity and function of genes and their control mechanisms has continuously changed and vastly expanded over the past forty years. It shifted from the suspected "handful" of oncogenes, to tumor suppressor genes, to specific mutations within the coding region of genes, to a specific temporal sequence of mutations in specific cancer genes (drivers), to "cancer context" inducing mutations in whole assemblies of gene networks, where the causal role of any single gene is entirely dependent on the network interactions and even to certain aspects of noncoding DNA. By maintaining the initial, linear logic of causal reasoning at the DNA level, *despite* the fact that its central definitions were constantly changing over time, generations of cancer researchers got used to "epistemological entrenchment" at the DNA level, while they witnessed an explosion of "complexity" in their attempts to match the causal paradigm with ever improving, more detailed multilevel descriptions of the malignant phenotype. This mind-set then generated the conditions for a vicious circle that required asking for ever more science to be carried out at the same level (DNA) to increase knowledge on DNA sequence information, in the hope that complexity might be reduced.

The so-called war was joined by many countries throughout the 1970s and 1980s, and cancer research became the global leading driver for biological discovery. Yet today, after almost fifty years, cancer remains "a key public health concern and a tremendous burden" on societies,[2] and cancer incidence is increasing worldwide at an alarming rate.[3] This situation brings the question of how to assess progress in cancer research into focus with increasing urgency. A lot has been written in the past about this apparent lack of progress with respect to the initial goal of finding a cure for cancer. In 2009, the European Partnership "Commission Communication on Action Against Cancer" set the ambitious goal to reduce cancer incidence by 15 percent by 2020 and launched several initiatives. At the World Oncology Forum held in Lugano, Switzerland, in 2012, the question was asked: are we winning the war on cancer, forty years on? The conclusion back then was, in general, no.[4] Pal et al.[5] already some time ago quantified the lack of any substantial improvement of patient survival from 1985 to 2007 computed on a large population, while Langer et al.[6] looked into measures for the rate of progress and showed that the pace of advancement is rather slow (a two-month increase in patient survival being considered a remarkable result). Despite the introduction of hundreds of new anticancer drugs based on genomics data, the current consensus is that for most forms of cancer, lasting disease-free treatment outcomes are still very rare and outright cures even rarer,[7] with some notable exceptions for certain types of leukemia and colorectal cancer, and in the 1980s for testicular tumors.[8]

Voices from within the scientific community have been expressing for some time the view that we might be completely missing something in our efforts to fight cancer, questioning the causal foundations of the science framework applied in cancer research—in particular, the ideas that the only actor of cancer is the cancer cell, that the "targets" for the magic bullets to come must be master genes causing the cancer phenotype, and that "enemy bases" are located at the molecular level.[9] While empirical evidence increasingly shows that these hypotheses are incomplete at best, the pharmaceutical industry continues to produce "new-generation" drugs based on discoveries of a supposed "central operative" of the enemy, namely, single genes or mutations (e.g., MAP kinases, TP53, KRAS). Moreover, Huang[10] and others have questioned for a long time the assumption that the primary focus in cancer therapy should be on *killing* cancer cells, and some cautious attempts at testing other strategies are just emerging with so-called adaptive therapies[11] (and see chapters 4, 8, and 10, this volume). Other scientists have pointed out a lack of meaning and causal understanding in many studies that are based on the current DNA sequencing technology-driven research practice that also dominates organizational structures, academic reward systems, and financial incentives.[12]

Against this tension between past and current enormous investment and research efforts, as well as the slow acknowledgment of a much lower than anticipated success rate with respect to the initial goal (eradicating cancer), the question remains, to what extent are the conceptual changes that *did* take place in cancer research over the past forty years able to influence future research practice, despite the continuing persistence of an outdated paradigm? Is their impact negligible, or can they enable a true paradigm shift and new developments? In other words, can cancer research move into new directions, even though at the moment it is still not clear who is winning the war?

2.1.2   The Philosophical Context: Why We Need a Working Notion of Progress

The public image of cancer research is still that of a long and difficult war characterized by slow but incremental progress and rare methodological and therapeutic breakthroughs, holding great promise in the future. Enthusiastic and progressionist claims about future prospects, however, are counterpointed by scientists and organizations declaring that we are losing the war. Other observers have pointed to a permanent oscillation between a hectic search for the few "key factors" that cause cancer and subsequent disappointment in the face of emerging "endless complexity" or the overwhelming failure of rationally designed "targeted" drugs in clinical trials. These contradictory messages from within the scientific community reflect an unresolved tension between, on the one hand, the epistemological evolution of the scientific foundations and, on the other hand, the results of empirical therapeutic approaches that are based on these foundations, as measured in treatment success. This tension is amplified by the fact that media attention in the public debate is usually focused on hypothetical future benefits for cancer patients due to technical advancements, as promised, for example, by personalized medicine, precision oncology, or immunotherapy. What is much less discussed, however, even among cancer scientists, is what conceptual progress in the mechanistic understanding of carcinogenesis beyond technological innovation has been made in the past decades, despite a persistent lack of a generalized cure for cancer. As this "progress gap" is increasingly noticed by the general public, it becomes a matter of urgency to understand *whether and how* advancement actually takes place in cancer research and at what levels. Being more aware of its dependency on certain epistemological positions and paradigms and the resulting internal relationships becomes increasingly important for cancer research when defending and communicating its legitimacy to the public. This legitimacy is currently still based on a presumed linear relationship between fundamental science and treatment outcomes based on it. In this time of instant access to information, public scrutiny of that relationship will continue to increase, and medical decision making will increasingly be influenced by patient choices based on information available "out there." Access to huge amounts of information, especially through the Internet and social networks,[13] combined with a lack of skills and possibilities to discern its scientific validity, can lead to a major negative impact on health outcomes in populations.[14] This is playing out, for example, in the current public battles around vaccination campaigns against childhood infectious diseases.

One result of this trend in the cancer field is the ever-increasing market for "alternative therapies" that grows on any flaws and uncertainties that can be found within the science framework of standard Western science/medicine. Although some elements of complementary and alternative medicine systems can be helpful and even get adopted at mainstream cancer centers,[15] the proliferation of alternative cancer treatments can also become a threat to patient safety on a larger scale,[16] as demonstrated by episodes of mass convergence on allegedly miraculous therapies,[17] reinforced by improbable conspiracy theories about pharmaceutical companies and scientific research.

Thus, assessing notions of epistemological progress at the fundamental science level and understanding its relationship with causal reasoning in cancer research is an important starting point for planning future research strategies and health policies that can then help close the gap with the overall goal of eradicating cancer.

2.1.2.1   Finding a suitable notion of scientific progress for cancer research

Achieving any degree of "objective" understanding of progress in cancer research is by no means trivial: it requires a serious philosophical reflection on what scientific advancement/progress may be taken to mean in the cancer context. Linear views of science are still attractive for the public communications of cancer research, where the rhetoric of linear progress is used for obtaining continuing financial and societal support. However, philosophy of science has identified some time ago that the notion of progress in science requires clear specifications of what is actually measured,[18] what context applies, and, interestingly, that it is in general an underanalyzed aspect of science.[19]

Approaches for analyzing the long-term development of scientific knowledge have focused, for example, on progress viewed as the *cumulative knowledge account* (epistemic approach),[20] as *paradigmatic change*,[21] as *pluralism and integration*,[22] or as P. Kitcher's model of *scientific advancement*.[23] Major tensions in these discussions of a suitable measure of progress in the sciences are within philosophy often linked to fundamental *philosophical* positions, such as realism, antirealism, reductionism, and antireductionism. One example of such a tension is a definition of progress linked to an approximation toward a notion of "truth" (= increasing verisimilitude). In this framework, progress is only made when new scientific knowledge has moved science closer to a perceived truth—whatever that may be or whoever may define it. Another notion of progress is linked to the production of solutions to specific problems (the functional-internalist approach). In this view, progress only exists when knowledge has solved the problem (independent of whoever defines the problem and then declares it as solved). The epistemic approach to progress attempts to provide an account of progress only based on the accumulative increase of knowledge without relying on the abovementioned "qualifying" entities (truth, problem) that act like goals for the scientific endeavor and therefore imply a certain directionality of expectation when measuring progress.

Given these philosophically well-established contexts, what are the challenges for any notion of progress in cancer research? Clearly, as cancer research is a science by definition dedicated to solving a specific problem, it seems very difficult to step back and study other notions of progress, such as epistemological progress, in isolation. Conceptual progress is clearly impacting the development/advancement of the basic science and therefore does not necessarily need to be measured at first against the constituting goal of the whole enterprise (eradicating cancer). Arguments could be made that "cancer research" in fact does not even constitute a science at all but is rather a "project" within the natural sciences that applies whatever theories and methods appear useful from various biological science subdisciplines and, increasingly, from physics. (Likewise, one could say that "rocket science" has gathered what seemed useful from physics and chemistry to solve a very specific problem.) It could be argued from such a perspective that different contributions to the cancer project, say from, biology, genetics, biochemistry, bioinformatics, and so on, would bring each their own, specific epistemological "load." This could mean that a unified measure of progress would either be impossible to find or represent a "compound term" that might be fraught with instability and uncertainty, depending on developments within the contributing subdisciplines.

In addition, one could argue that even the problem that the project aims to solve (cancer) is in itself not very well defined, as cancer phenotypes can be wildly different, and the debate whether it is legitimate to include all the different types of cancer currently known under a single category of disease is as old as cancer research itself, implying that we also should reframe the concept of "cancer disease."[24] Cancer research since the 1970s has moved from a focus on specific genes, genetic mutations, and clonal expansion of such mutations in somatic cells to multiple classes of genes responsible for cancer hallmark cell behavior and lately different kinds of nongenetic factors that were considered to have causal relevance (such as epigenetic mechanisms, various species of RNAs). This led to the acknowledgment of the increasingly *wild heterogeneity* of cancer-related mutations, cellular phenotypes, and even metabolic states not only between different types of cancer but also in different transformed cells in tumors of the same patient or of the same cancer type between different patients.[25] Moreover, there have been constantly new attempts to reclassify disease phenotypes according to quite different biological categories, such as by histology, protein markers, DNA mutations, and changes in gene network interactions.

Thus, increasing knowledge about the subject of study has *not* led to clearer, simpler definitions; on the contrary, it has made it ever more difficult to find them. The history of cancer research also shows that the scientific community performs this search with a long view, by progressively altering the epistemological features of their subject of study and by adapting the *adequacy* of explanations to it, all the while its definitions keep shifting. Such epistemological aspects are rarely—if at all—perceived or debated by the scientific community. This is, however, a pivotal question, as many cancer scientists appear to proceed as if the philosophical underpinnings that influence their day-to-day activities and how experiments are carried out and models are applied would be negligible.

From a "purist" philosophical perspective, one might ask, how can progress of a scientific project that is gathered around a phenomenon that itself seems to resist a clear definition from within the science ever be measured rigorously—in particular, if we want to assess *conceptual* progress/advancement, and not progress/advancement with respect to its overall goal, to cure cancer? Thus, to make a meaningful contribution to assessing progress, we need to relax some expectations about the way we go about defining progress. First, we need to let go of the idea that a notion of "advancement" will have to encompass simultaneously all the epistemological dimensions of cancer research—such as understanding, explanation, and treatment of cancer. Second, we don't need to insist that such a measure needs to simultaneously reach through all these aspects by assuming a tight causal interconnectivity between them. We will thus be able to remain at the level of knowledge and look at how theories, concepts, and scientific methods have changed over time (accumulative, epistemic approach), which will give an accumulative account of knowledge advancement. We can then define progress based on whether any *conceptual* change has led to *novel possibilities of causal explanation* (interpretation) of existing data or whether it enables new kinds of experiments or research practice. The detailed discussion of whether such progress can or should also lead to an increase in treatment success is initially kept aside, as is the discussion of how societal factors govern the overall embedding of cancer research into a framework of societal goals that introduce their own measures of progress (a discussion of "mandate"). This analytical separation opens up the possibility to assess epistemological progress independent from the societal goal of curing cancer. In a different discussion, we can then look at how the progress gap between these two levels can be closed.

2.2 Conceptual Progress in Cancer Research Emerges from Scientific Practice

In this section, we identify a process of *collective convergence* currently taking place along four dimensions of cancer research toward specific *conceptual and explanatory* frameworks that have been emerging out of scientific practice, mainly in response to the perceived increasing complexity of the cancer phenotype. Empirical evidence and the corresponding trends in explanatory models of carcinogenesis clearly point toward the fact that the complex organismic processes that are compromised in cancer are highly dynamic, nonlinear, and reciprocal. This is highlighted also by the recent conceptualization of "tumor heterogeneity" as the defining characteristic of cancer that needs to be causally understood in any effective explanatory model and therapeutic approach. It represents the encapsulation of several trends within cancer research to find causal explanations that match epistemologically the

complexity of the malignant phenotype. (Interestingly, the "causal," mechanistic explanatory basis of tumor complexity at the genomics and gene regulatory level is not the same
as the one that is applied to the concept of tumor heterogeneity.[26] The latter includes the
physical three-dimensional interactions of tumor cells with local, noncellular/nongenetic
factors, such as oxygen levels, pH, metabolic environments, and mechanical forces—but
the former does not.) However, this observed convergence of certain conceptual notions
out of scientific practice is currently neither explicitly acknowledged by cancer scientists
nor has it been systematically studied with respect to its potential to initiate a true "paradigm shift" or novel research programs (which would be another indicator of progress).

2.2.1   The Notion of Scientific Practice

In our view, conceptual convergence was mostly an emergent by-product of evolving
scientific practice, rather than the result of any one theoretical proposition made with this
intent. (On the contrary, scientists who have over many decades been advocating similar notions
aiming for a paradigm shift have mostly been ignored or dismissed by "mainstream" cancer
research.)

As the study of scientific practice has become a specialist field in philosophy of science
with its own definitions and connotations, we need to briefly clarify what we mean here by
scientific practice. The practice of cancer research has changed over time, not primarily as a
consequence of conceptual advancements but rather thanks to innovation and scaling up of
certain technologies and their wide availability at lower cost. This resulted in the dominance
of certain experimental methods, in turn shaping disciplinary and interdisciplinary organizational structures and collaborations. As a consequence, overall scientific practice has become
widely uniform, making it more resistant to conceptual change while becoming the main
measure of any notion of scientific advancement/progress. This progress notion resulting from
scientific practice has its own dynamics and epistemological outcomes that appear partially
independent from the overall "intellectual," conceptual frameworks applied, such as "the
current paradigm" of cancer research (the somatic mutation theory). This partial independence
needs to be highlighted here, as after the "practice turn" mainstream philosophy of science
has developed methods and categories to deal with several aspects of scientific practice that
are now considered essential for understanding how science as a whole works and how it
changes over time.[27] In particular, it was stated that practice "consists of organized or regulated
activities aimed at the achievement of certain goals, and its consideration is inseparable from
the philosophical study of models, theories, knowledge and their relationships with the
world."[28] This view of science also implies that any measure or concept of "adequacy" as
part of any notion of progress will naturally incorporate many human dimensions that characterize scientific work. These are all the contributions of the "human factor" in scientific
practice,[29] such as values, norms, and ideals inherent in the pursuit of scientific knowledge.
Including them into the analysis of scientific practice also considers the metaphysical and
ontological assumptions underlying these practices.

Such "inclusive" notions of practice, however, bring an element of "goal directedness" or "problem-solving directedness" into the discussion, which we are excluding here from our analysis of epistemological progress in cancer research. Though well aware of these aspects of current practice notions in the philosophy of science, we deliberately take the position here of remaining at the level of generated knowledge. In particular, we will now look at the advancement of concepts as a result of changing scientific practice that specifically enabled a change in *causal reasoning* when constructing explanations of carcinogenesis and of the biological processes involved.

2.2.2   Conceptual/Epistemological Progress along Four Fundamentals of Cancer Research

2.2.2.1   Definitions of cancer

Cancer research has been struggling to clearly define the pathology of the disease ever since it started focusing on the DNA level for constructing a causal narrative. The questions of what is cancer, and what is cancer a pathology of, have become more difficult rather than easier to answer the more knowledge became available in the past decades. Historically, earlier definitions were based on tissue-level characterization, mostly based on tumor histology, providing morphological descriptions of what the *effects* of carcinogenesis are with respect to normal cell and tissue architecture. The underlying causal mechanism was assumed a given, namely, uncontrolled, excessive cell proliferation. It was then the extent to which this process disrupted the surrounding tissue anatomy that was the basis of classifications, staging, and definitions of malignancies. In addition, detailed morphological descriptions of cancer cell types within tumors were part of definitions. Despite all the recent efforts to redefine cancer, based on various molecular and genetic markers, the organizational/anatomical definitions that have historically dominated are still the main working definitions in oncological clinical practice. For most cancers, the new markers appear overlaid onto the historic histology level, and many clinicians do not consider them on their own to be sufficient to guide therapeutic decision making (with the possible exception of HER+/HER– breast cancers and some forms of leukemia). However, treatment options are increasingly justified based on these additional molecular/genetic markers, even though their causal relevance in human cancers is often unclear.

Once top-level definitions of cancer, such as "Cancer is the name given to a collection of related diseases" (National Cancer Institute [NCI]), had been linked to the DNA level with the specific causal narrative: "Cancer is a genetic disease—that is, it is caused by changes to genes that control the way our cells function, especially how they grow and divide" (current NCI definition—note the absence of the term "DNA" or "mutation")—cancer has become a moving target, not just in the clinical sense[30] but also at the level of fundamental research.[31] With hindsight, this was to be expected as this definition depends on the knowledge of molecular processes that define the relationship between genotype and phenotype. This knowledge has increased/changed at a rapid pace over the past forty years all the while the fundamental causal assumption has remained the same (DNA changes cause cancer). As a historical aside, we would like to point out that the same phenotype to genotype causality issues have been extensively debated in the field of evolutionary developmental biology since the 1980s, based on earlier suggestions by C. H. Waddington, and have led between the late 1990s and mid-2000s to the formulation of the extended evolutionary synthesis (EES) of standard evolutionary theory. The tenets of the EES include epigenetic and generative physical mechanisms, causal reciprocity, and niche construction, among other nongenetic mechanisms, to explain the generation of phenotypes in the course of evolution and development.[32]

Within the current framework, cancer has been initially defined as the result of specific mutations in a limited number of genes in single cells that then expand clonally, but over time, other gene regulatory levels and most recently epigenetic levels have become defining factors of the neoplastic process. Once it became clear that regulatory levels that are associated with tissue-level cell interactions are at least as important as mutations in single cells, the construction of causal narratives based exclusively on specific DNA mutations became ever more difficult to justify. In addition, with the advent of whole-genome sequencing and more systematic genomic characterization of healthy tissues, it became established that most mutations thought to be cancer specific can be found in normal, disease-free tissues.[33] Ever increasing amounts of mutation data in cancer tissue samples have also shown that a large proportion of tumors does not show any expected "cancer mutations" at all and that there is in the majority of cases no good correlation between specific mutations or mutation signatures and patient prognosis and survival. Stronger correlations have been found in total copy number variations of unmutated genes, interestingly, irrespective of whether they translate into quantitative deregulation of normal gene products, or not.[34]

These findings have confirmed earlier speculations, made on purely conceptual grounds, that individual mutations found in a tumor biopsy are unlikely to be the causally relevant mutations that have initiated and driven carcinogenesis, simply because these are found at diagnosis in a tissue often decades after the cancer-initiating event. Cells carrying these mutations have by then undergone a long evolution within the abnormal tumor microenvironment that may have caused most of these mutations *after* the process was already well under way. This argument has been and still is considered somewhat heretical in the field, even though on the current webpage of the U.S. National Cancer Institute, one can find the following statement in the section "what is cancer": "In general, cancer cells have more genetic changes, such as mutations in DNA, than normal cells. Some of these changes may have nothing to do with the cancer; they may be the result of the cancer, rather than its cause" (NCI webpage: https://www.cancer.gov/about-cancer/understanding/what-is-cancer). This partial reversal of a fundamental causal narrative by a leading cancer research institution is certainly remarkable, and once directionality of cause and effect is not as clear any more as once believed, then a more thorough reassessment of the whole explanatory paradigm is probably in order.

Multilevel, organismic, or systemic definitions of cancer have been proposed, of course, all along, for example, by proposing that cancer is caused by "carcinogenetic fields." These would disrupt "morphogenetic" and "morphostatic fields" that were established during embryo development and maintain healthy tissue integrity through a set of cellular interactions in a three-dimensional, field-like manner (this view was termed *tissue organization field theory* or TOFT[35]). Definitions of cancer as development or organogenesis "gone awry" have been proposed repeatedly but not been taken up by mainstream cancer research as a framework for causal models. These quite different causal accounts offer a narrative of why and how tumor cells eventually show a *disintegrated functional activity* and a progressively aberrant genomic organization and the ability to metastasize in different organs.

While such organismic and systemic approaches have from the outset understood cellular processes, such as proliferation and differentiation, as outcomes of a dynamic network of cellular interactions heavily influenced and regulated by the tissue microenvironment, it took longer for reductionist genetic and molecular approaches based on DNA sequence data to converge on a processual view of cancer. This shift toward processual definitions was strongly influenced by the concept of the "hallmarks of cancer," which states that disruptions in six or eight *functional capabilities* of cells transform a healthy cell into a cancer cell[36]— see also next section. Even though proposed first in 2001 (revisited and extended 2011) by proponents of the mainstream paradigm, it was not acknowledged in the field that a shift toward *functional* causal definitions also changes the causal role of the DNA level.

Thus, in terms of epistemological advancement in defining cancer, we identify over the past fifteen years a clear shift in causal narratives. The neoplastic phenotype is not explained any more exclusively in terms of specific, single mutations or mutation signatures and the corresponding regulatory mechanisms but in terms of complex *cellular functions* and processes whose overall realization in cancer is no longer *integrated* with the normal/ healthy developmental and metabolic dynamics of the organism. Even though this epistemological shift has happened due to advancements at the knowledge level, it has not caused a change of explanatory paradigm in most of mainstream cancer research. Still most efforts of large-scale research projects and their funding streams focus on finding "the" causally relevant mutations in cancer cells, be it in whole-genome sequencing studies of ever larger numbers of patients or in personalized approaches.

2.2.2.2 Biological processes considered causally relevant to cancer

The suggestion that a number of complex and integrated processes should have equal importance in causal explanations of carcinogenesis, as proposed in the hallmarks concept, appears to be a direct consequence of the "complexity paradox" that has resulted from the epistemological entrenchment at the DNA level. The hallmarks concept was an attempt to create "some order" in the ever-increasing collection of genes and mutations that seemed to be causally relevant to cancer at the time.[37] Historically, causal explanations of the cancer pathology have linked specific causal mechanisms ultimately with excessive cell proliferation.

The current explanations based on the hallmarks idea are based on a number of simultaneously occurring, intricate reciprocal processes that all might play an equal causal role. This also means that the epistemological structure of such causal explanations would need to reflect the processual nature of the phenomena in question, but this is currently not the case.

What are these biological processes or cellular behavior "pathways" that are considered perturbed either simultaneously or in a certain sequential order to manifest a full cancer phenotype? Cancer hallmark behaviors as initially proposed are sustained proliferative signaling, evading growth suppressors, activating invasion and metastasis, enabling replicative immortality, inducing angiogenesis, resisting cell death, avoiding immune destruction, and deregulating cellular energetics. Tumor-promoting inflammation, increase of genomic instability, and mutations were proposed as additional, "enabling hallmarks." Each one of these cell behaviors is the result of hundreds of coordinated regulatory events that are highly dependent on functional integration not only within single cells but also within tissues and the organism as a whole.

Within this explanatory framework, causal complexity of the neoplastic process is expected, not an insolvable conundrum at the DNA sequence level. The causally relevant processes identified are dynamic and reciprocal, and what is eventually compromised is cellular differentiation status, or cell identity. (This is also the position of both the cancer stem cell model and the viewpoint of cancer as organogenesis gone awry.) Such a processual view of cancer represents a clear advancement of explanatory paradigm.[38] It also includes an explicit *cross-level phenomenology* of cancer that spans many levels of organization (gene, genome, epigenome, proteome, metabolome, cell, cell-cell interaction, tissue, organ, organism, etc.) and makes an important distinction between two processes that, despite both being intracellular processes, nonetheless encourage different levels of description, namely, proliferation and differentiation. In this framework, cancer is seen as a consequence of *interlevel* dysregulation due to the regulatory uncoupling of reciprocal processes (differentiation, apoptosis, and proliferation). This view also counters the earlier great emphasis on cell proliferation as the primary causal feature of cancer and holds that a perturbation of differentiation state is the primary disruption in the chain of cross-level regulation and long-range interactions that maintain normal organismic organization. Long-range interactions include systemic immune interactions with locally perturbed tissue conditions/tumors, and therefore cancer has also been described as a chronic medical condition maintained by immune-modulatory factors.[39] Descriptions of cancer as "wounds that do not heal," going back at least to the mid-nineteenth century,[40] can therefore be underpinned with our current mechanistic understanding of the immune system.[41] Parallels between embryonic development and cancer have been suggested repeatedly, proposing that "oncology recapitulates ontogeny."[42] The search for explanatory models based on systemic, integrative processes has also motivated a unified theory of development, aging, and cancer.[43]

One of the main reasons why such process-based views of cancer causation were not explicitly adopted by mainstream cancer research after the publication of the hallmark concept was the fact that each of the hallmark processes was studied by their own respective fields of science specialization. Within each of the disciplines, scientists continued to follow a linear, DNA sequence-based causal model, looking for the "most relevant" mutation within their pathway of study, rather than looking for an integrative, systems view of their respective hallmark under investigation.

2.2.2.3 Levels of biological organization that are considered causally relevant

From a processual point of view, causal explanations based exclusively on single mutations and deterministic selective advantages of single cells are clearly not sufficient. In particular, increasingly detailed knowledge about the heterogeneity of tumor cell populations has challenged earlier depictions of tumors as a simple clonal expansion of a single transformed cell type. In order to understand this heterogeneity, the complex three-dimensional structure (topological features) of tumors is now considered an important causal factor of cancer cell behavior as cells become functionally heterogeneous as a *consequence* of local differentiation events induced by very specific local conditions within the three-dimensional topography of the tumor (i.e., metabolic environments, oxygen stress, low pH, and other "stress factors" that emerge during tumor evolution). That tumor geometry has a profound effect on overall tumor response to therapeutic agents has been observed in the clinic for a long time and been well described by mathematical models (see chapter 10, this volume).

A small minority of scientists thus adopted the *emergent* properties of tissues as an explicit starting position to explain the specific causal characteristics of cancer, proposing a perspective that considers carcinogenesis as a developmental process gone awry or a result of a perturbation of tissue maintaining field-like properties of tissues (TOFT[44]). Explanations from this point of view initially appear to reverse the causal structure that is eventually compromised in carcinogenesis, be it genetic, epigenetic, topological, or metabolic, by assuming the causally relevant events for cellular phenotypes would reside in the respective higher levels of organization (top-down causality).

However, from a truly processual/relational point of view, the explanatory narrative of carcinogenesis is based on *interlevel* regulatory dynamics (*reciprocal* top-down *and* bottom-up/relational causation). Carcinogenic cell behavior arises in this causal model due to the perturbation of the very processes that ensure normal cell and tissue function by *integrating* different organizational levels (e.g., cell polarity, short- and long-range signaling within the tissue, microenvironmental factors, mechanical forces, and systemic factors such as immune response). In this line of reasoning, the notion of *developmental causality* has been presented[45] and an evo-devo (evolutionary developmental biology) notion of causality has been proposed by others in order to align explanations of the neoplastic process with the logic of explanations of normal developmental processes.[46]

What became clear through these conceptual advancements was that a single causally privileged level of organismic organization might not exist in cancer research and that the question of which level might be *more* fundamental in causal explanatory terms has eventually to change into a question of which *inter*level processes between levels of organization are causally relevant. Naturally, such judgment of causal relevance has to submit to pragmatic criteria (what can be technically studied, what do we define as an interlevel process, etc.) and specific societal scientific agendas (what is supported by funding, ethics, social consensus on goals, etc.). Currently, it is not clear how to describe perturbations of interlevel processes in a robust fashion that can be translated easily into scientific practice.[47] Thus, despite the actual possibility of a paradigm shift towards interlevel causality models, well supported by data on well-described biological processes, it is currently not clear how this shift can be incorporated into research practice without a wider acceptance of the need for a fundamental paradigm shift in causal explanations.

2.2.2.4 The causal relevance of environments and microenvironments

A large body of evidence in basic cell biology and clinical results have led to the wide acceptance of the insight that physiological cellular processes cannot be understood by studying cells in isolation. It became clear that the functional tissue context, including the noncellular components of the tissue microenvironment, provides key regulatory cues for cellular function with clear causal implications for the neoplastic process.[48] In particular, contextual parameters, such as physical, topological, chemical, and mechanical factors, as well as short- and long-range signaling interactions are becoming increasingly acknowledged for their role in *stabilizing* the structural and functional properties of intracellular molecular parts. Cellular tissue architecture *together* with its specific microenvironment is a three-dimensional organizational system that carries positional and historical information with "instructive" properties that define the cellular phenotype (conceptually similar to a "morphogenetic/-static field"). Both local cellular association patterns and cellular differentiation states become established as tissues and organs are formed during development and differentiation in a three-dimensional, contextual manner. The main instructive component of the microenvironment is the extracellular matrix (ECM), a tissue-specific, complex mixture of secreted macromolecules that controls gene expression either via ligand receptor interactions or through its physical properties, such as stiffness and tissue-specific mechanical forces.[49] The importance of these interactions has been demonstrated also in defined experimental systems, such as engineered three-dimensional cell culture systems.[50] These insights, after many decades of results supporting such a notion, have led to a wider acceptance of the tumor microenvironment as a major causally relevant factor for tumor-promoting *as well as* tumor-suppressing signals.[51]

This has important implications for causal explanations of carcinogenesis. From the perspective of the (mostly epithelial) tissue/tumor cell, changes in the microenvironment trigger changes in gene expression, irrespective of whether these changes were caused by some external factor (e.g., some chemical in food or drinking water, irradiation, or tissue trauma caused by infection, surgery, or a change of mechanical properties within a tissue) or by other cells, such as stromal fibroblasts. Importantly, long-term changes in the microenvironment can be induced via nongenetic mechanisms, which also explains why many carcinogens do not need to be necessarily mutagens that directly cause DNA sequence alterations while still causing abnormal signaling and gene misregulation by altering ECM properties. Once tissue-specific gene expression is altered through microenvironmental changes, affected cells may change their polarity, cell junctions, and proliferative and migratory behavior and thus begin to show "hallmarks" of cancer, irrespective of whether they carry "cancer mutations" or not. Cells showing hallmark behaviors also secrete in addition abnormal ECM, which over time will create an increasingly abnormal, stressful microenvironment that will also induce DNA mutations *as a consequence* of these conditions. This dominant role of the microenvironment has been repeatedly demonstrated by experimental reversion of tumor cell phenotypes by placing tumor cells into a normal microenvironment and thus restoring normal tissue phenotype and function irrespective of the presence of "cancer mutations"[52] as well as by evidence of tumor reversion obtained when cancerous cells were placed into a three-dimensional environment provided by eggs/oocytes.

With regards to causal explanations of carcinogenesis, it appears in such a scenario that the directionality of causality operates, from a single-cell perspective, outside–in, and most mutations found in late-stage tumors would be indeed the consequence of microenvironmental changes rather than their cause. Such an inverse causality with respect to mutations in tumor cells is supported also by findings that show that other levels of gene expression control, such as epigenetic mechanisms or regulatory RNAs, can cause a cellular cancer phenotype in the absence of mutations in the affected cells. However, from a tissue perspective, causality is of course reciprocal, as the components and properties of the ECM are continuously monitored as well as actively modified by tissue cells in response to changes in the microenvironment.[53] Despite the fact that these reciprocal relationships have been studied for a long time, their causal importance for carcinogenesis has only recently become acknowledged, although mostly through a mutation centric view of causation.

2.3   Molecular, Cellular, and Spatiotemporal Heterogeneity in Cancers Call Out for a New Explanatory Paradigm

Our understanding of the cellular and molecular composition as well as the physical properties of tumor tissue itself has changed fundamentally over the past decade due to a number of technological advancements in radiological imaging techniques, molecular biology, and single-cell genomics technologies as well as in evolutionary modeling applied to cancer genomics data. These methods have shown that tumors are composed of highly heterogeneous groups of cells that show a spatially complex distribution of highly diverse genomic, epigenetic, metabolic, and differentiation states.[54] This recent increase in resolution of tissue analysis methods has established that tumor heterogeneity is a key, defining characteristic of cancer (a "master hallmark," one might say) that needs to be understood as a basis for causal explanations as well as for developing treatment strategies. This insight further emphasized that the *temporal* evolutionary trajectory of cancer cells is the crucial aspect of malignancy that determines treatment success or failure.

The phenomenon of tumor heterogeneity in itself indicates that causality is unlikely to be unidirectional and simple. It is rather the morphological and molecular embodiment of the fact that what is perturbed in cancer cells is their ability to *integrate* cell behaviors *across* levels of spatial *and* temporal organization. Plausible scenarios feature interlevel circular causality; for example, as cell polarity is lost due to some transient carcinogenic insult (not necessarily causing mutations), three-dimensional tissue organization is lost, and as three-dimensional organization is lost, proliferative control is lost, which leads to more disorganization, changes in ECM composition, increased tissue stress factors, and so forth. As a result, functionally relevant tissue units cannot be maintained and what remains in terms of cell behaviors is a switch to basic cell survival mechanisms (representing cancer hallmarks) that are triggered and propagated by the abnormal, increasingly hostile tumor microenvironment. Under these conditions, cells *compete* against each other for resources (rather than cooperate), and increasing diversity/heterogeneity is just the expected fallback "evolutionary strategy" that has been the most successful in evolution whenever challenging external conditions have threatened eukaryotic cells, be it at the organism, the tissue, or the single cell level (see also chapter 9, this volume).

The very recent acceptance of tumor heterogeneity as a fundamental feature of the cancer phenotype will potentially support a shift away from linear "cancer mutation"–based models and might open up a discussion of the causality issues raised in the abovementioned other dimensions of cancer research. It will also help to refocus on the fact that cancer is foremost a perturbation of the integrating maintenance processes that generate and maintain three-dimensional *and* temporal organization in a tissue. Temporal organization stems from the fact that proliferation, cell aging, and normal cell death rates, as well as stem cell differentiation rates in tissue pools, are regulated in space *and* time. Thus, in a reciprocal manner, spatial order provides the correct context for temporal mechanisms such as cell division control to be maintained. For example, the four- to six-week renewal cycle in adult human skin, or the complete turnover of the intestinal lining by stem cell proliferation and differentiation every few days, can only occur when tissue morphology is normal. Only the cell type–specific, normal, three-dimensional location within the tissue provides the correct cues for normal proliferation control. Vice versa, these "timed" divisions in development and tissue maintenance provide the components for any organized tissue to form and be maintained in the first place.

The importance of the time dimension in carcinogenesis has been realized earlier and has led to the formulation of the multistage model of carcinogenesis[55] and, subsequently, the "multiple hit" hypothesis of carcinogenesis, postulating that a certain number of sequential mutations (between three and seven), which may have different phenotypic "weight" (passenger and driver mutations), need to accumulate in the same cell in order to give rise to cancer. These assumptions became necessary as it was clear already in these early attempts to model the progression of mutations during carcinogenesis that for most cancers, a single mutation would statistically not be sufficient to cause a fully malignant phenotype. Though conceptually important, at the time, technical limitations did not allow mapping DNA sequence data to the spatial complexity of tumor tissue.

How heterogeneity at the DNA level arises over time during tumor development is now studied specifically by the field of cancer evolution that applies theoretical concepts from evolutionary modeling to clonal groups of cells within tumor tissue to understand how genomic characteristics of cancer cells are changing over time.[56] However, most current cancer evolution analysis is based on DNA *mutation* data. Although mutations are one part of the evolutionary process that creates variation in cells, it is the selection forces that determine which phenotypes are to survive, as is the case in the evolution of species. Models including concrete information about what these specific selection forces are within tumor tissues are currently still lacking as it is acknowledged that a realistic mapping of selective forces and genotype-phenotype correlations in cancer tissue is not yet obtainable with the methods available[57] (see also chapter 8, this volume). In addition, mechanistic evolutionary arguments concerning the creation of phenotypic variation need to include epigenetic mechanisms and a way of describing highly selective physical properties of the tumor microenvironment, such as pH, oxygen concentrations, and mechanical forces. A better understanding of how the temporal dynamics of selective mechanisms generate wildly heterogeneous cell populations will help address issues such as the inevitable emergence of drug-resistant cells after exposure to the selective forces of cancer treatments as well as the fact that targeted treatment approaches against "specific" cancer mutations have so far mostly failed. In addition, one would expect that systemic factors, such as the aging of the immune system[58] or long-term hormonal and tissue microenvironmental changes that affect cell function, need to be included in any truly organismic view of cancer evolution.

These insights gained from the study of tumor heterogeneity are consistent with the notion of absolute context dependency of any cancer cell behavior, including the context of the time dimension. Thus, in earlier philosophical work on cancer, it has been proposed as a working definition that cancer is a "disease of life history."[59] Very recent evolutionary modeling that factors in differential aging and life history trait-dependent somatic selection also confirms the validity of such a definition.[60] It is the space- *and* time-dependent processes that integrate cell behaviors across levels of complexity and thereby structure biological development, growth, and maintenance of tissue phenotypes. Along these lines, it also makes sense to ask, instead of *why* cancer has emerged in a particular case, why we don't get *more* cancer[61]—with a view to better understand normal tissue *maintenance*. Dealing with cancer is to deal with the continuously ongoing, well-integrated structuring and maintenance processes of the organism, rather than the single, specific, and local changes at the molecular level that are detected upon diagnosis as a *consequence* of their disruption.

The conceptualization of tumor heterogeneity has further elicited the use of theoretical modeling approaches from complex nonlinear systems theory and network theory to study causally relevant processes in carcinogenesis. Describing cancer in terms of landscapes of gene expression attractors and complex gene and protein interaction networks demonstrated clearly that causal relationships with respect to the malignant phenotype are highly context-dependent, reciprocal, and emergent properties of complex cellular systems (see chapters 4, 5, and 6, this volume). Causal narratives formulated in this kind of modeling approaches not only describe the systemic nature of the neoplastic phenotype much better but also show that emerging causal relations can be transient, reciprocal, and reversible.

2.4 New Ways of Explaining Cancer

Explanation is a major topic of reflection in philosophy of science and has been specifically studied with regards to explanations of biological phenomena, including cancer.[62] Biological phenomena share fundamental ontological aspects, such as multilayer complexity, that influence the logical structure of explanations thereof. However, the available philosophical insights into this influence have not yet been integrated into any specific framework, such as a "theory of explanation for biological phenomena." From a historic perspective, it appears that what we are explaining (the **explanandum**) in cancer research and what has explanatory function (the **explanans**) have been continuously modified over the past fifty years. Extensive debates in philosophy of science have focused on the formal nature of the explanans, such as laws, generalizations, models, and simulations, or on the epistemic nature of explanations (reductionist, systemic, emergent, etc.) and the specific relationship between explanandum and explanans. This relationship determines, for example, whether an explanation is considered a causal, an ontological, or a functional explanation, distinctions that go back as far as Aristotelian thinking. Pragmatic models of explanation have emphasized the importance of contextual elements in determining the adequacy of an explanation, and Mitchell's integrative pluralism[63] has highlighted the possibility of a permanent coexistence of multiple explanations in science. Sometimes, explanations may lose the explanandum along the way. A change of definitions (and therefore the explananda) in scientific practice is common, for example, in operationalization, where the research question is changed according to what is experimentally feasible and to what is testable, or in reduction,[64] in which the phenomenon to be explained is modified/adapted during the explanatory process.

As for other fields of science, it does not seem meaningful to represent epistemological advancement/progress in explanations as competitive selection between explanantia[65] or as an accumulation of answers to different questions that can then be integrated to explain individual, concrete cases.[66] Explanatory advancement in cancer research has been characterized by a continuous modification of the explanans along with a more refined understanding of the explanandum. So far, cancer has resisted reduction to single, causally definitive molecular features[67]: there are no common genes or molecular targets identified as causally exclusive for the neoplastic phenotype of cells. Moreover, we are witnessing a growing multiplicity of levels of organismic organization, as well as of "molecular mechanisms" that are considered causally relevant in explaining carcinogenesis. But once the original and persistent explanandum is brought back into scientific practice, the explanantia that have been found often appear deprived of explanatory relevance (or power). On the other hand, trying to force-fit an explanans onto observational data eventually undermines the explanandum, and a revision of both becomes necessary. This has happened in cancer research when the definitions of cancer started shifting as a consequence of the failure of certain kinds of reductionist explanations.

In explanations of biological, nonlinear phenomena, there is a complex circular feedback between explanandum and explanans, so that advancement cannot simply be defined as an improvement of the explanandum-explanans fit.[68] Living entities typically embody their explananda over different levels of biological organization (e.g., genes, metabolic networks, cells, groups of cells, tissues, organisms) *as well as* through different organismic *functional processes* (e.g., replication, proliferation, differentiation, auto-stabilization). Despite a general acknowledgment of these aspects of multicellular life, monofunctional explanations are still ubiquitous in accounting for its complex properties. Thus, also in cancer research, single functional explanations cannot be adequate, because what is actually compromised in cancer is not one single, specific function (hence targeted molecular treatment strategies mostly fail). What is compromised is the tissue-specific dynamic and reciprocal *functionality* that integrates the cell within the overall functional dynamics of the organism *across* levels of organization. This happens all the way down from the tissue level to the molecular and genomic stability of single cells. What has causal explanatory relevance for the pathology of cancer is the change in the reciprocal, *dynamic dependency relations* of cells that maintain the normal, physiological space-time integrated state. Importantly, as the analysis of tumor heterogeneity has shown,[69] these dependency relations can be *asymmetrical*. For example, a loss of cell polarity (a defining feature of spatial organization) causes not only the loss of three-dimensional tissue integrity (= the higher level of spatial organization) but also changes in cell division control and ECM composition, which in turn cause changes in metabolic state, which in turn may cause mutations and so forth, all of which feed back to cellular functions that are causally very different from the level at which the original perturbation has occurred. This reciprocal *asymmetry in relations* at the cellular level also explains why we always need functional tests in order to make a causal inference about the role of any mutated gene in the neoplastic process and why tumor reversion strategies might eventually be the best way forward to cure cancer. Asymmetry of relations in the neoplastic process also affects the integration of time-dependent processes within an organism and therefore the constitutive *timescale integration* among different levels of the normal biological organization. It is this integration process that connects causally relevant elements over several organizational levels of the explanandum and enables the progressive refinement of the explanans in a dialectic/reciprocal process.

Such conceptual frameworks have been used in cancer research so far only by a small minority with the explicit aim to develop a novel kind of explanans to better match the causal properties of the neoplastic phenotype. These approaches often apply theoretical foundations from complex dynamical systems theory and network science and have recently gained some prominence not only for allowing a more realistic description of complex molecular network interactions but also for enabling analysis at higher levels of organization above the cell.[70] Applying this kind of models to organismic phenomena leads to descriptions that will not have causally privileged levels/units, which is perfectly consistent with the observational data on cancer phenotypes.

Thus, new methods and tools for explaining have been used for some time on cancer data and have as a "side effect" introduced a certain theory background to cancer research that contains formalized statements about fundamental biological processes that are much better suited to describe the phenomena in question. What is currently still lacking, however, are formal tools, such as a comprehensive mesoscopic language to deal with causal entities beyond the single cell/gene level that would allow translation of insights from this kind of models into new experimental approaches that can lead ultimately to a novel scientific practice.

2.5   A Relational Ontology to Inspire Scientific Practice

Given the above outlined shift in perception of cancer as a complex multilevel phenomenon, one might ask what kind of overarching epistemological framework needs to underpin the construction of any new theory of the neoplastic phenotype. From what we have learned about other complex biological phenomena over the past century, it appears appropriate to apply the logic of a relational ontology to cancer. By *relational ontology*, we refer to the epistemological framework that accounts for the *inter*level, reciprocal regulatory processes that maintain the dynamic space-time organization of the organism and are characterized by an intrinsic asymmetry of their overall dynamic relations. Such a framework explicitly includes the time dimension at all organizational levels, from cellular and organismic behavior to its interactions with the external environment, all together constituting the "life history" of the organism.

Such a relational ontology is based on the following fundamentals, encompassing essential aspects of a systemic perspective of biological phenomena, including cancer:

• Relations are prior to entities—both ontologically and conceptually.

• Biological/organismic entities are defined through the relational context they are embedded within—and hence by the web of relations they are part of.

• Causal relevance of entities therefore emerges from their contextual relations and is not intrinsic to them as such.

• The time dimension is acting on the system as a whole, not only on isolated parts of the system.

• Therefore, the generative and the generational dimensions of hierarchical biological organization are an inseparable continuum (germ cells—organisms, stem cells—tissues, etc.). This is the organizational basis of the temporal *historicity* of the organism and any of its emergent phenomena, such as embryonic development or carcinogenesis.

These fundamentals help construct suitable explanations for phenomena such as cellular differentiation and development or carcinogenesis, which are emergent phenomena precisely because of space- *and* time-integrated processes: they cannot naturally exist outside the time-dependent, cellular functional context and the spatial hierarchical organization of the organism. Multicellular organisms are the embodiment of this simultaneous coordination of ontogenetically and historically linked entities that generate what we perceive as the whole of which they are a part. A relational ontology is not simply an ontology of relations,[71] nor is it just a *replacement* of systems of parts with relations. A relational ontology holds that relations are what maintain and dynamically transform all we can observe; as such, relations are also the way to epistemologically access and define our observables. This is why a relational ontology offers a particularly suitable epistemology to explain organismic phenomena (see also chapter 15, this volume).

A relational ontology framework defines cancer as a disease caused by a perturbation of the continuously ongoing systemic relations of an organism, of its natural reciprocal dynamism. In the neoplastic process, cells lose their integrated functional properties and become disconnected from the relational network that initially constituted their identity, *as a consequence* of a perturbation in the relational network. They assume new functional states and organize into tumors within a relational constellation that is defined by the emerging properties of the tumor microenvironment as a result of cellular functionality that becomes enabled by these new functional states. These new morphological structures have lost their relational connectivity with the organism as a whole and begin to stabilize their newly emergent properties by establishing and maintaining their own local, short-range relational context (constructing their own niche, in evolutionary terms).

Thus, with a view to therapeutic strategies, a relational ontology approach would attempt to reestablish relations *across* the divide between the constituting relational network of the organism and the newly emerging, pathological local entity. Instead of looking for therapeutic solutions *within* the perturbation, one needs to understand which relations maintain the normal, healthy space-time integrated state of the whole and how these can be used to reverse the local perturbation. This means acknowledging that local perturbations can be caused by a disruption of systemic relations that act across many different levels of space-time organization, such as in the immune and endocrine systems. Conversely, cancer treatments that only address the local pathology might not be successful, exactly because the cancer-initiating, cross-level relational disruption may still persist. Testing such a relational framework for cancer research ultimately requires different experimental approaches and a novel research practice to produce clinically relevant results.

2.6   Conclusions and Outlook

In this chapter, we have proposed that epistemological progress has indeed occurred in cancer research despite the persistent lack of a new paradigm that would be more successful in explaining and treating cancer. This assessment was based on the following observations.

1) The large cancer genomics studies of the recent decade that were based on the current paradigm have shown (certainly not with this intent) that causal specificity of certain genes and mutations is very limited for most cancers, as is their diagnostic/prognostic value with respect to patient survival. In addition, most "targeted" anticancer drugs designed based on genomics data have failed in the clinic and have overall performed worse than other approaches. This is increasingly perceived as evidence for the necessity of a paradigm change.

2) The perception of ever increasing "complexity" resulting from attempts to match DNA sequence data with the observational data of carcinogenesis has ultimately led to the proposal of a more processual view of cancer causation. This view was initially conceptualized in a number of integrated cellular behaviors, termed "hallmarks of cancer," but has not motivated a change in causal reasoning. Instead of emphasizing the processual nature of the hallmarks concept, it was rather used to further entrench the DNA sequence level as the prime level of causal analysis. However, the acknowledgment that a number of integrated complex cell behaviors/processes might have equal causal relevance for the neoplastic phenotype has prepared the ground for investigating the phenotypic complexity of tumors with novel high-resolution approaches facilitating a "systems view" of cancer.

3) Phenotypic heterogeneity itself at many levels (cellular, molecular, genetic, metabolic, physical etc.) has been identified as the key property of malignancies, conceptualized as "tumor heterogeneity." The acceptance of cellular heterogeneity as an important "causal" factor in explaining cancer has also led to an increased acceptance of "contextual" causal factors, such as the tissue/tumor microenvironment and the importance of the time dimension in terms of evolutionary processes that shape cancer initiation and progression. Though cancer evolutionary modeling is still mostly mutation/cell centric, lacking currently theoretical formalisms to account for selective forces in carcinogenesis, it has introduced the notion of microenvironmental feedback loops (niche construction) and a "tissue ecology" view to the cancer field.

4) Modeling approaches from areas such as dynamical complex systems theory and network science were introduced over the past two decades to the cancer field to better describe the contextual emergent heterogeneity and complexity of tumors. These conceptual frameworks have carried over a certain theory background into cancer research that emphasizes a dynamic, reciprocal, multi- and cross-level, time-dependent relational nature of causality that clearly points to the possibility that a single privileged level of causality might not exist in carcinogenesis.

Although these insights are now acknowledged to some extent within the cancer research community, most research efforts and funding institutions are still perpetuating the gene/

mutation centric paradigm in a "doubling-down" mentality in the face of increasing paradigm instability. One reason why "insights in principle" are usually not sufficient to cause a paradigm change is the lack of a concrete theory framework and a language that enables the translation of novel concepts into research practice. As scientists, we certainly believe that conceptualization and theory construction are necessary not only for day-to-day planning and interpreting of experiments but also, more important, for deciding what evidence to look for in the first place. Thus, any novel conceptual framework will contain some epistemological innovation that requires new ways of generating data in the first place, before the overall robustness of the theory can be assessed experimentally.

A new paradigm based on the view that cancer is the consequence of a disruption of relational interactions between different levels of organismic organization requires that we understand the normal tissue integration and maintenance processes much better than we currently do. This requires currently lacking proper definitions of causally relevant "functional units" above the single cell level and an understanding of the rules that dictate their interactions/relations. It would also require that explanations of the cancer phenotype follow the logic structure of a mesoscopic way of reasoning, represented by a cancer mesosystem that exhibits regularities above the single cell level and in which the properties of the parts themselves emerge due to the overall properties of the system after it has been defined. Conversely, properties that are causally relevant for explanations may fail to emerge if the mesosystem is not correctly defined. This *"essentiality-by-location" principle* will be an unavoidable starting point for any subsequent reduction that may possibly be performed in studying complex problems in biology.[72] In such a mesoscale framework, explanation has to start with the identification of "middle levels" that maximize determinism while preserving the *explanandum* and always *several levels in the same explanation* to avoid losing causal relevance and meaningfulness. Such a mesoscopic approach is closely related to the approach introduced by Sydney Brenner and others[73] called "middle-out." In a system of multilevel interactions that involves both regulatory feed-forward and feedback pathways, as well as environmentally defined parameter constraints, there is really no alternative to breaking in at one level (the "middle" part of the metaphor) and then reaching "out" to neighboring levels using appropriate, experimental, and analytical methods. Middle-out avoids any rigid either bottom-up or top-down stance and conceptualizes insight at whichever level there is a good understanding of data and processes, and then connects this to higher and lower levels of structural and functional integration.

Some recent changes in clinical practice can be viewed as symptoms of a search and readiness for novel causal explanations—for example, the insight that not all types of breast cancer benefit from available chemotherapy, and therefore *not* attempting to kill as many "cancer cells" as possible can improve patient survival. Also, improved outcomes of combination therapies consisting of a mixture of lower-concentration drugs rather than "maximum-kill" concentrations point to the success of strategies that tacitly embrace a notion of a systemic approach possibly making use of the "normalizing" potential of the normal cell compartment within tumors. Manipulating the cross-level functionality of the immune system enables therapeutic interventions that can produce more promising results than approaches specifically targeting unique mutation signatures within a tumor. Very recent attempts to anticipate the evolutionary response of the "tumor system," with the aim to prevent the emergence of therapy resistance as a response to treatment, point toward the increasing acceptance of a more open-minded position with regards to which causal levels to consider for treatment approaches. Such therapeutic strategies are implicitly based on a different causal logic that does not necessarily involve detailed genomic information of the tumor tissue itself but attempts to make use of systemic, dynamical interactions of the organism to "target" the local, malignant perturbation. To formulate appropriate explanations for successful outcomes of such approaches that have mainly resulted from empirical observations in clinical practice would require a new language to describe the inherent logic of their mechanistic basis.

Here and in the other chapters of this volume, we propose some elements, some vocabulary, for such a language based on empirical evidence. Moreover, we suggest that the grammar of this new language may show features of a relational ontology.

Notes

1. Bertolaso, M. & Dupré, J. A processual perspective on cancer. In *Process Philosophy of Biology* (eds. Dupré, J. & Nicholson, D.) (Oxford University Press, 2017).

2. European Commission. Cancer. (2010). Available at: http://ec.europa.eu/health/major_chronic_diseases/diseases /cancer. (Accessed: 16 October 2016)

3. Bray, F. et al. Global cancer statistics 2018: GLOBOCAN estimates of incidence and mortality worldwide for 36 cancers in 185 countries. *CA. Cancer J. Clin.* 68, 394–424 (2018).

4. Hanahan, D. Rethinking the war on cancer. *Lancet* 383, 558–563 (2014).

5. Pal, S. K. et al. Lack of survival benefit in metastatic breast cancer with newer chemotherapy agents: the City of Hope experience. *ASCO Annu. Meet. Proc.* 26, 17510 (2008).

6. Langer, C. J. et al. Survival, quality-adjusted survival, and other clinical end points in older advanced non-small-cell lung cancer patients treated with albumin-bound paclitaxel. *Br. J. Cancer* 113, 20–29 (2015).

7. Hanahan, D. Rethinking the war on cancer. *Lancet* 383, 558–563 (2014). Friedman, A. A., Letai, A., Fisher, D. E. & Flaherty, K. T. Precision medicine for cancer with next-generation functional diagnostics. *Nat. Rev. Cancer* 15, 747–756 (2015). Letai, A. Functional precision cancer medicine—moving beyond pure genomics. *Nat. Med.* 23, 1028–1035 (2017).

8. Peckham, M. J. et al. The treatment of metastatic germ-cell testicular tumours with bleomycin, etoposide and cis-platin (BEP). *Br. J. Cancer* 47, 613–619 (1983).

9. Sonnenschein, C. & Soto, A. M. The death of the cancer cell. *Cancer Res.* 71, 4334–7 (2011). Huang, S. The war on cancer: lessons from the war on terror. *Cancer Mol. Targets Ther.* 4, 293 (2014).

10. Huang, S. The war on cancer: lessons from the war on terror. *Cancer Mol. Targets Ther.* 4, 293 (2014).

11. Maley, C. C. & Greaves, M. *Frontiers in Cancer Research.* (Springer New York, 2016). doi:10.1007/978 -1-4939-6460-4. Degregori, J. *Adaptive Oncogenesis.* (Harvard University Press, 2018).

12. Geman, D. & Geman, S. Opinion: Science in the age of selfies. *Proc. Natl. Acad. Sci.* 113, 9384–9387 (2016).

13. Du, L., Rachul, C., Guo, Z. & Caulfield, T. Gordie Howe's "Miraculous treatment": Case study of Twitter users' reactions to a sport celebrity's stem cell treatment. *JMIR Public Heal. Surveill.* 2, e8 (2016).

14. American Cancer Society. Cancer Information on the Internet. (2014). Available at: http://www.cancer.org /cancer/cancerbasics/cancer-information-on-the-internet. (Accessed: 16 October 2016)

15. Vickers, A. J. & Cassileth, B. R. Unconventional therapies for cancer and cancer-related symptoms. *Lancet Oncol.* 2, 226–232 (2001). Mulkins, A. L., McKenzie, E., Balneaves, L. G., Salamonsen, A. & Verhoef, M. J. From the conventional to the alternative: exploring patients' pathways of cancer treatment and care. *J. Complement. Integr. Med.* (2015). doi:10.1515/jcim-2014–0070.

16. Brigden, M. L. Unproven (questionable) cancer therapies. *West J. Med.* 163, 463–469 (1995). Werneke, U. et al. Potential health risks of complementary alternative medicines in cancer patients. *Br. J. Cancer* 90, 408–13 (2004).

17. Bifulco, M. & Gazzerro, P. The right to care and the expectations of society—controversial stem cell therapy in Italy. *EMBO Rep.* 14, 578 (2013).

18. Bird, A. *Philosophy of Science.* (Routledge, 1998). Bird, A. What is scientific progress? *Nous* 41, 64–89 (2007).

19. Niiniluoto, I. Scientific progress as increasing verisimilitude. *Stud. Hist. Philos. Sci.* 46, 73–7 (2014). Dellsén, F. Scientific progress: knowledge versus understanding. *Stud. Hist. Philos. Sci. Part A* 56, 72–83 (2016).

20. Bird, A. *Philosophy of Science.* (Routledge, 1998). Bird, A. What is scientific progress? *Nous* 41, 64–89 (2007). Niiniluoto, I. Scientific progress as increasing verisimilitude. *Stud. Hist. Philos. Sci.* 46, 73–7 (2014). Dellsén, F. Scientific progress: knowledge versus understanding. *Stud. Hist. Philos. Sci. Part A* 56, 72–83 (2016).

21. Ankeny, R. A. & Leonelli, S. Repertoires: a post-Kuhnian perspective on scientific change and collaborative research. *Stud. Hist. Philos. Sci. Part A* 60, 18–28 (2016). Bizzarri, M. & Cucina, A. SMT and TOFT: why and how they are opposite and incompatible paradigms. *Acta Biotheor.* 64, 221–239 (2016). doi:10.1007/s10441 -016-9281-4. Bizzarri, M., Cucina, A., Conti, F. & D'Anselmi, F. Beyond the oncogene paradigm: understanding complexity in cancerogenesis. *Acta Biotheor.* 56, 173–196 (2008). O'Malley, M. A. & Boucher, Y. Paradigm change in evolutionary microbiology. *Stud. Hist. Philos. Biol. Biomed. Sci.* 36, 183–208 (2005).

22. Mitchell, S. D. *Biological Complexity and Integrative Pluralism.* (Cambridge Studies in Philosophy and Biology, 2003). doi:10.1007/s10539-004-5896-y. Mitchell, S. D. *Unsimple Truths: Science, Complexity, and Policy.* (University of Chicago Press, 2009). Brigandt, I. Beyond reduction and pluralism: toward an epistemology of explanatory integration in biology. *Erkenntnis* 73, 295–311 (2010). Plutynski, A. Cancer and the goals of integration. *Stud. Hist. Philos. Biol. Biomed. Sci.* 44, 466–476 (2013).

23. Kitcher, P. *The Advancement of Science. Science without Legend, Objectivity without Illusions. Statewide Agricultural Land Use Baseline 2015* 1, (Oxford University Press, 1993).

24. Bizzarri, M., Minini, M. & Monti, N. Revisiting the concept of human disease: rethinking the causality concept in pathogenesis for establishing a different pharmacological strategy. In *Approaching Complex Diseases: Network-Based Pharmacology and Systems Approach in Bio-Medicine* (Springer, 2020).

25. Di Filippo, M. et al. Zooming-in on cancer metabolic rewiring with tissue specific constraint-based models. *Comput. Biol. Chem.* 62, 60–69 (2016).

26. Bizzarri, M., Cucina, A., Conti, F. & D'Anselmi, F. Beyond the oncogene paradigm: understanding complexity in cancerogenesis. *Acta Biotheor.* 56, 173–196 (2008). Plutynski, A. & Bertolaso, M. What and how do cancer systems biologists explain? *Philos. Sci.* 85, 942–954 (2018). Bertolaso, M. *Philosophy of Cancer: A Dynamic and Relational View.* (Springer, 2016). Bizzarri, M. et al. A call for a better understanding of causation in cell biology. *Nat. Rev. Mol. Cell Biol.* 20, 261–262 (2019).

27. Bertolaso, M. *The Future of Scientific Practice: 'Bio-Techno-Logos'.* (Pickering & Chatto Publishers, 2015).

28. Ankeny, R., Chang, H., Boumans, M. & Boon, M. Introduction: Philosophy of science in practice. *Eur. J. Philos. Sci.* 1, 303–307 (2011).

29. Bertolaso, M., Di Stefano, N., Ghilardi, G. & Marcos, A. Bio-techno-logos and scientific practice. In *The Future of Scientific Practice: 'Bio-Techno-Logos'* 179–192 (Pickering & Chatto Publishers, 2015).

30. Komarova, N. L. Cancer: a moving target. *Nature* 525, 198–199 (2015).

31. Francipane, M. G., Chandler, J. & Lagasse, E. Cancer stem cells: a moving target. *Curr. Phatobiol. Rep.* 18, 1199–1216 (2013).

32. Laland, K. N. et al. The extended evolutionary synthesis: its structure, assumptions and predictions. *Proc. R. Soc. B* 282, 20151019 (2015). Pigliucci, M. *Evolution: The Extended Synthesis.* (MIT Press, 2010).

33. Martincorena, I. et al. High burden and pervasive positive selection of somatic mutations in normal human skin. *Science (80-.).* 348, 880–886 (2015). Martincorena, I. et al. Somatic mutant clones colonize the human esophagus with age. *Science (80-.).* 362, 911–917 (2018). Kato, S., Lippman, S. M., Flaherty, K. T. & Kurzrock, R. The conundrum of genetic "drivers" in benign conditions. *J. Natl. Cancer Inst.* 108, djw036 (2016).

34. Smith, J. C. & Sheltzer, J. M. Systematic identification of mutations and copy number alterations associated with cancer patient prognosis. *Elife* 7, e39217 (2018).

35. Soto, A. M. & Sonnenschein, C. The tissue organization field theory of cancer: a testable replacement for the somatic mutation theory. *BioEssays* 33, 332–340 (2011).

36. Hanahan, D. & Weinberg, R. A. The hallmarks of cancer. *Cell* 100, 57–70 (2000). Hanahan, D. & Weinberg, R. A. Hallmarks of cancer: the next generation. *Cell* 144, 646–674 (2011). Cavallo, F., De Giovanni, C., Nanni, P., Forni, G. & Lollini, P. L. The immune hallmarks of cancer. *Cancer Immunol. Immunother.* 60, 319–326 (2011).

37. Weinberg, R. A. Coming full circle—from endless complexity to simplicity and back again. *Cell* 157, 267–271 (2014). Hanahan, D. Rethinking the war on cancer. *Lancet* 383, 558–563 (2014).

38. Bertolaso, M. & Dupré, J. A processual perspective on cancer. In *Process Philosophy of Biology* (eds. Dupré, J. & Nicholson, D.) (Oxford University Press, 2017).

39. Greten, F. R. et al. IKKbeta links inflammation and tumorigenesis in a mouse model of colitis-associated cancer. *Cell* 118, 285–96 (2004). Condeelis, J. & Pollard, J. W. Macrophages: obligate partners for tumor cell migration, invasion, and metastasis. *Cell* 124, 263–266 (2006).

40. Dvorak, H. F. Tumors: wounds that do not heal—redux. *Cancer Immunol. Res.* 3, 1–11 (2015).

41. De Vita, V. T., Lawrence, T. S. & Rosenberg, S. A. *Cancer. Principles and Practice of Oncology.* (Lippincott Williams & Wilkins, 2008).

42. Grier, D. et al. The pathophysiology of HOX genes and their role in cancer. *J. Pathol.* 205, 154–171 (2005). Skakkebaek, N. E. et al. Germ cell cancer and disorders of spermatogenesis: an environmental connection? *APMIS* 106, 3–11; discussion 12 (1998). Bizzarri, M. et al. Fractal analysis in a systems biology approach to cancer. *Semin. Cancer Biol.* 21, 175–82 (2011).

43. Finkel, T., Serrano, M. & Blasco, M. A. The common biology of cancer and ageing. *Nature* 448, 767–74 (2007). Soto, A. M., Maffini, M. V. & Sonnenschein, C. Neoplasia as development gone awry: the role of endocrine disruptors. *Int. J. Androl.* 31, 288–293 (2008). Levin, M. Morphogenetic fields in embryogenesis, regeneration, and cancer: non-local control of complex patterning. *Biosystems* 109, 243–261 (2012).

44. Soto, A. M. & Sonnenschein, C. Emergentism as a default: cancer as a problem of tissue organization. *J. Biosci.* 30, 103–118 (2005).

45. Bertolaso, M. & Dupré, J. A processual perspective on cancer. In *Process Philosophy of Biology* (eds. Dupré, J. & Nicholson, D.) (Oxford University Press, 2017).

46. Liu, K. E. Rethinking causation in cancer with evolutionary developmental biology. *Biol. Theory* 13, 228–242 (2018).

47. Bertolaso, M., Caianiello, S., Serrelli, E. (Eds.) *Biological Robustness: Emerging Perspectives from within the Life Sciences.* (Springer, 2018).

48. Bizzarri, M. & Cucina, A. Tumor and the microenvironment: a chance to reframe the paradigm of carcinogenesis? *Biomed Res. Int.* 2014, 934038 (2014).

49. Bissell, M. J. The differentiated state of normal and malignant cells or how to define a 'normal' cell in culture. *Int. Rev. Cytol.* 70, 27–100 (1981). Bissell, M. J., Hall, H. G. & Parry, G. How does the extracellular matrix direct gene expression? *J. Theor. Biol.* 99, 31–68 (1982). Muncie, J. M. & Weaver, V. M. The physical and biochemical properties of the extracellular matrix regulate cell fate. *Curr. Top Dev. Biol.* 130, 1–37 (2018).

50. Huh, D., Hamilton, G. A. & Ingber, D. E. From 3D cell culture to organs-on-chips. *Trends Cell Biol.* 21, 745–754 (2011). De Ninno, A., Gerardino, A., Girarda, B., Grenci, G. & Businaro, L. Top-down approach to nanotechnology for cell-on-chip applications. *Biophys. Bioeng. Lett.* 3 (2) (2010).

51. Roskelley, C. D. & Bissell, M. J. The dominance of the microenvironment in breast and ovarian cancer. *Semin. Cancer Biol.* 12, 97–104 (2002). Kenny, P. A. & Bissell, M. J. Tumor reversion: correction of malignant behavior by microenvironmental cues. *Int. J. Cancer* 107, 688–695 (2003). Bhat, R. & Bissell, M. J. Of plasticity and specificity: dialectics of the micro- and macro-environment and the organ phenotype. *Wiley Interdiscip. Rev. Membr. Transp. Signal.* 3, 147–163 (2014). Ricca, B. L. et al. Transient external force induces phenotypic reversion of malignant epithelial structures via nitric oxide signaling. *Elife* 7, e26161 (2018). Walker, C., Mojares, E.

& del Río Hernández, A. Role of extracellular matrix in development and cancer progression. *Int. J. Mol. Sci.* 19, 3028 (2018).

52. Ricca, B. L. et al. Transient external force induces phenotypic reversion of malignant epithelial structures via nitric oxide signaling. *Elife* 7, e26161 (2018). Kenny, P. A. & Bissell, M. J. Tumor reversion: correction of malignant behavior by microenvironmental cues. *Int. J. Cancer* 107, 688–695 (2003).

53. Bhat, R. & Bissell, M. J. Of plasticity and specificity: dialectics of the micro- and macro-environment and the organ phenotype. *Wiley Interdiscip. Rev. Membr. Transp. Signal.* 3, 147–163 (2014).

54. Hinohara, K. & Polyak, K. Intratumoral heterogeneity: more than just mutations. *Trends Cell Biol.* 29, 569–579 (2019).

55. Nordling, C. O. A new theory on the cancer-inducing mechanism. *Br. J. Cancer* 7, 68–72 (1953). Armitage, P. & Doll, R. A two-stage theory of carcinogenesis in relation to the age distribution of human cancer. *Br. J. Cancer* 11, 161–169 (1957).

56. Maley, C. C. & Greaves, M. *Frontiers in Cancer Research.* (Springer New York, 2016). doi:10.1007/978-1-4939-6460-4.

57. Martincorena, I. et al. Universal patterns of selection in cancer and somatic tissues. *Cell* 171, 1029–1041. e21 (2017). Turajlic, S., Sottoriva, A., Graham, T. & Swanton, C. Resolving genetic heterogeneity in cancer. *Nat. Rev. Genet.* 20, 404–416 (2019). Rozhok, A. & DeGregori, J. A generalized theory of age-dependent carcinogenesis. *Elife* 8, e39950 (2019).

58. Palmer, S., Albergante, L., Blackburn, C. C. & Newman, T. J. Thymic involution and rising disease incidence with age. *Proc. Natl. Acad. Sci.* 115, 1883–1888 (2018).

59. Bertolaso, M. La complessità del cancro. Verso nuove categorie concettuali per comprendere la rilevanza del contesto. *Med. Stor.* 5, 97–120 (2014). Bertolaso, M. Disentangling context dependencies in biological sciences. In *New Directions in Logic and Philosophy of Science* (eds. Felline, L., Paoli, F. & Rossanese, E.) (College Publications, 2016).

60. Rozhok, A. & DeGregori, J. A generalized theory of age-dependent carcinogenesis. *Elife* 8, e39950 (2019).

61. Bissell, M. J. & Hines, W. C. Why don't we get more cancer? A proposed role of the microenvironment in restraining cancer progression. *Nat. Med.* 17, 320–329 (2011).

62. Bertolaso, M. On the structure of biological explanations: beyond functional ascriptions in cancer research. *Epistemologia* 36, 112–130 (2013). Braillard, P.-A. & Malaterre, C. Explanation in biology: an enquiry into the diversity of explanatory patterns in the life sciences. In *History, Philosophy and Theory of the Life Sciences* (Springer, 2015). Silberstein, M. Reduction, emergence and explanation. In *The Blackwell Guide to the Philosophy of Science* (eds. Machamer, P. & Silberstein, M.) 80–107 (Blackwell Publishers Ltd, 2008). doi:10.1002/9780470756614.ch5. Plutynski, A. Explaining how and explaining why: developmental and evolutionary explanations of dominance. *Biol. Philos.* 23, 363–381 (2008). doi:10.1007/s10539-006-9047-5. Woodward, J. Causation in biology: stability, specificity, and the choice of levels of explanation. *Biol. Philos.* 25, 287–318 (2010). Germain, P.-L. Cancer cells and adaptive explanations. *Biol. Philos.* 27, 785–810 (2012). Richardson, R. C. & Stephan, A. Mechanism and mechanical explanation in systems biology. In *Systems Biology: Philosophical Foundations* (eds. Boogerd, F. C., Bruggeman, F. J., Hofmeyr, J. S. & Westerhoff, H. V.) 123–144 (Elsevier, 2007). Bechtel, W. Mechanism and biological explanation. *Philos. Sci.* 78, 533–557 (2011). Mitchell, S. D. Explaining complex behavior. In *Philosophical Issues in Psychiatry: Explanation, Phenomenology, and Nosology* (eds. Kendler, K. S. & Josef Parnas) 19–38 (Johns Hopkins University Press, 2008).

63. Mitchell, S. D. *Biological Complexity and Integrative Pluralism. Cambridge Studies in Philosophy and Biology.* (Cambridge University Press, 2003). doi:10.1007/s10539-004-5896-y.

64. Schaffner, K. F. Reduction: the Cheshire cat problem and a return to roots. *Synthese* 151, 377–402 (2006). Bertolaso, M. La dimensione non riduzionista della ricerca sperimentale dai modelli molecolari a quelli sistemici nella ricerca sul cancro. *Riv. di Neo-Scolastica* 4, 687–705 (2012). Bertolaso, M. Semantic roots of reductionism's limits. *Riv. di Neo-Scolastica* 3, 481–500 (2014).

65. Beatty, J. Natural selection and the null hypothesis. In *The Latest on the Best: Essays on Evolution and Optimality* (ed. Dupré, J.) 53–76 (The Linnean Society, 1987). Kitcher, P. The division of cognitive labor. *J. Philos.* 87, 5–22 (1990).

66. Mitchell, S. D. *Biological Complexity and Integrative Pluralism. Cambridge Studies in Philosophy and Biology.* (Cambridge University Press, 2003). doi:10.1007/s10539-004-5896-y.

67. Bertolaso, M. The neoplastic process and the problems with the attribution of function. *Riv. Biol.* 102, 273–295 (2009). Bertolaso, M. On the structure of biological explanations: beyond functional ascriptions in cancer research. *Epistemologia* 36, 112–130 (2013).

68. Bertolaso, M. & Campaner, R. Scientific practice in modelling diseases: stances from cancer research and neuropsychiatry. *J. Med. Philos.* 45, 105–128 (2019).

69. Bertolaso, M. *Philosophy of Cancer: A Dynamic and Relational View.* (Springer, 2016).

70. Bertolaso, M. *Philosophy of Cancer: A Dynamic and Relational View.* (Springer, 2016). Cherubini, C., Gizzi, A., Bertolaso, M., Tambone, V. & Filippi, S. A bistable field model of cancer dynamics. *Commun. Comput. Phys.* 11, 1–18 (2012). Loppini, A. et al. On the coherent behavior of pancreatic beta cell clusters. *Phys. Lett. A* 378, 3210–3217 (2014). Bertolaso, M. et al. The role of coherence in emergent behavior of biological systems. *Electromagn. Biol. Med.* 34, 138–40 (2015). Saetzler, K., Sonnenschein, C. & Soto, A. M. Systems biology beyond networks: generating order from disorder through self-organization. *Semin. Cancer Biol.* 21, 165–174 (2011). Plankar, M., Del Giudice, E., Tedeschi, A. & Jerman, I. The role of coherence in a systems view of cancer development. *Theor. Biol. Forum* 105, 15–46 (2012).

71. Bertolaso, M. & Ratti, E. Conceptual challenges in the theoretical foundations of systems biology. In *Systems Biology, Series: Methods in Molecular Biology* 1702, 1–13 (Springer, 2018).

72. Palumbo, M. C., Colosimo, A., Giuliani, A. & Farina, L. Functional essentiality from topology features in metabolic networks: a case study in yeast. *FEBS Lett.* 579, 4642–4646 (2005). Palumbo, M. C., Colosimo, A., Giuliani, A. & Farina, L. Essentiality is an emergent property of metabolic network wiring. *FEBS Lett.* 581, 2485–2489 (2007). Giuliani, A. Collective motions and specific effectors: a statistical mechanics perspective on biological regulation. *BMC Genomics* 11 Suppl 1, S2 (2010). Bertolaso, M. *How Science Works. Choosing Levels of Explanation in Biological Sciences.* (Aracne, 2013).

73. Brenner, S. et al. Understanding complex systems: top-down, bottom-up or middle-out? *Novartis Found. Symp. Complex. Biol. Inf. Process.* 239, 150–159 (2001).

3 Cancer as a System

Hard Lessons from Physics and a Way Forward

Thea Newman

Overview

My aim in this chapter is to provide a scientific rationale for a significant shift in the research strategy of cancer biology and more generally in much of the research on the complex diseases that continue to defy medical science. I use examples from physics, a discipline in which many complex phenomena have been thoroughly understood, to illustrate some hard-won lessons from two centuries of research in those areas. The overarching lesson is that reductionism is almost entirely helpless in understanding complex phenomena in physics, even though these problems appear extraordinarily simple compared to the problems one encounters in biology. On a positive note, complex systems have been understood in physics using a conceptual approach centered on "mesoscale constructs." I propose that this approach is adopted as a new research strategy in the cancer field.

To set the scene, section 3.1 defines some terms. Through being abstract, these terms are applicable to both physical and biological settings. Section 3.2 then examines two everyday physical systems: water and metals. I explain how these systems are understood in a modern physics setting and how this understanding is achieved, necessarily, by abandoning the reductionist approach. This section will contain a number of physics terms, which may be unfamiliar to the reader. I have tried to flag these clearly so that jargon does not obscure the underlying arguments.

In section 3.3, I apply the lessons learned from physical systems to a critique of reductionism in medical research, with an emphasis on cancer, although the lessons are more widely applicable.

In section 3.4, I revisit the physics examples with the aim of framing a more effective biological research strategy and discuss how this might be applied to cancer. I believe a conceptual systems approach used in physics has the potential to transform cancer research, yet will require an upheaval in how this research is organized and conducted. The chapter ends with a summary of the main points and some concluding remarks.

3.1   Life as a Hierarchy of Systems

I will use the word "system" to denote an ensemble of a large number of entities, which I call "components." Generally, the system does not have any additional material existence beyond the components and their mutual interactions, although the system may be subject to external influences (e.g., through the boundary that separates the system from its surroundings).

The following question may be posed:

Can the system (i.e., the ensemble of components) be understood directly in terms of its components?

Or, in other words:

Is an arbitrarily detailed knowledge of the components sufficient to directly achieve an understanding of the properties and behavior of the system?

At first glance, one might respond:

Of course. How else is one to understand the ensemble if not in terms of its components—one has nothing else to go on.

This commonsense reply, while natural, has been known from many years of practice in physics to be incorrect for a vast array of systems, each of which is far simpler than most systems one encounters in biology. Indeed, this is the crux of the matter and the subject of this chapter, and it leads to the following question:

To what degree is the component-centric approach of modern biology justifiable given two centuries of experience in physics, which has seen outstanding successes in understanding and manipulating systems through an overt and necessary dismissal of such an approach?

In subsequent sections, I will aim to explain the failure of the component-centric (i.e., reductionist) approach in understanding physical systems and describe the successful approach based on what I am calling here, for the sake of generality, "mesoscale constructs" and then ponder how such an approach could be applied in a cancer-related context.

In general terms, each system is itself a component of a larger system. Thus, one has a ladder, or hierarchy, of systems. (As an aside, this hierarchy in the natural world is a primary cause of subdisciplinary compartmentalization in the sciences, a deficiency that leads to overspecialization and lack of cross-fertilization between research areas.) It is helpful to provide two preliminary examples, one from physics and one from biology, to illustrate the system hierarchy concept. Let us consider a glass of water and, separately, a human body (figure 3.1).

It is obvious that the hierarchy of systems for water exists only below the molecular scale. The hierarchy spanning atomic, nuclear, and subnuclear levels will be common to

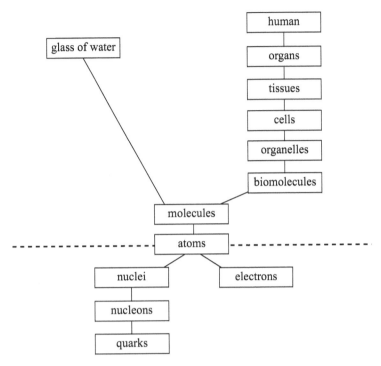

Figure 3.1
A representation of the hierarchy of systems for a glass of water and a human. The dashed line demarcates the subatomic hierarchy that is assumed for our purposes to be common to all systems considered.

all forms of matter and in most cases will be irrelevant to the observable properties of the systems at macroscopic scales. For this reason, we will restrict our attention to those scales of 10^{-10} m and above (above the dashed line in figure 3.1). We see then, in the first example, an absence of hierarchy between the H_2O molecules and the glass of water. In stark contrast, we see a dense hierarchy of systems in the biological example of the human body, with structural systems at numerous scales spanning molecular to organismal. Life is a hierarchy of systems, and we will return to this point in section 3.4.

First, let us consider nonhierarchical physical systems and ask to what degree they can be understood directly in terms of their atomic/molecular components.

3.2   A Physics Perspective on Systems

Over the past two centuries, many physical systems have been understood. Here I choose two of the simplest as they are commonplace and will be familiar to all. When I say "simplest," I mean in terms of their components. As we shall see, their behavior as systems is very challenging to unlock and has required a succession of nontrivial insights.

3.2.1   Example I: Water

Water is an apposite choice, as it is essential to life. In its own right, water exhibits an astonishing range of system-level behaviors, yet at a molecular scale, as discussed above, it is nothing other than an ensemble of identical components, H_2O molecules, whose structure is known with great precision (figure 3.2a). The higher electronegativity of oxygen pulls charge from the hydrogen atoms, giving these molecules a strongly polar configuration, from which arises the oriented, molecular interactions of hydrogen bonding.

With this in mind, then, we can restate our question of the previous section as follows:

Given detailed knowledge of H_2O molecules, can one directly predict the behavior of a large ensemble of such molecules, in the presence of various external influences (e.g., gravity, or gradients of pressure and temperature)?

A small selection of these system behaviors is illustrated in figure 3.2b–g, including cresting waves, vortices, fractal frost patterns, ice crystals, mackerel clouds, and Rayleigh-Bénard convection cells. Among other fascinating phenomena, one could also mention the very existence of the liquid state of water; the sharp phase transitions between solid, liquid, and gaseous forms of water; and the eighteen known phases of ice.

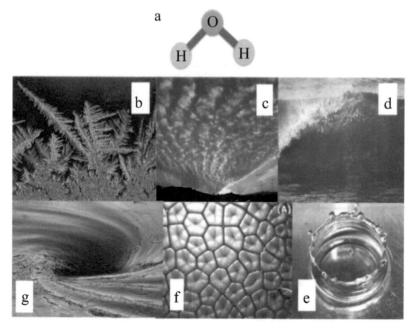

Figure 3.2
A collection of images representing a small subset of the many behaviors of a system of water molecules:
(a) the atomic structure of a water molecule and clockwise: (b) frost pattern on glass, (c) mackerel cloud,
(d) cresting wave, (e) splash pattern, (f) Rayleigh-Bénard convection cells (from Getling[1]), and (g) fluid vortex.

Borrowing from the language of biology, one could say, somewhat tongue-in-cheek, that the "genotype" of water (i.e., H_2O) gives rise to a dazzling array of "phenotypes" as one changes the "environment" in which the "genotype" is "expressed"—pleiotropy par excellence.

The deadly serious point, though, is that none of these behaviors can be directly predicted and understood from knowledge, alone, of the molecular structure of water. Why? Because it is too difficult.

Instead, our understanding of water at the systems level has arisen through experiment in tandem with abstract conceptual thinking over the past two centuries. This has required brilliant leaps of imagination and subsequent implementation of sophisticated mathematics.

For brevity, I will focus here on those behaviors in water that involve flow, for example, von Kármán vortex streets and Rayleigh-Bénard convection (arising, respectively, from large pressure and temperature gradients). These behaviors can be experimentally demonstrated with high levels of precision and control, but how does one achieve a predictive scientific understanding of them? The answer is through the derivation and manipulation of an equation that describes the spatial and temporal variations in the velocity and pressure of fluid. This was derived in the early nineteenth century and is called the Navier-Stokes equation. Such is this equation's continued importance that in 2000, its general solution was named by the Clay Mathematics Institute as one of its seven millennium problems, each of which comes with a prize of $1 million if successfully solved.

The Navier-Stokes equation was derived long before detailed knowledge of the molecular structure of water existed and yet has remained unchanged, since its form has nothing to do with the precise details of water molecules. (These details, however, will affect values of parameters that appear in the equation, such as the viscosity coefficient.) The equation is conceived to describe the velocity of an abstract mesoscale construct called a "fluid packet." One can motivate the fluid packet by considering a small volume containing billions of water molecules, at the scale of, say, one micron. This is small enough from the systems perspective to be considered infinitesimal. By considering the mechanical forces acting on the fluid packet and applying Newton's laws of motion, one derives the Navier-Stokes equation.

This equation is nonlinear and has proven too difficult to solve in complete generality. However, one can use mathematical methods to approximate the equation, yielding simpler theories that describe the behavior of water. This can be achieved under a variety of external conditions—such as high-velocity flow of water past an obstacle, which gives rise to downstream vortex streets,[2] or when a thin layer of water is heated from below, which gives rise to Rayleigh-Bénard convection patterns.[3] Such methods constitute the field of fluid mechanics,[4] a major activity in physics, applied mathematics, and engineering departments worldwide. The success of this methodology allows precise engineering of sea and air vessels, pipes, and turbines and provides an underpinning for predicting the effects of large-scale fluid flow in atmospheric physics and hydrology—all without a mention of the molecular components.

From a modern perspective, one might ask whether the Navier-Stokes equation can be derived directly from a molecular scale. The answer is essentially no. The equation can be derived rigorously for low-density fluids (i.e., gases) under certain limiting assumptions, a major achievement that goes back to Ludwig Boltzmann's research on statistical mechanics in the 1870s.[5] However, as the density of the fluid increases in the liquid state, the bimolecular collisions of single molecules are overwhelmed by far more complex polymolecular collisions and interactions, and the systematic derivation of a flow equation becomes prohibitively difficult.

For this reason, I believe water provides an important lesson for our thinking about biological systems. Here is a problem for which we know everything we could wish about the components, but we are helpless in using this information directly to predict the extraordinary variety of system-level behaviors that water exhibits. Yet, at the same time, by abandoning the components and introducing constructs at the mesoscale (i.e., fluid packets), one can deduce a theoretical framework that is very successful in predicting and describing the system behaviors. I think it is helpful to visualize the fluid packet as a stepping-stone, which allows one to cross a river, from the "microscale side" to the "macroscale side," a river otherwise impossible to cross in a single bound (figure 3.3).

The understanding of other system behaviors of water, such as phase transitions and ice crystals, relies not on flow but on thermodynamics.[6] This is another area developed in the

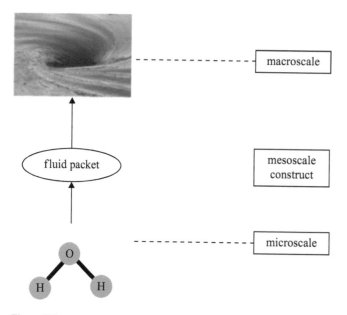

Figure 3.3
The fluid packet as a mesoscale construct; a stepping-stone across the "river of scales" separating H_2O molecules and the system-level behavior of water.

early to mid-1800s, prior to detailed knowledge of molecular components, and is based on abstract notions such as heat, entropy, and free energy, which required deep conceptualization to be formulated into the laws of thermodynamics, which have remained unchanged for nearly two centuries. Had the pioneers of thermodynamics had access to component-level "big data" of gases and liquids, and endless data sets of the coordinates and velocities of trillions of molecules, one has to doubt whether their progress would have been as successful or universally applicable. One can have too much information.

3.2.2   Example II: Metals

As a second example from the physical sciences, let us consider a piece of metal, say copper, at room temperature. In our everyday lives, we encounter copper in the form of household wiring, since it is, like most metals, an excellent conductor of electricity. And so, in the spirit of our systems discussion, a natural question might be as follows:

Can we directly predict the electrical and thermal properties of a piece of copper from an arbitrarily detailed knowledge of copper atoms? Can we understand why copper is a good conductor of electricity, while carbon, for example, is a poor conductor when in the form of diamond, and silicon has weak but nonlinear conducting properties (i.e., is a semiconductor)?

These systems appear more sedate than the dazzling behaviors of water and may not entice the reader, but we should recall that the understanding of metals, and the subsequent understanding of semiconductors, has changed our world perhaps more than any other scientific advance since World War II; I mean, of course, the silicon revolution. These advances occurred in an oft-overlooked field called "solid state physics,"[7] which has quietly collected twenty Nobel prizes since 1956.

There were reductionist theories of metallic conduction going back to the early twentieth century, based on a model of electrons as particles flowing along the wire, under the forcing of an external voltage, and bumping now and again into the array of metallic ions that constitutes the bulk material of the wire. This picture is unable, however, to account for a range of properties of conductors and in particular overestimates their thermal conductivity by a factor of several hundred. It is also helpless in explaining phenomena such as semiconductivity. With the advent of quantum theory in the 1920s, the field of solid-state physics was shortly thereafter established through the application of quantum mechanics to ensembles of very large numbers of atoms, with the aim of explaining their electrical, magnetic, optical, and thermal properties. The type of science employed here is extreme systems analysis, where only a very few properties of the atomic components are required, but then a series of mesoscale constructs are introduced that eventually allow one to make predictions at the macroscale, with tremendous success.

Due to limitations of space, I cannot provide a detailed explanation of this series of constructs, but I will aim to sketch out the stepping-stones allowing us to understand the link between copper atoms at the microscopic component scale and a metallic conductor at the macroscale. I believe some of the lessons learned are instructive in considering biological systems.

A reductionist approach, based directly on the components, focuses on individual electrons in the wire being pushed through an array of ions (i.e., the pre-quantum model mentioned earlier). Everything changes once one considers the quantum mechanical nature of the problem, and this despite the fact that we are considering a macroscopic system at room temperature.

First, since electrons are indistinguishable, one cannot speak of individual electrons but must rather describe electrons via a "many-electron wavefunction."

Second, it turns out that the electronic wavefunctions, on experiencing the regular lattice of ions, cease to be particle-like and become delocalized wave-like states (called "Bloch states"), which extend over hundreds of atomic distances.

Third, the thermal motion of the ions (which becomes more agitated with increasing temperature, thereby causing metallic resistance to increase with temperature) is also described quantum mechanically and by a spectrum of highly extended wave-like states called "phonons."

Fourth, because of the Pauli exclusion principle, each Bloch state can only be occupied at most by one electronic degree of freedom, and so most electronic states are "frozen" and not sensitive to temperature changes. Put bluntly, the vast majority of "electrons" (strictly speaking, electronic degrees of freedom) in a wire at room temperature are oblivious to temperature and act as if the wire were at absolute zero. This explains why the pre-quantum approach vastly overestimates the thermal conductivity of metals.

This last point then provides the key to the conceptual stepping-stones. Since only those Bloch states at the highest energy levels are able to respond to temperature or external forcing, the theoretical strategy is to recast the problem purely in terms of these states. This is the abstract notion that unlocks the problem and provides the mesoscale stepping-stones, which have names such as "Fermi surface," "quasi-particle," and "band gap" (figure 3.4). Modern solid-state physics describes the behavior of metals and semiconductors in terms of interactions between quasi-particles and phonons and is able to successfully predict a diverse range of properties, even in more exotic solid states at very low temperatures, such as superconductivity.[8]

3.3   Implications for Cancer Research

What messages can be distilled for biology, and cancer research in particular, from this whistle-stop tour of the physics of water and metals?

The overarching lesson is this: despite the simplicity of the components of water and metals, it is, in human terms, essentially *impossible* to predict their system properties directly from knowledge of their components. On the contrary, surprisingly little information of the components is required for a powerful system-level understanding. This may

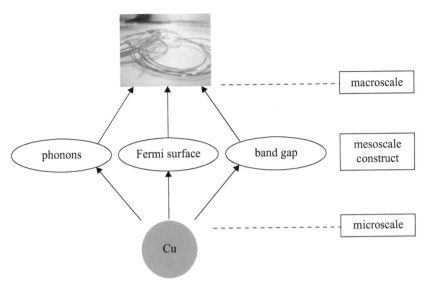

Figure 3.4
Some of the key mesoscale stepping-stones linking copper atoms to metallic conductivity in copper wire.

seem counterintuitive at first but arises due to an important and deep concept: new laws emerge when a large number of components form an ensemble. This was emphasized nearly fifty years ago in the beautifully entitled essay by Phil Anderson: *More Is Different.*[9] This idea is articulated widely in physics through terms like "emergent behavior," "effective theories," and "symmetry breaking."[10]

I have tried to lead the reader, who is most likely not a physicist, through the physics perspective with as little of the jargon as possible, but I realize that the fields of fluid mechanics and solid-state physics, while highly successful and with important lessons for other disciplines, are also technical and mathematical. Fortunately, the lessons we distill for other disciplines are not dependent on these technicalities.

Arising from our discussion so far, a question one might pose to the molecular biology community is as follows:

Given that the system properties of water and metals are not amenable to a reductionist approach, what grounds do we have in biology and medicine for expecting success in attempting to understand systems (e.g., human patients) directly from components (e.g., genes) as heralded in phrases such as "cancer is a genetic disease"? After all, is not water a molecular liquid and copper wire an atomic metal?

Obviously, biological systems are far more complex than physical ones. As we observed earlier, they are densely hierarchical in nature. In addition, they have huge heterogeneity—not one type of component as in water and copper but hundreds (e.g., cell types in an organism) or tens of thousands (e.g., protein types in a cell). They are also strongly out of equilibrium (by necessity in order to be living): a state that would discourage most physicists, as nonequilibrium systems have for many years remained a dimly understood research frontier in the physical sciences.[11] And, most fundamental of course, biological systems are the product of a process of evolution. Could it be that these significantly complicating aspects (hierarchy, heterogeneity, disequilibrium, evolution) conspire to breathe life back into the reductionist approach? Perhaps, but in the light of the physics perspective, I believe it is highly doubtful. A pragmatic view is that these aspects only make the reductionist approach, as it were, "more impossible."

To counter my view, one can look for historical evidence of success from the reductionist approach to system-level understanding in biology. Do we have solid bridges that allow successful prediction between the levels of organizational hierarchy in living systems? Is the cell understood in terms of its constituent biomolecules? Are organs understood in terms of their cell populations? Is the patient understood in terms of their malfunctioning tissues? Despite our wealth of knowledge in biology, the answer to all of these questions is presently "no." Moreover, it appears that the ethos in much of current biology is not to build bridges between levels of the hierarchy but to anchor all understanding at a molecular level. One might call this "ultra-reductionism." It is ambitious to be sure—the idea that once all of the "omics" have been exhaustively mapped and all details are collected and organized, then "everything will become clear." On the basis of the foregoing discussions about physical systems, I do not believe that the strategy of ultra-reductionism can lead to success and indeed appears to be optimally poor.

The hard evidence from the medical field would appear to support this pessimistic view. To what degree have fifty-plus years of molecular biology transformed medicine? The vast majority of biologists would have to say, reluctantly, that it has been disappointing—some successes, of course, but not transformational and not world changing as was hoped for in the early days of molecular biology or in the hubris preceding the human genome project. In fact, the outcomes are very much what one might have expected from the physics perspective. For those aspects of disease where there is a direct, causal one-to-one link from an aberrant gene to a specific pathology, then, of course, genetics is supreme and provides the vital link to understanding the pathology and providing a pathway to treatment. But, for those pathologies caused by polygenic influences and where strongly pleiotropic genes are acting, namely, where the components are strongly interacting, then the reductionist approach has been weak in explaining and treating such pathologies. The poster child for this lack of advance is cancer, but one can also make the same arguments for diabetes and neurological disorders. The response of most in the biomedical community to this lack of success within the reductionist approach has been to double down on the molecular approach, rather than to step back and vigorously search for other strategies. Such a response is pursued in good faith but from the outside appears to be one of desperation, like the gambler placing their last few chips on the same high-stakes game that has already ruined them.

One important point cannot be overstressed here. The physics perspective is *not* stating that system components are unimportant per se. One only has to look at physics to realize how passionately that discipline has pursued molecular, atomic, nuclear, and subnuclear phenomena. The component details are there to be researched and understood, for their own sake and for our own understanding of how the fabric of Nature has been woven. But, as I have emphasized repeatedly, a deeper understanding of component details does not necessarily enable an understanding of higher-level phenomena. The science of systems requires different intellectual and conceptual approaches—it is not Lego where one simply arranges the building blocks into different patterns. It could not be more different than that.

In the context of biology, the exploration of genetics, molecular biology, and biochemistry is beautiful and necessary to gain a deeper understanding of Nature's fabric.[12] However, two centuries of experience in physics teaches us not to expect transformational impact in trying to build directly from these components to an understanding of systems, such as cells, tissues, and organisms. Rather, these components will provide parametric inputs into independent lines of research, which will formulate new theories and concepts to understand life at various levels of the system hierarchy. This bears similarity to the modus operandi of biology prior to the 1970s, before the advent of the molecular biology revolution. What underpowered that era of biology, I believe, was the lack of theoretical work to complement the experimental work. Engineers, physicists, mathematicians, and computer scientists were not working in significant numbers alongside their biologist colleagues; hard-won insights from other areas about how systems work were not finding widespread application to biology. Ironically, strong physics input did help to drive the molecular biology approach, both through theoretical insights (e.g., inferring the structure of DNA) and technological innovation (e.g., X-ray crystallography of proteins). One would like to think interdisciplinary working is more common nowadays. What is still lacking, though, is a culture of intellectual partnership between disciplines; rather, disciplines tend to be co-opted into the life sciences based on their technical value rather than their conceptual value.

There is a yet harder lesson about reductionism that we learn from the example of metals—what one might call "the problem of entity identity." In the case of metals, we saw that the electron components that one has to hand, when viewed singly, cease to exist when one considers an ensemble of metal atoms, being replaced by the many-electron wavefunction, which then manifests itself in terms of Bloch states.

Thus, we have to consider two classes of difficulty when attempting to explain systems in terms of their components. In the first class, the components do not change their character when in the ensemble (e.g., water molecules in a glass of water), but it is beyond human ability to directly relate the components to the system. In the second class, the components themselves cease to exist when immersed in the ensemble, at least in the form we ascribe to them when they are single entities. In such a case, the reductionist approach cannot even get out of the starting blocks. It is meaningless to try to work from the bottom up as the components' identities depend on the system environment. In such a case, bottom-up and top-down are eerily entangled, and this must be accounted for in even defining the identity of the component entities.

An example of the entity identity problem in cancer biology occurs when considering cancer cells as components of a system such as a tissue or tumor. One might study those cancer cells individually in a petri dish to determine their properties, their identity. But, in the tissue or tumor, the cells will be exposed to a different and far more complex biochemical and biomechanical microenvironment than in a petri dish and will adapt their gene expression accordingly, effectively changing their identity. This has been known for many years through experiments investigating the effects of different tissue architectures and microenvironments on cancer cells.[13] To go one step further, it might be that the very concept we have of discrete, autonomous cells in such environments is insufficient or even plain wrong. After all, the lesson from metals is that the system can profoundly change the very nature of the components. The entity identity problem can be handled in the case of metals through the formalism of quantum mechanics, where very strict rules apply to how an environment affects atomic-scale components. As yet, there is no such set of rules in cancer biology, but the logical conundrum is very similar and deserves far greater theoretical study.

3.4   Toward a New Approach

In the previous section, I have enunciated a set of lessons from the physics perspective, which challenge the reductionist mind-set of cancer research. In the following, I will come to a constructive conclusion from this same perspective, which can motivate a new approach. The clear positive message from the studies of water and metals is that, despite the failure of reductionism, these are solved problems! And, that is all that matters.

This last point must be emphasized. Understanding a system scientifically is not equivalent to causally connecting the system behavior to its components. This latter aim is more of a subjective preference or a psychological state. It is typically held by those scientists whose expertise is in the study of components and who, quite naturally, wish to show that such study can yield information about not only the components but also the systems comprising these components.

One finds the most extreme example of this ironically in physics, where the community has to a large degree bifurcated into reductionists (e.g., atomic, nuclear, particle physicists) and "systematists" (e.g., solid-state, material, and biological physicists). We have for decades heard from particle theorists, stating with sincere conviction that they are searching for "a theory of everything," by which they mean a theory that can unify the standard model of particle physics and Einstein's theory of general relativity (the modern theory of gravity). A unifying theory would describe all four fundamental forces. Such a theory would be a marvelous step forward in our understanding of the underlying structure of the universe, but it would be strictly a theory of the fundamental forces, not "a theory of everything." It would provide not one iota more insight into our understanding of water or metals or cells or cancer or the human brain, since these are systems whose behavior is insensitive to nuances in the subnuclear fabric of the universe. Perhaps the "theory of everything" is a marketing tool for sales of popular science books rather than a widespread delusion among a large group of very able scientists, but it is certainly not what it literally purports to be. Fortunately for physics, this is a mind-set affecting only a small subset of the community. Unfortunately for biology, a similar philosophy, encapsulated in phrases such as "cancer is a genetic disease" or "without a gene there is no mechanism," affects great swathes of the community, across fields whose supposed areas of study are complex systems, such as cancer patients or developing embryos—human systems that most likely are governed by scientific principles and mechanisms that are inaccessible from knowledge alone of molecular details.

On this note, it is worth repeating an idea that was mentioned in the previous section. I stressed that the negative statement "understanding components is not the right strategy for understanding systems" does not equate to the negative statement "components do not matter." System behavior *will* be sensitive to the nature of the components to a greater or lesser degree. It is just that one cannot *understand* the system in terms of its components. The periodic table of elements is a good example to illustrate this. As we add one proton to the nucleus of an element, it changes its identity to the next element along in the table, and we can use quantum mechanics to work out the effect that has on electron levels and so on. However, this firm grasp of the atomic details distinguishing one element from another does not allow us to directly understand the properties of ensembles of these different atoms—why some substances are hard, others brittle, some liquid at room temperature, others solid, some dull, others shiny, some conductors, others insulators, and so forth. For this one needs the systems approaches described in section 3.2. In the same manner, different alleles of a particular gene may have greatly different influences on the likelihood of a particular cancer arising. But this does not mean that understanding the nature of that cancer is necessarily possible through detailed examination of that particular gene and its immediate interactions.

Returning to the positive message: understanding the behavior of water and metals are solved problems, and the successful strategy has been to find the appropriate mesoscale constructs. For water, this was the fluid packet. For metals, the constructs were Bloch states, Fermi surfaces, and band gaps. These constructs provide stepping-stones from the microscale to the macroscale, providing the conceptual connection from the molecular complexities of the components to the often relatively simple and deterministic behavior of the system.

I propose that the approach of identifying mesoscale constructs would be a fruitful strategy for cancer research: to identify the stepping-stones.

This approach is almost certainly not equivalent to joining the dots between the hierarchy of systems in figure 3.1 (biomolecule to cell to tissue to patient). If that were the case, if the scales in the biological hierarchy were equivalent to the mesoscale constructs, then I think we would have identified clear paths from genes to cancer patients by now and be using these to design effective treatments.

Using the discrete levels themselves as stepping-stones implicitly assumes that none of the intermediate systems in the hierarchy are harder problems to solve than the final system at the top of the hierarchy. For example, is the detailed understanding of a cell a simpler or harder problem than understanding the cause and therapeutic intervention for a particular cancer? I do not believe the answer is obvious. I would wager that a detailed understanding of the cell is one of the very hardest problems in biology, and so placing the understanding of the cell before understanding certain diseases imposes a tremendous bottleneck on medical progress.

One way to think of this is through an analogy from ecology. One could write down the system hierarchy for, say, a herd of wildebeest on their annual migration across the African plains. I think it is clear that achieving a good understanding of the population dynamics of the herd is a far simpler problem than understanding one of the herd's components (i.e., the anatomical, neurological, and physiological properties of a single wildebeest).

On taking the view that the structural biological hierarchy of systems does *not* provide the mesoscale constructs, there is a need then for a new kind of intellectual journey. This requires courage as one must put aside a great deal of detailed work on biological structures to search instead for what will appear at first to be quite abstract constructs.

Recall that this is how Nature has arranged the simpler systems of water and metals. Through a collusion of bottom-up and top-down, the mesoscale constructs can be said to be "those effective components through which the system allows itself to be understood."

The search for these constructs is not blind—they are highly constrained in that they must be deducible (although not necessarily derivable) from the microscale components, and they must, by definition, be predictive of system behavior.

I would add, more as an intuitive comment, that such constructs should also lead to relatively simple models of system behavior. Experience shows that when Nature is revealed, it is most often through a lens of great simplicity. Examples of this abound, from Einstein's theory of relativity to Darwin's theory of natural selection. Theories and models that are dense in parameters and unknown constants are not what is being sought; in fact, they are the antithesis.

The strategy proposed here would not be created from a vacuum; far from it. There is a wealth of experimental and theoretical work in the biological and biomedical sciences that is not component-centric. However, it is often starved of resources and often speaks a reductionist language in order to survive.

In terms of the theoretical work, there is also much to be done to elevate this to the status of experimental work in biology and medicine. Theoretical work in biology has generally played "second fiddle" over the years, in part because its scientific impact has been modest, on the whole. This lack of impact is either because living systems are by their very nature not amenable to theoretical and conceptual approaches or because the theoretical approaches employed to date are not honed to the task at hand. I believe it is the latter. Only a small subset of theoretical work in biology follows the approach described in this chapter, despite its demonstrable success in the physical sciences. Recent work on stem cell dynamics[14] from the group of Ben Simons in Cambridge stands out in this regard, with expertise spanning solid-state physics, cell biology, and medical physics. Rather, we often find mathematical models in biology teeming with parameters, which can be fit to any data set and therefore have little predictive power. Or we find theoretical approaches based very loosely on the philosophy of mesoscale constructs but that are not sufficiently sophisticated in their conception. For example, there is an entire industry of theoretical work using concentrations or densities as the variables for models that are subsequently analyzed in a handle-turning fashion. This falls well short of the bold conceptual approach that is required to unlock the problems of biology.

There is not space here for a review of existing theoretical work in biology and its suitability to the task at hand. Instead, I would like to briefly report some observations based on personal experience. Over the period from 2011 to 2017, I encouraged my research group to pursue an overt strategy of simplicity and "unrealistic ambition" in our theoretical work: to assume that a simple explanation for a phenomenon existed and to craft models with the minimum of parameters ("no more than two") that could clearly encapsulate a biological hypothesis. We consistently found that this approach was "unreasonably effective." In two cases, I ascribe this to our identification of mesoscale constructs that framed the subsequent theoretical description: (1) long closed feedback loops in gene networks[15] and (2) an immune escape threshold in cancer incidence,[16] summarized in figure 3.5. In

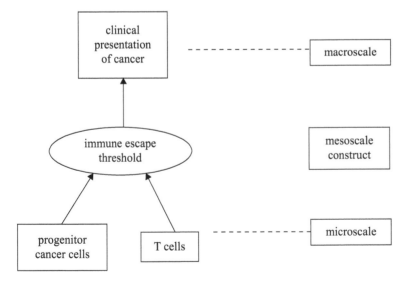

Figure 3.5
A mesoscale construct connecting the interactions of progenitor cancer and immune cells with the likelihood of subsequent clinical cancer presentation (cf. Palmer et al.[17]).

both cases, the resulting theoretical frameworks had high predictive power and were able to explain large amounts of data, thereby providing new biological insights.

I will expand on the second case mentioned above,[18] to help give some insight into how the approach described in this chapter can be carried out in practice. This project was initiated by observing the extraordinary fact that the known exponential decrease in thymic T-cell output with age mirrors perfectly the data on age-related increase in risk across a number of cancers. A reductionist approach to connect immune system decline and increasing cancer incidence with aging, in terms of the myriad cell types involved and their various gene expressions, is a task requiring thousands of scientists and enormous budgets. Our approach was to overtly ignore "mechanism," as defined in a reductionist sense, and to posit that there was a causal relation between immune decline and cancer incidence, but to then frame this within a mesoscale construct, so that we could derive some hard predictions and then test these with data. The mesoscale construct was based on the hypothesis that cancer progenitors are constantly arising at a rate independent of age but are eliminated by the immune system. However, very rarely a neotumor may evade the immune system long enough to exceed a threshold and become protected against immune control—this immune escape threshold was our mesoscale construct. It encapsulates an enormous amount of molecular and cell biology but does not require those details to be known in order to be defined and for some consequences of its existence to then be calculated. Since the immune system weakens with age, the immune escape threshold inexorably lowers with age, and thus the chance of a neotumor becoming established inexorably increases with age. The outcome of this project was a new prediction for cancer incidence with age based on immune decline that strongly outperforms the traditional model of cancer incidence based on sequential mutations. Our theory also straightforwardly explains the strong gender bias in cancer incidence, something that has evaded reductionist explanation for decades.

3.5   Summary and Concluding Remarks

Briefer forms of these arguments have made an appearance in previous essays of mine on the cultural divisions between physics and biology, as well as the power of simplicity in understanding biology.[19] My own understanding of the relative strengths and weaknesses of reductionism in biology has continued to evolve throughout this time.

In this chapter, I have challenged the reductionist modus operandi, which since the 1970s has held sway in much of biology and medicine, particularly cancer research. I have taken two examples from physics, water and metals, where two centuries of effort has concluded that it is essentially impossible to explain system-level behavior from the molecular components. Despite this, these systems, along with many others in physics, are extremely well understood through an entirely different research strategy, which I have explained in terms of "mesoscale constructs." I have argued that there is little reason to suppose that biological systems are more amenable to a reductionist approach than physical systems. In fact, one has every reason to suppose the reductionist approach is even weaker in biology. I have therefore proposed that given the lack of transformational progress in treating diseases such as cancer, we should engage in a significant new strategy for medical research, importing the valuable lessons and strategies from two centuries of study of physical systems and honing them for application in human medicine.

For convenience, I summarize below the lessons of physics as they pertain to systems:

1. Generally, it has proven impossible to predict a system's behavior directly from knowledge of the system's components.

2. This is because ensembles of components have their own emergent laws.

3. A component-centric or reductionist approach may succeed where there is a direct one-to-one relationship between a system behavior and a component.

4. Such reductionist approaches will generally fail where there is a many-to-many relationship between system behaviors and components.

5. Despite the failure of the reductionist approach in physics, many systems have been understood in great depth.

6. This success is due to conceptualization and utilization of "mesoscale constructs," which act like stepping-stones, connecting the microscale to the macroscale.

7. The mesoscale constructs require relatively little detail from the microscale components.

8. There are two classes of system:

    a. those for which the components do not change their intrinsic character through being in the ensemble, but nevertheless the system behavior can only be inferred through mesoscale constructs (e.g., water), and

    b. those for which the components do change their intrinsic character when in the ensemble, thereby requiring a confluence of bottom-up and top-down thinking to frame the problem and establish the mesoscale constructs (e.g., metals).

And, I append three statements relating to application of these lessons to biology, which have more the flavor of opinions of mine than lessons learned from physics:

9. The mesoscale constructs are highly unlikely to be the structural components intermediate between microscale and macroscale in the hierarchy of biological systems.

10. The problem of "entity identity": most problems in biology and medicine will fall under the second of the two classes defined above in lesson 8.

11. Theories using mesoscale constructs should be expressible in simple terms, requiring at most a handful of unknown parameters.

A lesson that was not intended and that emphatically should not be drawn from this chapter is that physicists have been more successful than biologists in understanding systems because they are better scientists. First, it is impossible to compare the disciplines as biological systems are far more complex in nature than physical systems. Second, one reason I think why physics has succeeded in understanding systems is because there is no hierarchy (cf. figure 3.1); that is, there is "nowhere to hide." This is perhaps more a statement on the psychology and sociology of science. For a community of physicists studying, say, water, they are literally faced with an ultimatum: "You have only H_2O molecules, on the one hand, and a dazzling array of phenomena, on the other. There is nothing in between. Either you re-create the system behavior from the bottom-up, starting with molecules, or you ignore the molecules and figure out the system behavior some other way." As we have seen, the first approach is too difficult, and the second is therefore the only option. This ultimately led to the Navier-Stokes equation and the laws of thermodynamics, which have been very successful. The mesoscale constructs approach has found subsequent success in a wide range of physical systems, its apogee being the development of the renormalization group,[20] a beautiful theoretical framework for certain critical phenomena, in which one can use mathematical methods to systematically derive mesoscale constructs from the microscale.

On the other hand, the community of biologists has always been confronted with not only a dazzling array of phenomena but also a dazzling array of structures, at every scale. I think this may have been profoundly distracting from a conceptual viewpoint, as instead of two strategies to choose from, one has potentially hundreds, trying to connect various scales in the hierarchy to one another directly. As a result, there has been little effort to seek conceptual approaches at the mesoscale that do not relate to the structural hierarchy. And, given the failure to bridge the hierarchical scales in a direct manner, the community has ended up painting itself into the corner of ultra-reductionism, convincing itself that in fact, understanding at all levels must ultimately come from the microscale, from genes and related biomolecular components. This logic is compelling but is, in my opinion, fatally flawed.

And where to from here? For the foreseeable future, I imagine the molecular biology research community will continue its work, in good faith, seeking direct connections between molecules and medicine, and will continue to garner the lion's share of resources, both financial and human. I wish them the very greatest success. Nothing would make me happier than to be entirely wrong and for a succession of "molecular magic bullets" to be developed in the next few years that once and for all remove the misery of cancer from the human race. From a physics perspective, though, I think this outcome is highly unlikely and so implore the scientific communities and the funders to vigorously pursue alternative strategies in parallel with the reductionist approach. In this vein, I believe the strategy I have sketched here, based on mesoscale constructs, to be worthy of closer attention.

Ultimately, transformation of our understanding of cancer and other diseases requires new generations of scientists who are able to think across disciplinary knowledge sets, connecting disparate ideas to create entirely new concepts and approaches. This requires a revolution in the way we train scientists. Within our current system, as soon as eighteen-year-olds make a choice between, say, physics and biology, they are on a one-way track to overspecialization and, ultimately, the irreversible acquisition of disciplinary belief systems. Broader education of young scientists is essential, throughout their training, in the spirit of an overarching natural philosophy rather than being pigeonholed into a science comprising separate disciplines.

Adopting new strategies for understanding and treating cancer requires large-scale change in our educational and research environments, change that will be highly disruptive. Most of us dislike change, particularly if it is rapid and large scale; our instincts, in universities and funding agencies, is to avoid this scale of disruption. But we must countenance it. Our discomfit as a scientific community is irrelevant when compared to our responsibility to the public—those who we work for and who pay for the research. The cancer deaths continue in their millions, year on year, each death a drawn-out agony for the patient and their family. They deserve far better from science than they have received to date. Personally, I remain entirely positive that science and medicine will ultimately prevail in managing (not necessarily curing) cancer, making it a benign condition. But the research community must be far more intellectually courageous and innovative in order to achieve this in decades rather than centuries.

Acknowledgments

The author gratefully acknowledges discussions with Julian Blow, Myles Byrne, Paul Davies, Stuart Lindsay, Adrian Saurin, Bernhard Strauss, Alastair Thompson, Emmy Verschuren, and Kees Weijer, along with all the members of her research group over the years.

Notes

1. Getling AV 1998, *Rayleigh-Bénard convection: structures and dynamics* (World Scientific).

2. von Kármán T 1963, *Aerodynamics* (McGraw-Hill).

3. Getling AV 1998, *Rayleigh-Bénard convection: structures and dynamics* (World Scientific).

4. Landau LD, Lifshitz EM 2003, *Fluid mechanics*, 2nd edition (Butterworth-Heinemann).

5. Keizer J 1987, *Statistical thermodynamics of nonequilibrium processes* (Springer).

6. Adkins CJ 1983, *Equilibrium thermodynamics*, 3rd edition (Cambridge).

7. Ashcroft NW, Mermin ND 1976, *Solid state physics* (Saunders College). Rosenberg HM 1978, *The solid state*, 2nd edition (Clarendon Press).

8. de Gennes PG 1999, *Superconductivity of metals and alloys* (Westview Press).

9. Anderson PW 1972, More is different, *Science* 177 393–396.

10. Laughlin R 2006 *A different universe: reinventing physics from the bottom down* (Basic Books).

11. Prigogine I 1980, *From being to becoming* (W. H. Freeman).

12. Alberts B, Johnson A, Lewis J, Raff M, Roberts K, Walter P 2012, *Molecular biology of the cell*, 5th edition (Garland Science).

13. Nelson CM, Bissell MJ 2006, *Of extracellular matrix, scaffolds, and signaling: tissue architecture regulates development, homeostasis, and cancer, Ann. Rev. Cell Dev. Biol.* 22 287–309.

14. Rulands S, Simons BD 2017, *Emergence and universality in the regulation of stem cell fate*, *Curr. Opin. Systems Biol.* 5 57–62

15. Albergante L, Blow JJ, Newman TJ 2014, *Buffered qualitative stability explains the robustness and evolvability of transcriptional networks*, *eLife* 3 e02863.

16. Palmer S, Albergante L, Blackburn CC, Newman TJ 2018, *Thymic involution and rising disease incidence with age*, *PNAS* 115 1883–1888.

17. Palmer S, Albergante L, Blackburn CC, Newman TJ 2018, *Thymic involution and rising disease incidence with age*, *PNAS* 115 1883–1888.

18. Palmer S, Albergante L, Blackburn CC, Newman TJ 2018, *Thymic involution and rising disease incidence with age*, *PNAS* 115 1883–1888.

19. Newman TJ 2011, *Life and death in biophysics*, *Phys Biol* 8 010201. Newman TJ 2014, *Water is a molecular liquid*, *Phys Biol* 11 033001. Newman TJ 2015, Biology is simple, *Phys Biol* 12 063002. Albergante L, Liu D, Palmer S, Newman TJ 2016, Insights into biological complexity from simple foundations, in *Biophysics of infection* (ed. Leake MC, Springer) 295–305.

20. Goldenfeld, N 1992, *Lectures on phase transitions and the renormalization group* (Addison-Wesley).

II THE SYSTEMS DIMENSION OF CANCER

4 The Logic of Cancer Treatment

Why It Is So Hard to Cure Cancer

Treatment-Induced Progression, Hyper-Progression, and the Nietzsche Effect

Sui Huang

What does not kill me makes me stronger.
—Friedrich Nietzsche

How wonderful that we have met with a paradox. Now we have some hope of making progress.
—Nils Bohr

Overview

Logically, given the crisp mechanistic rationales behind modern cancer drugs, we should be able to cure cancer. Yet there is a substantial deficit in the therapeutic benefit of any new cancer treatment compared to what one might expect given their mechanism of action. Could this discrepancy be due not to insufficient activity in suppressing tumors but rather to a hidden tumor-promoting activity of treatment itself that partially cancels its antitumor effect? Then, the perceived result of treatment is the net effect of two opposing forces: on the one hand, the desired cancer-reducing effect of the therapeutic intervention and, on the other hand, tumor progression *promoted by* the very treatment, as has been observed with various drugs, irradiation, surgery, and immunotherapy. The latter response may sometimes dominate treatment outcome and then appear as paradoxical. The epistemic habit in modern medicine of thinking in the terms of linear causality and molecular cascades keeps us defiant to accepting any such "double-edged sword" scenarios. Here, we present, in a nontechnical narrative, a conceptual framework based on principles of the theory of nonlinear dynamical systems that will facilitate the scientific comprehension of why cancer treatment can "backfire"—a response that epitomizes Friedrich Nietzsche's aphorism: "What does not kill me, makes me stronger." Understanding and applying this logic in cancer treatment may guide the development of a novel therapeutic modality that directly addresses an inherent but overlooked limitation of current cancer treatment.

4.1   Introduction: The Inherent Inefficacy of Current Treatments

Treatment failure in cancer therapy, even with the newest modalities, is a reality—much more so than is commonly acknowledged. The stream of celebratory reports on "breakthrough" therapies has numbed our awareness of the systemic failure of current cancer treatment. As a consequence, the insufficiency of therapies as such has remained rarely studied as a phenomenon in its own right. Life span of patients diagnosed with invasive cancer is typically extended by less than six months[1] even with the most recent therapies, and the overall response rate in terms of durable remission as of 2018 turned out to be less than 20 percent even in the case of immunotherapy, the therapeutic modality that after a long break has readmitted the word "cure" into the vocabulary of oncology.[2] The almost inevitable recurrence after remission following initial tumor regression is *the* major cause of therapy failure. Yet, although—or because—a new treatment "must work" in view of the rationale behind its mode of action, unfavorable outcomes rarely trigger scientific scrutiny. Instead, they simply fuel the ceaseless quest for ever more powerful ways to *kill* cancer cells. This chapter is an attempt at challenging this rigid mentality of "doubling down" and aims to motivate a rethink.

4.1.1   The Low Bar for "Treatment Success" Masks the Limitations of Latest Treatment Strategies

Unexplainable responses to treatment, or lack thereof, are considered threats to existing paradigms that have produced the mechanistic rationales behind the new drugs. They are shunned by research funding agencies that operate in a climate of epistemological rigidity. The Cancer Moonshot program, launched in 2015 by U.S. Vice President Joe Biden,[3] does not consider "treatment failure" a priority area, while it prioritizes yet more efficient (more "precise") ways to attack cancer cells. The universally disappointing performance of new therapies has over time lowered the bar for success. Tweaks to existing concepts, not new concepts, are considered innovation, and incremental prolongation of time to progression after therapy suffices to declare improvement. In fact, the most commonly used statistical measure for an effect of a new treatment remains the "hazard ratio" (ratio of rate of events, such as relapse or death between the treated and control groups) in the Cox proportional hazards model.[4] It quantifies the extension of the "time to progression" (or death) under the new therapy; thus, it tacitly assumes eventual failure of treatment and is by design agnostic of a long-term cure rate.[5] The goal of this chapter is to expose the epistemological and logical inadequacy of the current attitude toward treatment outcomes and to elucidate the fundamental theoretical and biological principles that explain why it is so hard to cure advanced cancer, which means to eradicate all tumor cells (as detectable) *and* prevent relapse over the remaining patient life span. The presentation of fundamental principles in turn will set the stage for proposing alternative therapeutic approaches based on a rationale that fundamentally departs from current thinking.

The two latest innovations in cancer drugs with the greatest impact in the clinic (at least as is currently thought) are (1) the target-selective drugs and (2) the immune checkpoint inhibitors. *The target-selective drugs* block molecular pathways that are considered to drive tumor cell proliferation. Such drugs typically bind to specific kinases involved in growth signaling pathways and are now often applied in combination with "companion diagnostics" that identifies the mutation in the tumor genome that activates the respective pathway.[6] The *immune-checkpoint inhibitors* (ICIs),[7] which unleash the power of the body's own immune defense against cancer cells by blocking the natural breaks in the cytotoxic arm of the immune system, have now surpassed target-selective drugs in glamour and promise for long-term cure.

Both innovations have been celebrated as revolutionary in cancer treatment. But inherent limitations have emerged soon after their rapid introduction to the clinic.[8] Targeted therapy in the majority of cases eventually fails: the tumor progresses after a certain period of treatment ("time to progression") or relapses in the form of invasive cancer and distant metastasis.[9] A rarely asked question is why the tumor not only recurs but also invariably does so in a more advanced, more malignant, therapy-resistant form.

While drug resistance can be explained by the selection of resistant cells, a principle well known from the study of microbial antibiotic resistance, there is no selection pressure imparted by treatment of the primary tumor that would promote the development of the sophisticated machinery for invasion and colonization of distant tissues, which requires the subverting of a robust tissue homeostasis.[10] A glimmer of hope is now offered by the ICIs: among the small fraction of patients who do respond to such immunotherapy, which is less than 20 percent,[11] a still smaller fraction exhibits hitherto unseen, sustained control of tumor suppression for several years, hence resurrecting the notion of the long-term "cure rate" that has been ignored in traditional statistical models.[12] Yet, this remarkable qualitative advance for a fraction of patients comes with a twist that pertains to the central theme of this chapter: not only are such cases of potential "cures" seen only in a minor subset of patients, but around 10 percent (or more) of patients receiving ICIs suffer "hyperprogression": the treatment itself seems to accelerate progression.[13]

4.1.2 Inadequate Explanation of Treatment Failure Perpetuates a Wrong Strategy

Unlike in other failures of technological and societal undertakings, such as plane accidents, building collapses, or social unrest, cancer research has never developed a sense for the need of thorough investigation to explain unwanted outcomes that defy their scientific rationale with the same scientific scrutiny that has underpinned its enterprise in the first place.

In the case of targeted therapy, explanation of treatment failure defaults to the broad concept of therapy resistance. A recurrent tumor in general (but not always[14]) does not respond to the same therapy any longer. Mutations in the targeted oncoprotein that prevent the target-selective drug from binding or utilization of alternative signaling pathways that obviate the cell's dependence of the protein being blocked by the drug can be readily identified.[15] Most prosaically, cellular detoxification systems that eliminate the drug from the cell are commonly found.[16] In the case of immune-checkpoint inhibitors, therapy failure is commonly explained by the tumor's *"escape"* from the immune system's surveillance activity.[17]

Cancer cells achieve such immune evasion in the simplest case by downregulating the expression of specific antigens used by certain clones of T cells to recognize and kill them. Alternatively, instead of reducing immunogenicity, tumor cells may actively modify the tumor microenvironment and co-opt alternative ways to enhance the natural immunosuppression that serves to keep T cells in check.

These concepts of drug resistance and immune escape epitomize the current thinking of how to overcome the tumors' unfathomable resources to survive therapy no matter its mode of action: that we face a *zero-sum game* of an arms race that we must win. Thus, we would need more powerful weapons to outrun the tumor's ability of developing resistance and kill the resistant cells or to block the many ways through which the tumor keeps tricking the immune system.

But why is a recurrent tumor so much more sophisticated as to be able to neutralize the cytotoxic compound of any imaginable chemical structure and perform complex biological functions, such as coordinating the formation of new blood vessels for its own supply, migrating into and colonizing new locations, restructuring the extracellular matrix, and emitting specific signals to keep the immune system at bay?

The unfathomably fast acquisition of these complex capabilities cannot be readily explained by the standard model of multistep tumor progression.[18] This model posits that random genetic mutations and natural selection of newly generated cell genotypes drive a Darwinian evolution of cancer cells in the affected tissue. However, the vast majority of somatic mutations in cancer cells are what evolutionary biologists call "nonadaptive" (or "neutral") because there is little firm evidence of "selective sweeps" that enrich these mutations beyond qualitative interpretations of mutation frequencies.[19]

Moreover, one cannot deny that the "novel" functions that the progressing cancer acquires, classically known as the "hallmarks of cancer,"[20] are already existing physiological functions utilized in normal reproduction, development, wound healing, and tissue homeostasis and repair and therefore don't need to be first summoned into existence within tumor tissue by random mutation and selection. Thus, tumor progression is a result of ongoing misregulation of existing cellular functions, leading to the misappropriation of functions in the wrong context, rather than a de novo somatic evolution of destructive capabilities under selection pressure. (The abnormal pH, oxygen saturation, etc. of the tumor microenvironment, however, have of course the potential to exert selection pressure, in addition to inducing stress responses, as we shall see, but such selection would only explain the *biochemical* capacity to survive and proliferate in this environment, a cellular function readily achieved by a single or few genetic mutations, and could not account for the coordinated cell behaviors that underlie coherent tumor tissue programs.)

That the notion of a zero-sum game and the attitude of "doubling down" in the face of a resilient opponent cannot be the winning strategy is a lesson that politicians, military strategists, and social scientists have long learned[21]: attacks with "partial success," which means short of total destruction of an adversary, will most often *backfire*. And most attacks have outcomes that qualify as "partial success" or "nonlethal." Bombing the land of terrorists will only strengthen their resolve and promote recruitment. Economic sanctions only cement the stronghold of dictators. And attacking your political opponent will increase their popularity among their supporters. A wounded animal is more dangerous.

4.1.3   Content and Goal of This Chapter

The tacit assumption of a zero-sum game and the desire to double down on the same strategy are the result of one-dimensional, linear, and deterministic thinking. But reality is often high-dimensional, nondeterministic (stochastic), and nonlinear. The latter means that the effect of interventions is not additive: after a partial success, one cannot simply come back to "finish off" because the initial action may have backfired. The enemy has responded to the nonlethal attack. The linear mode of thought has over the past decades become the default in biology and medicine, fostered by the abundance of simple cause-effect relationships embodied by molecular pathways. Departure from such linear thinking will entail the embrace of much neglected scientific concepts, including

- High dimensionality
- Nonlinearity (of interactions and resulting dynamics)
- Stochasticity (nondeterminism)
- Heterogeneity

We will see in this chapter what these abstract concepts, which are hallmarks of complex systems, concretely mean in the cancer context and show the value of embracing abstract principles alongside mechanistic explanations.

Specifically, we postulate, argue by deduction, and demonstrate that a treatment may not only kill the cancer cells, but that the intervention itself can make the tumor more aggressive and more resilient. In other words, the central thesis is that treatment is inherently a ***double-edged sword***: sometimes one of its two edges dominates and we eradicate the entire tumor, achieving "cure." At other times, the opposite edge that "cuts the wrong way" dominates, and treatment accelerates tumor progression. However, most often the net result of treatment is an intermediate (thus slightly better than the untreated control group) representing the superposition of these two opposing effects, each at its relative strength.

This chapter starts with an epistemological discussion of the collapse of traditional linear causality in the face of observed tumor behaviors, followed by an introduction to general systems dynamics concepts to satisfy the new epistemological needs. We then move to the concrete: the preclinical manifestations of the principle that treatment failure is not simply due to insufficient strength of the therapeutic intervention but in part actively caused by it, in what is called a nonlinear fashion. We explain the ontological necessity for this phenomenon as a consequence of core principles of nonlinear dynamical systems, without using technical language and mathematical formalisms, building the narrative toward the overarching idea: *that progression is the unleashing of an intrinsic potential that is a fundamental property of cancer cells and tumor tissues and can be triggered by the therapeutic intervention,* much like the forceful unfolding of a spring-loaded device when the box is opened by the therapeutic perturbation. We conclude with a proposal for the therapeutic control of the double-edged sword effect by uncoupling the intrinsic link between suppression and acceleration of tumor growth.

4.2   Epistemological Considerations

Our intention is to go beyond the usual type of mechanistic explanation of cancer recurrence. This type of explanation invokes genetic mutations that would confer cell survival capacity in the presence of cytotoxic drugs, suppress the antitumor immune response, or (re) express alternative cellular signaling pathways that the target-selective drug cannot block. Such *proximate explanations,*[22] which in cancer biology typically involve molecular mechanisms, are plenty and varied, and are readily found in review articles, are in a strict epistemic sense not satisfactory. They do not pin the process to be explained (the *explanandum*) to a *primordial logical necessity* in view of observed facts and known principles. In other words, they do not explain its *inevitability* given some set of first principles and conditions that warrant the invoking of a mechanism as explanation. Proximate explanations merely *describe* the process at a lower (molecular or cellular) level instead of *articulating why* a process (such as development of drug resistance) *has* to occur given the circumstances. It may satisfy the pragmatic mind to know that a tumor is "driven" by the mutated BRAF gene via the oncogenic protein BRAF^{mut} that it encodes because this explanation offers a molecular rationale for a *specific* targeted therapy. However, it does not provide a reason for *why* it is specifically BRAF^{mut} and not another oncoprotein that drives a (class of) tumor(s). We do in fact not know why the activating BRAF mutation is a "driver" for melanoma or colorectal cancer but rarely in other cancers, such as leukemia, and why deletion of the tumor suppressor protein Rb mainly causes retinoblastoma but, for instance, is not a driver in colorectal cancers, although these proteins play ubiquitous cellular roles.[23] Such cases illustrate the lack of generality and logical necessity in the descriptive proximate explanations. Although the identity of the *specific* culprit (the oncoprotein) is sufficient to account for a *given* observed process (tumor growth in a particular cancer subtype), it is *not necessary*, because an alternative oncoprotein might do the same job.

4.2.1   From Proximate Mechanism to Logical Necessity in Explanations

What does it mean to reach beyond proximate explanations and to anchor the *explanandum* to a logical necessity? What we seek here is to connect the observation to be explained to a set of pertinent biological facts or conditions, using steps of reasoning that comply with logic and mathematical principles, and demonstrate that given these facts, the occurrence of said observation is *inevitable*. The pertinent facts or conditions encompass both "first principles" as well as "ontological features" that are immanent to the system at study. They constrain its behavior and thereby generate its defining characteristics that hence are, given said conditions, inescapable. It is in doing so that the explanation meets the standard of logical necessity. We seek the *primum movens* for a system in a particular (instance-specific) internal and external condition that, given a set of valid general principles, provides a reason for *why* an observed process *must* take place. (Side note: The argument about the logical necessity is independent of the dualism between determinism and stochasticity since the stochastic behavior, as a physical phenomenon, can be part of a necessary behavior of a system; moreover, a prediction that relies on the necessity of a process can well be probabilistic.)

To be concrete: given a particular constellation (temperature, material composition, material properties, geometry of objects and surrounding, etc.), a particular chemical reaction *has* to proceed and a stone *has* to roll down the valley. Merely *describing* the specific underlying material changes (e.g., the kinetics of a reaction or the physical path of the rock) in detail does not provide the reason for why these processes *have* to happen in a particular direction, that is, are "spontaneous" given the circumstances (the initial conditions). Proximate explanations do not invoke the fundamental principles that govern these systems and that underlie the necessity of processes: the laws of thermodynamics and chemical kinetics, or of gravity and kinematical principles.

4.2.2   Toward "First Principles" of Cancer by Embracing
Dynamical Systems Theory

In current biomedical reasoning, the forces that drive an observed process are not connected to the logic of elementary governing principles, unlike in other domains of biology, such as evolutionary and ecological dynamics. Explanations that seek a logical necessity in terms of first principles are the basis for predictions in the physical sciences (as opposed to predictions in "data science" using statistics)[24] because they permit the computation of a predestined path—a *trajectory* in some abstract or in a concrete physical space. This is most prosaically epitomized by predictions of the impact site of a meteorite or even the course of a hurricane—within the accuracy range of some "probability cone." Biologists and clinicians of course welcome prediction, but prediction in biomedicine typically relies merely on empirical *statistical associations* (correlation between a biomarker and a treatment or disease outcome).[25] This does not mean that they are not useful.

Anchoring a process to the fundamental driving forces in biology does not mean reducing biological processes to chemistry to then use the framework of thermodynamics to derive the fundamental formalism of a driving force, which explains the spontaneity of the process and, hence, its inevitability. This would amount to the much-ridiculed naïve reductionism that reduces complex phenomena to classical thermodynamics to then invoke energy and entropy to explain what drives biological processes in one direction. This approach fails because biological systems operate far from thermodynamic equilibrium. The complexity of biological processes, including tumorigenesis, is characterized by emergent features, "higher-level" phenomena that cannot be reduced to their constituent parts. Their existence entails that we accordingly consider "higher-level" governing principles or "entirely new laws" in the words of the physicist Phil Anderson. ("Psychology is not applied biology, biology is not applied chemistry, and chemistry is not applied physics," to paraphrase Anderson.)[26] While not elementary in the same sense as the laws in physics and chemistry that cannot be violated, these new laws still can act as first principles within a certain formal framework that defines the elementary constraints of behaviors in a class of systems (including tumors). Applying these new laws, we seek to establish the logical necessity of processes that change the state of a complex system in a particular direction. This framework is essentially that of the theory of *complex dynamical systems* or, more technically, *nonlinear dynamics*. With a formal (mathematical) framework of nonlinear dynamical systems and its set of (relatively speaking) *first principles*, then necessities and apparent paradoxes can be identified.

Constraints defined by some higher-level principles in biology ("the cell cycle cannot turn backward") are certainly not as absolute as in physics ("the marble cannot spontaneously roll uphill"). Nonetheless, they provide a solid basis for explanation, because these constraints can in a first approximation be considered undefiable. But if they are violated in some well-characterized circumstances, then we have a paradox, which represents another well-defined observation in need for an explanation. The violation of a "higher-level" rule governing the dynamics of a class of complex systems can subsequently lead to a refined understanding of these principles for the class of systems in question. As Nils Bohr stated, paradoxes are key to progress. Thus, we will treat biological processes as processes that alter the state of a complex, high-dimensional system in a particular orderly way and reduce them to the fundamental mathematical laws of *dynamical systems theory.* In doing so, we seek here to expose the logical necessity of observable processes that are represented by the tumors' response to treatment.

In summary, we need to apply a broader category of reasoning that reaches beyond and is independent of the realm of specific, proximate molecular mechanisms. This wider and more encompassing epistemic category will explain the *inherent inevitability* of the forces that drive the development of resistance to treatment. It will complement the current default approach of identifying molecular pathways as the sole explanation (which, by identifying specific targets, is still important for drug development).

In the next section, we will introduce important basic concepts from systems theory, without using mathematical formalisms. We will take a first step beyond the epistemic habits of molecular biologists and toward general formalisms that will help in thinking about biological systems within the framework of general complex dynamical systems theory. After that, we will apply it to cancer to explain some of its inevitable properties.

4.3   General Principles: System States, Constraints, and the Landscape Metaphor

We seek a general explanatory framework in which when certain elementary conditions are met, development of drug resistance, immune evasion, and even progression are necessary treatment responses that are immanent to the tumor as a complex system. But we have to start with basic concepts first for understanding the behavior of complex systems.

4.3.1   A Complex System as a Network of Interactions of Its Components

A first step toward this goal is to consider a therapeutic intervention as a perturbation applied to a complex system. A complex system is a system made of *components* (cells, biomolecules) that *interact* with each other and in doing so produce the "emergent properties"—the characteristics associated with that system's *state*. In cancer biology, for instance, proliferation of cancer cells is an emergent property that results from the totality of regulatory activities that genes, proteins, and metabolites exert on each other to produce the cell division cycle. Other systems properties are the autonomy and robustness of the cell society in the tumor tissue that results from the communication between the various cell types in it. Thus, in the cancer context, we consider two levels at which a system and its behavior can be defined (figure 4.1).

(1) At the level of the *cell* as a system, the list of all genes N and their gene products (mRNA, proteins, and metabolites) are the interacting components: they regulate each other—according to a fixed scheme written into the genome that defines which transcription factor can bind to which regulatory element—thereby establishing the gene regulatory network (GRN) of the genome.[27] Each gene (or locus) is a node of the network and can be assigned a value x_i that represents the changing activity of the genomic locus i. A change in the value x_i thus translates into the change in the cellular abundance of the protein it

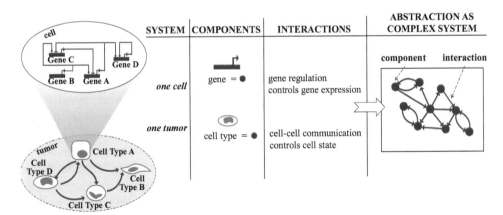

Figure 4.1
Formalizing the complex system for two levels: the cell (whose components are interacting genes) and the tumor tissue or cell population (whose components are the interacting cell types). The interacting components abstracted as black solid circles are the nodes (vertices) of a network in which the edges/arrows represent the interactions. The nodes are symbols that represent genes (or biomolecules) or cell types. They thus represent an ensemble of identical components (a named species of objects, such as a gene or a cell type), not individual components, and can take a value (abundance), represented by the variable x_i for node i.

encodes. Mathematically, x_i is a *variable*. The N gene loci (or nodes in the GRN) and their x_i values then collectively produce the *gene expression profile* of the entire genome in a particular cell, which defines the *state* of that cell as a system. This is of course a cartoon-ish, simplified view but will turn out to be instructive for our explanations.

(2) At the level of the tumor as a system, the components comprise the list of all N "cell types" within the tumor (figure 4.1). We take here the term "cell types" to be general and to refer to all possible phenotypic states, from a nominal cell type to cell subtypes, such as cancer-associated fibroblasts, to the biological cell states, such as the (cancer) stem cell state, the proliferative state, the senescent state, and the stressed cell state—each defined by a characteristic biological behavior and determined by a specific gene expression profile. These distinct cell types then are the nodes of the network and interact with each other, again following predetermined rules: the mediators (ligands) that a cell type produces and the receptors it expresses determine which cell type interacts with which, thus forming a cell-cell communication network. The relative fractions x_i of each cell types i in the tumor cell population (e.g., the actual number of cancer stem cells of a certain type) collectively determine the *tumor cell composition* and thus the *state* of the tumor as a system.

Note that for both levels of systems, the interaction network—the GRN or the cell-cell communication network—has a fixed "wiring diagram" determined by the molecular specificities that define each interaction. Hence, the architecture of the network of all possible (not necessarily all used) interactions is ultimately written in the genome. One genome maps into one wiring diagram—this operational assumption is central for the theory (explained in detail in Huang[28]). What changes and establishes the *dynamics* of a given system (with its fixed genome) are the *values* of the variables x_i: the expression (e.g., manifest as mRNA or protein abundance) of gene locus i in the cell or the fraction of cell types i present in the tumor, respectively. (This formalism also establishes a frame-work to better understand how mutations that alter the genome's identity, and thereby rewire the GRN, affect the dynamics of systems.)

With the recent arrival of single-cell molecular profiling, such as single-cell RNA-sequencing[29] that measures the transcriptome of each individual cell in a tumor cell popula-tion, we can determine the cell state of each cell (at least as defined by the mRNA profiles). But the same single-cell analysis, typically performed on a sample of tens of thousands of cells of a tumor, also offers a glimpse of the state of a given tumor by providing the relative fraction of cells in the various types or cell states.[30] For simplicity, we shall not consider here the physical position of the cells relative to each other in the physical space of the tissue, although considering tissue geography will be most interesting in the near future.

4.3.2   Basic Terminology: The System State *S* and the State Space

How do the interactions between the N system components (gene loci or cell types) give rise to a system state S (of cells or the tumor, respectively) that displays the "emergent properties"? Each of the large number N of distinctly interacting components i ($i = 1, 2, \dots N$)

can be assigned a variable value x_i (concentration, activity, number), and the resulting configurations of all the values $[x_1, x_2, \ldots x_N]$ define a state S at time t. The essence of the interactions between the components that form the network is then to change the N values x_i of each component i of the network (= node) in a coordinated manner to establish a particular configuration $S = [x_1, x_2, \ldots x_N]$. Thus, the state S in some way manifests the collective effect of all interactions. If the network of nodes and interactions is the blueprint of a system or its hardwired internal architecture, then the state S, defined by all the N values x_i, is its instantaneous appearance at a given time t (a snapshot), and the change of S in time is the dynamics of the system. Figure 4.2 illustrates all the basic theoretical concepts introduced here and below.

With the definition of a system state S as one specific configuration of the N values x_i among the almost innumerable number of combinatorially possible configurations $[x_1, x_2, \ldots x_N]$, we can define the important mathematical concept of the "**state space**": the abstract space of N dimensions that contains all the possible configurations of $[x_1, x_2, \ldots x_N]$. With the notion of a "space," we get the notion of a "location": in the state space, every configuration or state S maps into one point since every feature i of the system, with its corresponding value x_i contributing to defining the state S, represents one dimension in the N-dimensional state space. The configuration $S = [x_1, x_2, \ldots x_N]$ is thus a vector whose components x_i represent the coordinates for the position of S. As N is a large number, we call this a *high-dimensional* system. Two configurations S that are similar to each other are also near each other in state space. A pair of states S_A and S_B that are neighbors in this space would require few changes in gene expression values to shift into each other.

Back to biology: in the case of the cell as a system, the system state S is the gene expression profile, which encompasses the N activities x_i of all the N gene loci of the genome. It is approximately measured by assessing the transcriptome or even the proteome of a cell. In the case of the tumor as a system (figure 4.1), the N values that define the tumor tissue state are the relative abundance of the N cell types i (including cell states). For instance, some tumor may contain more of a certain subtype ($i = 1$) of cancer stem cells (high value for x_1), less differentiated cells of type $i = 2$ (lower value for x_2), a minimal proportion of senescent cells ($i = 3$) ($x_3 =$ near zero), and so on. As single-cell transcriptomics now reveals, the number of distinct cell (sub)types N in a tumor is larger than previously appreciated. Even among the neoplastic cells themselves (without the tumor stroma), many cell subtypes can be distinguished in "snapshot" measurements, although some may exist only transiently.[31]

The next point is a major step for understanding a central concept and will be illustrated again in a cartoon-like fashion. Consider now the level of the cell as a system: if each of the N gene loci had only two activity values x, say $x = $ ON and $x = $ OFF, we would already have $(2 \times 2 \times \ldots = 2^N)$ distinct configurations S of gene expression across the N genes. For $N = 10,000$ genes, this would amount to the hyper-astronomical number of $2^{10,000} \approx 10^{3,000}$ possible gene expression profiles S, each one mapping to a distinct position in the N-dimensional state space.

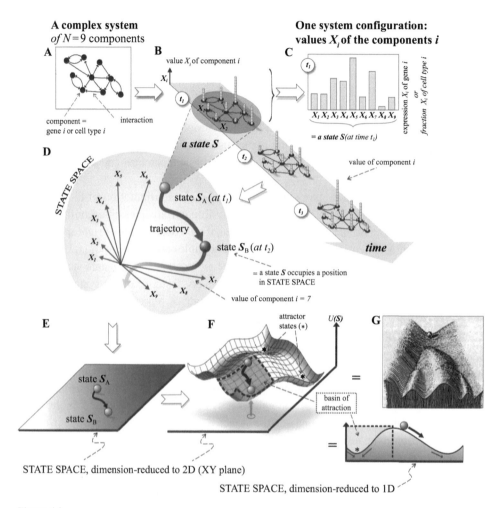

Figure 4.2
Basic concepts from dynamical systems theory applied to a gene regulatory network, GRN (for cell as a system), or to a cell-cell interaction network (tumor tissue as a system). For details, see text. (A) The fixed network architecture of a system of $N=9$ components $i=1, 2, 3 \ldots N$ as an example of a network. (B) The configuration of x_i activities of each network component (black dots) displayed as column. The entire configuration defines a state S and changes over time t, dictated by the interaction network. (C) Example of a configuration of the vector $[x_1, x_2, \ldots x_N]$ as a bar graph for the values of x_i at t_1. (D) The N-dimensional state space (indicated by the nine coordinate axes). (E) Dimension reduction of the nine-dimensional state space to a two-dimensional XY-plane. (F) The quasi-potential landscape plotted by assigning each position (= state S) in the XY-plane an elevation, the quasi-potential $U(S)$. (G) Waddington's epigenetic landscape (a 1957 version[32]) and below a simple landscape visualization for the one-dimensional state space, often used as pedagogical scheme.

But here is the rescue from a near-infinite continuum of cell states: because of the *interactions* between the components, the vast majority of these configurations logically cannot be realized—this will be explained in detail below. Even among those states that can be realized, because they have to comply with the logic of all the interactions, only a fraction is actually used by the organism and represent physiological cell states, such as the nominal cell types: a liver cell, lymphocyte, and so on. Cancer cells, then, as we shall see later in this chapter, may simply represent the pathological occupation of unused but realizable cell states by aberrant cells: they implement configurations $(x_1, x_2, \ldots x_N)$ of S not encountered in the cancer-free organism. How cells get there is essentially the question of tumorigenesis.

4.3.3 Systems Dynamics: A Change of System State Is a Movement in State Space

A system, be it a cell or a tumor, is in a given state S at a given time t as defined above (figure 4.2): it has a particular configuration or activation pattern of its genes or its cell types, respectively (figure 4.1). The unique mapping of a system state to a location in state space means that a different system state would occupy a different position. Thus, a change of the system state S (for the same genome and, hence, the same GRN) results in its displacement because of changes in the x_i values of its state vector: The system *moves* from state S at time t_1, or $S(t_1)$, to another state at time t_2, or $S(t_2)$. This process is referred to as a *state transition* (figure 4.2B,C). This shift from one location to another in state space defines a "path" or a "trajectory." For instance, when a cell in the progenitor state transitions (= differentiates) into the state of a mature differentiated cell, its gene expression profile shifts to that of the differentiated state.

An ***intervention*** is an action applied to a complex system and is aimed at causing a directed change, from the current system state to another (desired) one (e.g., from the proliferative state of a cancer cell to the quiescent or apoptotic state), as intended in treatment. In current biology, such an induced state change is often understood in terms of linear causality, like in "to make the water hot, add heat to it": to cause a cell in state S_A to shift its state to state S_B, which exhibits the trait Y that is not manifest in S_A, we typically activate or "cause" in the cell the expression of trait Y that is characteristic of state S_B. But in reality, an intervention in a complex system and its consequences can often not be understood in terms of such simple linear causality because the network interactions among the system components impose convoluted constraints on the response. This idea will be central in explaining why therapeutic interventions can backfire.

4.3.4 Interactions between Components Constrain the Realization of System States and Their Dynamics

We now expand on the defining crux of a complex system: the interactions. Since components within the system continuously *interact* with each other, they narrow down the hypothetically *possible combinatorial choices* of the N values x_i (e.g., gene expression levels x_1, x_2, x_3, ... x_N) of the system's N components to only a small subset of configurations, or states, of the system.

More concretely, the vast majority of the hyper-astronomical, theoretically possible configurations are *logically* not allowed because they may be incompatible with the regulatory logic of interactions. If gene *1* inhibits gene *2*, then any configuration *S* in which both genes are simultaneously highly expressed (i.e., in the region of state space in which the coordinates x_1 and x_2 both have large values) would be "forbidden" terrain; the configuration of gene expression there would be inconsistent with the (hardwired) interactions of the network. By contrast, a state *S* in which we have the configuration $x_1 \gg x_2$ would be "allowed." At the level of the tumor state, the various cell types communicate with each other to control cell division and death as well as cell state transitions. Such mutual influences between the cells constrain the cell population dynamics of the tumor tissue, similarly allowing only for particular cell-type compositions to be homeostatically stable.

In the hypothetical absence of interactions between the component parts, any of the combinatorially possible configurations, or system states *S*, would be equally possible, free of constraints: then steering the system to shift *S* in any direction to another (desired) state would be as easy as drawing any abstract figure on a vast, empty canvas. But the system's interactions impose *constraints* that embody the system's characteristic intrinsic architecture. The constraints force state changes through a particular path that minimize "violating the constraints." Violating constraints takes effort because it must counteract the regulatory interactions.

The fact that intrinsic constraints channel system state changes is the deeper reason why a complex system responds to interventions in a particular manner, not simply by mirroring our intervention as when nudging a satellite to bring it onto a desired trajectory. Constraints determine ease and resistance of change in a given direction. Thus, often, an inappropriate intervention may, instead of shifting a system state into a desired direction, unleash "energy" stored in the constraining interactions and subsequently trigger a self-propelling process, moving the system away from any behavior that is desired. (Note that "energy" is not to be taken literally here as in the meaning of physics, as detailed later.) A self-propelling process not intended by the intervention thus uses the internal "energy" stored in the interactions and dissipates it until when the logic of the interactions is satisfied.

Metaphorically, an unintended self-propelling response could be envisioned as accidentally taking a *wrong step during a walk on a mountain ridge*: the result is not simply a fall to the ground but into a valley: propelled by gravity, the reduction in altitude continues until a spot with zero slope is reached. This process is almost irreversible, and most important, the descent is along either one of the two slopes on one or the other side of the mountain ridge. The intrinsic geometry of the system (landscape with the ridge) is such that the state has high "potential energy" that can propel the process and that the two possible directions of the fall from the "high-energy state" are predetermined and opposite of each other. In which valley one falls depends on minuscule details of the misstep. The same misstep on a flat plane would simply slightly nudge the hiker's direction in a predictable manner; here the linear relationship between cause and effect is maintained.

4.3.5   Constraints Imposed by Interactions Give the State Space a Topographic Structure

How do we connect this imagery of a metaphoric landscape whose topography visualizes the channeling of state change by some internal structure with the abstract high-dimensional state space introduced above? Since it is difficult to think in N-dimensional space, one can, at the cost of some information loss, but with a gain of intuition, project the high-dimensional state space into a two-dimensional space. In it a state is placed at a position in a YX-plane, defined by its XY-coordinates (figure 4.2E). This dimension reduction (which sacrifices some information on neighbor relationships) now affords the following useful mental cartoon: if there were no interactions between its components, the system could in principle effortlessly travel on the *flat* XY-plane from any point, which represents a state S, to any other. But the system's internal *interactions* constrain the system's behavior, such that the "ease" and "resistance" of displacement from any one state S_A to another S_B is vastly affected.

The central idea is thus that the interactions that constrain state changes "fold up" the flat plane into a three-dimensional topography, with regions of high or low elevation, hills and valleys (figure 4.2F). Each state S (at its XY-position) is thus assigned an "elevation" U. This third dimension represents the ***quasi-potential*** $U(S)$ of the system. The landscape is no longer a loose metaphor but a mathematical construct. All elevation values $U(S)$ at all points S in the plane collectively establish a landscape over the flattened state space, giving rise to a characteristic topography: the ***quasi-potential landscape.*** (The term "quasi" is to indicate that these potentials are not actual *energy* potentials in the classical sense; see Zhou et al.[33] and section 4.6 for more details.) The slope at the XY-position that represents state S_A is determined by how much the system state at S_A is "driven" by the interactions of its components to change and in which direction.

Take the above example of the regulatory interaction by which gene $i = 1$ inhibits gene $i = 2$; then at any state S_A that is in regions of the state space representing high values of both x_1 and x_2, which defies the logic of the regulatory interactions, the slope at S_A will indicate that it will be driven away ("downhill") into a neighboring state S_B with higher x_1 *and lower* x_2 (which better complies with the rules of the regulatory interactions). The slope at S_B then represents the driving force there and so forth. Thus, the difference in quasi-potential U between two positions S_C and S_D, $\Delta U = U(S_D) - U(S_C)$, graphically depicted as an elevation difference on the landscape, corresponds to a sort of "gradient" of a "gravitational pull" (in a metaphorical sense) that produces the ***driving force*** for the state change at that state S_C toward state S_D.[34] States experience no "force" anymore when all network interactions are "balanced," or logically satisfied. Then, they occupy points in state space that have no slope (i.e., no gradient). In the concrete application to the GRN, a gene expression profile of state S that "complies" with all the gene regulatory interactions prescribed by the GRN will not change its gene expression profile: it is not driven and thus exposed to no downhill force, or slope, in the landscape. Such states are referred to as *steady states* and correspond to a point in the quasi-potential landscape that is flat: exactly at the bottom of a valley, or exactly on the hilltop (or a ridge).

With this mathematical landscape, we have now erected a formal framework that not only describes net changes of states S, which correspond to a shift in geographic position, but also describes the "ease" or "difficulty" of such state changes and the paths that incur the least effort (uphill climb in working against the regulatory constraints). Thus, the specific landscape topography captures the effort and the paths of changes of system states. The intuition that a marble would roll downhill represents the force emanating from the interactions in the network that drive the system toward states that comply with the interactions.

4.3.6 Nonlinearity: Stability, Instability, and the Coexistence of Multiple Attractor States

Now we need to introduce a second critical feature of complex dynamical systems on top of the mere existence of interactions, namely, their property of nonlinearity. An interaction, for example, between "upstream" regulat*ing* genes and a "downstream" regulat*ed* gene is considered "*nonlinear*" if the equations that describe these interactions are nonlinear. These equations contain functions that govern how the *regulating* (upstream) genes jointly affect the rate of change of expression of the *regulated* (downstream) gene. The rate equations of biochemical reactions are often nonlinear mathematical functions due to the molecular properties of molecular binding events, such as cooperativity. The solutions of these rate equations represent the "states with rate change equals zero," thus the steady states of the system or the points on the landscape that are flat (have no slope). Nonlinear functions allow for multiple solutions (but need not do so), much as the quadratic equation $x^2 = 4$ has two solutions: $x = 2$ *and* $x = -2$.

Therefore, since a solution of rate equations represents a steady state and "nonlinear interactions" in a system refer to nonlinear rate equations that can have multiple solutions, a system with nonlinear interactions can give rise to *multiple* steady states. Since on the quasi-potential landscape, each steady state maps onto either the top of a hill or the bottom of a valley, the existence of multiple steady states in a system is manifest in a landscape with multiple hills and valleys. We can thus distinguish between two types of steady states.

A point at the bottom of valleys or a "local minimum" of the quasi-potential landscape represents a *stable* steady state to which unstable states in its neighborhood are "attracted" toward, much as the bottom of a valley would drain water toward it. A marble at such a stable steady state would return to it when slightly nudged out of its position, as if attracted back. The stable steady states at the bottom of valleys are thus called *attractor states*, an important concept that we will use later in section 4.6. The attractor state is a *self-stabilizing* state because the return following a perturbation is an "automatic" consequence of the interactions. In the case of a GRN, attractor states guarantee stability of distinct gene expression profiles.

A point exactly at a hilltop (or on a mountain ridge in the hiker analogy earlier) is, in the absence of any perturbations (or missteps), also a steady state: a marble (system) poised exactly on the hilltop would stay there. But this steady state is **unstable**: the slightest perturbation ("wind gust on the marble") will push it down. Once minimally perturbed, the system finds a slope and rolls down on one side of the hill, departing from the unstable steady state at the peak, and continues to roll down even after the perturbation has long subsided. This descent represents a **symmetry-breaking event** because the system has made a choice for one direction, either by chance or biased by the push of the perturbation or local topographic details of the landscape. A self-perpetuating shift of the system state ensues. The change continues in a self-propelling manner, an "automatic" consequence of the interactions in the system, until the marble comes to a halt at a stable steady state, a nearby valley somewhere else in state space.

In general, a network of interactions between the components of a complex system capable of high-dimensional dynamics will contain many nonlinear interactions, many of which are feedback loops. The resulting landscape, which represents its entire behavior repertoire, will then be a rugged mountainous terrain with multiple valleys and, by logical necessity, hills in between. The existence of multiple steady states allows for a system to have multiple alternative *stable* attractor states, the valleys separated by the hills. This property is a fundamental property of some classes of complex systems and is called **multistability**. It readily emerges in a complex system with nonlinear interactions between its component parts, irrespective of whether the system is composed of chemicals, genes, cells, species, or humans forming a society. It is what provides a system distinct, ordered macroscopic behaviors. The continuum of the hyper-astronomical number of "micro-states" in the absence of interactions is compartmentalized into a much lower number of realized or realizable stable attractor states by the interactions.

The slopes on the hilly landscape represent one of the "first-principle" explanations for the "necessity" or "spontaneity" of a process of state change in a complex system, an internal "urge." When seeking to change the state of a complex system from state S_A to S_B, we typically cannot freely roll the marble in just the desired direction (along direct geographical line $S_A \rightarrow S_B$), free of resistance. We may need to exert efforts against the interactions that pull the marble back into the attractor (bottom of valley) and move "uphill," or take detours around hills. Thus, the landscape topography visualizes how internal constraints of a system imposed by the interactions channel the consequences of our intervention in a specific, often nonobvious way.

We now have explained two key ingredients listed above that make a complex system: *high dimensionality* (N dimensions), or the fact that system states are configurations of the values of all of its N components, and the *nonlinear interactions* between these components laid down in the wiring diagram of a system, such as the GRN, that produce the hilly landscape topography. The two other related key ingredients, stochasticity and heterogeneity, will follow soon.

4.3.7   A Historical Note on the "Landscape" Concept and What It Means for the Double-Edged Sword Effect in Cancer Therapy

The metaphor of the landscape on which the system state is symbolized by the marble at a given position has a formal mathematical basis in the *quasi-potential landscape* and the theory of "least action" in stochastic dynamical systems.[35] But this mathematical landscape has also a stunning equivalent in the historical metaphor of *the epigenetic landscape* in biology (inset in figure 4.2G), proposed by Conrad Waddington in the 1950s.[36] In his view, the valleys represent the distinct, stable self-stabilizing cell types and tissues, whereas the slopes represent the self-propelling process of embryonic development with its predestined paths, the "chreods" (discussed in section 4.8). Waddington proposed the landscape metaphor because he recognized the quasi-discrete nature of the distinct cell types (cell states): there is no continuum between the cell types (which would be manifest in a gradually changing series of equally stable cell types between two cell types).[37] Even newly discovered "intermediate" types between two known cell types are discrete entities.[38] The landscape picture will come in handy later when we explain why any treatment of cancer may have unintended, opposing consequences.

In complex systems, due to the constraints imposed by the network of interactions, the windows of opportunity for causing a desired change are narrow. Many system components must change their values in a particular coordinated manner, prescribed by the interactions, to implement the desired configuration S_{final} of the destination phenotype. Because of the interactions, conflicts readily arise: what if the configuration S_{final} intended by our intervention, or intermediate states on the way to it, is not compatible with the network interactions? The system appears to "have its own mind" as it follows the urge to comply with the logic of its interactions. Such behavior is at the core of a formal explanation of the double-edged sword effect of cancer treatment—which is essentially an attempt to shift the (cancer's) system state. Moving into regions of instable states, or onto the hilltops, means loss of control, incurring the risk of "rolling down the wrong side of the mountain, to the opposite of the desired direction."

In less formal parlance: an action intended to change a complex system into a new state can take the form of a "catch-22" or a *dilemma*. There are often no universally good options; there are trade-offs. The promise of the good is inseparably linked to the risk of the bad, and both can be just adjacent possibilities, with a fine separating line in between, epitomized by the watersheds of the landscape.

In the next section, we discuss a central logical blind spot of habitual thinking about cancer treatment and present some clinical and preclinical examples that illustrate these omissions before coming back to theory in section 4.5 as we move toward an explanation of why treatment can backfire.

4.4 Paradoxes and the Double-Edged Sword Effect in Cancer

What do all cancer drugs have in common? There are two answers: first, they all are designed to kill cancer cells. Second, they all do not work as desired—considering the almost certain relapse after initial remission of an advanced (invasive) cancer following treatment, be it surgical, physical, chemical, or biological. (Rare exceptions do exist, as seen in some immunotherapies and targeted therapy, but they confirm our thesis because they are encountered in particular circumstances.) By simple logic, the two answers raise this unarticulated possibility: *Could it be that the very act of killing cancer cells eo ipso also plays a part in promoting the relapse?* Could a failed attempt of eradicating the entire tumor, despite killing most cancer cells, actually make the surviving cells and the residual tumor tissue more malignant? Such a seemingly paradoxical response by the tumor would precisely recapitulate the statement by Friedrich Nietzsche: *"What does not kill me makes me stronger."*[39]

4.4.1 Beyond the Naïve Logic of Causation of Cure and of Doubling Down in Treatment

It is a natural reaction to dismiss the suggestion that failure of treatment could have an active, unintended *counterproductive* component. Instead, one commonly associates treatment failure and relapse with a passive principle, the tacit assumption that treatment just has *not* been *potent enough*. To see beyond this entrenched paradigm, let us consider the naked logic of the relationships between treatment and improvement of clinical outcome: since improvement of outcome, if not even cure (in brief, "benefit"), requires treatment, then the observation of a benefit in a patient (A) implies that she has been treated (B): in brief, A (benefit) implies B (treatment was given). At the logical level, it is generally readily understood that from "A implies B" does not follow that "B implies A." Thus, if "benefit" (A) implies the "patient has received treatment" (B), it is clear that "patient has been treated" (B) does not automatically imply a "benefit" (A), since not all patients that have been "treated" (B) will exhibit a "therapeutic benefit" (A). A is a subset of B and not vice versa. But the biological implications are usually not considered, and we will do so below.

Whereas it is readily appreciated that if "A implies B," then it does not mean that one can say "B implies A" (that is, "treatment" (B) implies "benefit" (A) is logically incorrect), it is often erroneously concluded that if "$\underline{A}$ implies B," then "non-A ('no benefit') implies non-B ('no treatment')" or vice versa. This is perhaps because of a tacit notion that the premise "A implies B" is so elementary that it has absolute validity. Too much ingrained in our thinking is the notion that adequate treatment and benefit are tightly linked (equivalent) so that the negation of the entire premise is also valid.

However, we shall now think outside of the premise: namely, that "A implies B" may not even be correct, let alone that "non-B ('no treatment') implies non-A ('no cure')." In other words, A and B, while typically assumed to be tightly linked and interdependent, are not necessarily so. They are logically *orthogonal*, in that there is room for logically permissive but practically underappreciated combinations of these two elements, *A* and *B*. This leads us to two not considered yet possible scenarios: (*α*) There is the logical possibility that *A* does not require *B*, that is, non-*B* *does* allow *A*. Concretely, *A* (benefit) lies partly outside of *B* (treatment). Thus, improvement if not cure may *NOT* require treatment. This outlandish possibility has of course been recognized in the theoretical consideration of the rare cases of "spontaneous remission" (discussed below).

But another, complementary scenario from allowing for independence between *A* and *B* is the unconsidered scenario (*β*): *B* negates *A*; that is, "treatment" (*B*) may *prevent or reduce* a "benefit" (*A*), beyond just having no effect. In other words, treatment may even abrogate the theoretically possible benefit associated with nontreatment just discussed. This possibility is compatible with the accepted elementary notion discussed above that "benefit" (*A*) is a subset of "treatment" (*B*), but few have ventured into thinking about all the possibilities outside the subset (*A*) but within the set (*B*).

Taken together, with the acceptance of a logical separation between benefit and treatment, we shall also consider the two outlandish possibilities of (*α*) *improved outcome without treatment compared to with treatment* and (*β*) *poorer outcome with treatment than without*. Here we will focus on the second scenario because it is directly linked to actionable changes in the current practice of cancer treatment, which essentially is taken to mean "killing cancer cells." Concretely, although the default assumption that "killing tumor cells is beneficial" is correct and enjoys a mechanistically plausible rationale, by never questioning it, we deprive ourselves of an unexplored possibility that "not killing tumor cells" also could have a (hidden) benefit that could be enhanced and that, even more so, killing tumor cells may actually make the outcome even worse.

4.4.2 The Practical Challenge of Seeing beyond Association of Treatment and Cure

Clearly, treatment of a large tumor is important to avert a medical threat (the clinical necessity for reducing tumor burden) and we are not advocating skipping treatment. But we cannot logically exclude the possibility that treatment, while inducing clinical remission by killing cancer cells, *also* contributes to recurrence in an active manner (scenario *β*). We now need to demonstrate the underlying biological principle and the specific mechanisms.

Why do we not encounter evident manifestation of this seemingly outlandish scenario that we have so far considered solely on logical grounds and not based on direct observation? There are two reasons. First, for obvious ethical reasons, we cannot compare treated patients with a cohort of completely untreated patients within the same clinical setting in order to expose the occasional, likely conditional benefit of nontreatment (scenario *α*). The natural history of the course of tumor development is largely unknown given that nowadays, virtually every patient who is formally diagnosed with cancer also receives treatment. Thus, we cannot even measure the small, conditional probability of the much speculated upon phenomenon of spontaneous regression of untreated tumors, although ample anecdotal evidence exists. (Reported cases are riddled with controversies regarding proper diagnosis.[40])

Second, a possible tumor-promoting effect of treatment (scenario β) might be overwhelmed by the *aggregate* benefit of treatment: the effectiveness of various treatments is quantified at the level of an entire *patient cohort*. Study results (end points) are typically reported as percentage of patients who are alive after five years or as median survival time in the patient cohort. Such a reporting format ignores the *heterogeneity* of the cohort in which the behaviors of individual tumors may be very diverse in response to the same treatment. The focus on cohort averages masks the divergence of response trajectories of individual patients. The phenomenon of heterogeneity is not as trivial as commonly perceived and will be discussed at length later.

4.4.3   Hidden behind Survival Curves: Tumor Acceleration in "Treated but Not Responding" Patient Cohorts

Consider a numerical example of a report on the success of treatment, where the general term "response" implies a desired therapeutic effect, such as reduction of tumor size or extension of time to progression. Take, for instance, an outcome of a new treatment in which the treated group (T) has a response rate T_R (e.g., tumor regression revealed by imaging) of $T_R = 45$ percent compared to the control group C treated with existing therapy, with a response rate $C_R = 28$ percent[41]; this would be considered a success if the difference in $T_R \gg C_R$ is statistically significant.

Now let us dissect these aggregate responses: among the patients of the treated group T, in which $T_R = 45$ percent did respond, there are by definition $1 - T_R$ or $100 - 45 = 55$ percent *nonresponders*. We call them "treated nonresponders," hereafter T_{NR}, an important but much ignored subgroup (defined by observed result, not study group design). We have by definition $T = T_R + T_{NR} = 100$ percent. But who are these $T_{NR} = 55$ percent of the treated patients? In what sense do they not respond?

Recall the above logical analysis that if A ("improved outcome") implies B ("treatment"), then *non-B* does not necessarily imply *non-A* but might actually also result in A. As said above, it is in principle entirely possible that a however small portion p_α of patients, under some yet to be specified circumstances, may fare better if left untreated. But we shall now also consider the scenario β in that for a small fraction p_β of the treated patients T, the treatment may have actively increased the risk of relapse (despite inducing remission in the initial therapy), and therefore, *not* treating in the usual manner (non-B) this subset of patients actually would have been more beneficial (A).

It may even be that for some tumors and treatments, the fraction of patients in which treatment promotes progression, p_β, is larger than the fraction of untreated patients who experience spontaneous remission, p_α. In fact, the fraction of patients with advanced invasive tumors that are actually cured (complete eradication, no relapse) by any treatment is currently vanishingly small, perhaps in the same order of magnitude as the fraction of spontaneous cure without treatment p_α. We simply do not know.

Thus, in our example, among the $T_{NR} = 55$ percent patients who have been treated but considered "nonresponding," some may actually have an outcome worse than if they had not been treated. Because results are lumped together in treatment groups, this fraction p_β among the treated nonresponders in which the treatment has accelerated progression compared to patients in the control group C is not directly evident in reported data.

Results of clinical trials are commonly presented as Kaplan-Meier *survival curves* where the proportion (with 100 percent referring to the initial cohort size) of patients who at a given time point are either in remission (and do not progress) or are alive is plotted against time. The resulting curves then visualize either the "progression-free survival" (PFS) or the "overall survival" (OS), respectively, as a function of time since start of the trial (figure 4.3A). A simple theoretical simulation (in figure 4.3B) based on the widely used proportional hazard model (exponential decay of the surviving fraction, i.e., the hazard rate = death rate is constant over time) shows the survival curves of all treated patients T_R represented as one curve (dashed line) and of all patients in the control group C as another curve (solid line). We will use this model later.

Note that the survival curves represent entire patient cohorts since only for an ensemble of individuals can one compute a fraction of patients at any time that remain progression free or are alive, respectively. As mentioned, the curves are aggregates and represent the *nominal* treatment groups T and C defined *before* the study, independent of outcomes, and do not reveal individual fates. The group of treated but nonresponding patients, T_{NR}, is usually not separately displayed.

Even if among the concealed fraction, $T_{NR} = 55$ percent of patients in the treated group who were considered nonresponders, a small subset of patients exhibits the possibility β and thus has actually worsened under the new therapy more so than the patients in the control group C, then the survival curve of the treatment group can still indicate a *net* benefit for the treated group T.

4.4.4 Preclinical Studies Can Reveal Treatment-Induced Tumor Promotion

Only in animal studies can we reveal the size of fraction p_β: we can expose the phenomenon of accelerated progression caused by treatment within the fraction T_{NR}—the treated but nonresponding animals—because, unlike with human patients, we can compare the treated animals to entirely untreated ones. Since mice can be monitored individually in parallel and longitudinally under exact conditions, we can break down the nominal study cohorts not just by treatment groups (T vs. C) but also according to the types of response. Figure 4.4A shows a study on the therapeutic effect of the HDAC inhibitor drug VPA on a murine xenograft model for renal carcinoma.[42] The authors not only recorded tumor growth curves of all treated animals as aggregate, as is commonly done, but also graphed the responding and nonresponding animals in the treatment group separately (i.e., we have the breakdown

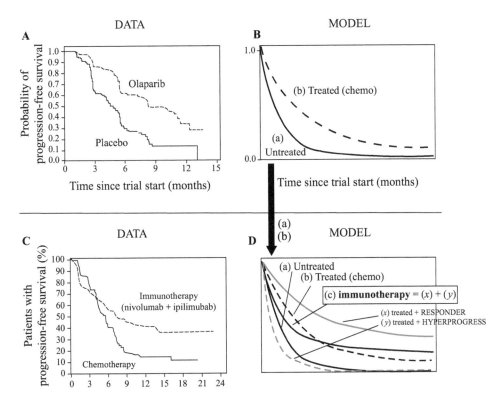

Figure 4.3

Kaplan-Meier survival curves and how aggregating the outcomes can conceal treatment-induced progression. (A) Example of a typical Kaplan-Meier survival curve comparing the treatment group (Olaparib) and the placebo group (from Ledermann et al.[43]). (B) Mathematical model of survival based on proportional hazard (exponential decrease of disease-free survivors at constant rate). Dashed lines represent the treated group (a) and solid line the control group. (C) "Crossover" survival curves as often observed for immune-checkpoint inhibitors (ICIs) compared to chemotherapy, indicating benefit of immunotherapy with the two ICIs, nivolumab and ipilimubab: the initially fast descending survival curve (dashed) of the treated group bends and crosses the curve of the control group (chemotherapy, solid line) toward a higher plateau indicating a higher rate (fraction) of progression-free survival at the end of the study at two years (from Hellmann et al.[44]). (D) In a simulation, we compare the two groups (a) and (b) from panel B with a third group (c) of patients (thick solid black line) treated with ICIs for which, again, as in C, long-term survival with a plateau has been observed. In the mathematical model, the immunotherapy group (c) can be broken up into groups (x and y) presented by the two curves (finely dashed) that represent responders (x) and hyper-progressors (y). These two dashed lines add up to the solid line: $(x) + (y) = (c)$. Thus, the aggregate (c) displays the net effect, showing a benefit that "appears later" than the control and recapitulates the typical curve for ICI treatment with the crossover seen in panel C.

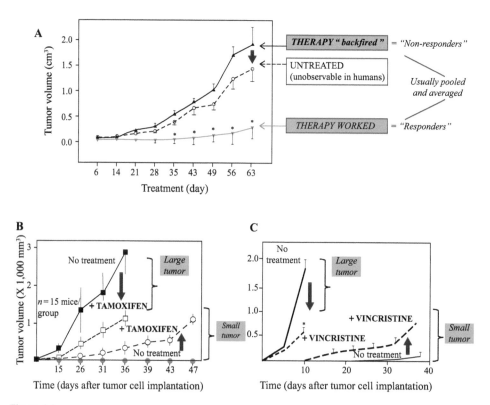

Figure 4.4
Animal experiments exposing the existence of treatment-induced progression. (A) Comparison of the treated group with an entirely untreated group reveals that the treated but nonresponding animals (top curve) displayed faster tumor growth than animals in the untreated control group, thus treatment "backfired." Adapted from Juengel et al.[45] (B, C) Apparently paradoxical treatment effect in two examples of grafted tumor mouse models (B, breast cancer and C, lymphoma). Depending on the number of subcutaneously implanted tumor cells, they establish either a large, fast-growing tumor or a small, slow-growing tumor. Drug treatment (dashed curves) with tamoxifen or vincristine, respectively, slows tumor growth of the large tumor but accelerates growth of the slow-growing tumor. From D. Panigrahy and S. Huang, unpublished observations, and Sulciner et al.[46]

of treatment group T into T_R and T_{NR}). The curves show an acceleration of tumor growth in the treated but nonresponding animals T_{NR}. The tumor grew over time to a significantly larger size than the tumor in the untreated control C (dashed lines). Had the authors combined the treated, nonresponding cases of T_{NR} with the treated responding tumors of T_R, the aggregate curve T would still have suggested that the drug has an effect in slowing tumor growth, albeit not as much as when the responders are displayed separately. In clinical trials, the curve for the T group cannot be readily separated into T_R and T_{NR} since the survival rate (an aggregate) is measured. Moreover, the T group is compared with a control cohort C in which patients are treated with an older (standard) drug and, for ethical reasons, cannot be compared with a "treatment-naïve" control group that reveals the natural course.

A starker demonstration of a direct tumor-promoting effect of therapy is the treatment of small dormant tumors. This can be studied in animal models by grafting of an atypically small number of tumor cells. Such small tumors stay for extended time in a quasi-dormant state without "taking off." Only until after a long latency period of months do they either spontaneously or upon experimentally induced stress, such as wounding and surgery,[47] escape the dormancy; they then exhibit the typical rapid, near-exponential growth.[48] Therefore, such small tumors are used for studying tumor dormancy. However, preclinical drug testing is performed on the most aggressive tumors in which a large number of cancer cells (in the millions) are grafted. This practice in industry is meant to set the bar high for the capacity of the experimental treatment to suppress a vigorously growing tumor. Such aggressive tumors either respond to treatment, as evident by a flattening of the tumor growth curve, as in figure 4.4B,C, or are resistant to the experimental treatment and grow at the biologically possible maximal rate—like that of control (vehicle-treated) animals. The third outcome, a *further* acceleration of the already maximal tumor growth by treatment, cannot be observed.

By contrast, we and others have shown that a dormant tumor can be triggered by the same anticancer drug that suppresses large tumors to exit dormancy and enter near-exponential macroscopic growth (figure 4.4B,C). We have observed this dramatic tumor-stimulating effect of antitumor treatment in animal models for a variety of tumor types (including lung cancer, breast cancer, and other carcinomas, as well as melanoma and lymphoma) and for both chemotherapy as well as target-selective therapy (Huang and Panigrahy, unpublished observations and Sulciner et al.[49]). These results establish a positive causal relationship between treatment and tumor progression. Moreover, this effect of treatment is sustained: treatment not only accelerated the growth rate of the (small or dormant) primary tumor but also greatly enhanced metastasis as well as metastatic relapse. These experiments also teach us that only in particular circumstances, such as in slow-growing tumors, can the paradoxical tumor-promoting effect of treatment be observed.

4.4.5   Hyperprogression following Treatment with Immune-Checkpoint Inhibitors

In clinical trials, the identification of cases of treatment-accelerated tumor growth requires that patients are individually monitored by imaging (or biomarker, if available) and also reported individually, not as an aggregate. However, in a cohort, a heterogeneous response could manifest in the particular shape of the survival curves. This appears to be the case in immunotherapy: the existence of T_{NR} patients that not only do not respond but also do worse under therapy has emerged as a problem in immunotherapy with ICIs. Such patients are referred to as *hyperprogressors*. As mentioned earlier, the promise of immunotherapy lies in the observation that for a small fraction c percentage of patients, sustained tumor-free survival over several years is possible ("cure fraction").[50] This outcome is manifest in the characteristic "bending" of the descending survival curve toward a plateau that will (likely) not further descend to the line of 0 percent survivors (see figure 4.3C, an example for ICI treatment of lung cancer).

But the curve does not always simply begin as a flatter curve that exhibits less decline compared to the curve of the control group treated with conventional therapy (e.g., chemotherapy). Instead, one consistently observes a characteristic initial steep decline of the survivor rate in the ICI-treated cohort, often even steeper than that of the conventional treatment group. Only later does the curve bend and flatten by *crossing* the survival curve of the control group, toward the plateau of long-term survivors, whose level defines the cure fraction c that is significantly higher than that of the control group.

We propose that the initial steep drop of the fraction of survivors, followed by the flattening and characteristic "crossing over" of the curve for the control treatment group (figure 4.3C), could be the result of a *superposition* of multiple curves rather than simply reflecting a "late effect" of ICIs, the current default interpretation by clinicians. Computational simulation of such a scenario shows that the shape of a composite survival curve can arise from the additive combination of hyperprogressors and responders in the same treatment group (figure 4.3D).

Hyperprogression in response to ICIs could be explained by different types of mechanisms, including an overshooting response to the activation of the adaptive immune system by cells that normally control (suppress) immune defense, or the misdirection of the inherent proliferation-stimulating activity of the ICIs from the intended target, the T cells, to the tumor cells. Independent of the mechanism, we have reason to assume that hyperprogression is only the most visible expression of a more general, perhaps universal phenomenon. Treatment either suppresses the tumor or induces its progression because the perturbation of a complex system (here consisting of interacting cancer and stromal cell populations that stimulate and/or inhibit each other and form a robust self-stabilizing system) can have unintended consequences. These are channeled by the internal dynamics of the system, and as a whole, its response can go in either one of two opposite directions.

Could the split of the cohort response of the treatment group into the T_R and T_{NR} groups and the presence of a fraction p_β of patients in the T_{NR} group that do worse than the control group be at the heart of the inherent limitation of cancer treatment? If acceleration of progression is an intrinsic behavior of a tumor as a self-sustained stable system, could a gentle noncytotoxic drug blunt this undesired effect of cell-killing-based treatment? This would increase the overall efficacy of treatment—an underappreciated room for improvement in cancer therapy that does not depend on enhancing the potency of treatment in killing tumor cells.

4.4.6 From Patient-Cohort Heterogeneity to Cell-Population Heterogeneity: The Core of the Paradoxical Response to Treatment

The treatment-induced adoption of a more malignant behavior is not only a phenomenon at the tumor tissue level but can also be observed at the *cellular level*. In cell culture, an almost universal behavior has come to light with the arrival of single-cell resolution analysis methods.[51] Increasing the dose of a drug will kill an increasing percentage of cells in the cell culture—creating the characteristic sigmoidal "kill curve" that displays the dose dependence of the cell death fraction and is used to establish the dose that kills 50 percent of the cells, the EC_{50} ("half-maximal effective concentration"), as a measure for the cell-killing potency of a compound (figure 4.5). The fact that the proportion of cells that succumb to the drug increases gradually with the dose of the drug, and not suddenly at a characteristic dose at which all cells die, indicates that each cell has its own distinct "death threshold" that varies in the cell population. Thus, we need to consider *ensembles* (populations) of individual complex systems (cells).

This opens the path to the profound notion of *heterogeneity* not only of patient populations (as discussed above) but also of the cell populations that constitute a tumor that will be discussed in the theory of section 4.5. Heterogeneity is one of the four central concepts underlying the key idea of this article mentioned in the Introduction: the variability of nominally identical individual cells in the cell population with respect to their cell state S that determines drug susceptibility of each cell (figure 4.5). Such phenotypic heterogeneity

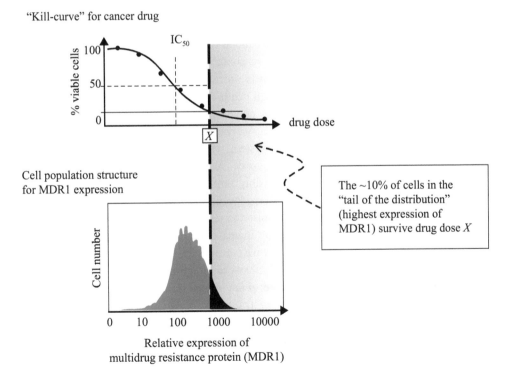

Figure 4.5
Relationship between heterogeneity of cancer cell populations with respect to drug susceptibility and the slope of the dose-response curve. At a very high drug dose X, still ~10 percent of cells survive because of the heterogeneity of the cell population, for instance with respect to the expression of resistance genes, such as MDR1.

of cells exists even in a clonal cell population (consisting of genetically identical cells), but this finding is often taken for granted.[52] At any dose, many cells do not die, even if they are all genetically identical: at a high dose that kills 90 percent of the cells, 10 percent survive. What happens to the cells that survive?

While one fraction of the cell population undergoes apoptosis triggered by chemotherapy, target-selective compounds, or irradiation, the nonkilled cells do not just survive as unaffected "innocent bystanders." It is plausible that the nonkilled cells have escaped death narrowly and thus have encountered a massive cell stress, which must have affected their cellular state S. As has long been suggested,[53] this can now be demonstrated with single-cell resolution analysis of the transcriptome of individual cells, which allows the measurement of the high-dimensional cell state[54]: the surviving cells, which have been exposed to the nonlethal cytotoxic action of treatment, are pushed into various states characterized by the expression of *stress-response* genes, including those encoding proteins for xenobiotic detoxification, as well as repair genes and stem cell–related genes[55] (more details in section 4.9). These surviving but stressed cells exhibit tolerance to many stressors beyond toxins, such as radiation, hypoxia, and low pH, and can better cope with genome damage compared to the more differentiated, unstressed cells. Sometimes they express regulators of multipotency normally found only in early embryonic cells, such as Oct4 or Bmi.[56] Because this stress/stem cell response confers increased resistance and even "rejuvenation" (in view of the stemness), it is warranted to consider such a "response of the nonresponders" as being *opposite* to cell death. Thus, a stressful, toxic but nonlethal perturbation of an ensemble of systems may achieve the opposite in a fraction of systems.

In concluding this section on empirical observations, we propose that a fundamental property of a class of complex, resilient systems can exhibit multiple alternative, self-stabilizing modes of behavior. As a consequence, such a system can respond to the same perturbation in either one of two (or possibly more) opposite ways by entering these alternative states. Then, in a composite (heterogeneous) ensemble of such systems (patients, cells), members of the ensemble may be poised to respond differently to the same perturbation. Thus the perturbation would further increase the diversity of the ensemble. This phenomenon applies to both levels, whether we consider the cancer cell or the entire tumor as a system. At both levels, the double-edged sword effect of an intervention is apparent, or following the logic of Nietzsche's quote: *If you intervene and do not manage to kill a cancer cell with your cytotoxic perturbation, you trigger its adoption of a more malignant (stem cell–like) phenotype, and if you do not manage to eradicate the entire tumor, you may cause it to progress and relapse later.* We have an "either-or" binary choice response: both responses can be triggered by the same nonlethal therapeutic perturbation, and systems responding in either way can coexist in an ensemble of systems. The problem is that such nonlethal intervention, that is, an intermediate success or a partial antitumor effect, is the norm because of elementary logistic reasons (diminishing returns of hitting a rare cell population, drug accessibility, systemic toxicity).

In the next section, back to theoretical concepts, we will relate this counterintuitive observation to the general principles of dynamical systems introduced in the previous section.

4.5   Stochasticity and Poised Instability: Linking the Double-Edged Sword Effect to the Landscape Concept

Why can a system, the apparently same type of tumor or the same type of tumor cell, respond to the same perturbation in one way or in the opposite way? Shouldn't the same perturbation of the same system produce the same one response (figure 4.6A)? And likewise, why, if that system is present as an *ensemble* of multiple similar or (near) identical replicates (patients in patient cohorts, cells in cell populations), would some of them respond in one way, while others would respond in the opposite way to the very same intervention?

4.5.1   Apparent Stochasticity: The Binary Decision on the Hilltop

One obvious starting point to address these questions is to assume that the system's internal state, technically, the "initial state," when receiving the therapeutic perturbation (be it defined

A   LINEAR (PROXIMATE) CAUSATION

(No underlying formalism beyond the ad hoc "causation" by an intervention of a transformation a → B)

B   NONLINEAR COMPLEX SYSTEM WITH MULTISTABILITY

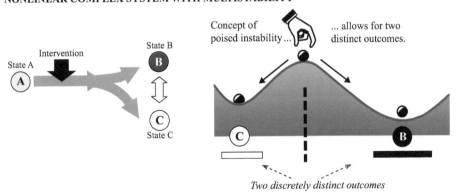

Two discretely distinct outcomes

Figure 4.6
"Poised instability" due to the nonlinear dynamics in a system can account for the unexpected binary outcome of an intervention aimed at shifting the system state. (A) Traditional linear process in simple systems. (B) In reality, in complex systems, the desired transition [A][B] is complicated by the appearance of state [C]. Such splitting of outcome into two opposites can be intuitively framed by placing the state initial [b] (black marble) on an unstable steady state or hilltop on a landscape, where the state exhibits "poised instability." The two outcomes manifest in two adjacent valleys, which naturally explains the discrete separation of the two outcomes.

by the molecular profile or by phenotypic traits, such as the tumor growth rate), is slightly different in each individual system (a cell in a tumor or a tumor in a patient) that is a member in an ensemble of systems (the tumor or the patient cohort). These differences in the system configuration of individual systems may be so small as to barely be noticeable, yet sufficiently large to determine the "macroscopic" outcome: whether the cancer *cell* as a system will undergo apoptosis or become a stem cell–like, resistant cell that drives later relapse. (The same principle applies to the *tumor* as a system: whether a tumor will progress or shrink in response to the same treatment may depend on fluctuations in the tumor tissue state defined by relative cellular abundances.)

If no apparent and consistent difference between individual systems (cancer cells or tumors) can be measured, then it may appear as if a system selects one of the two intrinsic possible responses randomly when facing a choice between the two distinct available outcomes (figure 4.6A,B). This is effectively the same as if the internal initial state of each system fluctuates randomly at the microscopic scale (i.e., unnoticeably) but sufficiently to produce the macroscopic *apparent* randomness of the binary outcomes, as if chosen by tossing an internal coin. If this indeed is the case, it would explain the historical difficulty of identifying molecular biomarkers that can predict the therapeutic benefit.[57]

The microscopic random fluctuations are caused by random fluctuations in the abundance of regulatory proteins in the cell, and thus they are elementary and inevitable—they embody the apparent internal coin tossing. These fluctuations are the physical consequences of the inherently stochastic nature of chemical reactions that depend on random molecular collisions and become manifest in small reaction volumes with low numbers of molecules, such as in cell compartments, and are prosaically manifest in the phenomenon of *gene expression noise*.[58] These random fluctuations in the values of x_i for all network components result in a random walk-like "wiggling" of the trajectories of the state S in state space, with micro-steps that are not in accordance with the constraints imposed by the regulatory interactions.

4.5.2   The Landscape Topography Amplifies Molecular Noise

To explain the counterintuitive but robust feature of the possibility of a system displaying one of two distinct macroscopic responses to the apparently same intervention, we map the cell (or tumor) behavior onto the quasi-potential landscape. Imagine again the idealized cartoonish depiction for now: upon some intervention (to be specified), the cancer cell or the tumor, as a system represented by the marble, finds itself placed on an unstable steady state, a hilltop (if projected) or, better, a mountain ridge (figure 4.6C). As discussed earlier, the system will roll down in one of two opposite directions upon the slightest perturbation, thereby undergoing a symmetry-breaking event. (*Why* a system is placed on an unstable hilltop state will be discussed in the next section.) Either one of the two diverging outcomes will persist even after the perturbation is long gone. The hilltop thus acts like a watershed to the water droplet: the descent into one but not the other valley implies that a binary choice *must* be made. Hereafter, the system follows the self-propelling divergence of the trajectories dictated by the slopes of the landscape. Each valley represents the state of the tumor or of the tumor cell after the perturbation: destined either to grow and progress or to regress and disappear.

When a system is poised at a hilltop of a landscape, it cannot hold the unstable steady state when perturbed, which is why it "cannot *not* respond"; it has to choose one of the two alternative options as response. We call this property ***poised instability***. The quasi-potential landscape, a mathematical construct introduced in section 4.3, permits the use of metaphors, such as the hilltop, or equivalently, the mountain ridge or the watershed (figure 4.6C). The topographical structures illustrate intuitively why minuscule temporary perturbations, such as the aforementioned tiny random fluctuations of the system state, constitute imbalances that push the system away from the unstable steady state (hilltop) and lock in a specific choice between the two opposite possibilities.

More generally, because the state space is not a flat plane but structured by a landscape with separating hills, the noise-driven microscopic fluctuations of S are not automatically averaged out but can occasionally be amplified and have macroscopic consequences. This is the case when the system is poised at an unstable steady state, where a tiny departure from the hilltop steady state, typically due to noisy fluctuations, pushes the system down one of the valleys. This amplification of micro-fluctuations by the landscape topography explains why complex systems, such as the cell or the tumor, which use homeostatic mechanisms to maintain a robust phenotype far above the microscopic scale of molecular fluctuations, still can undergo switching between such high-level (macroscopic) phenotypes that appears to occur at random. Thus, we can link the probabilistic nature of an observable (macroscopic) behavior to the inherently stochastic nature of the (microscopic) molecular processes. We thus introduce the third fundamental feature of tumors as a complex system that we have listed: *stochasticity*—even at the macroscopic levels of cell fates and tumor outcomes.

In a deeper theoretical consideration, I would like to note that stochasticity is fundamental and at the heart of the quasi-potential landscape. It is not merely measurement noise. Without stochasticity of the dynamics, the landscape would not exist. It is only owing to the intrinsic, "unmotivated" stochastic fluctuations immanent to a system state that a system can, without apparent external cause, move spontaneously away from an unstable steady state, the poised instability, and descend into one attractor state. Although the landscape topography may appear as a static, deterministic structure that unambiguously dictates the trajectories of descent, its very existence depends on how the interactions in the system "play out" in the presence of stochastic noise but frozen *in the mathematical limit of zero noise*. Thus, the landscape topography is an invariant, fundamental property of a given system with its system-defining, fixed diagram of the interactions between its N types of components. The existence of the landscape is not unlike the "instantaneous velocity" of an object at a specific time point ("snapshot"), which by definition has a duration of zero, yet the object has a velocity, which is generally defined as distance traveled over a time period (e.g., 100 miles per hour), namely, for the "limit" as the length of the time period approaches infinitesimally close to zero. In the same sense, the state space of a system folds into a defined landscape because of the very notion of stochastic noise but for the limit of zero noise.

4.5.3 Poised Instability Can Divide and Push Apart a Heterogeneous Ensemble of Systems

We have encountered the fourth fundamental property of cells or tumors as complex systems, namely, *heterogeneity*, and have seen how it is concealed by aggregating data in the clinical survival curve and the cell kill curve (section 4.4.5). With a looking glass that resolves details below that level of data aggregation, we can now more formally state that heterogeneity arises from stochastic diversification of individual systems (e.g., cells) with respect to their states in a statistical *ensemble* of replicates of these systems (e.g., a cell population). Heterogeneity is thus a property of an ensemble of (complex) systems, which in our case are either the cells in a tumor or the tumors in a patient *cohort*, and must not be confounded with stochasticity.

More specifically, cells, even within the same nominal type and within a clonal cell population, are always more or less slightly different from each other, resulting in "*intra-tumor* heterogeneity."[59] These differences of cell phenotype may either be short-lived and random, caused by stochastic gene expression noise or, alternatively, may represent distinct, predetermined, and durable differences due to multistability as specified by the network. Multistability may appear as the presence of multiple biological subtypes of cells within one population of one cell type, such as stem cells, the transient amplifying cells, or immune-activated states. Both types of heterogeneity are produced even within cells of one clone and thus is referred to as *nongenetic heterogeneity*. If, by contrast, we consider somatic mutations, which alter the GRN and hence the landscape of each individual cell, the diversity of possible cell states in a tumor further increases, adding the layer of *genetic heterogeneity*.[60] At the level of the entire tumor in different patients, even if classified under the same diagnostic type with respect to the tissue of origin (e.g., squamous cell carcinoma), there is an additional layer of heterogeneity: the tumor may differ from each other with respect to their detailed cellular composition, for instance, due to variation in the relative abundance of immune cells and of cancer cells, resulting in "*intertumor* heterogeneity" within a cohort of patients with the same nominal tumor type (i.e., arising from the same cell type of origin).

Heterogeneity and poised instability jointly produce macroscopic behaviors at the level of ensembles of systems that defy the effect of "averaging out" the behaviors of individual systems: this results in diverging behaviors of a composite system, the tumor as an ensemble of cells, or the patient cohort as an ensemble of tumors (figure 4.7A). In a tumor at a state of poised instability, the individual cells of a cell population that as a whole is on a hilltop of the epigenetic landscape are placed at slightly distinct spots around the

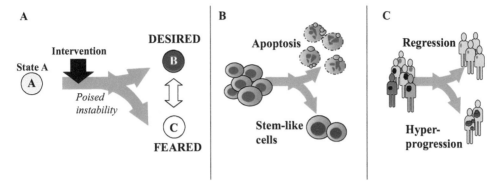

Figure 4.7
Heterogeneity of ensembles at the two levels can be subjected to poised stability and lead to the two discrete outcomes, one desired, and one feared. (A) Because of the branching paths, in heterogeneous systems (cells or patient tumors), the ensemble is split in the two outcomes: (B) at the level of cells in cell populations and (C) at the level of tumors in patient cohorts.

hilltop, as single-cell transcriptome measurements now amply demonstrate. It is then obvious from the hilltop metaphor that individual cells will descend in one direction or another and move to opposite states: they can enter the quiescent or the proliferative state in response to the same perturbation (figure 4.7B). Similarly, at the level of tumors as systems, if a heterogeneous population of patients with the same nominal type of cancer is at poised instability, their tumors are on the hilltop of the landscape of tumor states at slightly distinct positions, depending on minor differences with respect to the relative abundances of the various cell types. Thus, an individual tumor as a cell society within a heterogeneous cohort of patients could be biased to descend into one direction or the opposite: the valley of tumor regression or the valley of accelerated progression, as seen in hyperprogressors (figure 4.7C). Hence, patients with nominally the same tumor type may exhibit qualitatively distinct, macroscopically opposite responses due to microscopic differences in cell-type composition.

Taken together, members of an ensemble of complex systems of the apparently same (macroscopic) type can under some circumstances exhibit distinct, *opposing* behaviors in response to almost identical perturbations. This principle explains why the human mind's habit of a 1:1 linear mapping between cause and effect (figure 4.6A) can fail when explaining living systems in medical research that is still dominated by a culture of linear thinking. Poised instability and binary outcomes are related to the oft-cited principle of *sensitivity to initial conditions* in chaos theory,[61] explaining the diverging outcomes of two *almost* (but not entirely) *identical* systems to the same perturbation (figures 4.6B, 4.7A). In the landscape framework, the duality in the Nietzsche response to therapeutic intervention or, equivalently, the double-edged sword property is captured by placing the system state at a spot of poised instability represented by the hilltop.

But why does the therapeutic (or other external) perturbation create the situation of poised instability for the cancer cell, or for the tumor, in the first place? In the next two sections, we will address this question. To do so, we will first recapitulate the notion of attractors representing cell (pheno)types and ask how cells transition from one attractor (cell type) to another.

4.6 Process Spontaneity: The Fundamental Inevitability in Multistable Systems

To better understand the *directionality* or, equivalently, the *spontaneity* of biological processes, let us now apply the basic concepts of nonlinear dynamical systems theory introduced in section 4.3 to biology. First, we shall develop a more precise notion of what is a "*biological state*." We then reason about the transition from one such characteristic state to another and what drives it. These *state transitions* are the elementary processes that we need to understand. To illustrate these principles, we consider here the level of the cell (not of the tumor) as a complex system. We again regard the phenotype of the cell, or the cell state S, as the emergent property of the interactions of the gene regulatory network (GRN) that collectively control which genes in the genome are expressed into active proteins, thereby generating the gene expression configuration $[x_1, x_2, \ldots x_N]$ (figure 4.2) across all the gene loci of the genome (measured as "*transcriptome*" and "*proteome*"). These genome-wide molecular patterns, to a first practical approximation, determine the observable, characteristic *cell phenotype*, which thus is a biological state.

4.6.1 Cell Types Correspond to Attractor States

A cell type is a particular form of a cell state that can be any functional, observable biological state of a cell: a cell of a given cell type can be in the proliferative, quiescent, activated, senescent, or apoptotic state. The epithelial or mesenchymal phenotypes, as well as the more stem cell–like phenotypes of cells, including the *cancer stem cell*, are also cell states. In a coarser (higher level) view of the whole organism, the specific cell types in the metazoan body, such as muscle cells, liver cells, and lung cells, which are defined by specific configurations S of the activities of genomic loci, are also cell states.

As first presented in section 4.3.7, biologically characteristic phenotypes, such as cell types, are discretely distinct cell states that Conrad Waddington equated with the valleys on his epigenetic landscape. These phenotypic states are, in a macroscopic approximation, not only discrete but also stable: they maintain and restore their identity even in the presence of gene expression noise and minor perturbations in the tissue. An essential corollary of these two properties is that cells can undergo switch-like (all-or-none) phenotype conversions between these states given appropriate (major) perturbations. Thus, cell states are self-stabilizing but not rigid. Such behavior is naturally captured by mapping the cell states to the bottom of the valleys in the quasi-potential landscape, or to the ***attractor states***, or in brief "attractors," as introduced in section 4.3.

The concept that *attractor states of molecular interaction networks* represent distinct biological cell phenotypes had been proposed in the late 1940s independently by Max Delbrück (in a mathematical way)[62] and by Waddington (metaphorically, using his pictorial *epigenetic landscape*)[63] to explain cellular differentiation, and then again by Monod and Jacob shortly after their discovery of gene regulatory interactions in 1961. Later, in 1969, Stuart Kauffman[64] extended these ideas, along with a series of fundamental insights,[65] to high-dimensional gene regulatory networks consisting of thousands of interacting genes, postulating that the high-dimensional attractors correspond to cell types.

The attractor concept captures a dynamical feature of biological reality that is often taken for granted: because attractor states are self-stabilizing, once a cell is placed in a specific attractor state, the factor that has "caused" the associated phenotype can vanish, yet the phenotype persists. The attractor state thus represents a memory of itself—it maintains its molecular identity once it has adopted it. (There is no need to invoke some molecular "epigenetic marks" in the form of DNA methylation and chromatin modification, which are dynamic and reversible anyway.[66])

On the other hand, many paths in state space can lead to the same attractor: it suffices to reach any point within the valley that "drains" to the attractor—or the "***basin of attraction***" (see figure 4.2F); the downhill stretch toward the new attractor state is then taken care of by the "attraction." In this segment of the path, the process is goal-directed toward the (new) attractor state and is said to be "spontaneous" because it follows the gradient (slope) in the quasi-potential energy U. In the metaphor of the landscape and water flow, the basin of attraction of an attractor is defined by the totality of the region (of state space) in which all "initial states" will drain into that attractor state. Importantly, a cell in the attractor, while remembering its state following minor perturbations (as long as its departure from the attractor state stays within the basin), "forgets" its past, that is, from where it entered its attractor basin. This loss of information on history is the cost that the system pays for the stability of a steady state.

4.6.2 Cell Type Switching as Transition between Attractor States on the Quasi-Potential Landscape

An essential characteristic of complex systems is the coexistence of multiple attractors (multistability), and hence, one is also interested in their relationships with each other, which are captured by the topography of the quasi-potential landscape. It is on this landscape that biology takes place. Then the relationships between the attractors are evoked by the following question: how do cells switch between the discretely distinct stable phenotypes, that is, "move" between the attractors? Such an attractor state transition is the formal representation of cell-biological processes, such as differentiation, or, as is the topic here, treatment-induced cell phenotype switching. What happens in detail when an *intervention* causes a particular attractor transition in view of the stability of a cell state? How does the intervention overcome the attracting forces that seek to restore the cell state to that of the attractor state following a perturbation? Let us consider the cancer cell as a complex system with its landscape containing all possible cell states and examine the process of cell-killing therapy as the intervention. The goal of drug treatment is to cause a shift of the cancer cell state: from the proliferative state to the apoptotic state, both of which are distinct attractor states.

In the simplest formal model, the drug impacts a set of signaling pathways that are biochemically connected to the drug's molecular target. This event alters the expression and activities of a specific set of proteins in the cells. This multipoint perturbation of the molecular network results in the change of the configuration of the values x_i of the components i (expression levels of genes, protein activities; see section 4.3) that collectively define the cell state S and thus causes a displacement of S in the landscape. In the case of an anticancer drug that induces apoptosis, this would shift the cell from the proliferation attractor to *some* place in the basin of the apoptosis attractor, taking advantage of the aforementioned multiplicity of ways to enter a new attractor. As long as the cell is placed in the basin of attraction of the destination phenotype, it will reach it, descending in a "self-propelling" way (downhill) to the lowest point of the new attractor state that executes the apoptosis program.

This default model of attractor transition faces a challenge: mathematically, if we take into account the landscape topography, moving the state S out of the attractor of proliferation involves a *climb* out of it. This uphill path embodies the effort required to overcome the regulatory constraints that stabilize the proliferative state. This effort is represented by the quantity ΔU, the depth of the attractor—or height of the hill or "energy barrier" separating two attractors. By contrast, the descent within the basin of the new attractor is spontaneous, or "effortless," once the "climb" out of the proliferation attractor valley has been successful. Such crossing of a hilltop (climb and descent into new attractor) appears to have the feature of a threshold property or a "point of no return" that can be experimentally observed.

4.6.3   Transition between Attractor States Can Be Facilitated by Bifurcations

The uphill climb to overcome the elevation ΔU to exit the basin of attraction requires enormous effort to defy the high-dimensional homeostatic forces of the regulatory network of thousands of genes: a very particular combination of alterations of the components in the interaction network would be required, one that is difficult to implement by external controlled manipulation. It is also extremely unlikely to occur by chance fluctuations of the state S as introduced in section 4.5. The steepness of ascent and size of the basin are the mathematical expression of how low the "first exit" probability[67] is for the event of exiting a given attractor basin by chance, that is, driven solely by the randomly wiggling course of the state S.

Yet, experimental observations of cell state transitions suggest that in the presence of intermediate to low doses of a signal known to cause a phenotype switch, a state transition is possible: it may occur not that rarely but is probabilistic in nature. In fact, if we consider an entire cell population sitting in the attractor, with each cell in a slightly different state because of gene expression fluctuations, it is conceivable that some cells will be more susceptible to the intervention and to exiting the attractor than others.[68] This would be consistent with the notion of a statistical distribution of response thresholds among the individual systems in the population explained in figure 4.5.

The relative ease by which a signal can, even at low dose, already push some cells into the new attractor prompted us to propose an alternative principle to explain the ubiquity of such probabilistic and partial responses: a *change of landscape topography*. In the theory of nonlinear dynamical systems, this type of change is part of a **bifurcation** process: we postulate[69] that the external perturbation that causes a cell to switch to a new phenotype lowers the hills or "energy barrier" between the attractors. Mathematically, a bifurcation process is driven by altering the value of just one (or two, not many more) *parameter(s) μ* that define(s) the strength and modality of a given interaction in the network (figure 4.8). Since these interactions are responsible for folding the state space into a landscape in the first place, a change in the nature of interactions will alter the landscape topography. However, mapping the alteration in the interactions, represented by the change of value of the parameter μ, to a distortion of the topography of the rugged high-dimensional landscape, is not mathematically straightforward. What is clear is that we cannot use the intuition of

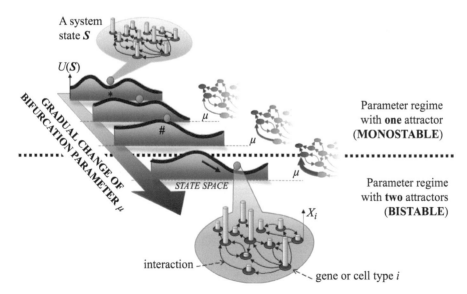

Figure 4.8
Basic terms for the key principle of bifurcations. Gradual change of the bifurcation parameter μ, which represents the "strength" of a critical interaction in the network (right) and has an impact on the quasi-potential landscape: the attractor (*) flattens and suddenly disappears, creating an unstable state (hilltop, poised instability #).

linear causality, as epitomized in the countless "arrow-arrow" schemes of textbooks on drug mechanisms and pathogenesis, to predict which of the "microscopic" (molecular) regulatory interactions must be tweaked to cause a certain landscape distortion so as to trigger a particular "macroscopic" cell state transition.[70]

One characteristic of a bifurcation process is that it causes not simply a *quantitative* distortion in which hills become higher or attractors lower, but it triggers at some point a *qualitative* (discontinuous) alteration of the quasi-potential landscape; this is the actual bifurcation event. One illustrative type of bifurcation involves the destabilization ("flattening") of an attractor (stable steady state) until it ceases to be attracting. Then, even as the relevant bifurcation parameter μ gradually changes, the valley of the original attractor state (* in figure 4.8) at one critical value of μ flips into a hilltop (unstable steady state). In this abrupt, qualitative change, the attracting forces are reversed and become "repelling" forces. The attractor becomes a hilltop, which is still a steady state, but it is unstable and thus exhibits now poised instability (# in figure 4.8). Globally, bifurcations can change the number of attractors (in a certain subregion) of the landscape and hence confer a qualitative change of the behavioral repertoire of a system. In the example of figure 4.8 (a so-called pitchfork bifurcation), the system shifts from a mono-stable to a bistable dynamics.

Destabilizing an attractor or, equivalently, lowering the separating hills that act as "energy barriers" is reminiscent of catalysis in chemical reactions in the world of classical equilibrium systems: catalysis increases the reaction rates for a transition from one stable steady state (the substrate in chemical equilibrium) to another (the product) by lowering the "energy barrier" that otherwise would prevent a spontaneous reaction along the free energy difference to a lower energy state and that would demand "activation energy" to start the reaction. Once the barrier is sufficiently low, the reaction can, driven by thermal fluctuations, move to the alternative chemical state at a lower "free energy." Without the "energy barrier," the reaction is now an inevitable "spontaneous" reaction.

4.6.4 The Rough Equivalence to Energy Landscapes of Classical Equilibrium Thermodynamics

For those who wish to relate the above introduced theoretical framework to the elementary concepts from physics, one could summarize the relationship in the following way: in dynamical systems theory, notably when applied to biology, we do not deal with classical equilibrium systems as studied in physics or chemistry, with free energy as the potential function that indicates the intrinsic "urge" of a system to shift its state toward thermodynamic equilibrium—the lowest energy state. Instead, we have a nonlinear complex system (the living organism) that is far from thermodynamic equilibrium and maintains itself, thanks to the internal constraints, in some state of orderly fashion away from thermodynamic equilibrium. This is possible, and its apparent spontaneous order does not defy the laws of thermodynamics, because it is an *open* system that takes up from outside free energy (nutrients) that fuels the interactions of its component parts, which in turn impose constraints that "frustrate" a descent into thermodynamic equilibrium. The system is "stuck" in attractor states away from thermodynamic equilibrium while consuming energy. Its orderly behavior, far from thermodynamic equilibrium, consists of state shifts between the multiple local minima, the attractor states of the system produced by the network of inter-actions between the system's components. Examples of such organized behaviors are development and homoeostasis of living organisms. Transitions between ordered, stable states herein are governed by the "quasi-potential" energy U that is visualized as the eleva-tion of the "quasi-potential" energy landscape and is not the same as the free energy in classical potential landscapes. With due caution, one can articulate meaningful equivalence between these two worlds of equilibrium and far-from equilibrium dynamics.

In summary, we have dissected the inevitability of a process within a complex system: rather than saying an intervention (a treatment) "causes" a change, a more accurate state-ment would be as follows: a new constellation, defined by interaction parameters altered by the intervention, establishes a new condition for spontaneous change of the system and thereby the spontaneity of change from one stable state to another. This picture profoundly differs from the "arrow-arrow" cartoons depicting causal chains of molecular events used to explain the molecular mechanism of action of drugs.

Equipped with the generic formal concepts of attractor transitions, we can now add back the specific biological content and explain how the attempt to kill cancer cells may make the stressed, nonkilled cancer cells more malignant or how enforcing a reduction of tumor size may make the tumor tissue more aggressive. We return to the question raised in section 4.5: why does the therapeutic intervention place the cells at a hilltop, conferring poised instability that can explain the two alternative outcomes? And why do the two outcomes opposite to each other specifically take the form of regression versus strengthening of the tumor?

4.7   Why the Nietzsche Effect Is Immanent to Cancer Cells and Tumors

Tweaking an interaction between components of a complex system to drive a specific bifurcation that destabilizes an attractor will flatten the attractor basin—or equivalently, reduce the "energy barriers" around an attractor—and thereby initiate an attractor transi-tion. We now combine these ideas with the notion of stochastic noise and heterogeneity (explained in sections 4.5 and 4.6). Together these concepts, which are all derived from dynamical systems theory (introduced in section 4.3), allow us to present the general principle that explains why an intervention in a complex system can trigger diverging responses and backfire.

4.7.1   Flattening of the Attractor Basin before a Bifurcation Event Increases the Diversity of System States in the Ensemble of Systems

In the process of reducing an "energy barrier" around an attractor as the bifurcation param-eter changes, the attractor basin first goes through the phase of gradually flattening before the qualitative step of its sudden disappearance at the bifurcation point. Therefore, as conditions that alter the critical interaction in the network nudge the system closer to a bifurcation point, the attracting forces of the basin of attraction decrease, the homeostasis weakens, and the fluctuations of the system state (driven by gene expression noise in the case of the cell state) grow in amplitude. Individual cells, although still in the original valley, no longer exhibit the exact gene expression profile of the attractor and appear more different from each other at any given time point because the fluctuations are not synchronized, even if on average the cell population still displays the characteristic gene expression profile of the attractor state. This manifestation of destabilization has now been experimentally demonstrated using single-cell resolution gene expression profiling: a diversification of the "micro-states" (transcriptomes) of the individual cells of a cell population *before* binary cell fate decisions.[71] This phenomenon of increased variance of cell states around the same mean state is illustrated in the cartoon of figure 4.9.

Let us now apply the principle of attractor destabilization and exit from it to a therapeutic intervention: a treatment seeks to cause a transition from the attractor of the proliferative state

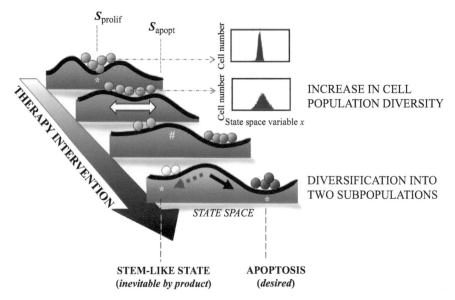

Figure 4.9
The principle of bifurcation events explains the paradoxical generation of stem-like cells during cancer treatment. Each ball is a cell on the quasi-potential landscape. The intervention aims at shifting cancer cells from their attractor in the basal proliferative state (*) to that of the apoptotic state on the right. The therapeutic intervention is modeled as the change of a bifurcation parameter (large arrow from back to front)—consistent with the action of drugs on molecular interactions. As the system approaches the bifurcation point at which the original attractors vanish, the diversity of cell states increases—as measurable in one state space dimension x and displayed as histograms for the value of x across the cell population *before* the bifurcation event. Because the bifurcation process generates a poised instability (#), control is lost and some cells in the heterogeneous cancer cell population can, against therapeutic intention, "spill" over into the neighboring attractor encoding for an abnormal stem-like state (left), which are equivalent to regenerative states in normal cells.

S_{prolif} to that of the apoptotic state S_{apopt}. This attractor transition $S_{\text{prolif}} \rightarrow S_{\text{apopt}}$ follows the bifurcation process that destabilizes the state of origin, S_{prolif}. But unlike in chemical catalysis or ecological state shifts,[72] where lowering the energy barrier in a two-potential-well system (figure 4.10A) will ensure drastically accelerated transition into the potential well of the (sole) destination state, we have now a multistable system—the rugged quasi-potential landscape. Here the destabilization and ensuing poised instability will open up access to two attractors on both sides of the watershed (figure 4.10B). In fact, high-impact perturbation of complex high-dimensional systems can trigger not one but a sequence of multiple such symmetry-breaking bifurcations, at each of which there can be two possible outcomes. The ensuing hierarchy of binary decisions eventually creates new "kinetic paths," branching out to facilitate access to a large number of lower-energy attractors on the landscape. Driven by stochastic fluctuations of the system state S and starting from a destabilized system at a single original state with relatively high quasi-potential energy, the system now has the potential to enter any of these previously inaccessible or newly created stable phenotypic states (figure 4.10C), a set of nonphysiological attractors in uncharted or ancient territory of the landscape.

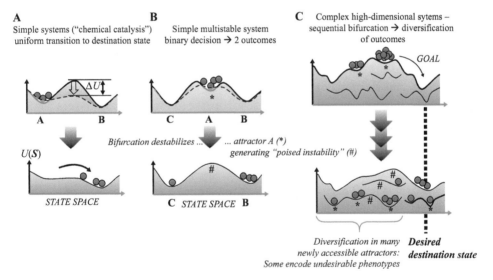

Figure 4.10
Bifurcations on increasingly complex ("rugged") landscapes. (A) In the classical case, the original attractor A disappears, the "energy barrier" is lowered, and the system uniformly (with all the members of the ensemble) moves to the new state [B]. This case corresponds to chemical catalysis (in the realm of thermodynamic equilibrium) or of the "fold bifurcation" (in simple nonlinear dynamical systems). There is no separation of the trajectory into two discretely distinct outcomes. (B) The case of binary branching in multistable systems after the attractor [A] is converted into a poised instable steady state (#). (C) For complex high-dimensional systems with rugged landscapes, a series of bifurcations and passages through multiple unstable poised instabilities eventually lead to the diversification of cells into a variety of quasi-stable states (small attractors)—reflecting the heterogeneity of cell types among cancer cells.

4.7.2 Bifurcation Events Are Low-Dimensional and Create Access to Opposite Fates

Why are the two outcomes of the double-edged sword effect functional (biological) opposites? Specifically, why does the same intervention induce either stemness and rejuvenation, or senescence and death? In section 4.5, we saw how the landscape image forces a system in poised instability to make a binary decision and move in opposition directions. This is why the watershed picture, or our earlier image of the walk on the mountain ridge, is so powerful.

At a closer look, the actual consequence of an attractor vanishing is dictated by the detailed "geometry" of the bifurcation: in the general case, although the dynamics of complex networks takes place in a high-dimensional system, bifurcations are low-dimensional. In GRNs, the bifurcations that create two new cell fates are driven by a single bifurcation parameter μ and do not affect simultaneously all the variable x_i. Thus, the destabilization can be projected into a few main dimensions (genes) of the system state. The bifurcation represents a low-dimensional event within high-dimensional dynamics. (Technically, one interesting manifestation of this behavior is the rapid increase in correlation of the expression levels x_i across all the cells for many key regulatory genes as the network comes close to the bifurcation event.[73]) Therefore, we can focus on describing a single dimension $x_i = X$ for the type of bifurcation discussed here: the forces that attract the system with respect to the dimension X, the horizontal axis in figure 4.9 or 4.10 (expression level of gene X), toward the attractor state at X^* from a state near it, namely, at $X^* - X$ or at $X^* + X$, are converted into repelling, opposing forces along the same axis X. Hence, as stability at stable state X^* is converted into poised instability, the system is pushed away from the *value* X^* in the direction of increasing or decreasing values of X relative to the attractor value X^* (toward the stable state B or C in figure 4.10B). The system thus will descend into opposite directions by either increasing or decreasing the value of feature X.

The change of expression of gene X may trigger specific downstream regulatory events, resulting in qualitatively distinct outcomes depending on whether the value of X is high or low. If the axis X, along which the qualitative separation takes place, also maps into an important biological function, then we have a binary choice between two alternative biological fates that are at opposite ends in the spectrum of a phenotypic feature implemented by X: in development, a binary cell fate decision of a stem cell might be, for instance, about choosing between the two "opposing" sister cell fates, such as "white" (myeloid) blood cell versus "red" (erythroid) blood cell; in the case of cell fates within one cell type, such as a cancer cell, the binary branching can be between increased survival mechanisms and the apoptotic program, between stemness and terminal differentiation, and so on. This concludes the mathematical-geometrical explanation for the double-edged sword effect of treatment.

4.7.3   A Heterogeneous Ensemble of Systems Undergoing Bifurcations Will Diversify

Having explained the diversification of the potential paths of state change for an individual cell as a system, we now recall the concepts of section 4.5 and place the theoretical principles in the context of the reality of cell *populations*. A tumor is, as explained in section 4.5, a heterogeneous population of millions of cells. Thus, a cell population is not ONE single marble on the quasi-potential landscape but a "cloud of points" where each point is a cell, previously represented by the marble (figure 4.9). This cloud is spread over some extended region of the landscape, representing the heterogeneity of the cell states. Upon converting an attractor state S^* into a poised unstable state of a hilltop during the bifurcation event, the population that previously occupied the attractor will be *partitioned* by the new watershed, and cells disperse in opposing directions (figure 4.9). In which valley a cell ends up with respect to the new separating watershed is thus not solely "determined" by pure chance in the moment of poised instability described in section 4.5 (figure 4.6) but is also biased by the position in the original attractor cloud that a cell held when the bifurcation progressed.

This phenomenon has been repeatedly observed in single-cell resolution analyses (flow cytometry) of embryonic and adult stem or progenitor cells that are multipotent. These cells, when exposed to conditions known to induce them to differentiate, make the fate decision in a way that is heavily biased by their position in the original attractor at the time just before the bifurcation event, manifest in the measurable value of expression of one or few informative marker X even if the differentiation signal is known to promote a particular fate.[74]

In summary, the heterogeneity of an ensemble of individual systems underlies the partitioning of its individual members into two subpopulations that take the two fates following an intervention that causes a bifurcation, as illustrated in figure 4.7. Thus, the two opposing outcomes of a system are not theoretical *potentialities* of one system that has to make a binary decision but in an ensemble of systems they can be *realized in parallel* by its members. If Mary faces the choice to study medicine or law, she may state that her preferences are 60 percent for medicine and 40 percent for law. But in a group of 100 students with the exact same preference distribution in the same career decision situation, one will observe that 60 percent of her colleagues would go into medicine and 40 percent into law. To observe such partitioning of a population of systems, the resolution of analysis must be able to expose the fates of its individual member systems and, even better, measure the extent to which a fate is realized.

At the level of cells as systems, future high-resolution single-cell transcriptomics and proteomics will undoubtedly expose the full extent of cell state diversification induced by nonlethal treatment stress to cancer cells. But already the inspection by fluorescence microscopy of individual cancer cells following chemotherapy reveals the generation of stem cells and senescent cells side-by-side.[75] At the level of tumors as systems, the case in the animal tumor treatment model already reveals the accelerated progression in the treated nonresponders (section 4.4.4, figure 4.4). Importantly, the specific fate of an individual tumor is, while probabilistic, greatly biased by its "microstate" (= position in the attractor basin) within the ensemble at the time just before the intervention or decision. This phenomenon could be harnessed for developing predictive biomarkers based on a scientific rationale as opposed to blind statistical associations.

4.7.4   Treatment Stress Causes Bifurcations That Diversify Tumor Cell States

We can now return to cancer treatment: following the conversion of the attractor state of robust proliferation by cytotoxic treatment into a state of poised instability in the intended push into the apoptotic fate, not all cells will automatically shift, as desired, into the attractor of the apoptotic state. Instead, many cells in the population will experience one or perhaps a series of bifurcations due to the harsh perturbation of the near (but non) lethal therapeutic intervention that affects multiple interactions of the network and thus drastically remodels the landscape. The result is a series of destabilizations of attractors and their conversions into new ridges that further and further split the cloud of cells into new attractors, thereby pushing some of the cancer cells further apart in state space as they adopt new stable phenotypic states (figure 4.10C). In fact, intratumor heterogeneity of the cell phenotypes in a tumor has long been known to be associated with poorer outcome.[76] As the millions of cancer cells swarm out into phenotypic space guided by the landscape surface, the cancer cell population becomes phenotypically more and more heterogeneous even before accumulating genetic mutations (which they will inevitably do since the pathological attractors are not endowed with efficient DNA repair systems). In the multi-valleyed landscape, only a fraction of the diverse cells reaches the attractor state of apoptosis and dies as intended by treatment.

Cancer is a disease of cell populations. As treatment with cytotoxic agents that perturb the system state destabilizes and disperses the cells into new niches in phenotype space, the treatment is more akin to chasing cats, not herding sheep. Many cancer cells will be lost in phenotype space, stuck in states without a downward trajectory ("drainage"), neither toward normal differentiation nor toward apoptosis. (Only occasionally may abnormal cancer cells "spill" past the barriers and still reach normal chreods and differentiate, likely much less frequently than in normal development,[77] as single-cell analyses now increasingly show. Such spontaneous, normal differentiation of tumor stem cells is supported by detection of oncogenic mutations in differentiated, apparently normal cells.

Of course, many of the newly occupied attractors do not necessarily represent phenotypes of more aggressive cells, let alone viable cells. Among the surviving tumor cells, some will switch the cell type, as amply seen in the treatment of glioblastoma where under therapy, the proportion of the cell types (mesenchymal, proneural, etc.) in the heterogeneous tumor shifts, not only due to differential survival but because of actual cell phenotype conversions.[78] Similarly, in melanoma[79] or in prostate cancer, treatment enriches for neural crest– or neuroendocrine-type cells, immature cell states that confer therapy resistance.[80] Given the variety of fates adopted under broadly cytotoxic treatment, cancer (stem) cells can also differentiate[81] into a postmitotic quiescent state, which is desirable[82] even if it most often occurs insufficiently so to be of therapeutic benefit. As mentioned, and consistent with the diversification into opposites, a response of cancer cells to nonlethal stress is not only stemness but they can also become senescent.[83] Importantly, like treatment-induced stemness, treatment-induced senescence can promote resistance via release of cytokines,[84] creating tissue fields as discussed below.

In the larger picture of resilience to therapy, it suffices that just a few of the nonkilled but perturbed cancer cells enter attractors encoding more aggressive, stem-like phenotypes capable of repopulating the tumor society and creating a cancer tissue field to pave the way for recurrence. If a tumor contains 10^{10} cells (roughly a 1- to 3-cm^3 size tumor), then even if 99 percent of the cells were killed (successfully steered into the apoptosis attractor), still 10^8 cells would survive. But these survivors are highly diverse.

As more and more distant and "dead-end" attractors far away from the normal developmental trajectories are occupied, many of these encode phenotypes compatible with the living state and may map to states near normal developmental states, for reasons discussed in the next section. Yet these immature cells are incapable of descending into the attractor of physiological mature cells at a rate that would preclude progression. This is why tumors invariably contain immature cells, recognized by pathologists as "dedifferentiation" or anaplasia, and perhaps why the degree of dedifferentiation scales with malignancy.

The standard view that the emergence of stemness and resistance is driven by a version of Darwinian evolution in the cell population (somatic mutation theory), which is discussed in the chapter 8 in this volume, does not contradict the nongenetic mechanism that is at the center of this chapter. In fact, many of the abnormal immature states, being implemented by normally unoccupied attractors that have never been exposed to natural selection for tuning physiological functionality, exhibit genome instability due to aberrant cell cycle and DNA repair, conferring them the "mutator phenotype"[85] and thus facilitating the accumulation of genetic mutations as a substrate for natural selection. One could then argue that the nongenetic induction of stemness allows for short-term survival and jump-starts and lubricates the ensuing Darwinian process.[86]

4.7.5   Treatment Stress Also Reconfigures the Cell-Cell Interaction Network

The same dynamics as described for cell states also operates at the level of the entire tumor as a complex system. Altering the cellular composition of the tumor tissue by a therapeutic intervention, with respect to nature and abundance of cell types, including the conversion of stromal cells to tumor-associated fibroblasts and the influx of inflammatory cells, will also change the cell-cell interaction network's interaction parameters. For instance, the production of secreted growth factors and chemokines by the new diversity of cell types, such stressed cells or the abovementioned senescent cells, may increase and add connections to the cell-cell interaction network, which in turn alters cell states, such as promoting the stem-like state.[87]

But the larger picture beyond such cell biology effects is that the changes in the cell-cell interaction network will alter the landscape for the tumor as a complex system and can also lead to a bifurcation at the level of tissue dynamics. The stability of the entire tumor can go through a bifurcation event of *tissue* behaviors, destabilizing the state of slow tumor growth and converting the tumor cell society to enter either into dormancy or into the macroscopic outgrowth state of uncontrolled wound healing, as illustrated in the animal model in section 4.4.

Cancer research has barely scratched the surface of the tissue-level dynamics that is conveniently summarized as "non-cell-autonomous" effects in current mainstream cancer biology, which essentially is a tacit reductionist embrace of the older, profound, but more abstract idea of "field cancerization."[88] A specific example that is emblematic of the principle of treatment backfiring is our recent observation that apoptotic cells and their debris send signals that further promote the stemness of the surviving cancer cells as well as promote influx and proliferation of inflammatory cells that secrete mitogenic signals.[89]

Here we have seen that the landscape concept, itself anchored in the formalisms of dynamical systems theory, naturally explains how a nonspecific (nonbiological) perturbation of tissues and cells induces, with relative ease, a variety of new cell phenotypes, represented by the newly occupied attractor states. But why are there so many empty attractors in the state space in the first place that are accessed by cancer cells under massive stress? Why can some of these pathological attractors encode specifically stem-like, less mature cells, characteristic of the therapy-resistant tumor?

4.8   The Inherent Inevitability of Cancer Attractors

The above analytic-geometric explanation offers the formal underpinning for the logical necessity of the Nietzsche response: the general dynamics of unleashing two opposite fates after destabilization of the system state. However, it is not sufficient to account only for the *inevitability* of the double-edged sword effect as a *generic* response to an intervention. We also need to connect such dynamics of destabilization and poised instability with the *specific biology* if we are to continue with our requirement of logically necessary explanations. Hence, we should ask, *Why do the aberrant attractors into which cells are unintentionally pushed by the nonlethal perturbation intended to be lethal encode specifically stem-like cell states?* Equivalently, at the level of the whole tumor as a system, *why does the tumor, if not fully destroyed by the treatment, enter a stable tissue attractor state that governs the coherent, even if unrestrained, behavior of development or regeneration?* Obviously, neither highly specific pathway-targeting agents nor the broadly acting nonbiological chemotherapeutics or radiation provide the specific biological *instruction* to guide the cells to adopt biological functions such as stemness, development, and regeneration. But cancer produces in a seemingly accidental, yet consistent, if not inevitable manner these sophisticated specialized phenotypes with apparent ease and at high frequency when treated.

4.8.1   Cancer as Access to a Hidden, Inevitable State: The Cancer Attractor

Stuart Kauffman in 1971 proposed that complex GRNs generate more attractor states than are used and hence more potentially stable cell phenotypes than are actually found in normal tissues. This observation prompted him to postulate that occupation of the unused attractors is the cause of cancer.[90] If there exist "*cancer attractors,*" this would offer a first general account for the very possibility of an accidental generation of a highly specific stable phenotype without a specific instructive signal, let alone specific genetic mutations, as we have argued before.[91] A broad class of perturbations could then "tip" the cell into the basin of a preexisting attractor state S immanent to the GRN that would encode for a particular developmental, malignant phenotype and allow for the spontaneous and robust implementation of that state.

Without invoking attractors, Paul Davies and Charles Lineweaver (see chapter 9, this volume) have proposed that cancer cells are "atavistic cells" that would reexpress features of evolutionary ancestral cells that resemble autonomous protozoa: more resilient, mobile, invasive, and not subjugating themselves to the needs of a cell society.[92] This hypothesis tacitly assumes hidden "genetic programs" in the genome that can be reactivated; it also provides some explanation for the accidental acquisition of the specific nature of the complex cancerous phenotype. However, additional reasons are needed to account for why Kauffman's unused attractors would typically encode not only developmentally immature but also evolutionarily archaic phenotypes.

If the rugged quasi-potential landscape of the GRN or of cell-cell interaction networks, in which attractors are stable cell or tissue states, is littered with attractors that are either never used or that encode evolutionarily or developmentally *ancestral phenotypes* not used anymore in the extant adult organism,[93] how have these excess valleys in the quasi-potential landscape of interaction networks been produced during evolution in the first place?

4.8.2   Evolution Shapes the Quasi-Potential Landscape

A network of interactions *inevitably* produces a multiplicity of stable attractors if its dynamical behavior belongs to the particular class of not too densely connected networks, referred to as "critical networks" according to Kauffman.[94] Such networks have the rather stunning property of producing a dynamics that is optimal for processing information and for balancing order (robustness) and chaos (flexibility).[95] Interestingly, most biological networks, including the GRN, belong to this class of critical networks (see also chapter 6, this volume). Such types of networks, connecting thousands of genes in the case of the GRN, even in a rather random fashion, will by mathematical necessity produce a large number of stable attractor states in its immense state space. The number of attractors scales in a particular manner with the number of components (genes) in the network, as Kauffman and others have shown in computer simulations since 1970.[96] Importantly, the combinatorial constraints and complexities in the mapping from network diagram to landscape topography are such that a "landscaping by design" by appropriate wiring of the network is sheerly unfathomable: even evolution with its proverbial potency to minimize wasted effort cannot generate network architectures exactly so as to create *just those* attractors that encode meaningful stable states of cells and tissues that will be used.

To understand this, recall that order in living systems emerges from "chaos" through the tight synergism between constraints erected by physical laws, including structure and self-organization, and natural selection of adaptive random variants.[97] Self-organization accounts for the just mentioned intrinsic ease of biological networks to generate scores of attractors even by random wiring of the connections. In fact, in the case of GRN governing cell state dynamics, mutations (in coding and noncoding regions) randomly "rewire" the regulatory interactions between the genes. Concretely, the network interactions that determine the network dynamics and thereby shape the landscape topography are mediated by the specific protein-protein and protein-DNA interactions and the regulatory binding sites in genomic DNA, which are all manifestations of the DNA sequence of the genome. Thus, the landscape topography is ultimately encoded in the genome and is altered by genetic mutations. It is therefore that random mutations can shape the quasi-potential landscape: they create (by chance) the new stable attractors in a class of networks that is intrinsically likely to produce attractors in the first place; then, the new phenotype that an attractor encodes can be tested by selection but only if the attractor is occupied by cells (or tissues), which thereby expose the associated network state S to natural selection.

We can thus state that the "wiring diagram" of the GRN (or the cell-cell communication network) is continuously redrawn by evolution, which in doing so controls the stability of cell states and tissue states seen by selection. Evolution then selects the adaptive phenotypes. In the finalistic anthropomorphic short-cut formulation, it would thus appear as if evolution "generates" new useful phenotypic states S by "making" them attractor states in the landscape and by encoding the landscape topography that has these attractors through mutational wiring of the network. (This is not to be confused with the fine-tuning of local topography during bifurcations through variation of the parameter values that determine the interaction properties and are subjected to external conditions.)

4.8.3   Unused Attractors as By-product of the Evolution of the Quasi-Potential Landscape

Now a critical feature of such sculpting of the landscape topography by evolution is that evolution has only limited leverage for changing the topography of the landscape as it rewires it by selecting the adaptive genomic alterations. There is no obvious "direct" mapping between *topology* of the interaction network (wiring diagram) and the *topography* of its quasi-potential landscape that the former produces. This fundamental complication explains the failure of approaches that assume a simple, linear mapping between independently acting genes and phenotypes. Recalling that the landscape topography is the mathematical expression of the interplay of all genetic interactions, it is worth mentioning that the tacit assumption of independence and linearity of gene loci activity that underlies the genome-wide association studies (GWAS) is a major oversight that may account for the limited success of such studies to explain the genetic basis of complex diseases.[98]

But more relevant here is the observation in computational simulations of virtual GRNs in which the topology is altered by random rewiring of the regulatory interactions to mimic mutations or genomic rearrangements. These analyses show that network rewiring in large networks is not likely to drastically alter the attractor structure. Not surprisingly, for complex networks, the landscape topography is in general rather robust to rewiring of individual interactions. Randomly deleting, adding, or reconnecting an interaction is much more likely to affect basin sizes and barrier heights than eliminating or creating entirely new attractors.[99] (Again, this holds at least for the class of "critical" networks.) Thus, while mathematical constraints explain the inevitability of attractors in complex networks, these same constraints entail that *evolution cannot easily eliminate unused attractors*. But it need not do so. As evolution rewires the GRN as it grows (generally gene numbers have increased over time in most organisms) and modifies the expanding quasi-potential landscape, it suffices to simply shut down, by elevating energy barriers, access to the many unused attractors that are created as by-products of the expansion of the landscape.[100] Unused attractors are hence "shielded" from the path of normal development, or the chreods.

But evolutionary innovation not only depends on new phenotypes (stable attractors); the latter also must be incorporated in the process of development, much as the engineer designing a more complex machine with new functionality must ensure that it still can be assembled in the first place. Thus, to create new phenotypes, for instance, in the innovation of new cell types, evolution must also carve developmental paths such that the immature (stem and progenitor) cells can reach the new attractors encoding new cell phenotypes at the appropriate rate. At the same time, accidental entry into "wrong attractors" must be avoided. This is one reason why the concept of accessibility of attractors, quantified by the quasi-potential value U, is central.

As a consequence of this principle, the many attractors that are not accessible ever since their creation can be left "unused" on the landscape, and those encoding ancestral or developmental phenotypes not used any more in the extant adult need not be completely deleted; rather, it suffices for evolution to reshape the landscape such that they are not accessed any longer. This requires much less genetic alterations than would be required for eliminating entire undesired attractors.

4.8.4   The Unused Attractors Are Evolutionary and Developmentally Older and Dysfunctional

From the above, it then follows that the quasi-potential landscape that has evolved to guide the development of all the metazoan cell and tissue types and to implement their specific physiological gene expression or cell state configurations is littered with countless unused (and normally inaccessible) attractors. Some of them represent ancient phenotypes that have been used in ancestors. Others are used only during embryonic development, not in the adult organism. Yet others will only be occupied after extreme, unnatural perturbations, such as when exposed to cytotoxic drugs or radiation, as we will see later.

Due to their utilization, the physiological attractors are exposed to natural selection, ensuring continuous functional improvement of the phenotype that they encode. For instance, through evolutionary fine-tuning, regulation of their proliferation rates must be adjusted to meet the tissue needs, and fidelity of their genome replication must be ensured to optimize organismal fitness. By contrast, the unused excess attractors, whether evolutionary remnants or never ever occupied attractors, will eventually erode, akin to unoccupied, unmaintained homes. They may, since they are not "maintained" by natural selection, encode imperfect biological states: they lack robust viability or may have imprecise DNA replication, as mentioned earlier, or they may still be endowed with ancestral functions not adapted to life in the cell society of present-day metazoan organisms.

Natural selection for efficient development has carved the epigenetic landscape as such to ensure safe and timely descent of cells into the physiological attractors. The term "descent" is important here: it represents the process of development and captures its global directionality. Because the driving force is determined by the "decrease in elevation" of the quasi-potential $U(S)$, the younger, useful attractors selected for encoding advanced phenotypes and for their accessibility will tend to have lower elevation and deeper valleys—so to speak, added to the bottom of the landscape. Perhaps trajectories of normal development, which are shielded from the aberrant paths that lead into unexplored regions in the rugged landscape, are what Waddington meant when he coined the term "chreod" (see section 4.3.7). Mechanisms of homeostasis, including "tumor suppressor genes" or the antiproliferative cell-cell interactions ("contact inhibition"), may have evolved to serve the stabilization of chreods.

In summary, attractors survive evolutionary selection and are used as part of the organism if they met these two conditions: (1) first, they can be reached from existing attractor states or developmental paths (as explained in sections 4.5 and 4.6 via extrinsically induced bifurcations with the help from gene expression noise). (2) Second, they represent "physiological attractors" in that they encode cell and tissue phenotypes that benefit organismal development and fitness.

4.8.5   The Use of Developmental Attractors for Evolving the Adult Physiological Regeneration and Tissue Repair Programs

The above conditions explain why unused cell attractor states are at higher elevation of the landscape and encode phylogenetic as well as ontogenetic immature phenotypes. Here we propose that as part of the unrealized repertoire of genetic programs in the organism's regulatory network, they play a role in adaptive evolution: with their similarities to developmental states and given the fact that they may become accessible under nonphysiological perturbations, these unoccupied immature attractors are predestined to be the raw material for the evolution of regenerative programs. If these attractor states are consistently induced by nonlethal injuries and happen to encode functionalities of development (such as stem cell–like cell phenotypes or developmental tissue plasticity), which is inherently likely given the way the landscape has evolved, it is then plausible that they could be coopted by adaptive evolution to evolve a **wound-healing** and **tissue regeneration** response (after some further fine-tuning by mutations and natural selection).

In fact, whether regeneration is an innate ancestral capacity, representing a remnant of asexual propagation, or had to be acquired by adaptive evolution under selection pressure imparted by recurrent nonlethal injuries,[101] the diversity in the animal kingdom of generative capacities, both in mechanism and extent, suggests that the capacity of regeneration, or at least tissue repair, is deeply related to development and was relatively easy to evolve (and to lose). Our proposal that entry into the developmental or ancestral unused attractor may also be associated with tumorigenesis is consistent with the widely held idea that the limited regenerative capability of higher mammals is a result of evolutionary protection against the danger of neoplasia.

The opportunistic use of by-products of evolution, such as unused attractors, that are inevitable due to fundamental constraints of complex systems, be it due to principles of network dynamics or any morphological arrangement, is a common feature of adaptive evolution by natural selection and may in part account for cases of stunningly rapid acquisition of complex functions.[102] Such evolutionary paths represent "short-cuts" since mutation and selection need not sculpt the new functional structures from scratch but can take over what the intrinsic system constraints have already produced but not used. This idea does not contradict neo-Darwinian natural selection but is also not central to mainstream thinking in evolutionary theory and is loosely referred to as "structuralism."[103]

4.8.6 Cells "Stuck" in an Unused Attractor Could Become Cancer Cells or Increase Malignancy

A corollary of the above principles feeds a central idea regarding the origin of carcinogenesis. Cancer might be initiated when developing cells are "stuck" in nonphysiological attractor states either on their way down from an elevated position, such as following a "false" decision in a state of poised instability during normal tissue differentiation from stem cells, or after tissue stress due to imperfect shielding of cells in the chreods and mature state attractors that prevents them from entering the pathological unoccupied attractors. Developing cells and tissues, once derailed into pathological attractors by stress, are literally "stuck" because the latter have not evolved an outflow to the low-energy, stable physiological attractors of the mature state. This picture naturally explains why cancer cells invariantly display an immature, "dedifferentiated" phenotype, similar to that of stem or progenitor cells, and why tumor tissues have cellular compositions reminiscent of development or regeneration, such as mesenchymal cells, matrix remodeling, and inflammation, warranting the old idea of cancer as a disease of "maturation arrest."[104]

The above conceptual framework offers a straightforward explanation for why a relatively random and "blunt" nonlethal perturbation that causes extensive disturbance of the neoplastic cell and tissue state results specifically in the entry into atavistic and embryonic attractors. The injury may trigger the molecular pathways for a physiological repair response. However, with the abnormal starting point of cancerous cells and tissue, this response may miss the physiological wound-healing attractors and instead steer cells and tissues into other pathological attractors of increasing immaturity that lurk in the uncharted regions of the landscape. By contrast, mutations caused by genotoxic agents may in parallel rewire the regulatory interactions and thus permanently distort the landscape, which could lower the quasi-potential energy barriers that normally prevent the cells on the chreods from veering into side valleys. If the latter become occupied and are pathological attractors that encode a rapidly proliferating state without an exit path back to a normal chreod, such aberrant cells may be selected for and become cancerous. Thus, it is easy to see how nongenetic and genetic processes synergize in tumorigenesis and tumor progression.

4.8.7 Cancer Cells Recapitulate the Evolution of Tissue Repair Mechanisms—but Lose Control

The mimicking of normal repair and regenerative processes may help to explain why incomplete killing of tumor cells under treatment triggers the complex response that involves not only the production of stem-like cells but tissue-level reactions, such as inflammation, angiogenesis, extracellular matrix remodeling, and immune suppression—all cell functions activated in normal wound healing and embryonic development, including for placental function.[105] Such recapitulation of programs of development, pregnancy, and regeneration also encompasses immune modulation after injury: the physiological reconstitution of normal tissue after damage goes through a phase in which developmental proteins, which may be antigenic, are reexpressed. Thus, during normal wound healing, the immune system is transiently suppressed via mediators secreted by inflammatory cells, injured epithelial cells,[106] much like in early pregnancy, where immune modulation by placenta cells serves to avoid fetus rejection.[107]

But there is a more profound twist than just neoplastic tissues copying physiological repair: if nature has exploited the development-like programs of unused attractors for the rapid adaptive evolution of physiological tissue repair and regeneration, a similar mechanism of co-opting existing functional modules may also be applied by stressed tumor cells. Tumor inflammation is thus not simply a pathological version of a tissue's intrinsic capacity to repair itself upon damage but may also, in a more fundamental way, expose the immanent evolvability of organisms that undergo a complex development: tumorigenesis may be a parodic manifestation of the innate capacity of living tissues to evolve repair functionality by exploiting by-products of ontogenesis for adaptive evolutionary innovation. If normal regeneration is the evolutionary response to repeated threat of nonlethal injury to tissues, treatment-induced progression might be nothing more than the unfortunate, equally innate, but *nonevolved* response to nonlethal injury of the tumor.

Thus, there is a critical, practical difference between normal tissue repair and the apparent repair functionalities co-opted by tumors. In cancerous tissues, the repair response is aberrant—it is a caricature of the normal well-orchestrated wound healing and regeneration. One reason for this difference to physiological repair is that unlike in the normal tissue, the stressed state of tumor tissues and tumor cells has not been exposed to selection pressure to maintain organismal integrity and thus has never been fine-tuned by evolution to serve the *entire organism*. The homeostatic self-limitation of normal regeneration does not exist in tumor tissues. The consequence is what observant pathologists have long suggested: "*A tumor is a wound that never heals*."[108] It is a regeneration that never stops. This property is at the core of what drives tumor progression and relapse. And it is amplified by treatment induced injury. The parallels to the normal wound-healing process would also naturally explain the stunningly specific capacity of tumors to evade antitumor immunity,[109] since immune suppression is also a central feature of normal tissue repair.

In summary, the formalism behind the quasi-potential landscape not only explains the evolutionary optimization of robustness of physiological function. By embracing the mathematical necessity of the existence of unoccupied attractors as inevitable by-products generated by evolving interaction networks, it can account for the response to injury and explain why oncogenesis appears to recapitulate both phylogenesis and ontogenesis[110] and why cancer cells appear to be maturation arrested.[111] Finally, principles of the evolution of the topography of the quasi-potential landscape also offer an explanation for why cancer treatment, which is nothing but a nonlethal injury to the tumor tissue, promotes paranormal regenerative processes that fail to terminate. Sadly, these deeply biological phenomena of cancer, long articulated by keen observers, have evaded the narrow attention of molecular geneticists who view cancer exclusively through the lens of mutations that need to be targeted by drugs.

4.9   Conclusion: Specific Mechanisms That Epitomize the Theoretical Principles

The goal of this chapter is to present a particular mode of scientific reasoning that can help to comprehend counterintuitive phenomena but is largely unknown in biomedicine. It is one that seeks the *logically necessary* explanation of phenomena and that anchors them to fundamental constraints. In doing so, it deemphasizes proximate mechanistic explanation, which typically is framed in terms of a cellular and molecular process. Therefore, we have so far barely mentioned specific molecular pathways that are usually celebrated as the chief *rerum causae* and the rational basis for therapeutic intervention.

Formulation of the constraints permits the prediction that in a particular set of circumstances, a particular process *has* to occur (thus is inevitable or spontaneous). If in physics and chemistry, the constraints that predict the spontaneity of a process are geometry, forces, energy, entropy, and so on, here the analogy of constraints are the first principles of the theory of dynamical systems.

Yet while exposing the inevitability of cancer behaviors by grounding them in "first principles" offers epistemological satisfaction, to be of practical use, these principles must be mapped to specific biological functions or "pathways." Even if guided by theory, detailed mechanistic knowledge of biological functions is required to identify the lever points for a therapeutic intervention. The diversity of molecular mechanisms that has been implicated in tumor progression, development of resistance, and evasion from the immune system is immense. Their descriptions are readily found in the myriads of review articles in cancer biology. To connect our theoretical arguments with the realm of specific proximate explanations, we now provide a few examples of canonical mechanistic pathways.

4.9.1 Specific Molecular Mechanisms That Embody the Inevitable First Principles

As discussed, the induction of a stem cell–like state in cancer cells by a cytotoxic, nearlethal perturbation is part of an aberrant and disorganized attempt of tissue repair: a caricature of a physiological wound-healing and regeneration program triggered by tissue injury. In this abnormal stress response of malignant cells, the cellular detoxification apparatus is a central component. Indeed, the cytotoxic stress imparted by a broad range of treatments induces or enhances the expression of **ABC transporter proteins**, a family of membrane pumps that protect the cells by promoting the efflux of toxic compounds from the cell[112] and thereby account for multidrug resistance after exposure to just one drug. Interestingly, ABC transporters are also prototypic markers for normal stem cells in the adult tissue, as well as for cancer stem cells.[113]

In several cancer cell lines, induction of ABC transporters in response to treatment is mediated by the **Wnt/β-catenin pathway.**[114] This pathway is critical in maintaining stemness and also governs the expression of major embryonic stem cell transcription factors, such as Oct4, which has also been detected in chemotherapy-stressed cancer cells and in cancer stem cells.[115] Since Wnt is a secreted mediator molecule that binds to its cognate receptors on neighboring cells to activate the canonical developmental signal transducer β-catenin, this response implies a non-cell-autonomous process that connects intracellular response with tissue-level response. Consistent with our overarching framework, the normal Wnt system also orchestrates stemness, inflammation,[116] and immunosuppression[117] during regeneration.

How exactly cell stress caused by anticancer therapy activates the stemness-promoting signaling pathways remains currently elusive. Again, a broad palette of possible mechanisms has been described.[118] One of the best-studied pathways for coping with cell stress is governed by the **protein p53** (*TP53* gene), whose synthesis was historically described to be induced by DNA damage. But we know now that a variety of other cell stresses can also activate the *TP53* gene.[119] The p53 protein causes cell cycle arrest and, concomitantly, either stimulates expression of the DNA repair apparatus or activates apoptosis or senescence. Both responses are interpreted as a way to prevent propagation of mutations in the cell population. Interestingly, p53 has also been implicated in stemness programs in embryonic stem cells, in part also via the induction of Wnt.[120] More broadly, it appears that following cytotoxic perturbation of the cell, however nonspecific, several cellular programs, including the heat-shock response (which is mediated by the ubiquitous stress-response transcriptional regulator NF-κB), the hypoxia response (mediated by the transcription factor hypoxia inducible factor [HIF]), metabolic stress, and endoplasmic reticulum (ER) stress, connect the nonspecific cell perturbation to specific pathways in the GRN to orchestrate repair, permit dormant survival associated with stemness, and promote wound healing and regenerative functions.[121]

Of note is that the cell injury that triggers the counterreaction need not be caused by cytotoxicity of broadly acting chemotherapy or radiation. Even target-selective compounds can display the double-edged sword effect. Imatinib (Glivec), employed to target and block the oncogenic fusion proteins Bcr-Abl and the cell signaling kinases c-Kit and the growth factor receptor PDGFR in several malignancies, has been the single most successful targeted therapy so far (at least in Bcr-Abl-driven chronic myelogenous leukemia). But it has also been shown to induce epithelial-mesenchymal transition (EMT), a process at the core of tumor progression that is associated with stemness and therapy resistance.[122]

At the tumor tissue level, the Nietzsche effect that strengthens the tumor in the wake of partial destruction is a coordinated wound-healing or regeneration-like response of the cell society to the damage inflicted by treatment. Much as inflammation is the first step of healing after damage to normal tissue, the stressed neoplastic tissue mounts an inflammatory response. But it is one that is pathological: the inflammatory process is not orchestrated to terminate when the tissue defect is repaired (perhaps also because the signal of restored tissue integrity that acts as negative feedback control is never produced).

A central trigger of inflammation is the sensing of cell death. This tissue response to tissue destruction in the tumor is mediated by a large array of signals in the tumor microenvironment and involves the tumor stroma. Dying tumor cells produced by cytotoxic treatment release "alarmins,"[123] a group of intracellular components that act as danger signals and induce a cell state shift toward stemness in the surviving tumor cells and also triggers the recruitment of inflammatory and immune cells to the site of injury, starting the signaling cascade of inflammation. The induction of regenerative processes by cell debris in tumors[124] is a prosaic example of how a potent homeostatic mechanism, which normally replaces cell loss by the transient reactivation of cell division in the adjacent postmitotic cells, is co-opted by the tumor in response to treatment injury. We recently found that apoptosis (which in healthy tissues by definition, in contrast to necrosis, does not evoke inflammation) is a potent stimulator of inflammatory macrophages in a tumor, which in turn can promote tumor progression. This response is in part mediated by the molecular hallmarks of apoptosis, such as activation of caspases and exposure of lipid phosphatidylserine on the outer side of the cell membrane.[125] But chemotherapy-induced senescence, as mentioned earlier,[126] can also trigger secretion of cytokines that cause inflammation and stemness.[127]

Among the many modes of inflammation, some of which stimulate the adaptive immune response to foreign material, cell death and debris-induced inflammation in tumors can cause immune suppression, in part via the central mediator of inflammation, prostaglandin E2 (PGE2).[128] This facet of cell death–induced inflammation may help explain the immune-suppressive potency of tumor tissues that contain cell debris.[129] It is then a small step to connect the dots and hypothesize that the immune-checkpoint inhibitors, if successful, may cause overt cell death (and debris not cleared by the immune system), which then could trigger the opposing response: inflammation-driven tumor acceleration accompanied by (transient) immune suppression that is manifest in the resistance to treatment with ICIs and may explain the hyperprogression discussed in section 4.4.

4.9.2   Outlook: The Unification of Two Distinct Epistemic Cultures Will Open Up New Ways to Control Cancer

The above examples of specific biomolecules and cellular processes behind the phenomenon of treatment-induced cancer progression epitomize the traditional mode of proximate explanation in mainstream cancer biology. While the concreteness of such proximate explanations has its practical appeal, only an epistemology that seeks generic first principles derived from a theory can anchor the counterintuitive behaviors of tumors to a fundamental necessity. And only by appreciating the logic of necessary constraints, dictated by principles of dynamical systems theory, can we one day utilize the vast amount of molecular data to predict the trajectory of an individual tumor to guide personalized treatment.

Thus, we need to unite two complementary epistemic cultures: the familiar endeavor of describing specific proximate mechanisms and the (still) unorthodox quest for the underlying governing principles, without which the apparent paradox of the Nietzsche effect, the backfiring of treatment, cannot be comprehended. To achieve this unification, first a change in the mode of thinking within the community of empirical practitioners must take place, which includes clinicians and translational researchers as well as editors and research policy makers. Practitioners eschew theory, but theorists espouse practical applications (of their theory). Therefore, this call for a rethink is mostly directed at the former group. A shift of thinking needs to involve the following measures:

• Dynamical systems theory, even just the qualitative concepts as presented here, must be taught in the early semesters of medical schools and undergraduate programs of all life sciences. At the very least, we must sensitize young minds to the utility of thinking in terms of abstract first principles and to familiarize them with nonlinear phenomena.

• Biomedical research funding agencies must support the development of theoretical principles and frameworks leading to new theories that can be tested. Unfortunately, the momentum behind the surge of quantitative biology and systems biology of the past decade has been almost completely absorbed by theory-free computational and statistical analyses. There is only so much that we can get out of "big data" analytics without a theory on the governing principles.

• Journal editors and managers of governmental research funding agencies, the two gate-keepers of innovation in science, must develop a working-grade epistemological aware-ness and learn to appreciate hypothesis-driven research and formal theory-based reasoning. Historically, such an approach was commonplace before the data-driven, technology-enabled quest for proximate molecular explanations has pushed aside thinking about first principles.

• Theorists can also contribute to the change by confidently participating with their mode of reasoning and by communicating their distinct epistemology in sensible ways to experi-mentalists while resisting the temptation of adopting quick, proximate explanations. They have to learn that pure science and pursuit of basic principles are not only at home in the province of fundamental questions, such as on the origin of life or evolution of complexity, but could be ported to the domain of open practical and translational questions, such as those on resilience and the arrow of progression of cancer.

Facilitated by the above unification of the two cultures, we must engage in new research programs that explicitly examine the mechanisms of the paradoxical phenomena discussed in this chapter with the same zeal as in the current quest for new molecular targets for ever more potent cell-killing drugs:

• Elucidate the molecular pathways of stress-induced stemness, dissect inflammatory mechanisms to distinguish those that fight and those that fuel tumor progression, and develop bioassays in drug discovery not for identifying tumoricidal activity but to find "gentle" (state-modifying) drugs that do not backfire or even can specifically block the deleterious stress response.

• Design and analyze experiments in a refined, quantitative, and theory-guided manner to uncover dose regimes of intervention that minimize the "backfiring."

• At the clinical front, examine the exceptional responders both in the positive (cure) as well as negative sense (treatment-induced progression) as real biological phenomena and not as statistical flukes.

• We also must be open to new schemes for trials that currently limit the early studies of novel classes of treatment to patients who have failed all therapies—precisely because previous treatment may have altered and diversified cancer cells on the epigenetic land-scape, rendering them essentially untreatable.

Our failure in the management of cancer cannot be overcome with the current mind-set of brute-force molecular profiling and by doubling down on killing cancer cells more efficiently. Comprehending the counterintuitive paradoxical phenomenon of the Nietzsche effect will require appreciation for logical reasoning and theoretical principles that have been sidelined in the past decades of experimental and technology-driven molecular biology. Putting theory to practice still entails the traditional systematic characterization of the proximate molecular mechanisms; however, this endeavor must be guided by logic and the theory of the first principles. Uniting these two epistemic cultures will unleash unprecedented synergism in developing new ways to control cancer.

Notes

1. Fojo, T., Mailankody, S. & Lo, A. Unintended consequences of expensive cancer therapeutics-the pursuit of marginal indications and a me-too mentality that stifles innovation and creativity: the John Conley Lecture. *JAMA Otolaryngol Head Neck Surg* 140, 1225–1236, doi:10.1001/jamaoto.2014.1570 (2014). Davis, C. et al. Availability of evidence of benefits on overall survival and quality of life of cancer drugs approved by European Medicines Agency: retrospective cohort study of drug approvals 2009–13. *BMJ* 359, j4530, doi:10.1136/bmj. j4530 (2017). Haslam, A. & Prasad, V. Estimation of the percentage of US patients with cancer who are eligible for and respond to checkpoint inhibitor immunotherapy drugs. *JAMA Netw Open* 2, e192535, doi:10.1001/jama networkopen.2019.2535 (2019). West, H. J. No solid evidence, only hollow argument for universal tumor sequencing: show me the data. *JAMA Oncol* 2, 717–718, doi:10.1001/jamaoncol.2016.0075 (2016).

2. Haslam, A. & Prasad, V. Estimation of the percentage of US patients with cancer who are eligible for and respond to checkpoint inhibitor immunotherapy drugs. *JAMA Netw Open* 2, e192535, doi:10.1001/jamanet-workopen.2019.2535 (2019). The Lancet. Calling time on the immunotherapy gold rush. *Lancet Oncol* 18, 981, doi:10.1016/S1470–2045(17)30521–1 (2017).

3. Office of the Press Secretary, T. W. H. *Fact sheet: at Cancer Moonshot Summit, Vice President Biden announces new actions to accelerate progress toward ending cancer as we know it*, <https://www.whitehouse .gov/the-press-office/2016/06/28/fact-sheet-cancer-moonshot-summit-vice-president-biden-announces-new> (2016).

4. Cox, D. R. Regression models and life-tables. *J R Stat Soc B* 34, 187–220 (1972).

5. Asano, J., Hirakawa, A. & Hamada, C. Assessing the prediction accuracy of cure in the Cox proportional hazards cure model: an application to breast cancer data. *Pharm Stat* 13, 357–363, doi:10.1002/pst.1630 (2014). Maetani, S. & Gamel, J. W. Evolution of cancer survival analysis. *Surg Oncol* 19, 49–51; discussion 61, doi:10.1016/j.suronc.2010.03.002 (2010).

6. Klaeger, S. et al. The target landscape of clinical kinase drugs. *Science* 358, doi:10.1126/science.aan4368 (2017). Seebacher, N. A., Stacy, A. E., Porter, G. M. & Merlot, A. M. Clinical development of targeted and immune based anti-cancer therapies. *J Exp Clin Cancer Res* 38, 156, doi:10.1186/s13046-019-1094-2 (2019). Stoughton, R. B. & Friend, S. H. How molecular profiling could revolutionize drug discovery. *Nat Rev Drug Discov* 4, 345–350 (2005). Overington, J. P., Al-Lazikani, B. & Hopkins, A. L. How many drug targets are there? *Nat Rev Drug Discov* 5, 993–996 (2006). Jorgensen, J. T. & Hersom, M. Companion diagnostics: a tool to improve pharmacotherapy. *Ann Transl Med* 4, 482, doi:10.21037/atm.2016.12.26 (2016).

7. Topalian, S. L., Drake, C. G. & Pardoll, D. M. Immune checkpoint blockade: a common denominator approach to cancer therapy. *Cancer Cell* 27, 450–461, doi:10.1016/j.ccell.2015.03.001 (2015). Nishino, M., Ramaiya, N. H., Hatabu, H. & Hodi, F. S. Monitoring immune-checkpoint blockade: response evaluation and biomarker development. *Nat Rev Clin Oncol* 14, 655–668, doi:10.1038/nrclinonc.2017.88 (2017).

8. Kim, C. & Prasad, V. Cancer drugs approved on the basis of a surrogate end point and subsequent overall survival: an analysis of 5 years of US Food and Drug Administration approvals. *JAMA Intern Med* 175, 1992–1994, doi:10.1001/jamainternmed.2015.5868 (2015).

9. Fojo, T., Mailankody, S. & Lo, A. Unintended consequences of expensive cancer therapeutics-the pursuit of marginal indications and a me-too mentality that stifles innovation and creativity: the John Conley Lecture. *JAMA Otolaryngol Head Neck Surg* 140, 1225–1236, doi:10.1001/jamaoto.2014.1570 (2014). Davis, C. et al. Availability of evidence of benefits on overall survival and quality of life of cancer drugs approved by European Medicines Agency: retrospective cohort study of drug approvals 2009–13. *BMJ* 359, j4530, doi:10.1136/bmj

.j4530 (2017). Rosenzweig, S. A. Acquired resistance to drugs targeting receptor tyrosine kinases. *Biochem Pharmacol* 83, 1041–1048, doi:10.1016/j.bcp.2011.12.025 (2012). Marquart, J., Chen, E. Y. & Prasad, V. Estimation of the percentage of US patients with cancer who benefit from genome-driven oncology. *JAMA Oncol* 4, 1093–1098, doi:10.1001/jamaoncol.2018.1660 (2018).

10. Sonnenschein, C. & Soto, A. M. The aging of the 2000 and 2011 Hallmarks of Cancer reviews: a critique. *J Biosci* 38, 651–663 (2013). Bernards, R. & Weinberg, R. A. A progression puzzle. *Nature* 418, 823 (2002).

11. Haslam, A. & Prasad, V. Estimation of the percentage of US patients with cancer who are eligible for and respond to checkpoint inhibitor immunotherapy drugs. *JAMA Netw Open* 2, e192535, doi:10.1001/jamanetworkopen.2019.2535 (2019).

12. Maetani, S. & Gamel, J. W. Evolution of cancer survival analysis. *Surg Oncol* 19, 49–51; discussion 61, doi:10.1016/j.suronc.2010.03.002 (2010). Berkson, J. & Gage, R. P. Survival curve for cancer patients following treatment. *J Am Stat Assoc* 47, 501–515 (2012).

13. Kato, S. et al. Hyperprogressors after immunotherapy: analysis of genomic alterations associated with accelerated growth rate. *Clin Cancer Res* 23, 4242–4250, doi:10.1158/1078–0432.CCR-16–3133 (2017). Fuentes-Antras, J., Provencio, M. & Diaz-Rubio, E. Hyperprogression as a distinct outcome after immunotherapy. *Cancer Treat Rev* 70, 16–21, doi:10.1016/j.ctrv.2018.07.006 (2018). Frelaut, M., Le Tourneau, C. & Borcoman, E. Hyperprogression under immunotherapy. *Int J Mol Sci* 20, doi:10.3390/ijms20112674 (2019).

14. Riely, G. J. et al. Prospective assessment of discontinuation and reinitiation of erlotinib or gefitinib in patients with acquired resistance to erlotinib or gefitinib followed by the addition of everolimus. *Clin Cancer Res* 13, 5150–5155, doi:10.1158/1078–0432.CCR-07–0560 (2007).

15. Rosenzweig, S. A. Acquired resistance to drugs targeting receptor tyrosine kinases. *Biochem Pharmacol* 83, 1041–1048, doi:10.1016/j.bcp.2011.12.025 (2012).

16. Gottesman, M. M. & Ling, V. The molecular basis of multidrug resistance in cancer: the early years of P-glycoprotein research. *FEBS Lett* 580, 998–1009 (2006).

17. O'Donnell, J. S., Teng, M. W. L. & Smyth, M. J. Cancer immunoediting and resistance to T cell-based immunotherapy. *Nat Rev Clin Oncol* 16, 151–167, doi:10.1038/s41571-018-0142-8 (2019).

18. Vogelstein, B. & Kinzler, K. W. The multistep nature of cancer. *Trends Genet* 9, 138–141 (1993).

19. Buisson, R. et al. Passenger hotspot mutations in cancer driven by APOBEC3A and mesoscale genomic features. *Science* 364, doi:10.1126/science.aaw2872 (2019). Williams, M. J., Werner, B., Barnes, C. P., Graham, T. A. & Sottoriva, A. Identification of neutral tumor evolution across cancer types. *Nat Genet* 48, 238–244, doi:10.1038/ng.3489 (2016).

20. Hanahan, D. & Weinberg, R. A. Hallmarks of cancer: the next generation. *Cell* 144, 646–674, doi:10.1016/j.cell.2011.02.013 (2011).

21. Huang, S. The war on cancer: lessons from the war on terror. *Front Oncol* 4, 293, doi:10.3389/fonc.2014.00293 (2014).

22. Tinbergen, N. Derived activities; their causation, biological significance, origin, and emancipation during evolution. *Q Rev Biol* 27, 1–32 (1952). Mayr, E. Cause and effect in biology. *Science* 134, 1501–1506 (1961).

23. Tiong, K. L. & Yeang, C. H. Explaining cancer type specific mutations with transcriptomic and epigenomic features in normal tissues. *Sci Rep* 8, 11456, doi:10.1038/s41598-018-29861-1 (2018). Yamamoto, H. et al. Paradoxical increase in retinoblastoma protein in colorectal carcinomas may protect cells from apoptosis. *Clin Cancer Res* 5, 1805–1815 (1999).

24. Shmueli, G. To explain or to predict? *Stat Sci* 25, 289–310, doi:10.1214/10-STS330 (2010). Huang, S. The tension between big data and theory in the "omics" era of biomedical research. *Perspect Biol Med* 61, 472–488, doi:10.1353/pbm.2018.0058 (2018).

25. Huang, S. The tension between big data and theory in the "omics" era of biomedical research. *Perspect Biol Med* 61, 472–488, doi:10.1353/pbm.2018.0058 (2018).

26. Anderson, P. W. More is different. *Science* 177, 393–396 (1972).

27. Davidson, E. H. & Erwin, D. H. Gene regulatory networks and the evolution of animal body plans. *Science* 311, 796–800 (2006).

28. Huang, S. The molecular and mathematical basis of Waddington's epigenetic landscape: a framework for post-Darwinian biology. *Bioessays* 34, 149–155 (2012).

29. Kolodziejczyk, A. A., Kim, J. K., Svensson, V., Marioni, J. C. & Teichmann, S. A. The technology and biology of single-cell RNA sequencing. *Mol Cell* 58, 610–620, doi:10.1016/j.molcel.2015.04.005 (2015).

30. Patel, A. P. et al. Single-cell RNA-seq highlights intratumoral heterogeneity in primary glioblastoma. *Science* 344, 1396–1401, doi:10.1126/science.1254257 (2014). Shaffer, S. M. et al. Rare cell variability and drug-induced reprogramming as a mode of cancer drug resistance. *Nature* 546, 431–435, doi:10.1038/nature22794 (2017). Hovestadt, V. et al. Resolving medulloblastoma cellular architecture by single-cell genomics. *Nature* 572, 74–79, doi:10.1038/s41586-019-1434-6 (2019).

31. Waddington, C. H. *The strategy of the genes.* (Allen and Unwin, 1957).

32. Patel, A. P. et al. Single-cell RNA-seq highlights intratumoral heterogeneity in primary glioblastoma. *Science* 344, 1396–1401, doi:10.1126/science.1254257 (2014). Shaffer, S. M. et al. Rare cell variability and drug-induced reprogramming as a mode of cancer drug resistance. *Nature* 546, 431–435, doi:10.1038/nature22794 (2017). Hovestadt, V. et al. Resolving medulloblastoma cellular architecture by single-cell genomics. *Nature* 572, 74–79, doi:10.1038/s41586-019-1434-6 (2019). Li, Q. et al. Dynamics inside the cancer cell attractor reveal cell heterogeneity, limits of stability, and escape. *Proc Natl Acad Sci U S A* 113, 2672–2677, doi:10.1073/pnas.1519210113 (2016). Pisco, A. O. et al. Non-Darwinian dynamics in therapy-induced cancer drug resistance. *Nat Commun* 4, 2467, doi:10.1038/ncomms3467 (2013). Su, Y. et al. Single-cell analysis resolves the cell state transition and signaling dynamics associated with melanoma drug-induced resistance. *Proc Natl Acad Sci U S A* 114, 13679–13684, doi:10.1073/pnas.1712064115 (2017).

33. Zhou, J. X., Aliyu, M. D., Aurell, E. & Huang, S. Quasi-potential landscape in complex multi-stable systems. *J R Soc Interface* 9, 3539–3553, doi:10.1098/rsif.2012.0434 (2012).

34. Zhou, J. X., Aliyu, M. D., Aurell, E. & Huang, S. Quasi-potential landscape in complex multi-stable systems. *J R Soc Interface* 9, 3539–3553, doi:10.1098/rsif.2012.0434 (2012).

35. Zhou, J. X., Aliyu, M. D., Aurell, E. & Huang, S. Quasi-potential landscape in complex multi-stable systems. *J R Soc Interface* 9, 3539–3553, doi:10.1098/rsif.2012.0434 (2012). Freidlin, M. & Wentzell, A. *Random Perturbations of Dynamical System.* (Springer-Verlag, 1984).

36. Waddington, C. H. *The Strategy of the Genes.* (Allen and Unwin, 1957).

37. Waddington, C. H. The epigenotype. *Endeavour* 1, 18–20. (1942). Waddington, C. H. *Principles of Embryology.* (Allen & Unwin Ltd, 1956).

38. Grosse-Wilde, A. et al. Stemness of the hybrid epithelial/mesenchymal state in breast cancer and its association with poor survival. *PLoS One* 10, e0126522, doi:10.1371/journal.pone.0126522 (2015). Jolly, M. K. et al. Stability of the hybrid epithelial/mesenchymal phenotype. *Oncotarget* 7, 27067–27084, doi:10.18632/oncotarget.8166 (2016).

39. Nietzsche, F. *Twilight of the Idols.* (Oxford: Oxford University Press, 1998).

40. Tokunaga, E. et al. Spontaneous regression of breast cancer with axillary lymph node metastasis: a case report and review of literature. *Int J Clin Exp Pathol* 7, 4371–4380 (2014). Bramhall, R. J., Mahady, K. & Peach, A. H. Spontaneous regression of metastatic melanoma—clinical evidence of the abscopal effect. *Eur J Surg Oncol* 40, 34–41, doi:10.1016/j.ejso.2013.09.026 (2014). Chang, W. Y. Complete spontaneous regression of cancer: four case reports, review of literature, and discussion of possible mechanisms involved. *Hawaii Med J* 59, 379–387 (2000). Challis, G. B. & Stam, H. J. The spontaneous regression of cancer: a review of cases from 1900 to 1987. *Acta Oncol* 29, 545–550, doi:10.3109/02841869009090048 (1990). Cole, W. H. Spontaneous regression of cancer and the importance of finding its cause. *Natl Cancer Inst Monogr* 44, 5–9 (1976).

41. Reck, M. et al. Pembrolizumab versus chemotherapy for PD-L1-positive non-small-cell lung cancer. *N Engl J Med* 375, 1823–1833, doi:10.1056/NEJMoa1606774 (2016).

42. Juengel, E. et al. Resistance after chronic application of the HDAC-inhibitor valproic acid is associated with elevated Akt activation in renal cell carcinoma in vivo. *PLoS One* 8, e53100, doi:10.1371/journal.pone.0053100 (2013).

43. Ledermann, J. et al. Olaparib maintenance therapy in platinum-sensitive relapsed ovarian cancer. *N Engl J Med* 366, 1382–1392, doi:10.1056/NEJMoa1105535 (2012).

44. Hellmann, M. D. et al. Nivolumab plus ipilimumab in lung cancer with a high tumor mutational burden. *N Engl J Med* 378, 2093–2104, doi:10.1056/NEJMoa1801946 (2018).

45. Juengel, E. et al. Resistance after chronic application of the HDAC-inhibitor valproic acid is associated with elevated Akt activation in renal cell carcinoma in vivo. *PLoS One* 8, e53100, doi:10.1371/journal.pone.0053100 (2013).

46. Sulciner, M. L. et al. Resolvins suppress tumor growth and enhance cancer therapy. *J Exp Med* 215, 115–140, doi:10.1084/jem.20170681 (2018).

47. Sieweke, M. H. & Bissell, M. J. The tumor-promoting effect of wounding: a possible role for TGF-beta-induced stromal alterations. *Crit Rev Oncogenesis* 5, 297–311 (1994).

48. Sulciner, M. L. et al. Resolvins suppress tumor growth and enhance cancer therapy. *J Exp Med* 215, 115–140, doi:10.1084/jem.20170681 (2018). Rashidi, B. et al. Minimal liver resection strongly stimulates the growth of human colon cancer in the liver of nude mice. *Clin Exp Metastasis* 17, 497–500 (1999). Panigrahy, D. et al. Preoperative stimulation of resolution and inflammation blockade eradicates micrometastases. *J Clin Invest* 129, 2964–2979, doi:10.1172/JCI127282 (2019).

49. Sulciner, M. L. et al. Resolvins suppress tumor growth and enhance cancer therapy. *J Exp Med* 215, 115–140, doi:10.1084/jem.20170681 (2018).

50. Berkson, J. & Gage, R. P. Survival curve for cancer patients following treatment. *J Am Stat Assoc* 47, 501–515 (2012).

51. Brock, A., Chang, H. & Huang, S. Non-genetic heterogeneity—a mutation-independent driving force for the somatic evolution of tumours. *Nat Rev Genet* 10, 336–342, doi:nrg2556 [pii] 10. 1038/nrg2556 (2009).

52. Brock, A., Chang, H. & Huang, S. Non-genetic heterogeneity—a mutation-independent driving force for the somatic evolution of tumours. *Nat Rev Genet* 10, 336–342, doi:nrg2556 [pii] 10. 1038/nrg2556 (2009). Brock, A. & Huang, S. Precision oncology: between vaguely right and precisely wrong. *Cancer Res* 77, 6473–6479, doi:10.1158/0008-5472.CAN-17-0448 (2017).

53. Brock, A. & Huang, S. Precision oncology: between vaguely right and precisely wrong. *Cancer Res* 77, 6473–6479, doi:10.1158/0008-5472.CAN-17-0448 (2017). Pisco, A. O. & Huang, S. Non-genetic cancer cell plasticity and therapy-induced stemness in tumour relapse: 'what does not kill me strengthens me'. *Br J Cancer* 112, 1725–1732, doi:10.1038/bjc.2015.146 (2015).

54. Shaffer, S. M. et al. Rare cell variability and drug-induced reprogramming as a mode of cancer drug resistance. *Nature* 546, 431–435, doi:10.1038/nature22794 (2017). Neftel, C. et al. An integrative model of cellular states, plasticity, and genetics for glioblastoma. *Cell* 178, 835–849 e821, doi:10.1016/j.cell.2019.06.024 (2019).

55. Pisco, A. O. & Huang, S. Non-genetic cancer cell plasticity and therapy-induced stemness in tumour relapse: 'what does not kill me strengthens me'. *Br J Cancer* 112, 1725–1732, doi:10.1038/bjc.2015.146 (2015).

56. Amini, S., Fathi, F., Mobalegi, J., Sofimajidpour, H. & Ghadimi, T. The expressions of stem cell markers: Oct4, Nanog, Sox2, nucleostemin, Bmi, Zfx, Tcl1, Tbx3, Dppa4, and Esrrb in bladder, colon, and prostate cancer, and certain cancer cell lines. *Anat Cell Biol* 47, 1–11, doi:10.5115/acb.2014.47.1.1 (2014). Trosko, J. E. From adult stem cells to cancer stem cells: Oct-4 Gene, cell-cell communication, and hormones during tumor promotion. *Ann N Y Acad Sci* 1089, 36–58, doi:10.1196/annals.1386.018 (2006). Bhattacharya, R., Mustafi, S. B., Street, M., Dey, A. & Dwivedi, S. K. Bmi-1: At the crossroads of physiological and pathological biology. *Genes Dis* 2, 225–239, doi:10.1016/j.gendis.2015.04.001 (2015). Rowbotham, S. P. & Kim, C. F. Don't stop re-healin'! Cancer as an ongoing stem cell affair. *Cell* 169, 563–565, doi:10.1016/j.cell.2017.04.030 (2017). Blanco, S. et al. Stem cell function and stress response are controlled by protein synthesis. *Nature* 534, 335–340, doi:10.1038/nature18282 (2016). Hung, K. F., Yang, T. & Kao, S. Y. Cancer stem cell theory: are we moving past the mist? *J Chin Med Assoc* 82, 814–818, doi:10.1097/JCMA.0000000000000186 (2019).

57. Diamandis, E. P. The failure of protein cancer biomarkers to reach the clinic: why, and what can be done to address the problem? *BMC Med* 10, 87, doi:10.1186/1741-7015-10-87 (2012). Kern, S. E. Why your new cancer biomarker may never work: recurrent patterns and remarkable diversity in biomarker failures. *Cancer Res* 72, 6097–6101, doi:10.1158/0008-5472.CAN-12-3232 (2012). Poste, G. Bring on the biomarkers. *Nature* 469, 156–157, doi:10.1038/469156a (2011).

58. Kaern, M., Elston, T. C., Blake, W. J. & Collins, J. J. Stochasticity in gene expression: from theories to phenotypes. *Nat Rev Genet* 6, 451–464 (2005). Raj, A. & van Oudenaarden, A. Nature, nurture, or chance: stochastic gene expression and its consequences. *Cell* 135, 216–226, doi:S0092-8674(08)01243-9 [pii] 10.1016/j.cell.2008.09.050 (2008). Raser, J. M. & O'Shea, E. K. Noise in gene expression: origins, consequences, and control. *Science* 309, 2010–2013 (2005).

59. Brock, A., Chang, H. & Huang, S. Non-genetic heterogeneity—a mutation-independent driving force for the somatic evolution of tumours. *Nat Rev Genet* 10, 336–342, doi:nrg2556 [pii] 10. 1038/nrg2556 (2009). Marusyk, A., Almendro, V. & Polyak, K. Intra-tumour heterogeneity: a looking glass for cancer? *Nat Rev Cancer* 12, 323–334, doi:10.1038/nrc3261 (2012).

60. Huang, S. Genetic and non-genetic instability in tumor progression: link between the fitness landscape and the epigenetic landscape of cancer cells. *Cancer Metastasis Rev* 32, 423–448, doi:10.1007/s10555-013-9435-7 (2013).

61. Strogatz, S. H. Exploring complex networks. *Nature* 410, 268–276 (2001).

62. Delbrück, M. *Unités biologiques douées de continuité génétique Colloques Internationaux du Centre National de la Recherche Scientifique* 33–35 (CNRS, Paris, 1949).

63. Waddington, C. H. *The Strategy of the Genes.* (Allen and Unwin, 1957).

64. Kauffman, S. Homeostasis and differentiation in random genetic control networks. *Nature* 224, 177–178 (1969).

65. Kauffman, S. A. *The Origins of Order.* (Oxford University Press, 1993).

66. Huang, S. The molecular and mathematical basis of Waddington's epigenetic landscape: a framework for post-Darwinian biology. *Bioessays* 34, 149–155 (2012). Kubicek, S. & Jenuwein, T. A crack in histone lysine methylation. *Cell* 119, 903–906 (2004).

67. Aurell, E. & Sneppen, K. Epigenetics as a first exit problem. *Phys Rev Lett* 88, 048101 (2002).

68. Chang, H. H., Hemberg, M., Barahona, M., Ingber, D. E. & Huang, S. Transcriptome-wide noise controls lineage choice in mammalian progenitor cells. *Nature* 453, 544–547 (2008).

69. Mojtahedi, M. et al. Cell fate decision as high-dimensional critical state transition. *PLoS Biol* 14, e2000640, doi:10.1371/journal.pbio.2000640 (2016).

70. Wells, D. K., Kath, W. L. & Motter, A. E. Control of stochastic and induced switching in biophysical networks. *Phys Rev X* 5, doi:10.1103/PhysRevX.5.031036 (2015).

71. Mojtahedi, M. *Single-Cell Analysis for Cell-Fate Decision Studies.* PhD thesis, University of Calgary, (2014). Richard, A. et al. Single-cell-based analysis highlights a surge in cell-to-cell molecular variability preceding irreversible commitment in a differentiation process. *PLoS Biol* 14, e1002585, doi:10.1371/journal.pbio .1002585 (2016).

72. Scheffer, M. et al. Anticipating critical transitions. *Science* 338, 344–348, doi:10.1126/science.1225244 (2012).

73. Mojtahedi, M. et al. Cell fate decision as high-dimensional critical state transition. *PLoS Biol* 14, e2000640, doi:10.1371/journal.pbio.2000640 (2016).

74. Hough, S. R. et al. Single-cell gene expression profiles define self-renewing, pluripotent, and lineage primed states of human pluripotent stem cells. *Stem Cell Reports* 2, 881–895, doi:10.1016/j.stemcr.2014.04.014 (2014). Martinez Arias, A. & Brickman, J. M. Gene expression heterogeneities in embryonic stem cell populations: origin and function. *Curr Opin Cell Biol* 23, 650–656, doi:10.1016/j.ceb.2011.09.007 (2011). Rotem, A. et al. Single-cell ChIP-seq reveals cell subpopulations defined by chromatin state. *Nat Biotechnol* 33, 1165–1172, doi:10.1038 /nbt.3383 (2015). Canham, M. A., Sharov, A. A., Ko, M. S. & Brickman, J. M. Functional heterogeneity of embryonic stem cells revealed through translational amplification of an early endodermal transcript. *PLoS Biol* 8, e1000379, doi:10.1371/journal.pbio.1000379 (2010).

75. Jackson, T. R. et al. DNA damage causes TP53-dependent coupling of self-renewal and senescence pathways in embryonal carcinoma cells. *Cell Cycle* 12, 430–441, doi:10.4161/cc.23285 (2013).

76. Gay, L., Baker, A. M. & Graham, T. A. Tumour cell heterogeneity. *F1000Research* 5, doi:10.12688/f1000 research.7210.1 (2016).

77. Rastrick, J. M., Fitzgerald, P. H. & Gunz, F. W. Direct evidence for presence of Ph-1 chromosome in erythroid cells. *Br Med J* 1, 96–98, doi:10.1136/bmj.1.5584.96 (1968). Takahashi, N., Miura, I., Saitoh, K. & Miura, A. B. Lineage involvement of stem cells bearing the Philadelphia chromosome in chronic myeloid leukemia in the chronic phase as shown by a combination of fluorescence-activated cell sorting and fluorescence in situ hybridization. *Blood* 92, 4758–4763 (1998).

78. Neftel, C. et al. An integrative model of cellular states, plasticity, and genetics for glioblastoma. *Cell* 178, 835–849 e821, doi:10.1016/j.cell.2019.06.024 (2019). Halliday, J. et al. In vivo radiation response of proneural glioma characterized by protective p53 transcriptional program and proneural-mesenchymal shift. *Proc Natl Acad Sci U S A* 111, 5248–5253, doi:10.1073/pnas.1321014111 (2014).

79. Su, Y. et al. Single-cell analysis resolves the cell state transition and signaling dynamics associated with melanoma drug-induced resistance. *Proc Natl Acad Sci U S A* 114, 13679–13684, doi:10.1073/pnas.1712064115 (2017).

80. Rickman, D. S., Beltran, H., Demichelis, F. & Rubin, M. A. Biology and evolution of poorly differentiated neuroendocrine tumors. *Nat Med* 23, 1–10, doi:10.1038/nm.4341 (2017).

81. Prabhakaran, P., Hassiotou, F., Blancafort, P. & Filgueira, L. Cisplatin induces differentiation of breast cancer cells. *Front Oncol* 3, 134, doi:10.3389/fonc.2013.00134 (2013). Brambilla, E. et al. Cytotoxic chemotherapy induces cell differentiation in small-cell lung carcinoma. *J Clin Oncol* 9, 50–61, doi:10.1200/JCO.1991.9.1.50 (1991).

82. Sell, S. Stem cell origin of cancer and differentiation therapy. *Crit Rev Oncol Hematol* 51, 1–28 (2004).

83. Jackson, T. R. et al. DNA damage causes TP53-dependent coupling of self-renewal and senescence pathways in embryonal carcinoma cells. *Cell Cycle* 12, 430–441, doi:10.4161/cc.23285 (2013). Prabhakaran, P., Hassiotou, F., Blancafort, P. & Filgueira, L. Cisplatin induces differentiation of breast cancer cells. *Front Oncol* 3, 134, doi:10.3389/fonc.2013.00134 (2013). Herr, R. et al. B-Raf inhibitors induce epithelial differentiation in BRAF-mutant colorectal cancer cells. *Cancer Res* 75, 216–229, doi:10.1158/0008-5472.CAN-13-3686 (2015).

84. Ewald, J. A., Desotelle, J. A., Wilding, G. & Jarrard, D. F. Therapy-induced senescence in cancer. *J Natl Cancer Inst* 102, 1536–1546, doi:10.1093/jnci/djq364 (2010). Guillon, J. et al. Chemotherapy-induced senescence, an adaptive mechanism driving resistance and tumor heterogeneity. *Cell Cycle* 18, 2385–2397, doi:10.1080/15384101.2019.1652047 (2019).

85. Beckman, R. A. Efficiency of carcinogenesis: is the mutator phenotype inevitable? *Semin Cancer Biol* 20, 340–352, doi:10.1016/j.semcancer.2010.10.004 (2010). Bielas, J. H., Loeb, K. R., Rubin, B. P., True, L. D. & Loeb, L. A. Human cancers express a mutator phenotype. *Proc Natl Acad Sci U S A* 103, 18238–18242 (2006).

86. Brock, A., Chang, H. & Huang, S. Non-genetic heterogeneity—a mutation-independent driving force for the somatic evolution of tumours. *Nat Rev Genet* 10, 336–342, doi:nrg2556 [pii] 10. 1038/nrg2556 (2009). Huang, S. Tumor progression: chance and necessity in Darwinian and Lamarckian somatic (mutationless) evolution. *Prog Biophys Mol Biol* 110, 69–86 (2012).

87. Korkaya, H., Liu, S. & Wicha, M. S. Regulation of cancer stem cells by cytokine networks: attacking cancer's inflammatory roots. *Clin Cancer Res* 17, 6125–6129, doi:10.1158/1078-0432.CCR-10-2743 (2011). Fordyce, C. A. et al. Cell-extrinsic consequences of epithelial stress: activation of protumorigenic tissue phenotypes. *Breast Cancer Res* 14, R155, doi:10.1186/bcr3368 (2012). Hangai, S. et al. PGE2 induced in and released by dying cells functions as an inhibitory DAMP. *Proc Natl Acad Sci U S A* 113, 3844–3849, doi:10.1073/pnas.1602023113 (2016). Kurtova, A. V. et al. Blocking PGE2-induced tumour repopulation abrogates bladder cancer chemoresistance. *Nature* 517, 209–213, doi:10.1038/nature14034 (2015).

88. Rubin, H. Fields and field cancerization: the preneoplastic origins of cancer: asymptomatic hyperplastic fields are precursors of neoplasia, and their progression to tumors can be tracked by saturation density in culture. *Bioessays* 33, 224–231, doi:10.1002/bies.201000067 (2011). Soto, A. M. & Sonnenschein, C. The tissue organization field theory of cancer: a testable replacement for the somatic mutation theory. *Bioessays* 33, 332–340, doi:10.1002/bies.201100025 (2011). Dotto, G. P. Multifocal epithelial tumors and field cancerization: stroma as a primary determinant. *J Clin Invest* 124, 1446–1453, doi:10.1172/JCI72589 (2014). Simple, M., Suresh, A., Das, D. & Kuriakose, M. A. Cancer stem cells and field cancerization of oral squamous cell carcinoma. *Oral Oncol* 51, 643–651, doi:10.1016/j.oraloncology.2015.04.006 (2015).

89. Sulciner, M. L. et al. Resolvins suppress tumor growth and enhance cancer therapy. *J Exp Med* 215, 115–140, doi:10.1084/jem.20170681 (2018). Brock, A. & Huang, S. Precision oncology: between vaguely right and precisely wrong. *Cancer Res* 77, 6473–6479, doi:10.1158/0008-5472.CAN-17-0448 (2017).

90. Kauffman, S. Differentiation of malignant to benign cells. *J Theor Biol* 31, 429–451 (1971).

91. Huang, S., Ernberg, I. & Kauffman, S. Cancer attractors: a systems view of tumors from a gene network dynamics and developmental perspective. *Semin Cell Dev Biol* 20 869–876, doi:S1084-9521(09)00149-9 [pii] 10 .1016/j.semcdb.2009.07.003 (2009). Huang, S. On the intrinsic inevitability of cancer: from foetal to fatal attraction. *Semin Cancer Biol* 21, 183–199, doi:S1044-579X(11)00032-0 [pii] 10.1016/j.semcancer.2011.05.003 (2011).

92. Davies, P. C. & Lineweaver, C. H. Cancer tumors as Metazoa 1.0: tapping genes of ancient ancestors. *Phys Biol* 8, 015001, doi:10.1088/1478-3975/8/1/015001 (2011).

93. Huang, S. Genetic and non-genetic instability in tumor progression: link between the fitness landscape and the epigenetic landscape of cancer cells. *Cancer Metastasis Rev* 32, 423–448, doi:10.1007/s10555-013-9435-7 (2013). Huang, S. On the intrinsic inevitability of cancer: from foetal to fatal attraction. *Semin Cancer Biol* 21, 183–199, doi:S1044-579X(11)00032-0 [pii] 10.1016/j.semcancer.2011.05.003 (2011).

94. Kauffman, S. A. *The Origins of Order.* (Oxford University Press, 1993).

95. Bornholdt, S. & Rohlf, T. Topological evolution of dynamical networks: global criticality from local dynamics. *Phys Rev Lett* 84, 6114–6117 (2000). Nykter, M. et al. Gene expression dynamics in the macrophage exhibit

criticality. *Proc Natl Acad Sci U S A* 105, 1897–1900 (2008). Ramo, P., Kesseli, J. & Yli-Harja, O. Perturbation avalanches and criticality in gene regulatory networks. *J Theor Biol* 242, 164–170 (2006). Torres-Sosa, C., Huang, S. & Aldana, M. Criticality is an emergent property of genetic networks that exhibit evolvability. *PLoS Comput Biol* 8, e1002669, doi:10.1371/journal.pcbi.1002669 (2012).

96. Kauffman, S. A. *The Origins of Order.* (Oxford University Press, 1993). Bagley, R. J. & Glass, L. Counting and classifying attractors in high dimensional dynamical systems. *J Theor Biol* 183, 269–284 (1996). Huang, S. & Kauffman, S. In *Encyclopedia of Complexity and Systems Science* (ed. R. A. Meyers) 1180–1213 (Springer, 2009).

97. Huang, S. The molecular and mathematical basis of Waddington's epigenetic landscape: A framework for post-Darwinian biology. *Bioessays* 34, 149–155 (2012).

98. Janssens, A. & Joyner, M. J. Polygenic risk scores that predict common diseases using millions of single nucleotide polymorphisms: is more, better? *Clin Chem* 65, 609–611, doi:10.1373/clinchem.2018.296103 (2019).

99. Torres-Sosa, C., Huang, S. & Aldana, M. Criticality is an emergent property of genetic networks that exhibit evolvability. *PLoS Comput Biol* 8, e1002669, doi:10.1371/journal.pcbi.1002669 (2012). Aldana, M., Balleza, E., Kauffman, S. & Resendiz, O. Robustness and evolvability in genetic regulatory networks. *J Theor Biol* 245, 433–448 (2007).

100. Huang, S. Genetic and non-genetic instability in tumor progression: link between the fitness landscape and the epigenetic landscape of cancer cells. *Cancer Metastasis Rev* 32, 423–448, doi:10.1007/s10555-013-9435-7 (2013). Huang, S. On the intrinsic inevitability of cancer: from foetal to fatal attraction. *Semin Cancer Biol* 21, 183–199, doi:S1044–579X(11)00032–0 [pii] 10.1016/j.semcancer.2011.05.003 (2011).

101. Maden, M. The evolution of regeneration—where does that leave mammals? *Int J Dev Biol* 62, 369–372, doi:10.1387/ijdb.180031mm (2018). Slack, J. M. Animal regeneration: ancestral character or evolutionary novelty? *EMBO Rep* 18, 1497–1508, doi:10.15252/embr.201643795 (2017).

102. Gould, S. J. & Lewontin, R. C. The spandrels of San Marco and the Panglossian paradigm: a critique of the adaptationist programme. *Proc R Soc Lond B Biol Sci* 205, 581–598 (1979).

103. Aubin, D. Forms of explanations in the catastrophe theory of René Thom: topology, morphogenesis, and structuralism. In *Growing Explanations: Historical Perspective on the Sciences of Complexity* (ed. M.N. Wise) 95–130 (Duke University Press, 2004). Webster, G. & Goodwin, B. C. A structuralist approach to morphology. *Riv Biol* 92, 495–498 (1999).

104. Sell, S. On the stem cell origin of cancer. *Am J Pathol* 176, 2584–2494, doi:10.2353/ajpath.2010.091064 (2010).

105. Holtan, S. G. et al. The dynamic human immune response to cancer: it might just be rocket science. *Immunotherapy* 3, 1021–1024, doi:10.2217/imt.11.109 (2011).

106. Fordyce, C. A. et al. Cell-extrinsic consequences of epithelial stress: activation of protumorigenic tissue phenotypes. *Breast Cancer Res* 14, R155, doi:10.1186/bcr3368 (2012). Hangai, S. et al. PGE2 induced in and released by dying cells functions as an inhibitory DAMP. *Proc Natl Acad Sci U S A* 113, 3844–3849, doi:10.1073/pnas.1602023113 (2016). Kurtova, A. V. et al. Blocking PGE2-induced tumour repopulation abrogates bladder cancer chemoresistance. *Nature* 517, 209–213, doi:10.1038/nature14034 (2015). Kalinski, P. Regulation of immune responses by prostaglandin E2. *J Immunol* 188, 21–28, doi:10.4049/jimmunol.1101029 (2012). Stoecklein, V. M., Osuka, A. & Lederer, J. A. Trauma equals danger—damage control by the immune system. *J Leukoc Biol* 92, 539–551, doi:10.1189/jlb.0212072 (2012). Wang, D. & DuBois, R. N. Immunosuppression associated with chronic inflammation in the tumor microenvironment. *Carcinogenesis* 36, 1085–1093, doi:10.1093/carcin/bgv123 (2015). Xiao, W. et al. A genomic storm in critically injured humans. *J Exp Med* 208, 2581–2590, doi:10.1084/jem.20111354 (2011).

107. Holtan, S. G. et al. The dynamic human immune response to cancer: it might just be rocket science. *Immunotherapy* 3, 1021–1024, doi:10.2217/imt.11.109 (2011).

108. Dvorak, H. F. Tumors: wounds that do not heal. Similarities between tumor stroma generation and wound healing. *N Engl J Med* 315, 1650–1659, doi:10.1056/NEJM198612253152606 (1986).

109. O'Donnell, J. S., Teng, M. W. L. & Smyth, M. J. Cancer immunoediting and resistance to T cell-based immunotherapy. *Nat Rev Clin Oncol* 16, 151–167, doi:10.1038/s41571-018-0142-8 (2019).

110. Zhou, J. X. et al. Phylostratigraphic analysis of tumor and developmental transcriptomes reveals relationship between oncogenesis, phylogenesis and ontogenesis. *Convergent Sci Phys Oncol* 4, 025002 (2018).

111. Sell, S. On the stem cell origin of cancer. *Am J Pathol* 176, 2584–2494, doi:10.2353/ajpath.2010.091064 (2010).

112. Abolhoda, A. et al. Rapid activation of MDR1 gene expression in human metastatic sarcoma after in vivo exposure to doxorubicin. *Clin Cancer Res* 5, 3352–3356 (1999). Chin, K. V., Tanaka, S., Darlington, G., Pastan, I. & Gottesman, M. M. Heat shock and arsenite increase expression of the multidrug resistance (MDR1) gene in human renal carcinoma cells. *J Biol Chem* 265, 221–226 (1990). Correa, S. et al. Wnt/beta-catenin pathway regulates ABCB1 transcription in chronic myeloid leukemia. *BMC Cancer* 12, 303, doi:10.1186/1471-2407 -12-303(2012).Mickley,L.A.etal.Modulationoftheexpressionofamultidrugresistancegene(mdr-1/P-glycoprotein) by differentiating agents. *J Biol Chem* 264, 18031–18040 (1989). Pisco, A. O., Jackson, D. A. & Huang, S. Reduced intracellular drug accumulation in drug-resistant leukemia cells is not only solely due to MDR-mediated efflux but also to decreased uptake. *Front Oncol* 4, 306, doi:10.3389/fonc.2014.00306 (2014).

113. Donnenberg, V. S. & Donnenberg, A. D. Multiple drug resistance in cancer revisited: the cancer stem cell hypothesis. *J Clin Pharmacol* 45, 872–877 (2005).

114. Correa, S. et al. Wnt/beta-catenin pathway regulates ABCB1 transcription in chronic myeloid leukemia. *BMC Cancer* 12, 303, doi:10.1186/1471-2407-12-303 (2012). Hung, T. H. et al. Wnt5A regulates ABCB1 expression in multidrug-resistant cancer cells through activation of the non-canonical PKA/beta-catenin pathway. *Oncotarget* 5, 12273–12290 (2014). Stein, U. et al. Impact of mutant beta-catenin on ABCB1 expression and therapy response in colon cancer cells. *Br J Cancer* 106, 1395–1405, doi:10.1038/bjc.2012.81 (2012).

115. Trosko, J. E. From adult stem cells to cancer stem cells: Oct-4 Gene, cell-cell communication, and hormones during tumor promotion. *Ann N Y Acad Sci* 1089, 36–58, doi:10.1196/annals.1386.018 (2006). Jackson, T. R. et al. DNA damage causes TP53-dependent coupling of self-renewal and senescence pathways in embryonal carcinoma cells. *Cell Cycle* 12, 430–441, doi:10.4161/cc.23285 (2013). Hu, X. et al. Induction of cancer cell stemness by chemotherapy. *Cell Cycle* 11, 2691–2698, doi:10.4161/cc.21021 (2012).

116. Bastakoty, D. & Young, P. P. Wnt/beta-catenin pathway in tissue injury: roles in pathology and therapeutic opportunities for regeneration. *FASEB J* 30, 3271–3284, doi:10.1096/fj.201600502R (2016). Bielefeld, K. A., Amini-Nik, S. & Alman, B. A. Cutaneous wound healing: recruiting developmental pathways for regeneration. *Cell Mol Life Sci* 70, 2059–2081, doi:10.1007/s00018-012-1152-9 (2013).

117. Goldsberry, W. N., Londono, A., Randall, T. D., Norian, L. A. & Arend, R. C. A review of the role of Wnt in cancer immunomodulation. *Cancers (Basel)* 11, 771, doi:10.3390/cancers11060771 (2019).

118. Hu, X. et al. Induction of cancer cell stemness by chemotherapy. *Cell Cycle* 11, 2691–2698, doi:10.4161 /cc.21021 (2012).

119. Hafner, A., Bulyk, M. L., Jambhekar, A. & Lahav, G. The multiple mechanisms that regulate p53 activity and cell fate. *Nat Rev Mol Cell Biol* 20, 199–210, doi:10.1038/s41580-019-0110-x (2019).

120. Lee, K. H. et al. A genomewide study identifies the Wnt signaling pathway as a major target of p53 in murine embryonic stem cells. *Proc Natl Acad Sci U S A* 107, 69–74, doi:10.1073/pnas.0909734107 (2010).

121. Lee, E. et al. Metabolic stress induces a Wnt-dependent cancer stem cell-like state transition. *Cell Death Dis* 6, e1805, doi:10.1038/cddis.2015.171 (2015). Lettini, G. et al. Heat shock proteins in cancer stem cell maintenance: a potential therapeutic target? *Histol Histopathol* 35, 25–37, doi:10.14670/HH-18–153 (2019). Cubillos-Ruiz, J. R., Bettigole, S. E. & Glimcher, L. H. Tumorigenic and immunosuppressive effects of endoplasmic reticulum stress in cancer. *Cell* 168, 692–706, doi:10.1016/j.cell.2016.12.004 (2017). Lu, H. et al. Chemotherapy triggers HIF-1-dependent glutathione synthesis and copper chelation that induces the breast cancer stem cell phenotype. *Proc Natl Acad Sci U S A* 112, E4600–4609, doi:10.1073/pnas.1513433112 (2015). Ciocca, D. R., Arrigo, A. P. & Calderwood, S. K. Heat shock proteins and heat shock factor 1 in carcinogenesis and tumor development: an update. *Arch Toxicol* 87, 19–48, doi:10.1007/s00204-012-0918-z (2013). Sosa, M. S. et al. NR2F1 controls tumour cell dormancy via SOX9- and RARbeta-driven quiescence programmes. *Nat Commun* 6, 6170, doi:10.1038/ncomms7170 (2015).

122. Puissant, A. et al. Imatinib triggers mesenchymal-like conversion of CML cells associated with increased aggressiveness. *J Mol Cell Biol* 4, 207–220, doi:10.1093/jmcb/mjs010 (2012).

123. Stoecklein, V. M., Osuka, A. & Lederer, J. A. Trauma equals danger—damage control by the immune system. *J Leukoc Biol* 92, 539–551, doi:10.1189/jlb.0212072 (2012). Chan, J. K. et al. Alarmins: awaiting a clinical response. *J Clin Invest* 122, 2711–2719, doi:10.1172/JCI62423 (2012). McDonald, B. et al. Intravascular danger signals guide neutrophils to sites of sterile inflammation. *Science* 330, 362–366, doi:10.1126/science.1195491 (2010).

124. Sulciner, M. L. et al. Resolvins suppress tumor growth and enhance cancer therapy. *J Exp Med* 215, 115–140, doi:10.1084/jem.20170681 (2018).

125. Sulciner, M. L. et al. Resolvins suppress tumor growth and enhance cancer therapy. *J Exp Med* 215, 115–140, doi:10.1084/jem.20170681 (2018). Huang, Q. et al. Caspase 3-mediated stimulation of tumor cell repopulation during cancer radiotherapy. *Nat Med* 17, 860–866, doi:10.1038/nm.2385 (2011).

126. Ewald, J. A., Desotelle, J. A., Wilding, G. & Jarrard, D. F. Therapy-induced senescence in cancer. *J Natl Cancer Inst* 102, 1536–1546, doi:10.1093/jnci/djq364 (2010). Guillon, J. et al. Chemotherapy-induced senescence, an adaptive mechanism driving resistance and tumor heterogeneity. *Cell Cycle* 18, 2385–2397, doi:10.1080 /15384101.2019.1652047 (2019).

127. Ewald, J. A., Desotelle, J. A., Wilding, G. & Jarrard, D. F. Therapy-induced senescence in cancer. *J Natl Cancer Inst* 102, 1536–1546, doi:10.1093/jnci/djq364 (2010). Guillon, J. et al. Chemotherapy-induced senescence, an adaptive mechanism driving resistance and tumor heterogeneity. *Cell Cycle* 18, 2385–2397, doi:10.1080 /15384101.2019.1652047 (2019).

128. Hangai, S. et al. PGE2 induced in and released by dying cells functions as an inhibitory DAMP. *Proc Natl Acad Sci U S A* 113, 3844–3849, doi:10.1073/pnas.1602023113 (2016). Kalinski, P. Regulation of immune responses by prostaglandin E2. *J Immunol* 188, 21–28, doi:10.4049/jimmunol.1101029 (2012).

129. Liu, D. & Hornsby, P. J. Senescent human fibroblasts increase the early growth of xenograft tumors via matrix metalloproteinase secretion. *Cancer Res* 67, 3117–3126, doi:10.1158/0008-5472.CAN-06-3452 (2007).

5 The Cell Attractor Concept as a Tool to Advance Our Understanding of Cancer

Ingemar Ernberg

Overview

In this chapter, I explore how useful the concept of the "cellular attractor," adopted from complex dynamical systems theory, might be in the analysis and study of cellular identity, plasticity, and cellular phenotypes generated by the intracellular gene regulatory network in the cancer context. First, I provide a brief summary of the considerations that have led cancer scientists to realize that we need to go beyond the metaphor of "programing" of biological/cellular function by DNA in order to better understand cellular function. In short, the overarching question we need to answer is, "Who/what makes decisions in the cell?" or rather, "How are decisions made in a cell?" This might sound teleological or provoke unintended dangerous or misleading arguments, but handled with care, I hope to guide the reader to the essence of this metaphorical question—in brief, the enigma of how the same genome shared by approximately ten trillion cells in a human body can give rise to the few hundred widely divergent cell phenotypes with their specific functions that we can observe.

5.1 Can DNA Specify a Cell Type?

What is the most important feature of a cell in a multicellular organism? Basic functions, such as metabolism and cell division, are necessities and defining for all cellular life, but within the context of the whole organism, the most important feature, or the *overall function* of the cell, is *its contribution to the survival of the organism*. We usually represent cell functions, or at least groups of cellular functions, by defining cell types and their phenotypes. Therefore, the typical cell types crudely represent different functions or functional groups of cells (e.g., in a human body). The different cell types of a multicellular organism carry identical genomic information but execute widely divergent functions, such as electric signaling, contraction, mobility, laying down the skeleton, defense against invading organisms and molecules, translating photons into "meaningful" signals, or protein production as required for the release of enzymes, hormones, or antibodies. How can the individual organism's single genome operate in so many widely divergent contexts?

Conventionally, it is understood that this is possible due to the differentiation programs that are directing the development of an organism and/or epigenetic control mechanisms. How the actual cues for cell differentiation operate and how they are controlled is still the subject and focus of much of current research, and many of the biological processes involved are reasonably well understood. However, what has become clear by now is that they cannot simply be considered "programed algorithms" encoded in and executed by DNA. If this were the case, then the plethora of observed end points, different cell types, and functions that can emerge from the same DNA sequence cannot be explained. When it comes to other, additional levels of gene expression control, such as "epigenetic programming," we have currently only a very limited understanding of what is the chicken and what is the egg. How is specificity conferred for closing or opening a chromatin region or gene? How are specificities of this kind coordinated between many thousands of genes and hundreds of thousands of proteins? The close-to-end operator, for example, may be an acetylase, deacetylase, methylase, or possibly a demethylase, but how this enzyme's action becomes specific for a certain intracellular region or a specific gene within the context of the intracellular totality is not very clear. By retrograde, linear analysis, one inevitably will end up losing any clear notion of hierarchy with regards to who is in command of this specificity. In fact, from the study of complex systems, we have learned that it is not sufficient for solving the specificity problem to refer to cellular differentiation exclusively in terms of "genetic or epigenetic programming" in the way we study them today. To rephrase the problem, using a language metaphor, we could say: we do not understand *who* decides in the cell and *how* decisions are made, for instance, the decision to become a neuron, a B-lymphocyte, a retinal-rod cell, a muscle cell, or an osteoblast. Without having solved this problem, we must admit that we are still missing a major piece in understanding how a cell "works."

A major reason for this shortcoming might be that we have not really tried to explain the "decision-making" process of a cell. Since the seminal and ingenious discovery of the DNA structure in 1953 and the subsequent decades of unravelling the genetic code and the biochemistry of transcription and translation, we have been caught up in the rather limited perspective that all biology can be explained by understanding DNA and that it is the "DNA codes" ($\approx$ genes) that are running different programs within different cells. This reductionist viewpoint has been, and still is, instrumental and useful for the impressive progress in cell biology, biology, and biotechnology and serves a major task in, for example, applying biomedicine to human disease. The late Nobel laureate Sydney Brenner has succinctly commented on our generation's admiration for the significance of the genome:

I think the most important thing we know about living systems is that they've got genes. It is through the genes that one living system propagates descendants that look like it.

What science has accomplished to tell us is that this happens because organisms contain genes in them and the future organism is written in this somehow. And "somehow" is what we have to explain. We have to say not somehow but how.[1]

5.2 What Has Been Missing in Cell Studies?

DNA-centered reductionism has certainly led to a better understanding of the cell biology of diseases such as cancer and to the development of drugs that can successfully target mutated genes and their products. While busy harvesting its low-hanging fruits, we have been struck by hubris, believing that a biological system is controlled solely by its DNA. Despite much impressive, detailed knowledge about how the information held in DNA is used in a cell, we still cannot easily translate the operation of genes and genetic information into the complex functions of different cell types. Although somewhat aware of this deficiency, we have confined ourselves to a view that causality in biology is largely a linear process—a view that perceives biological systems in a manner similar to the perception of the (classical) foundations of the natural sciences at the end of the nineteenth century, before Einstein, Schrödinger, Heisenberg, and Bohr. Scientists are only now realizing that we cannot any longer ignore the fact that biological systems are complex, with tens of millions of interacting actors in a single cell. There is now growing acceptance of the fact that a cell represents a complex, adaptive system with emerging properties. The most pressing aspect of this complexity is that we have very little understanding of how "decision making" functions in such systems. One suggestion is that it takes place in a distributed manner, such that close neighbors influence each other locally, and that the sum of such local interactions adds up to a systems-level output that we perceive as a "decision," as is observed in self-organizing systems.[2] Such a framework has so far not been applied widely by biologists in the study of cells. Moreover, many basic features of cells that determine their physical properties have been neglected until now and would need much more attention to understand cellular life.

One such fundamental property of cells is the extremely high protein concentration, or molecular density, that exists at the micro- and nanoscales within cells. This and other properties clearly show that cells operate far from thermodynamic equilibrium, employing long chains of simultaneously occurring chemical reactions with many feedback and feedforward loops in a highly exceptional three-dimensional nanoenvironment. This type of high-density chemistry of a complex network of coupled chemical reactions has so far not been studied much due to technical limitations. What are the rules of such complex non-hierarchical networks of nonlinear chemical reactions? How can we understand cause and effect in such a system? The intracellular chemistry is consequently mainly surface chemistry, involving proton gradients and organized water molecules. It has been suggested that the densely packed molecular surfaces bumping into each other inside the cell represent states of continuous/constant "phase transitions." Moreover, such cellular systems are good in maintaining homeostasis over a range of change in external conditions, which is poorly understood. Once we understand these phenomena at the levels of cell biology, (cell) physics, and (cell) chemistry (as these might be quite different from "standard" physics and chemistry) applying a complex systems framework, we can then consider applying similar principles to the organism level.

5.3   The Role of DNA Is Not to Program Cells

Denis Noble, one of the founders of systems biology, in a review in 2013 referred to ten principles of systems biology, which he first presented in a meeting a few years earlier.[3] He provocatively argues that

- Biological functionality is multilevel.
- Transmission of information is not one way.
- The theory of biological relativity: there is no privileged level of causality.
- DNA is not the sole transmitter of inheritance.
- There is no genetic program.
- There are no programs at any other level.
- There are many more to be discovered; a genuine "theory of biology" does not yet exist.

If this is the case, the famous statement by Carl Sagan and Francis Crick participating in a conference, sponsored jointly by the Academies of Science of the United States and the USSR, on "Communication with Extraterrestrial Intelligence," or CETI, at the mountaintop observatory outside of Yerevan, Armenia, in 1971, was erroneous when they stated that "if we could send a string of DNA out into the universe so that another civilization could pick it up, then they would understand our form of life."[4] Clearly, much more is needed to understand life, first and foremost the physical and chemical constraints that appear to be highly specific for the forms of life known to us, and moreover the obvious impact of the inherited "plan" for a cell, the sophisticated cellular organizational structure that is not inherited through DNA, but through the (fertilized) egg itself. The egg cell as a whole preserves the genome within this overall structure and makes it work to maintain structural integrity.

In such a world of intracellular exceptional complexity, what is then the role of DNA if it is not the programing of cellular functions? It is nevertheless certainly very important. It provides the boundaries for the possibilities of functional expression. It is an important part of the intracellular network and the source of information in building new protein molecules. It also functions as a vital source memory for new molecules when the cell "needs" a new function over the course of evolution. DNA, so to speak, fixes and stores "memories" of evolutionary experiences for future generations.

5.4   The Intracellular Molecular Interaction Network

Considering the above arguments, to answer what ultimately defines and controls a cell type, a specific cell function, or phenotype, it appears that they must be laid down in the intracellular network of possible molecular interactions! This "network" emerges from the boundary conditions defined by genes and DNA *as well as* the structural organization of the cell but also independently of these. For example, local activities of the network in different parts and compartments of the cell can operate and interact, without any direct involvement of the genes coding for the proteins active at different locations. Obvious examples are muscle contraction, the circadian rhythm generators, or signaling resulting from photons hitting a rod cell in the retina. The metaphorical comparison of a cell with a city is therefore somewhat more useful and less misleading than the often-used machine or engine metaphor. There are many local activities resulting from the interactions of various social networks (today, real life and web based) in different parts of a city, which appear more or less autonomous. The boundaries (constraints) of these activities are provided by the tangible, physical structure (buildings, workplaces, restaurants, meeting halls, streets, etc.) of the city and its infrastructure, as well as by the "invisible," intangible social rules and norms typical for that specific community, as agreed upon and always dynamically changing and emerging from the interactions within the social network (official, static representations of these are found for example in the local and national laws). Importantly, these local activities are not emerging as a result of the existence of books of law or regulations or because of officials reading the law in the city hall. Equally, the intracellular network only visits the genome and asks for spare parts (proteins) when a local actor is lost, broken down, or needed in order to fulfill a specific task. Thus, the decision making resulting in cellular differentiation/function/phenotype takes place within the *whole of the intracellular network* of which the genome is just one but an important constituent part. As Denis Noble put it, DNA by itself does not program anything, as little as the city law cooks food or plays music in the concert hall.

What do we know about this intracellular network of molecular interactions? Well certainly not enough. Currently, the best available models for it are provided by so-called large, scale-free networks, as developed by complex dynamical systems theory and network theory. These networks are characterized by features, which at least superficially fit with observations made in intracellular molecular interaction networks. They can be extremely large networks but governed by a limited number of simple rules (see also chapter 6, this volume). They can show highly complex collective behavior, despite lacking an obvious central control level or leader entity. Their behaviors are also hard to predict, which are a fundamental characteristic of all complex systems studied so far. They use, produce, and integrate information and signals from both within the network and from the environment. They can change and adapt in order to improve survival or successful operations through learning and evolution and are able to respond quickly to changing external stimuli. Large scale-free networks exhibit nontrivial emergent and self-organizing behaviors.[5] Nodes and hubs are connected by different interactions (e.g., physical contacts or long-range signaling molecules), and the hubs that contact many other nodes and hubs have a tendency to further enhance their number of contacts ("the rich get richer"—a common principle of successful complex networks). In this network model, nodes and hubs correspond in a cell to individual proteins and molecules interacting with each other.

5.5 Meta-Analysis of the Phenotypic Effects of Intracellular Network Interactions Leads to the Concept of Cellular Attractors

The diversity of cell types in a multicellular organism results therefore from nongenetic switching between distinct phenotypes, despite the fact that each cell has the very same genome. The transient dynamic nature of multiple distinct phenotype states within a clonal cell population can be captured by complex nonlinear dynamical systems theory applied to gene regulatory networks (GRNs). A network of genes that influences the expression of each other can assume a very large number of theoretical (combinatorial) gene expression configurations (states of the network). Each one of these combinatorially possible gene expression patterns/profiles can be considered a position, a point, in a coordinate system with n dimensions, where n is the number of interacting genes. The dynamics of GRNs have been investigated by using Boolean algebra simulations as a conceptual model to represent fundamental features in the functionality of real GRNs. These simulations showed that not all states of the system are equally stable (probable to occur) but that some network states, as dictated by the GRN, represent stable, steady states, called the attractor states, to which the dynamics of all possible state changes will be "attracted" toward.

The concept of cellular attractors was originally introduced by Stuart Kauffman,[6] emerging from the application of Boolean algebra to gene regulatory networks. If governed by a few simple rules, the intracellular network of interacting genes or their products ultimately ends up with only a limited number of stable states of gene expression, irrespective of the starting conditions or the infinite amount of theoretically, combinatorially possible states. This limited number of actually realized stable states corresponds to the limited number of real-world cell types, although there may be some additional possible stable states which are not exploited by real-world cell types.

Real gene expression data show that cells representing one cell type will not be positioned in a singular attractor point (in a mathematical sense) but rather be localized within a more or less distributed cloud around a statistical attractor center, due to conspicuous heterogeneity/variability in each biological parameter. Earlier work has demonstrated that stochastic variations of levels of the same protein between cells of the same cell type are a fundamental feature of most cells studied so far.[7] Stochastic fluctuations due to molecular noise in gene expression contribute to the distribution range and shape of the cellular "phenotype cloud." The volume/area of a specific cloud can be visualized as a basin of attraction in a virtual landscape of all cell types.[8] This variability, in itself, together with the attractor represents the "prototypical" phenotype (= mode) of the cell type. Note that this view is in contrast to the conventional model where all cells in the whole population of cells of a given phenotype are assumed to behave in an identical manner as an abstract "average cell." The dynamic attractor concept, on the other hand, may be viewed as a modern version of the classical epigenetic landscape metaphor, intuitively introduced by C. H. Waddington seventy years ago.[9]

5.6   Cell Types According to the Attractor Concept

The distinct cell "types" observed in organisms represent therefore dynamical attractor states: "self-stabilizing" configurations of gene activities across the genome that arise due to the constraints emerging from the collective output of overall, total gene expression, imposed by gene-gene regulatory interactions *and* the overall structure of the GRN.[10] Therefore, also multipotent states or terminal cell types in normal tissues, or the stem-like (tumor-initiating) or metastatic states in cancer, would be such attractor states. Attractor states display robustness against stochastic fluctuations, such that a clonal population of cells appears as a coherent "cloud" of cells when the gene expression pattern of each cell is displayed as a point in a high-dimensional gene expression space.[11] This is the reason why cells can collectively be identified as a distinct phenotype representing what we know as "cell type," despite substantial cell-to-cell variability.

Yet, cells can, in the presence of sufficiently high levels of fluctuations or in response to a deterministic regulatory signal, switch between attractors, that is, undergo phenotype conversions.[12] Since attractor states are the result of distinct configurations of overall gene activity patterns, no genetic mutations at the DNA level with long-lasting effects need to be involved in these phenotype transitions, although mutations can facilitate state transitions by modifying the whole of the attractor landscape.[13]

5.7   Applying the Cell Attractor Model to Cancer Cells:
Experimental Observations

The cell attractor model was originally proposed to explain the maintenance of normal cell phenotypes.[14] The phenotype of a cell population emerges as the result of continuous dynamic state changes over time, depending on continuous variations in the internal molecular network (gene expression levels and patterns, protein interactions, metabolic fluxes, etc.) of single cells.

We and others have shown that any subpopulation of cells isolated and grown separately from a population that shows the original full spectrum of realized expression levels can reestablish the original parental distribution of phenotypic variation within days to a few weeks.[15] This is one predicted outcome from applying the attractor model. Furthermore, any subpopulation of cells remaining in the original, whole "parent" population also changes its expression level in a similar way like cells isolated from this population and grown in a separate new culture. Cells, with an expression phenotype corresponding to a position at the edge of the cloud of realized phenotypes, over time reposition themselves toward the middle of the distribution. Likewise, cells in the center move away from the middle toward the edges of the phenotype distribution, in both cases ending up statistically as being distributed all over the population cloud (thus representing the "average" phenotype). Intercellular signals and cell-cell communication do not seem to be the main determinant for this behavior in our and similar reductionistic experimental systems of cultured, cloned derivatives of cell lines in vitro.[16] It is inevitable that in a more complex environment, such as a tissue or a whole organism, the input of neighboring cell contacts, short-range signaling from the microenvironment, or long-range signaling by hormones resets the boundaries of an exposed cellular attractor, probably redefining both the attractor points and shapes of cell state clouds. The tools have just become available to study this balance between the basic set points of the intrinsic cellular network/GRN and the extent of modification due to its interactions with the microenvironment.

The distribution behavior of "phenotypic" states within a cell population and their relaxation dynamics within this population can be described by using the Fokker-Planck equation. Modeling based on this equation suggested that the dynamics of cell states around a cancer cell attractor is reasonably well represented by a deterministic drift force and a diffusion term representing the major driving forces of cell state dynamics, namely, the regulatory constraints of the GRN and molecular noise.[17]

One important feature of the virtual cell-type attractor landscape is that spaces between basins of attraction ("hills and ridges") are void of cells. These are positions in the state landscape that individual cells are unlikely to take as at these positions outside an attractor basin, the intracellular network state is unstable or does not allow a viable phenotype to exist due to the constraints posed by the existing network interactions.[18] So, is it possible for cells to (occasionally) escape from their native phenotype attractor basin and move beyond its edge to enter another (sub) attractor just due to the stochastic variations in the intracellular network? This would imply a temporary and probably very critical (risky, in terms of cell survival) passage, crossing over the void, or the ridge between two attractor valleys of "impossible" or "not allowed" states in the landscape. It would be an exceptional phenomenon as it would violate the ordered structure of cell phenotypes and tissues in a multicellular organism. Preliminary data from our in vitro cell model system suggest that cells at the edge of the phenotype variation distribution—and probably cells transgressing the ridge between two basins—are fragile, prone to apoptosis, and slowly proliferating. In a highly structured normal, healthy microenvironment, they might merely manage the transition, but just not survive, but in an environment experiencing tissue stress and disorder, such as in low-grade chronic inflammation, they might have a greater chance to survive and to enter a path to premalignancy, as has been also suggested by others.[19] The appearance (although at very low frequency) of cells that enter a neighboring phenotype attractor, which might even be a novel attractor in the landscape and thus physiologically illegitimate, could contribute to the initiation of tumors without the need for initiating mutations. Such a scenario represents an extension of the epigenetic progenitor cancer model suggested earlier by Feinberg et al.[20]

Another important observation in our experimental model was the existence of cells with a nontypical marker expression profile but at a low frequency. From the perspective of the attractor model, these cells could represent a transient shift toward another attractor or an incomplete, partial change of phenotype to a subattractor. When isolated and cultured, outlying edge cells (in relation to the central basin of attraction) in our experimental cell cultures proliferated slowly with a high apoptosis rate but ultimately returned to the parental status. Such cells, when isolated, showed distinct changes in messenger RNA expression patterns involving a large number of genes.

Cell attractors do not exist as such in the real world. They are just one way of representing a cell type, providing some (partly counterintuitive) insights into the behavior of cells in populations. I advocate that these insights about the nature of intercellular microheterogeneities within a cell population of one cell type can be generalized to all cell types in our biosphere.

5.8 Applying the Attractor Model for Finding New Treatment Approaches

With the successful unravelling of basic molecular cell biology, a great leap forward has been made in our understanding of cancer. The concept of "driver" versus "passenger" mutations in cancer biology has led to a new phase in developing drugs for cancer treatment, targeting one or several of the mutated genes found in tumors by a number of different approaches.[21] The development of such precise drugs that are expected to correct or circumvent genetic errors was a great achievement in translating basic science findings into clinical application. While precision cancer medicine (PCM) is still a worthwhile and overall promising approach, the success rates of this type of drugs in the clinic has so far been only modest. As pointed out also elsewhere in this volume, a more comprehensive understanding of the systems biology of cancer will be necessary to improve the treatment concepts and develop more successful ones.[22]

Applying the attractor model to cancer phenotypes may lead to important implications for cancer cell biology and the treatment of cancer. It is quite possible that the impact of the local microenvironment (including local exposures) on the cell network and the resulting phenotype may precede subsequent carcinogenic mutations, thus preconditioning the cell for tumor progression in line with the epigenetic progenitor cancer model mentioned. More frequently, mutated oncogenes or suppressor genes and their aberrant gene products represent the "original sin" in the tumors´ pathogenesis. However, the errors inflicted on the cell by the mutated genes have been compensated for by the resilience of the intracellular network since long in the manifest tumor. The complex cellular network carries the feature to "heal its wounds." Then the effects of these aberrant genes will no longer be instrumental to the phenotype. The progressing cell with mutated genes managed to survive due to adaptive and compensatory mechanisms offered by the intracellular network.[23] These compensatory mechanisms are probably the ones that should be sought out and targeted when conventional treatment approaches have failed.[24] They are likely the same that operate in normal homeostasis to maintain the "healthy" state.

Another important consequence for future cancer therapy can be derived from the attractor model. The wide distribution of single cells within the range of a cancer cell attractor may explain why some cells in the population are resistant to a given drug at a certain time point but not at another time point when they have shifted position in the phenotype state space[25] (see also chapter 10, this volume). When the selective pressures of treatment are relieved after a course of treatment, the original malignant population can be reestablished by a few temporarily resistant edge cells. If a malignant cell clone represents a stochastic distribution of different gene expression states, including rare variants, it may be necessary to identify combinations of treatments that not only reduce the average viability in the population but specifically eliminate rare edge cells that may be transiently drug resistant. This is a challenging complementary view on drug resistance, in addition to the issues of resistant cancer progenitor cells or the appearance of new mutants due to treatment selection.[26]

Can the cell attractor concept be useful for improving or developing new cancer therapies? This can only be the case when the observed behavior of single cells in a cancer cell attractor can be explained in terms of the "real" physicochemical network properties of the intracellular world. I can see no other way than to map out the intracellular network, its nodes and hubs, and the rules and principles for its interactions. In this understanding of the network, new principles can be derived for targeting of pathways and subcircuits of the network, principles currently still beyond our reach. Today, the efforts of Peter Csermely (see chapter 6, this volume) come closest to such an attempt, when he finds that it is more efficient to target neighbors of a strong hub rather than the strong hub itself.[27] Another approach would be to focus on targets/network nodes that are there to stabilize the network, should they exist. The aim would be to demolish (collapse) or freeze (hyperstabilize) the intracellular network, which would result in cell death due to disintegration of the network or to a "locking in" of a specific network state.

In order to be able to pursue such conceptually promising approaches, we need a much better understanding of the highly exceptional *physical* reality of the intracellular world. This requires novel ways to investigate the properties of the physical and chemical conditions inside a cell, which are likely to be quite extreme and very different compared to the everyday physical world and the textbook chemistry of linear, few-step linear reaction pathways. It will be a future challenge to integrate this necessary basic knowledge with that of intracellular/GRN network modeling that also incorporates nonlinear dimensions. For it to be productive, this challenge will require close collaboration between cell and cancer biologists, system biologists, computer scientists, biochemists, chemists, physicists, and mathematicians.

Acknowledgments

This author is supported by the Swedish Cancer Society, the Childhood Cancer Foundation, Radiumhemmets Research Funds, and Karolinska Institutet. The author is indebted to Qin Li, Harvard Medical School; Erik Aurell, Royal School of Technology, Stockholm; Anders Wennborg and Jie-Zhi Zou, Karolinska Institutet; and Sui Huang, ISB, Seattle, for collaborations on these topics and discussions leading to this review.

Notes

1. Brenner S & Wolpert L. 2001. *My Life in Science*. BioMed Central Ltd. ISBN-10: 0954027809.

2. Wolfram S. 2002. *A New Kind of Science*. Wolfram Media. ISBN 1-57955-008-8.

3. Bard J, Melham T, Werner E, & Noble D. 2013. Plenary discussion of the conceptual foundations of systems biology *Prog Biophys Mol Biol* 111:137–140.

4. Crick F. 1982. *Life Itself: Its Origin and Nature*. McDonald & Co, London and Sydney.

5. Mitchell M. 2011. *Complexity: A guided Tour.* Oxford University Press. ISBN-10: 0199798109.

6. Kauffman S. 1969. Metabolic stability and epigenesis in randomly constructed genetic nets. *J Theoretical Biol* 22(3):437–467. doi:10.1016/0022-5193(69)90015-0. PMID 5803332.

7. Sigal A, et al. 2006. Variability and memory of protein levels in human cells. *Nature* 444(7119):643–646.

8. Sisan DR, Halter M, Hubbard JB, & Plant AL. 2012. Predicting rates of cell state change caused by stochastic fluctuations using a data-driven landscape model. *Proc Natl Acad Sci U S A* 109(47):19262–19267.

9. Waddington CH. 1940. *Organisers & Genes*. Cambridge University Press. Huang S. 2011a. Systems biology of stem cells: three useful perspectives to help overcome the paradigm of linear pathways. *Philos Trans R Soc Lond B Biol Sci* 366(1575):2247–2259.

10. Huang S & Kauffman S. 2009. Complex gene regulatory networks—from structure to biological observables: cell fate determination. *Encyclopedia of Complexity and Systems Science*, ed Meyers RA (Springer), pp 1180–1213.

11. Huang S. 2011. Systems biology of stem cells: three useful perspectives to help overcome the paradigm of linear pathways. *Philos Trans R Soc Lond B Biol Sci* 366(1575):2247–2259.

12. Kalmar T, et al. 2009. Regulated fluctuations in nanog expression mediate cell fate decisions in embryonic stem cells. *PLoS Biol* 7(7):e1000149. Munoz-Descalzo S, de Navascues J, & Arias AM. 2012. Wnt-Notch signalling: an integrated mechanism regulating transitions between cell states. *Bioessays* 34(2):110–118.

13. Huang S. 2011. On the intrinsic inevitability of cancer: from foetal to fatal attraction. *Semin Cancer Biol* 21(3):183–199.

14. Kauffman S. 1993. *Origins of Order: Self-Organization and Selection in Evolution*. Oxford University Press.

15. Chang HH, Hemberg M, Barahona M, Ingber DE, & Huang S. 2008. Transcriptome-wide noise controls lineage choice in mammalian progenitor cells. *Nature* 453(7194):544–547. Gupta PB, et al. 2011. Stochastic state transitions give rise to phenotypic equilibrium in populations of cancer cells. *Cell* 146(4):633–644. Li Q, et al. 2016. Dynamics inside the cancer cell attractor reveal cell heterogeneity, limits of stability, and escape. *Proc Natl Acad Sci U S A* 113(10):2672–2677.

16. Andrecut M, Halley JD, Winkler DA, & Huang S. 2011. A general model for binary cell fate decision gene circuits with degeneracy: indeterminacy and switch behavior in the absence of cooperativity. *PLoS ONE* 6(5):e19358.

17. Li Q, et al. 2016. Dynamics inside the cancer cell attractor reveal cell heterogeneity, limits of stability, and escape. *Proc Natl Acad Sci U S A* 113(10):2672–2677.

18. Chang HH, Hemberg M, Barahona M, Ingber DE, & Huang S. 2008. Transcriptome-wide noise controls lineage choice in mammalian progenitor cells. *Nature* 453(7194):544–547.

19. Huang S. 2011. On the intrinsic inevitability of cancer: from foetal to fatal attraction. *Semin Cancer Biol* 21(3):183–199. Andrecut M, Halley JD, Winkler DA, & Huang S. 2011. A general model for binary cell fate decision gene circuits with degeneracy: indeterminacy and switch behavior in the absence of cooperativity. *PLoS ONE* 6(5):e19358.

20. Feinberg AP, Ohlsson R, & Henikoff S. 2006. The epigenetic progenitor origin of human cancer. *Nat Rev Genet* 7(1):21–33.

21. Hanahan D & Weinberg RA. 2011. Hallmarks of cancer: the next generation. *Cell* 144(5):646–674. Bernards R. 2012. A missing link in genotype-directed cancer therapy. *Cell* 151(3):465–468.

22. Huang S. 2012. Tumor progression: chance and necessity in Darwinian and Lamarckian somatic (mutationless) evolution. *Prog Biophys Mol Biol* 110(1):69–86.

23. Gupta PB, et al. 2011. Stochastic state transitions give rise to phenotypic equilibrium in populations of cancer cells. *Cell* 146(4):633–644. Roesch A, et al. 2010. A temporarily distinct subpopulation of slow-cycling melanoma cells is required for continuous tumor growth. *Cell* 141(4):583–594. Csermely P, Hódsági J, Korcsmáros T, Módos D, Perez-Lopez AR, Szalay K, Veres DV, Lenti K, Wu LY, & Zhang XS. 2015. Cancer stem cells display extremely large evolvability: alternating plastic and rigid networks as a potential mechanism: network models, novel therapeutic target strategies, and the contributions of hypoxia, inflammation and cellular senescence. *Semin Cancer Biol* 30:42–51.

24. Pisco AO, et al. 2013. Non-Darwinian dynamics in therapy-induced cancer drug resistance. *Nat Commun* 4:2467.

25. Roesch A, et al. 2010. A temporarily distinct subpopulation of slow-cycling melanoma cells is required for continuous tumor growth. *Cell* 141(4):583–594. Pisco AO, et al. 2013. Non-Darwinian dynamics in therapy-induced cancer drug resistance. *Nat Commun* 4:2467. Schadt EE, Friend SH, & Shaywitz DA. 2009. A network view of disease and compound screening. *Nat Rev Drug Discov* 8(4):286–295. Quintana E, et al. 2010. Phenotypic heterogeneity among tumorigenic melanoma cells from patients that is reversible and not hierarchically organized. *Cancer Cell* 18(5):510–523. Holzel M, Bovier A, & Tuting T. 2013. Plasticity of tumour and immune cells: a source of heterogeneity and a cause for therapy resistance? *Nat Rev Cancer* 13(5):365–376.

26. Bernards R. 2012. A missing link in genotype-directed cancer therapy. *Cell* 151(3):465–468.

27. Csermely P, Hódsági J, Korcsmáros T, Módos D, Perez-Lopez AR, Szalay K, Veres DV, Lenti K, Wu LY, & Zhang XS. 2015. Cancer stem cells display extremely large evolvability: alternating plastic and rigid networks as a potential mechanism: network models, novel therapeutic target strategies, and the contributions of hypoxia, inflammation and cellular senescence. *Semin Cancer Biol* 30:42–51.

6 Adaptation of Molecular Interaction Networks in Cancer Cells

Peter Csermely

Overview

Network theory as part of complex systems theory has over the past decades developed powerful tools to analyze and make sense of the enormous amounts of molecular interaction data obtained by modern molecular biology methods at the DNA, RNA, and protein levels. These complex network models of intracellular molecular interactions (from here on: cellular networks) can reveal aspects of cellular function and overall molecular interaction behavior that cannot be described by other means. We will give here a brief introduction into the general logic of such network models and how we can gain novel, biologically, and medically relevant insights from applying them to the cancer context. This chapter describes network-based adaptive mechanisms that bring about the "creativity" of cancer cells to survive and expand in the unpredictable, often hostile, environments of tumor tissues. First, we describe the dominance shift from "business-as-usual" processes driven by the core of cellular molecular interaction networks to changes in the network periphery that lead to "creative" shortcuts between distant network regions and thus allow the network to respond to novel challenges.[1] This forms a general adaptation/learning mechanism that characterizes the initial stages of cancer development.[2] Such adaptive changes may change the topology of cellular networks from a rigid to a plastic state. Rigid networks have a dense core, disjunct modules (network groups), prominent hierarchy, low network entropy and so-called sink dominance. Rigid networks have only a few dominant attractors (i.e., stable states to where the network converges). Plastic networks have a fuzzy core, overlapping modules, less hierarchy/more loops, high network entropy, and source dominance. Plastic networks have many attractors, which are often dispersed. Alternating changes of network plasticity and rigidity help to encode novel information into the network structure, thus remodeling the network core and developing novel system attractors.[3] Cancer stem cells are characterized by exceptionally high evolvability involving rapid alternations between plasticity and rigidity.[4] Plastic and rigid networks (characterizing early and late-stage tumors, respectively) require conceptually different drug design strategies. Plastic networks (which dissipate stimuli very well) should only be attacked with a "central hit," targeting hubs, bridges, and bottlenecks. If they were attacked at the network periphery, the effect of the drug would never reach the center of the network due to efficient stimulus dissipation. In contrast, rigid networks (which transmit stimuli without much dissipation) may become "overexcited" by "central hit" attacks, leading to unwanted side effects such as adverse drug reactions. Rigid networks require the "network influence drug design strategy" targeting the neighbors of their hubs and central nodes.[5] "Network influence targeting" of neighbors of key network nodes increases the precision of the intervention by targeting only certain functions of the key, neighboring network node. The chapter will conclude with the outline of network dynamics-based, personalized multitarget drug design strategies as a promising perspective for future therapies.

6.1 Network Science Provides Important Insights into Complex Cell Behaviors, Including Cancer

6.1.1 Definition of Cellular Networks

The term "cellular networks" encompasses many types of interaction networks inside a cell, such as protein-protein interaction networks (interactomes), signaling networks, gene transcription networks, and metabolic networks. Recently, additional types of intracellular networks have also been outlined, such as cytoskeletal networks, cellular organelle networks, and chromatin networks. However, currently we do not have enough information on most of these latter networks to include them into a detailed analysis of network adaptation processes of cancer cells.[6] Importantly, a rapidly emerging area of network science is the assessment of intercellular networks, which gives insight into the interactions of heterogeneous cancer cell types within the tumor, stromal cells, and immune cells.[7] The analysis of these networks has to date not yet yielded enough information to be included into this review, but their adaptation processes are an exceptionally interesting area of future study.

6.1.2 The Core-Periphery Learning Mechanism in Biological Systems

Three discoveries in the field of complex systems theory provided important insights into general adaptation mechanisms of complex systems.

a) Network core and periphery distinction. Starting with the work of Steve Borgatti and Martin Everett in 1999, a number of studies showed that most complex networks can be dissected down to a core and a periphery.[8] The network core refers to a central and densely connected set of a few network nodes, where connection density is often increased further by large edge (i.e., network node interaction) weights, which reflects the larger probability of the functional use of these interactions at the center of the network. In contrast, the network periphery consists of nodes that are noncentral, are sparsely connected, and attach preferentially to the core.[9] Importantly, some networks that have a well-developed modular structure and small module overlap possess multiple cores, which correspond to the cores of their modules; such module cores can be defined by several algorithms.[10] Nodes of a network core are (evolutionarily) conserved and shielded from the environment of the network by the periphery.[11] Peripheral nodes are often sources of innovation, since they have a large degree of freedom (which is, for example, described in social networks as a lack of social pressure[12]).

b) Attractors of complex systems are deepened by learning. In 1969, Stuart Kauffman described that random genetic control networks develop a surprisingly small number of attractors.[13] Later studies of William Little, Gordon Shaw, and John Hopfield showed that attractors are deepened, or stabilized, during learning processes of networks of real or artificial neurons.[14]

c) Attractors of complex systems are encoded by core nodes of their network representation. Recent studies by Reka Albert, Bernold Fiedler, Atsushi Mochizuki, and their coworkers showed that attractors are encoded by overlapping node subsets of the strongly connected network region, which is the core of directed, bowtie networks.[15] These node subsets are both necessary and sufficient to determine the given attractor of the network.

6.1.2.1 The core-periphery learning theory

Based on the above three key observations and on several other studies described in Csermely,[16] the following general core-periphery learning theory for complex networks was conceived (figure 6.1). In most cases, a stimulus first affects peripheral nodes, since they are much more numerous than core nodes, and core nodes are often shielded by peripheral nodes from the network environment. The stimulus propagates then rapidly from the periphery to the core, since peripheral nodes are preferentially connected to core nodes. Once the stimulus signal has reached one node within the network core, it becomes rapidly shared/distributed throughout the entire core of the network by a fast process, since core nodes are densely connected, and their connecting edges (interaction paths) have a large weight, or "importance" or "preference," compared to other interactions (see the solid lines of figure 6.1).

After these initial steps, one of the following three scenarios may happen.[17]

Scenario 1. Activation of a previously encoded attractor. If the incoming stimulus had been experienced by the complex system several times before, a subset of core nodes has already formed a subgroup within the core that is even more densely interconnected than the rest of the core and also has already well-established connections with the "sensory" nodes of the periphery. This subgroup of network core nodes drives the complex system quickly to an attractor (outcome system response) that gives an adequate response to the formerly experienced stimulus. If now the same stimulus is repeated again, it is channeled immediately to this subgroup of core nodes, which drive the system to the very same attractor/response (figure 6.1A). This mechanism results in a fast, reliable, and robust response of the whole complex system.[18]

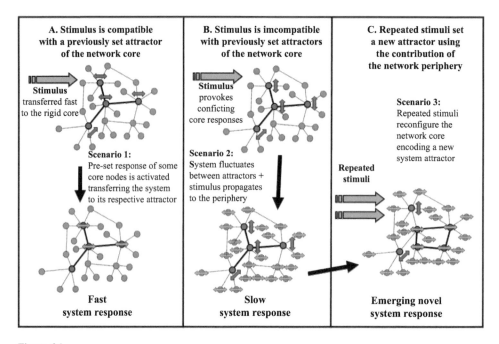

Figure 6.1
Description of the core-periphery learning mechanism of complex systems. (A) *Scenario 1.* The stimulus is rapidly channeled to the rigid core of the network (*dark gray nodes*) as a result of the central position of the core. It becomes "instantly" shared by core nodes due to their dense connections having large edge weights (*solid lines*). The stimulus (*large arrow*) is compatible with a previously set attractor of the complex system. This attractor is encoded by a subset of the core nodes (*top figure, two-headed arrows*) and provokes a fast, matching response (*bottom figure, two-headed arrows*), which rapidly dissipates the signal. (B) *Scenario 2.* The stimulus is incompatible with previously set attractors of core nodes (*highlighted*), provoking a fluctuation between attractors (*vertical two-headed arrows*). Consequently, the stimulus has enough time to spread back to the network periphery (*nonhighlighted nodes*), where it induces a slow, system-level, integrative response (*bottom figure horizontal two-headed arrows*). Here, a collective decision of the entire network emerges in a selection process of many, mostly stochastic steps (1). (C) *Scenario 3.* Repeated novel stimuli reconfigure the core (*highlighted nodes*) encoding a new system attractor (*horizontal two-headed arrows*). Reproduced with permission from Csermely.[19]

Scenario 2. Initial development of a new attractor. If the stimulus is a consequence of a novel, unexpected situation (figure 6.1B), it may be incompatible with any of the existing attractors encoded by the current core of the complex network. As a consequence, this novel stimulus may provoke conflicting core responses, inducing the complex system to fluctuate between its original attractors. This prolongs the time during which the stimulus cannot be dissipated by the system. During this extended period of time, the stimulus may have the chance to propagate back to the weakly connected peripheral nodes of the network, which form the majority of nodes in most networks and are not connected to each other, and therefore can only be accessed via the core. This process stabilizes the system (by modifying the position, size, saddles, or depth of the attractor basins in the complex system). The emergent periphery response is usually slow. This is partly because the reorganization of the periphery requires a large number of rather slow, mostly stochastic steps.[20] A key example of such a "learning step" of a complex system is the case of a "creative node,"[21] which has a dynamic position in the network (often acting as a "date hub"[22]), and creates a shortcut between previously distant network regions, allowing an entirely novel combination of the information previously encoded in these network regions.[23] In addition, the emerging system response is slow, because stimulus-driven periphery reorganization must often be attempted hundreds (if not thousands) of times before a new, adequate response is found.[24]

Scenario 3. Stabilization and encoding of the new attractor. In case the novel stimulus is repeated (many times), the peripheral network nodes, which were involved in "Scenario 2," may gradually reconfigure the network core adding nodes to it or exchanging its nodes (figure 6.1C). This process encodes the newly acquired response as a novel attractor of the system. Core reconfiguration may weaken or erase some of the earlier system attractors and thus may also serve as a "forgetting"/"erasure" mechanism.[25]

6.1.2.2 The core-periphery learning mechanism characterizes a wide range of complex systems

The core-periphery learning theory described above characterizes adaptation processes of a wide range of complex systems from protein structures to social networks.[26] In case of proteins, a rigid core is often surrounded by intrinsically disordered protein domains, which may become at least partially ordered during signaling processes when interacting with other proteins, hence forming a "conformational memory," which represents a learning process at the molecular level.[27] Individual cells may "learn" by the modification of signaling pathway dynamics and, most important, by developing epigenetic, chromatin memory.[28] (It will be a question of future studies whether these changes are initiated in the periphery of the signaling network and become part of its core.) Metabolic networks possess a reaction core containing all essential biochemical processes and have a large, adaptive periphery, which is switched on and off by transcriptional and regulatory processes driven by the flow of nutrients and emerging needs of the cell or by its environment.[29] In neuronal networks, peripheral nodes are becoming core nodes during the learning process. In social groups, "peripheral" individuals are not belonging to the social "elite," are free of social pressure, do not have the intrinsic need to maintain the "status quo," and thus may often become innovators. The collective action of peripheral (not well-connected) individuals is often called the "wisdom of crowds."[30]

6.1.2.3 The core-periphery learning theory applied to cancer

To date, we have already a number of good examples demonstrating that cell behaviors as described by the core-periphery learning theory may also drive the development of cancer. The following observations support this notion. Determinant nodes of the attractors of the epithelial-mesenchymal transition reside in the strongly connected region of the dynamic signaling network describing this process.[31] Expression patterns of the strongly connected region of microRNA-mediated intergenetic networks had an efficient prognostic potential for breast and colorectal cancer patients.[32] A recent study highlighted the importance of the first and second neighbors of cancer-related proteins in cancer development and their potential role in therapeutic approaches.[33] For example, this study could show that first neighbors of cancer-related (i.e., mutated or differentially expressed) proteins in interactomes and signaling networks have a degree of between-ness, centrality, and clustering coefficient at least as high as cancer-related proteins themselves, indicating a previously unknown central network position. Furthermore, there are 223 marketed drugs already targeting first neighbor proteins but applied mostly outside oncology, providing a potential list for drug repurposing against solid cancers.[34] It will be a task of further studies to prove or refute whether peripheral nodes of protein-protein interaction, signaling, or metabolic networks play a distinctive role in the development of novel responses of cancer cells to carcinogenic stimuli, stressful changes in the microenvironment, or cancer drugs.

6.1.3 Alternating Network Plasticity and Rigidity as a Hallmark of Cancer Cells

Complex systems often reside in one of two major configurations: either plastic or rigid. Plasticity and rigidity may be defined as a functional term of the complex system and as a structural term of the network description of the complex system. Functional and structural plasticity and rigidity are not describing the same phenomenon, but they largely correlate in their occurrence.[35] In the following, I will first introduce some general concepts concerning plastic and rigid networks and then give some examples of how these may apply to cancer.

6.1.3.1 Differences between functionally rigid and plastic complex systems

Functionally rigid systems have only few attractors, typically only one, and therefore have a very rough attractor landscape. (A rough attractor landscape means a set of attractors that are separated by large barriers.) A rigid object, such as a porcelain vase, is not able to change its state, unless it breaks, where this noncontinuous, nondifferentiable transition forms an entirely different system. In contrast, a functionally plastic system has a large number of attractors often associated with a smooth attractor landscape. (In a smooth attractor landscape, attractors are separated by very small barriers.) A plastic object, such as a paper clip, may adopt a large number of configurations, without an abrupt change. Consequently, rigid systems have a very poor adaptation (learning) potential, but they have an extremely good "memory" performing their dedicated task(s) with high precision and efficiency. On the contrary, plastic systems have an extremely good adaptation (learning) potential but have a very poor "memory," so they can perform specific tasks with only a low precision and efficiency.[36]

6.1.3.2   Differences between structurally plastic and rigid networks

Structurally plastic networks often have an extended, fuzzy core, where the network core cannot be easily demarcated and often contains most of the network nodes (instead of only a few). Plastic networks have fuzzy modules that also overlap to a large extent. Usually, plastic networks have little hierarchy, have more loops, and, if they have directionality, are source dominated. In contrast, structurally rigid networks have a small, dense core and disjoint, tightly organized, dense modules. Rigid networks are characterized by a strong hierarchy and, if they have directionality, by sink dominance (figure 6.2[37]). In summary, plastic networks are periphery dominated, while rigid networks are core dominated. This is in good agreement with the finding that network attractors are encoded by core nodes,[38] since the small and well-organized core of rigid networks encodes only a few attractors, where these attractors can be reached with a high probability and provide an optimized, highly efficient response. Plastic networks, on the other hand, have a large number of poorly defined attractors, which are encoded by a large number of poorly discriminated core nodes.

The novel stimulus, described in scenario 2 above, may "melt"/restructure part of the network core by decreasing the core edge weights. Note that this will also decrease the core rigidity, which leads to the destabilization of the original attractors and to an increase of learning potential to develop new attractors. Plastic network configurations can be induced and maintained by "soft spots," that is, nodes that are highly dynamic and have multiple, weak connections (figure 6.2). Note that these "soft spots" are the same as the "creative nodes" mentioned in scenario 2 above and are the ones that have a dynamic position in the network and can create shortcuts between previously distant network regions, allowing an entirely novel combination of nodes to encode the same information that was earlier already encoded in these same network regions.[39]

If the novel stimulus is repeated, as described in scenario 3 above, it may encode a novel set of constraints into the network structure, establishing a new region of the network core. This core extension makes the network more rigid again.[40] These rigid network configurations can be induced and maintained by "rigidity seeds," that is, nodes that increase the size of densely connected network clusters, for example, by completing a larger complete subgraph (i.e., clique, where every node is interacting with every other node) in the network or by joining two densely connected network regions (figure 6.2).

6.1.3.3   Alternating plasticity and rigidity is a general adaptation mechanism

Plastic-rigid transitions characterize a large number of complex systems from protein structures to social networks. An example of a protein-level alternation between plastic and rigid states are molecular chaperones that have an ATP (adenosine-tri-phosphate) hydrolysis-driven "chaperone-cycle," where they help the refolding of misfolded proteins by the physical extension of misfolded proteins, which is followed by their release from the chaperone cage. In their extended form, misfolded proteins become rigid, while after release, they are plastic again. If the misfolded protein folds to its native conformation, it

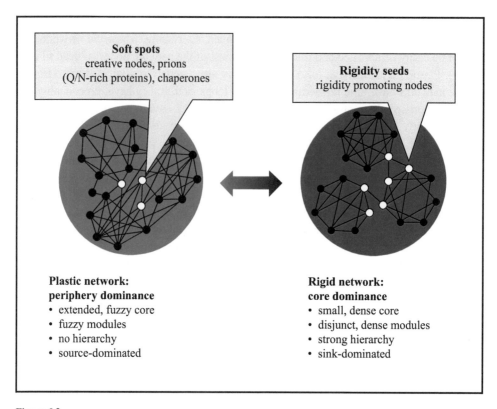

Figure 6.2
Properties of plastic and rigid networks. Network structures may adopt structurally plastic or rigid[41] network configurations. Plastic networks often have an extended, fuzzy core, where the network core cannot be easily discriminated and the core often contains most of the network nodes (instead of only a few). In addition, plastic networks have fuzzy modules with a large overlap. Usually, plastic networks display little hierarchy, have more loops, and, if they are directed, are source dominated.[42] In contrast, rigid networks have a small, dense core and disjoint, tightly organized, dense modules. Rigid networks are characterized by a strong hierarchy and, if they have directionality, by sink dominance.[43] In summary, plastic networks are periphery dominated, while rigid networks are core dominated. Plastic network configurations can be induced and maintained by "soft spots," that is, nodes that have highly dynamic and multiple, weak connections, such as creative nodes[44] exemplified by molecular chaperones, prions, or prion-like, Q/N-rich proteins.[45] In contrast, rigid network configurations can be induced and maintained by "rigidity seeds," that is, nodes that increase the size of densely connected network clusters, for example, by completing a larger complete subgraph (clique) in the network or by rigidly joining two densely connected network regions.

becomes more rigid, since it is stabilized in one conformation (attractor) instead of the competing many conformations (attractors) of the misfolded, at least partially disordered state. Such chaperone-driven extension-release (rigidity-plasticity) cycles iterate until the misfolded protein is refolded correctly again or becomes discarded by proteasomal degradation.[46]

Cell differentiation progresses via an initial "disorganization" phase of the gene expression networks in progenitor cells. This can be measured by (a) measuring the similarity of gene expression profiles using symmetrized Kullback-Leibler distances, (b) applying a hierarchical clustering algorithm and calculating the giant component of the network, and (c) comparing the size of the giant component to that of the complete gene expression network. This final measure shows the level of organization of transcriptional processes. The initial "disorganization" is followed by the development of the much more "organized" gene expression network of the differentiated cell. In agreement with a transient increase of system plasticity during the cell differentiation process, the heterogeneity of the cell population increases considerably after the start of the differentiation process compared to the progenitor cells. As differentiation advances further, the heterogeneity of the cell population then markedly decreases, usually much below that observed within the progenitor cell population.[47] In addition, most terminally differentiated cells are highly specialized, which means they have often simple, hierarchical, rigid networks.

There are several other studies showing that plasticity-rigidity changes within neuronal networks can be observed during a large number of learning processes, such as bird song learning or infant speech learning. Human creativity consists of alternating "blind variation" and "selective retention" processes corresponding to more plastic and rigid neuronal states, respectively. Plasticity-rigidity cycles also characterize organizational learning processes.[48]

6.1.3.4 Plasticity-rigidity changes in carcinogenesis and cancer progression

The initial stages of cancer are characterized by an overall increase in network entropy of cellular networks[49] due to an increased number of stochastic processes (noise[50]) and loops,[51] as well as by increased phenotypic plasticity.[52] All these changes contribute to the increase of cellular phenotypic heterogeneity of cancer cells within a developing tumor (figure 6.3[53]). Higher-degree entropy of signaling networks was found to correlate with lower survival of prostate cancer patients.[54] A detailed investigation of normalized local and intermodular signaling network entropies revealed increased entropies in benign adenomas when compared to that of healthy colon epithelial cells. Importantly, colon carcinoma cells showed decreased entropies when compared to that of benign adenoma cells.[55] Similar changes showing transiently higher entropy were observed in early stage B-cell lymphoma and early hepatocellular carcinoma,[56] as well as in the more plastic, early stage proliferative phenotypes, compared to lower entropy levels in gene expression signatures of remodeling phenotypes of various late-stage cancer types.[57] This is a pattern of changes in system disorder remarkably similar to the one observed during cell differentiation.[58] Cells, which start from healthy attractors, develop during cancer initiation and progression a specific set of attractors, called "cancer attractors."[59] The change of the attractor landscape from the initially, relatively "rough" surface, which defines the healthy attractor(s), through a much "smoother" attractor landscape, where novel attractors arise and/or may become accessible, to the final stage of advanced tumors, where a well-developed and relatively stable set of cancer attractors becomes occupied and stabilized, corresponds very well to the observed increase and then decrease of signaling network entropy.[60]

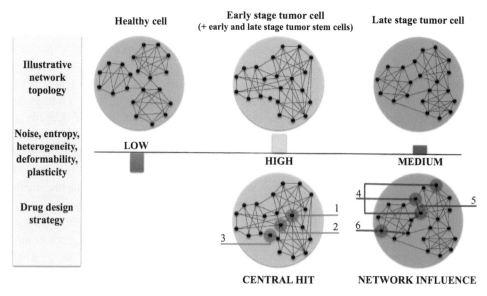

Figure 6.3

Conceptual summary: Cancer as an adaptation process of increasing and decreasing plasticity that requires corresponding, different drug-targeting strategies. The figure summarizes literature data[61] showing that cancer progresses by an initial increase of system plasticity followed by a late-stage decrease of plasticity. The plastic to rigid transition of network structure during cancer development requires distinctly different drug-targeting strategies in early versus late tumors. While at the early phase of carcinogenesis, central hits of hubs ("1"), intermodular bridges ("2"), or bottlenecks ("3") may be a winning strategy, at later stages of cancer progression, the more indirect means of a "network influence strategy," such as multitarget ("4"), edgetic ("5"), or allo-network drugs ("6"), should be used.[62] Unfortunately, most anticancer drug tests use cancer cell lines that have more plastic networks resembling those of the "early stage tumor-like" cells, while most patients are diagnosed with late-stage tumors with rigid cellular networks. Importantly, the heterogeneous cell populations of tumors may harbor early and late-stage cells at the same time.[63] Moreover, cancer stem cells may have the ability to change their plasticity from that of early to late-stage tumor cells and vice versa.[64] Therefore, multitarget, combinatorial, or sequential therapies using both central hit– and network influence–type drugs may provide a promising therapeutic modality. (Reproduced with permission from Gyurkó et al.[65])

6.1.3.5   Cancer stem cell–like cells display a high degree of evolvability of plasticity/rigidity changes

Cancer stem cells possess the capacity to self-renew and to repeatedly rebuild the heterogeneous lineages of cancer cells that comprise a tumor in a changing tumor microenvironment. They can assume both plastic and rigid network structures and cellular phenotypes. The plastic phenotype is rapidly proliferating and characterized by symmetric cell divisions. The rigid phenotype is characterized by not so frequent, asymmetric cell divisions and by increased invasiveness. A highly increased ability of plasticity modulation (which results in an increased level of evolvability) may prove to be a major discriminatory hallmark of cancer stem cells. In cancer development, cancer stem cells are repeatedly selected for high evolvability and become "adapted to adapt." Importantly, this increased plasticity modulation ability may be a key reason why anticancer therapies often induce new cancer stem cells, instead of killing or transforming them. Such behavior was observed in non–small cell lung carcinoma after conventional chemotherapy (paclitaxel) or targeted therapies (Erlotinib) in breast cancer after taxane or anthracycline treatment or in hepatocellular carcinoma after carboplatin treatment. In these examples, network plasticity was increased by induction of specific transcription factors, such as Sox2 or Oct3/4, SRC, or IGFR signaling and epigenetic changes.[66]

6.2   Different Drug Design Strategies Are Required against Early and Late-Stage Tumors

Plastic and rigid networks require completely different drug-targeting strategies. Plastic networks have a rich and rather undifferentiated contact structure, which is able to dissipate "unexpected" external stimuli rather well. Drug treatment can be perceived as an "unexpected" intervention toward which the cancer cell has not developed an adequate response yet. Targeting noncentral nodes in a periphery-dominant plastic network would result in a fast dissipation of the intervention. Thus, plastic networks require a "central hit" that targets their central nodes, such as hubs, intermodular bridges, or bottlenecks (see labels 13 in figure 6.3, respectively). Rapidly dividing bacteria are typical examples of more plastic cellular networks. Not surprisingly, many antibiotics target central nodes of bacterial networks (with the notable exception of "choke point drugs," which target enzymes producing a key molecule for bacterial survival).[67] Rapidly proliferating cells of early stage cancers, as well as the rapidly proliferating, symmetrically dividing phenotype of cancer stem cells, have plastic networks, since the continuous changes of rapid cell division can be more adequately served by a contact-rich, noncentralized network structure. But it is also possible that rapid cell division is often a *consequence* of rapid changes in the tissue microenvironment that may initiate the switching to a plastic network structure; once network plasticity is increased, then more proliferation becomes possible as restrictive context is reduced. It can then become one further strategy to adapt by generating more cells that produce a microenvironment that supports cancer cell survival. Thus, "central hit"–type drugs may be more efficient against plastic phenotypes of cancer cells of early stage tumors. In agreement with the "central-hit" strategy, targets of anticancer drugs are often hubs.[68] Moreover, intermodular interactome hubs were found to associate more often with carcinogenesis than intramodular hubs.[69]

Rigid networks have a well-differentiated, centralized, hierarchical, modular structure, which is specialized to perform certain functions very efficiently. Rigid structures do not dissipate unexpected, random signals very well, since they were optimized for the rapid and efficient dissipation of only certain, previously experienced signals. As a consequence, rigid structures transmit signals rather well. This may cause rigid networks that are exposed to "central hits" to "overshoot," whereby not only the intended reaction but also unintended side effects may emerge. Cells forming a stable, cooperating community, such as cells of a tissue, have most of the time rigid networks. This makes the network influence strategy a key strategy in most diseases, such as diabetes or neurodegenerative diseases. As an example, the p62/SQSTM1 protein, which is a neighbor of raptor, the regulatory protein of mTOR (the mammalian target of rapamycin protein complex), is emerging as a novel target in both diabetes and cancer.[70] Late-stage tumors contain often "highly experienced cells," which have already been organized as a part of a community either in the original tumor or in metastases. The "overshoot" of "central-hit" targeting of cancer cells with rigid networks may result in the secretion of molecules that increase drug resistance of neighboring cells or cause necrosis instead of apoptosis, inducing various survival programs in their neighboring cells. Thus, instead of "central hits," the more indirect means of the "network influence strategy" should be used when targeting the rigid networks of late-stage tumors. The "network influence strategy" may target (first or second) neighbors of key network nodes.[71] Drugs for such a targeting method have been called "allo-network drugs" (label 6 in figure 6.3). This may allow the excitation of only a subset of the signaling pathways related to the central network node, which gives a much larger specificity to the intervention. (Such fine-tuning is close to impossible in extremely plastic networks, where the rich interaction structure channels the intervention to any direction, and thus the "fine-tuned" intervention becomes soon dissipated). "Network-influence targeting" may also be achieved by multitarget or combination therapies, which may use a combination of submaximum doses and may reach their goal by superimposing two (or more) actions at specific nodes of the network in a specific way, mobilizing again only a subset of the signaling pathways related to that particular node (label 4 in figure 6.3). Both neighbor-targeting and combination targeting may actually behave as "edgetic drugs" (label 5 in figure 6.3), which are targeting not an entire node but only one of its interactions, that is, one edge (interaction) of the signaling network. Edgetic targeting was used in the case of the superhub mTOR[72] or by inhibiting the p53/MDM2 connection by nutlins.[73] Neighbors of cancer-related proteins were found as widespread targets of drugs mainly used in diseases other than cancer and were suggested as candidates for potential repurposing efforts.[74] Several candidates for potential combination therapies were initially discovered by network-based identification.[75]

Most anticancer drug tests are currently performed on cancer cell lines, which are rapidly proliferating cells that have developed a plastic network, and from this point of view resemble more the phenotype of early stage tumors. Unfortunately, most patients are first diagnosed with late-stage tumors that have already more rigid cellular networks. Importantly, the heterogeneous cell populations of tumors may at the same time harbor cells that have both plastic and rigid networks.[76] Moreover, as described in the previous section, cancer stem cells have the ability to change their networks from a plastic to a rigid state and vice versa.[77] Cancer stem cells follow Nietzsche's proverbial saying "what does not kill me makes me stronger." Thus, conventional anticancer therapies may actually provoke cancer stem cell development.[78] In such scenarios, multitarget, combinatorial, or sequential therapies using both central hit– and network influence–type drugs may provide a more promising therapeutic modality.

6.3   Conclusions and Perspectives: Toward Personalized Drug Design Based on Insights from Network Science

This chapter introduced two key network-based cellular adaptation mechanisms that might play important roles in cancer. Both mechanisms modulate the evolvability of cancer cells to help their survival in an unpredictable tumor tissue environment. The first network-based adaptation mechanism is based on the "core-periphery learning theory."[79] Responses to previously experienced stimuli are encoded by node sets in the core of the network, while peripheral nodes are needed to "invent" novel responses to unexpected environmental changes. Thus, peripheral nodes are expected to play a major role in early stages of cancer development. Late-stage tumor cells may have already encoded several successful survival mechanisms into the core of their networks. The exploration of these ideas needs future studies. The second network-based adaptation mechanism was the alternation between plastic and rigid network states.[80] Alternating changes in network plasticity and rigidity help to encode novel information to the network structure by remodeling the network core and therefore developing novel system attractors. Cancer stem cells utilize this mechanism to develop and maintain an exceptionally high evolvability.[81]

Importantly, plastic and rigid networks (mainly characterizing early and late-stage tumors[82]) require conceptually different drug design strategies. Plastic networks require "central hits" targeting their hubs, bridges, and bottlenecks. On the contrary, rigid networks require the "network influence drug design strategy" targeting the neighbors or edges of their hubs and central nodes.[83]

Although the above suggestions have been formulated as a result of integrating a number of individual experimental studies listed in references 1 through 6, they require further experimental evidence to establish their precise limits and possibilities. A few of these important areas of future research are as follows:

1. Further studies are needed to characterize the core-periphery mechanism and plastic/rigid alternations of progressing cancer cell and cancer stem cell networks.

2. Systematic studies must show differences in the efficacy of targeting various network positions in early and late-stage tumors.

3. Systemwide studies (such as total gene expression/complete exome data preferably at single cell level) are needed to clarify the network consequences of multitarget, combinatorial, or sequential targeting therapies.

4. The above areas require extension to intercellular network interactions as to date only a few studies have been performed.[84]

5. Both intra- and intercellular networks may be "personalized" and/or "localized," taking into account the location of the hosting cells within the tissue, building in the functional (e.g., signaling) consequences of the specific profile of mutations and epigenetic changes of the given tumor and subsequent modification of network nodes and edges according to the transcriptome and proteome of the given tumor. Importantly, due to the heterogeneity of tumors and the complexity of chromatin modifications, this task may be much more complex than initially thought.

6. Last but not least, most of the above considerations involved mainly structural changes of cellular networks and have not taken into account the dynamic analysis of cellular networks in order to determine, predict, and modify the changes in their attractor landscape structure. Several important recent studies[85] have established the novel area of "cancer attractor redesign." The aim of such approaches is to develop multitarget drugs and drug combinations, which (a) do not allow the dominance of proliferation, invasiveness, and so on attractors of cancer cells; (b) act as "differentiation therapies" guiding cancer cells back to their healthy attractors[86]; and (c) might lock cancer stem cells into their plastic or rigid phenotype.

The author is very optimistic that a paradigm shift is about to emerge that will lead to a change in the design strategies for anticancer therapies, such that the primary goal will be cancer cell "reeducation" and guidance toward the healthy state, instead of their mass murder, a strategy that does not work so well, as many decades of clinical outcome data have shown. The emerging knowledge on network adaptation mechanisms in complex molecular interaction networks will certainly be very useful in guiding and driving these efforts. From a network science perspective, it is quite clear that simplistic, linear, mutation-centric concepts, such as looking for "the" cancer genes and/or mutations, cannot solve the entire puzzle of successful cancer therapies. Such a "systems view of cancer" requires novel strategies that have to be drastically different from the current "mutation fishing" approaches. The author hopes that complex network theory and its applications, presented in this chapter, will contribute to a systems-level understanding of cancer and to the development of system-based anticancer therapies.

Acknowledgments

The author acknowledges the contribution of current and former LINK-Group (http://linkgroup.hu) members for the conceptualization of the ideas presented. The author is a founder and advisor of the Turbine startup (http://turbine.ai) that applies the attractor structure of personalized signaling networks to predict optimal intervention points of anticancer combination therapies. Work in the author's laboratory is supported by the Hungarian National Research Development and Innovation Office (OTKA K131458), by the Higher Education Institutional Excellence Program of the Ministry of Human Capacities in Hungary, within the framework of the molecular biology thematic programs of the Semmelweis University (Budapest, Hungary), and by the Artificial Intelligence Research Field Excellence Programme of the National Research, Development and Innovation Office of the Ministry of Innovation and Technology in Hungary (TKP/ITM/NKFIH).

Notes

1. Peter Csermely, "Creative elements: network-based predictions of active centres in proteins, cellular and social networks," *Trends Biochemical Sciences* 33, (2017): 569–576. Peter Csermely, "The wisdom of networks: a general adaptation and learning mechanism of complex systems. The network core triggers fast responses to known stimuli; innovations require the slow network periphery and are encoded by core-remodeling," *Bioessays* 40, (2018): 201700150.

2. Peter Csermely, "The wisdom of networks: a general adaptation and learning mechanism of complex systems. The network core triggers fast responses to known stimuli; innovations require the slow network periphery and are encoded by core-remodeling," *Bioessays* 40, (2018): 201700150.

3. Peter Csermely, "The wisdom of networks: a general adaptation and learning mechanism of complex systems. The network core triggers fast responses to known stimuli; innovations require the slow network periphery and are encoded by core-remodeling," *Bioessays* 40, (2018): 201700150. Peter Csermely, "Plasticity-rigidity cycles: a general adaptation mechanism," (2015): http://arxiv.org/abs/1511.01239

4. Peter Csermely, János Hódsági, Tamás Korcsmáros, Dezső Módos, Áron Ricardo Perez-Lopez, Kristóf Szalay, Dávid V. Veres, Katalin Lenti, Lin-Yun Wu, Xiang-Sun Zhang, "Cancer stem cells display extremely large evolvability: alternating plastic and rigid networks as a potential mechanism. Network models, novel therapeutic target strategies and the contributions of hypoxia, inflammation and cellular senescence," *Seminars in Cancer Biology* 30, (2015): 42–51.

5. Peter Csermely, Tamás Korcsmáros, Huba J. M. Kiss, Gábor London, Ruth Nussinov, "Structure and dynamics of biological networks: a novel paradigm of drug discovery. A comprehensive review," *Pharmacology and Therapeutics* 138, (2013): 333–408. Dávid M. Gyurkó, Dániel V. Veres, Dezső Módos, Katalin Lenti, Tamás Korcsmáros, Peter Csermely, "Adaptation and learning of molecular networks as a description of cancer development at the systems-level: potential use in anti-cancer therapies," *Seminars in Cancer Biology* 23, (2013): 262–269. Áron Ricardo Perez-Lopez, Kristóf Z. Szalay, Dénes Türei, Dezső Módos, Katalin Lenti, Tamás Korcsmáros, Peter Csermely, "Targets of drugs are generally, and targets of drugs having side effects are specifically good spreaders of human interactome perturbations," *Scientific Reports* 5, (2015): 10182. Dezső Módos, Krishna C. Bulusu, Dávid Fazekas, János Kubisch, Johanne Brooks, István Marczell, Péter M. Szabó, Tibor Vellai, Péter Csermely, Katalin Lenti, Andreas Bender, Tamás Korcsmáros, "Neighbours of cancer-related proteins have key influence on pathogenesis and could increase the drug target space for anticancer therapies," *NPJ Systems Biology and Applications* 3, (2017): 2.

6. Peter Csermely, Tamás Korcsmáros, Huba J. M. Kiss, Gábor London, Ruth Nussinov, "Structure and dynamics of biological networks: a novel paradigm of drug discovery. A comprehensive review," *Pharmacology and Therapeutics* 138, (2013): 333–408. Kivilcim Ozturk, Michelle Dow, Daniel E. Carlin, Rafael Bejar, Hannah Carter, "The emerging potential for network analysis to inform precision cancer medicine," *Journal of Molecular Biology* 430, (2018): 2875–2899. Paramasivan Poornima, Jothi Dinesh Kumar, Qiaoli Zhao, Martina Blunder, Thomas Efferth, "Network pharmacology of cancer: from understanding of complex interactomes to the design of multi-target specific therapeutics from nature," *Pharmacological Research* 111, (2016): 290–302. Yoo-Ah Kim, Dong-Yeon Cho, Teresa M. Przytycka, "Understanding genotype-phenotype effects in cancer via network approaches," *PLoS Computational Biology* 12, (2016): e1004747.

7. James S. Hale, Meizhang Li, Justin D. Lathia, "The malignant social network: cell-cell adhesion and communication in cancer stem cells," *Cell Adhesion and Migration* 6, (2012): 346–355. Yu Wu, Lana X. Garmire, Rong Fan, "Inter-cellular signaling network reveals a mechanistic transition in tumor microenvironment," *Integrative Biology (Cambridge)* 4, (2012): 1478–1486. Joseph X. Zhou, Roberto Taramelli, Edoardo Pedrini, Theo Knijnenburg, Sui Huang, "Extracting intercellular signaling network of cancer tissues using ligand-receptor expression patterns from whole-tumor and single-cell transcriptomes," *Scientific Reports* 18, (2017): 8815.

8. Steve P. Borgatti, Martin G. Everett, "Models of core/periphery structures," *Social Networks* 21, (1999): 375–395.

9. Peter Csermely, András London, Ling-Yun Wu, Brian Uzzi, "Structure and dynamics of core-periphery networks," *Journal of Complex Networks* 1, (2013): 93–123.

10. István A. Kovács, Robin Palotai, Máté S. Szalay, Peter Csermely, "Community landscapes: an integrative approach to determine overlapping network module hierarchy, identify key nodes and predict network dynamics," *PLoS One* 5, (2010): e12528. Máté Szalay-Bekő, Robin Palotai, Balázs Szappanos, István A. Kovács, Balázs Papp, Peter Csermely, "ModuLand plug-in for Cytoscape: determination of hierarchical layers of overlapping network modules and community centrality," *Bioinformatics* 28, (2012): 2202–2204.

11. Peter Csermely, András London, Ling-Yun Wu, Brian Uzzi, "Structure and dynamics of core-periphery networks," *Journal of Complex Networks* 1, (2013): 93–123.

12. Peter Csermely, András London, Ling-Yun Wu, Brian Uzzi, "Structure and dynamics of core-periphery networks," *Journal of Complex Networks* 1, (2013): 93–123.

13. Stuart Kauffman, "Homeostasis and differentiation in random genetic control networks," *Nature* 224, (1969): 177–178.

14. William A. Little, "The existence of persistent states in the brain," *Mathematical Biosciences* 19, (1974): 101–120. William A. Little, Gordon L. Shaw, "Analytic study of the memory storage capacity of a neural network," *Mathematical Biosciences* 39, (1978): 281–290. John J. Hopfield, "Neural networks and physical systems with emergent collective computational abilities," *Proceedings of the National Academy of Sciences of the USA* 79, (1982): 2554–2558.

15. Bernold Fiedler, Atsushi Mochizuki, Gen Kurosawa, Daisuke Saito, "Dynamics and control at feedback vertex sets. I: Informative and determining nodes in regulatory networks," *Journal of Dynamics and Differential Equations* 25, (2013): 563–604. Atsushi Mochizuki, Bernold Fiedler, Gen Kurosawa, Daisuke Saito, "Dynamics and control at feedback vertex sets. II: A faithful monitor to determine the diversity of molecular activities in regulatory networks," *Journal of Theoretical Biology* 335, (2013): 130–146. Assieh Saadatpour, Reka Albert, Timothy C. Reluga, "A reduction method for Boolean network models proven to conserve attractors," *SIAM Journal on Applied Dynamical Systems* 12, (2013): 1997–2011. Jorge G. Zañudo, Reka Albert, "Cell fate reprogramming by control of intracellular network dynamics," *PLoS Computational Biology* 11, (2015): e1004193.

16. Peter Csermely, "The wisdom of networks: a general adaptation and learning mechanism of complex systems. The network core triggers fast responses to known stimuli; innovations require the slow network periphery and are encoded by core-remodeling," *Bioessays* 40, (2018): 201700150.

17. Peter Csermely, "The wisdom of networks: a general adaptation and learning mechanism of complex systems. The network core triggers fast responses to known stimuli; innovations require the slow network periphery and are encoded by core-remodeling," *Bioessays* 40, (2018): 201700150.

18. Peter Csermely, "The wisdom of networks: a general adaptation and learning mechanism of complex systems. The network core triggers fast responses to known stimuli; innovations require the slow network periphery and are encoded by core-remodeling," *Bioessays* 40, (2018): 201700150.

19. Peter Csermely, "The wisdom of networks: a general adaptation and learning mechanism of complex systems. The network core triggers fast responses to known stimuli; innovations require the slow network periphery and are encoded by core-remodeling," *Bioessays* 40, (2018): 201700150.

20. Peter Csermely, "The wisdom of networks: a general adaptation and learning mechanism of complex systems. The network core triggers fast responses to known stimuli; innovations require the slow network periphery and are encoded by core-remodeling," *Bioessays* 40, (2018): 201700150.

21. Peter Csermely, "Creative elements: network-based predictions of active centres in proteins, cellular and social networks," *Trends Biochemical Sciences* 33, (2017): 569–576.

22. István A. Kovács, Robin Palotai, Máté S. Szalay, Peter Csermely, "Community landscapes: an integrative approach to determine overlapping network module hierarchy, identify key nodes and predict network dynamics," *PLoS One* 5, (2010): e12528. Jing-Dong J. Han, Nicolas Bertin, Tong Hao, Debra S. Goldberg, Gabriel F. Berriz, Lan V. Zhang, Denis Dupuy, Albertha J. M. Walhout, Michael E. Cusick, Frederick P. Roth, Marc Vidal, "Evidence for dynamically organized modularity in the yeast protein-protein interaction network," *Nature* 430, (2004): 88–93.

23. Peter Csermely, "Creative elements: network-based predictions of active centres in proteins, cellular and social networks," *Trends Biochemical Sciences* 33, (2017): 569–576.

24. Peter Csermely, "The wisdom of networks: a general adaptation and learning mechanism of complex systems. The network core triggers fast responses to known stimuli; innovations require the slow network periphery and are encoded by core-remodeling," *Bioessays* 40, (2018): 201700150.

25. Peter Csermely, "The wisdom of networks: a general adaptation and learning mechanism of complex systems. The network core triggers fast responses to known stimuli; innovations require the slow network periphery and are encoded by core-remodeling," *Bioessays* 40, (2018): 201700150.

26. Peter Csermely, "The wisdom of networks: a general adaptation and learning mechanism of complex systems. The network core triggers fast responses to known stimuli; innovations require the slow network periphery and are encoded by core-remodeling," *Bioessays* 40, (2018): 201700150.

27. Peter Csermely, "The wisdom of networks: a general adaptation and learning mechanism of complex systems. The network core triggers fast responses to known stimuli; innovations require the slow network periphery and are encoded by core-remodeling," *Bioessays* 40, (2018): 201700150. Peter Tompa, "The principle of conformational signaling," *Chemical Society Reviews* 45, (2016): 4252–4284.

28. Tanmay Mitra, Shakti N. Menon, Sitabhra Sinha, "Emergent memory in cell signaling: persistent adaptive dynamics in cascades can arise from the diversity of relaxation time-scales," (2018): https://arxiv.org/abs/1801 .04057. Augustina D'Urso, Jason H. Brickner, "Mechanisms of epigenetic memory," *Trends in Genetics* 30, (2014): 230–236.

29. Eivind Almaas, Zoltan N. Oltvai, Albert Lászlo Barabási, "The activity reaction core and plasticity of metabolic networks," *PLoS Computational Biology* 1, (2005): e68.

30. Peter Csermely, "The wisdom of networks: a general adaptation and learning mechanism of complex systems. The network core triggers fast responses to known stimuli; innovations require the slow network periphery and are encoded by core-remodeling," *Bioessays* 40, (2018): 201700150.

31. Steven Nathaniel Steinway, Jorge G.T. Zañudo, Wei Ding, Carl Bart Rountree, David J. Feith, Thomas P. Loughran Jr., Reka Albert, "Network modeling of TGFβ signaling in hepatocellular carcinoma epithelial-to-mesenchymal transition reveals joint sonic hedgehog and Wnt pathway activation," *Cancer Research* 74, (2014): 5963–5977.

32. Vladimir V. Galatenko, Alexey V. Galatenko, Timur R. Samatov, Andrey A. Turchinovich, Maxim Y. Shkurnikov, Julia A. Makarova, Alexander G. Tonevitsky, "Comprehensive network of miRNA-induced intergenic interactions and a biological role of its core in cancer," *Scientific Reports* 8, (2018): 2418.

33. Dezső Módos, Krishna C. Bulusu, Dávid Fazekas, János Kubisch, Johanne Brooks, István Marczell, Péter M. Szabó, Tibor Vellai, Péter Csermely, Katalin Lenti, Andreas Bender, Tamás Korcsmáros, "Neighbours of cancer-related proteins have key influence on pathogenesis and could increase the drug target space for anticancer therapies," *NPJ Systems Biology and Applications* 3, (2017): 2.

34. Dezső Módos, Krishna C. Bulusu, Dávid Fazekas, János Kubisch, Johanne Brooks, István Marczell, Péter M. Szabó, Tibor Vellai, Péter Csermely, Katalin Lenti, Andreas Bender, Tamás Korcsmáros, "Neighbours of cancer-related proteins have key influence on pathogenesis and could increase the drug target space for anticancer therapies," *NPJ Systems Biology and Applications* 3, (2017): 2.

35. Peter Csermely, "Plasticity-rigidity cycles: a general adaptation mechanism," (2015): http://arxiv.org/abs /1511.01239. Merse E. Gáspár, Peter Csermely, "Rigidity and flexibility of biological networks," *Briefings in Functional Genomics* 11, (2012): 443–456.

36. Peter Csermely, "Plasticity-rigidity cycles: a general adaptation mechanism," (2015): http://arxiv.org/abs /1511.01239. Merse E. Gáspár, Peter Csermely, "Rigidity and flexibility of biological networks," *Briefings in Functional Genomics* 11, (2012): 443–456.

37. Peter Csermely, "Plasticity-rigidity cycles: a general adaptation mechanism," (2015): http://arxiv.org/abs /1511.01239. Merse E. Gáspár, Peter Csermely, "Rigidity and flexibility of biological networks," *Briefings in Functional Genomics* 11, (2012): 443–456. Justin Ruths, Derek Ruths, "Control profiles of complex networks," *Science* 343, (2014): 1373–1376.

38. Bernold Fiedler, Atsushi Mochizuki, Gen Kurosawa, Daisuke Saito, "Dynamics and control at feedback vertex sets. I: Informative and determining nodes in regulatory networks," *Journal of Dynamics and Differential Equations* 25, (2013): 563–604. Atsushi Mochizuki, Bernold Fiedler, Gen Kurosawa, Daisuke Saito, "Dynamics and control at feedback vertex sets. II: A faithful monitor to determine the diversity of molecular activities in regulatory networks," *Journal of Theoretical Biology* 335, (2013): 130–146. Assieh Saadatpour, Reka Albert, Timothy C. Reluga, "A reduction method for Boolean network models proven to conserve attractors," *SIAM*

Journal on Applied Dynamical Systems 12, (2013): 1997–2011. Jorge G. Zañudo, Reka Albert, "Cell fate reprogramming by control of intracellular network dynamics," *PLoS Computational Biology* 11, (2015): e1004193.

39. Peter Csermely, "Creative elements: network-based predictions of active centres in proteins, cellular and social networks," *Trends Biochemical Sciences* 33, (2017): 569–576.

40. Peter Csermely, "Plasticity-rigidity cycles: a general adaptation mechanism," (2015): http://arxiv.org/abs /1511.01239. Merse E. Gáspár, Peter Csermely, "Rigidity and flexibility of biological networks," *Briefings in Functional Genomics* 11, (2012): 443–456.

41. Peter Csermely, "Plasticity-rigidity cycles: a general adaptation mechanism," (2015): http://arxiv.org/abs /1511.01239. Merse E. Gáspár, Peter Csermely, "Rigidity and flexibility of biological networks," *Briefings in Functional Genomics* 11, (2012): 443–456.

42. Justin Ruths, Derek Ruths, "Control profiles of complex networks," *Science* 343, (2014): 1373–1376.

43. Justin Ruths, Derek Ruths, "Control profiles of complex networks," *Science* 343, (2014): 1373–1376.

44. Peter Csermely, "Creative elements: network-based predictions of active centres in proteins, cellular and social networks," *Trends Biochemical Sciences* 33, (2017): 569–576.

45. Peter Csermely, "Creative elements: network-based predictions of active centres in proteins, cellular and social networks," *Trends Biochemical Sciences* 33, (2017): 569–576. Peter Tompa, "The principle of conformational signaling," *Chemical Society Reviews* 45, (2016): 4252–4284.

46. Peter Csermely, "Plasticity-rigidity cycles: a general adaptation mechanism," (2015): http://arxiv.org/abs /1511.01239.

47. Indika Rajapakse, Mark Groudine, Mehran Mesbahi, "Dynamics and control of state-dependent networks for probing genomic organization," *Proceedings of the National Academy of Sciences of the USA* 108, (2011): 17257–17262.

48. Peter Csermely, "Plasticity-rigidity cycles: a general adaptation mechanism," (2015): http://arxiv.org/abs /1511.01239.

49. Andrew E. Teschendorff, Simone Severini, "Increased entropy of signal transduction in the cancer metastasis phenotype," *BMC Systems Biology* 4, (2010): 104. James West, Ginestra Bianconi, Simone Severini, Andrew E. Teschendorff, "Differential network entropy reveals cancer system hallmarks," *Scientific Reports* 2, (2012): 802. Dylan Breitkreutz, Lynn Hlatky, Edward Rietman, Jack A. Tuszynski, "Molecular signaling network complexity is correlated with cancer patient survivability," *Proceedings of the National Academy of Sciences of the USA* 109, (2012): 9209–9212. János Hódsági, "*Network entropy as a measure of plasticity in cancer,*" MSc Thesis, UCL London (2013).

50. Elisabet Pujadas, Andrew P. Feinberg, "Regulated noise in the epigenetic landscape of development and disease," *Cell* 148, (2012): 1123–1131.

51. Luca Albergante, J. Julian Blow, Timothy J. Newman, "Buffered qualitative stability explains the robustness and evolvability of transcriptional networks," *eLife* 3, (2014): e02863.

52. Peter Csermely, János Hódsági, Tamás Korcsmáros, Dezső Módos, Áron Ricardo Perez-Lopez, Kristóf Szalay, Dávid V. Veres, Katalin Lenti, Lin-Yun Wu, Xiang-Sun Zhang, "Cancer stem cells display extremely large evolvability: alternating plastic and rigid networks as a potential mechanism. Network models, novel therapeutic target strategies and the contributions of hypoxia, inflammation and cellular senescence," *Seminars in Cancer Biology* 30, (2015): 42–51. Dávid M. Gyurkó, Dániel V. Veres, Dezső Módos, Katalin Lenti, Tamás Korcsmáros, Peter Csermely, "Adaptation and learning of molecular networks as a description of cancer development at the systems-level: potential use in anti-cancer therapies," *Seminars in Cancer Biology* 23, (2013): 262–269. Dongya Jia, Mohit Kumar Jolly, Prakash Kulkarni, Herbert Levine, "Phenotypic plasticity and cell fate decisions in cancer: insights from dynamical systems theory," *Cancers (Basel)* 9, (2017): E70.

53. Andriy Marusyk, Vanessa Almendro, Kornelia Polyak, "Intra-tumour heterogeneity: a looking glass for cancer?" *Nature Reviews of Cancer* 12, (2012): 323–334. Luonan Chen, Rui Liu, Zhi-Ping Liu, Meiyi Li, Kazuyuki Aihara, "Detecting early-warning signals for sudden deterioration of complex diseases by dynamical network biomarkers," *Scientific Reports* 2, (2012): 342. Luonan Chen, Rui Liu, Zhi-Ping Liu, Meiyi Li, Kazuyuki Aihara, "Detecting early-warning signals for sudden deterioration of complex diseases by dynamical network biomarkers," *Scientific Reports* 2, (2012): 342. Rui Liu, Meiyi Li, Zhi-Ping Liu, Jiarui Wu, Luonan Chen, Kazuyuki Aihara, "Identifying critical transitions and their leading biomolecular networks in complex diseases," *Scientific Reports* 2, (2012): 813. Elke K. Markert, Arnold J. Levine, Alexei Vazquez, "Proliferation and tissue

remodeling in cancer: the hallmarks revisited," *Cell Death and Disease* 3, (2012): e397. Stuart Kaufman, "Differentiation of malignant to benign cells," *Journal of Theoretical Biology* 31, (1971): 429–451. Sui Huang, Ingemar Ernberg, Stuart Kauffman, "Cancer attractors: a systems view of tumors from a gene network dynamics and developmental perspective," *Seminars in Cell and Developmental Biology* 20, (2009): 869–876. Sui Huang, "Tumor progression: chance and necessity in Darwinian and Lamarckian somatic (mutationless) evolution," *Progress in Biophysics and Molecular Biology* 110, (2012): 69–86. Wei-Yi Cheng, Tai-Hsien Ou Yang, Dimitris Anastassiou, "Biomolecular events in cancer revealed by attractor metagenes," *PLoS Computational Biology* 9, (2013): e1002920. Peter Csermely, Vilmos Ágoston, Sándor Pongor, "The efficiency of multi-target drugs: the network approach might help drug design," *Trends in Pharmacological Sciences* 16, (2005): 178–182. Quan Zhong, Nicolas Simonis, Qian-Ru Li, Benoit Charloteaux, Fabien Heuze, Niels Klitgord, Stanley Tam, Haiyuan Yu, Kavitha Venkatesan, Danny Mou, Venus Swearingen, Muhammed A. Yildirim, Han Yan, Amélie Dricot, David Szeto, Chenwei Lin, Tong Hao, Changyu Fan, Stuart Milstein, Denis Dupuy, Robert Brasseur, David E. Hill, Michael E. Cusick, Marc Vidal, "Edgetic perturbation models of human inherited disorders," *Molecular Systems Biology* 5, (2009): 321. Ruth Nussinov, Chung-Jung Tsai, Peter Csermely, "Allo-network drugs: harnessing allostery in cellular networks," *Trends in Pharmacological Sciences* 32, (2011): 686–693.

54. Dylan Breitkreutz, Lynn Hlatky, Edward Rietman, Jack A. Tuszynski, "Molecular signaling network complexity is correlated with cancer patient survivability," *Proceedings of the National Academy of Sciences of the USA* 109, (2012): 9209–9212.

55. János Hódsági, "*Network entropy as a measure of plasticity in cancer,*" MSc Thesis, UCL London (2013).

56. Luonan Chen, Rui Liu, Zhi-Ping Liu, Meiyi Li, Kazuyuki Aihara, "Detecting early-warning signals for sudden deterioration of complex diseases by dynamical network biomarkers," *Scientific Reports* 2, (2012): 342. Rui Liu, Meiyi Li, Zhi-Ping Liu, Jiarui Wu, Luonan Chen, Kazuyuki Aihara, "Identifying critical transitions and their leading biomolecular networks in complex diseases," *Scientific Reports* 2, (2012): 813.

57. Elke K. Markert, Arnold J. Levine, Alexei Vazquez, "Proliferation and tissue remodeling in cancer: the hallmarks revisited," *Cell Death and Disease* 3, (2012): e397.

58. Indika Rajapakse, Mark Groudine, Mehran Mesbahi, "Dynamics and control of state-dependent networks for probing genomic organization," *Proceedings of the National Academy of Sciences of the USA* 108, (2011): 17257–17262.

59. Stuart Kaufman, "Differentiation of malignant to benign cells," *Journal of Theoretical Biology* 31, (1971): 429–451. Wei-Yi Cheng, Tai-Hsien Ou Yang, Dimitris Anastassiou, "Biomolecular events in cancer revealed by attractor metagenes," *PLoS Computational Biology* 9, (2013): e1002920.

60. János Hódsági, "*Network entropy as a measure of plasticity in cancer,*" MSc Thesis, UCL London (2013).

61. Dávid M. Gyurkó, Dániel V. Veres, Dezső Módos, Katalin Lenti, Tamás Korcsmáros, Peter Csermely, "Adaptation and learning of molecular networks as a description of cancer development at the systems-level: potential use in anti-cancer therapies," *Seminars in Cancer Biology* 23, (2013): 262–269. Andrew E. Teschendorff, Simone Severini, "Increased entropy of signal transduction in the cancer metastasis phenotype," *BMC Systems Biology* 4, (2010): 104. James West, Ginestra Bianconi, Simone Severini, Andrew E. Teschendorff, "Differential network entropy reveals cancer system hallmarks," *Scientific Reports* 2, (2012): 802. Dylan Breitkreutz, Lynn Hlatky, Edward Rietman, Jack A. Tuszynski, "Molecular signaling network complexity is correlated with cancer patient survivability," *Proceedings of the National Academy of Sciences of the USA* 109, (2012): 9209–9212. János Hódsági, "*Network entropy as a measure of plasticity in cancer,*" MSc Thesis, UCL London (2013). Elisabet Pujadas, Andrew P. Feinberg, "Regulated noise in the epigenetic landscape of development and disease," *Cell* 148, (2012): 1123–1131. Luca Albergante, J. Julian Blow, Timothy J. Newman, "Buffered qualitative stability explains the robustness and evolvability of transcriptional networks," *eLife* 3, (2014): e02863. Dongya Jia, Mohit Kumar Jolly, Prakash Kulkarni, Herbert Levine, "Phenotypic plasticity and cell fate decisions in cancer: insights from dynamical systems theory," *Cancers (Basel)* 9, (2017): E70. Andriy Marusyk, Vanessa Almendro, Kornelia Polyak, "Intra-tumour heterogeneity: a looking glass for cancer?" *Nature Reviews of Cancer* 12, (2012): 323–334. Luonan Chen, Rui Liu, Zhi-Ping Liu, Meiyi Li, Kazuyuki Aihara, "Detecting early-warning signals for sudden deterioration of complex diseases by dynamical network biomarkers," *Scientific Reports* 2, (2012): 342. Rui Liu, Meiyi Li, Zhi-Ping Liu, Jiarui Wu, Luonan Chen, Kazuyuki Aihara, "Identifying critical transitions and their leading biomolecular networks in complex diseases," *Scientific Reports* 2, (2012): 813. Elke K. Markert, Arnold J. Levine, Alexei Vazquez, "Proliferation and tissue remodeling in cancer: the hallmarks revisited," *Cell Death and Disease* 3, (2012): e397. Stuart Kaufman, "Differentiation of malignant to benign cells," *Journal of Theoretical Biology* 31, (1971): 429–451. Sui Huang, Ingemar Ernberg, Stuart Kauffman,

"Cancer attractors: a systems view of tumors from a gene network dynamics and developmental perspective," *Seminars in Cell and Developmental Biology* 20, (2009): 869–876. Sui Huang, "Tumor progression: chance and necessity in Darwinian and Lamarckian somatic (mutationless) evolution," *Progress in Biophysics and Molecular Biology* 110, (2012): 69–86. Wei-Yi Cheng, Tai-Hsien Ou Yang, Dimitris Anastassiou, "Biomolecular events in cancer revealed by attractor metagenes," *PLoS Computational Biology* 9, (2013): e1002920.

62. Peter Csermely, Vilmos Ágoston, Sándor Pongor, "The efficiency of multi-target drugs: the network approach might help drug design," *Trends in Pharmacological Sciences* 16, (2005): 178–182. Quan Zhong, Nicolas Simonis, Qian-Ru Li, Benoit Charloteaux, Fabien Heuze, Niels Klitgord, Stanley Tam, Haiyuan Yu, Kavitha Venkatesan, Danny Mou, Venus Swearingen, Muhammed A. Yildirim, Han Yan, Amélie Dricot, David Szeto, Chenwei Lin, Tong Hao, Changyu Fan, Stuart Milstein, Denis Dupuy, Robert Brasseur, David E. Hill, Michael E. Cusick, Marc Vidal, "Edgetic perturbation models of human inherited disorders," *Molecular Systems Biology* 5, (2009): 321. Ruth Nussinov, Chung-Jung Tsai, Peter Csermely, "Allo-network drugs: harnessing allostery in cellular networks," *Trends in Pharmacological Sciences* 32, (2011): 686–693.

63. Andriy Marusyk, Vanessa Almendro, Kornelia Polyak, "Intra-tumour heterogeneity: a looking glass for cancer?" *Nature Reviews of Cancer* 12, (2012): 323–334.

64. Peter Csermely, János Hódsági, Tamás Korcsmáros, Dezső Módos, Áron Ricardo Perez-Lopez, Kristóf Szalay, Dávid V. Veres, Katalin Lenti, Lin-Yun Wu, Xiang-Sun Zhang, "Cancer stem cells display extremely large evolvability: alternating plastic and rigid networks as a potential mechanism. Network models, novel therapeutic target strategies and the contributions of hypoxia, inflammation and cellular senescence," *Seminars in Cancer Biology* 30, (2015): 42–51.

65. Dávid M. Gyurkó, Dániel V. Veres, Dezső Módos, Katalin Lenti, Tamás Korcsmáros, Peter Csermely, "Adaptation and learning of molecular networks as a description of cancer development at the systems-level: potential use in anti-cancer therapies," *Seminars in Cancer Biology* 23, (2013): 262–269.

66. Peter Csermely, János Hódsági, Tamás Korcsmáros, Dezső Módos, Áron Ricardo Perez-Lopez, Kristóf Szalay, Dávid V. Veres, Katalin Lenti, Lin-Yun Wu, Xiang-Sun Zhang, "Cancer stem cells display extremely large evolvability: alternating plastic and rigid networks as a potential mechanism. Network models, novel therapeutic target strategies and the contributions of hypoxia, inflammation and cellular senescence," *Seminars in Cancer Biology* 30, (2015): 42–51. Dávid M. Gyurkó, Dániel V. Veres, Dezső Módos, Katalin Lenti, Tamás Korcsmáros, Peter Csermely, "Adaptation and learning of molecular networks as a description of cancer development at the systems-level: potential use in anti-cancer therapies," *Seminars in Cancer Biology* 23, (2013): 262–269. Wanyin Chen, Jihu Dong, Jacques Haiech, Marie-Claude Kilhoffer, Maria Zeniou, "Cancer stem cell quiescence and plasticity as major challenges in cancer therapy," *Stem Cells International* 2016, (2016): 1740936. Mary R. Doherty, Jacob M. Smigiel, Damian J. Junk, Mark W. Jackson, "Cancer stem cell plasticity drives therapeutic resistance," *Cancers (Basel)* 8, (2016): 8. Marina Carla Cabrera, Robert E. Hollingsworth, Elaine M. Hurt, "Cancer stem cell plasticity and tumor hierarchy," *World Journal of Stem Cells* 7, (2015): 27–36.

67. Peter Csermely, Tamás Korcsmáros, Huba J. M. Kiss, Gábor London, Ruth Nussinov, "Structure and dynamics of biological networks: a novel paradigm of drug discovery. A comprehensive review," *Pharmacology and Therapeutics* 138, (2013): 333–408.

68. Takeshi Hase, Hiroshi Tanaka, Yasuhiro Suzuki, So Nakagawa, Hiroaki Kitano, "Structure of protein interaction networks and their implications on drug design," *PLoS Computational Biology* 5, (2009): e1000550.

69. Ian W. Taylor, Rune Linding, David Warde-Farley, Yongmei Liu, Catia Pesquita, Daniel Faria, Shelley Bull, Tony Pawson, Quaid Morris, Jeffrey L Wrana, "Dynamic modularity in protein interaction networks predicts breast cancer outcome," *Nature Biotechnology* 27, (2009): 199–204.

70. Peter Csermely, Tamás Korcsmáros, Huba J. M. Kiss, Gábor London, Ruth Nussinov, "Structure and dynamics of biological networks: a novel paradigm of drug discovery. A comprehensive review," *Pharmacology and Therapeutics* 138, (2013): 333–408.

71. Dezső Módos, Krishna C. Bulusu, Dávid Fazekas, János Kubisch, Johanne Brooks, István Marczell, Péter M. Szabó, Tibor Vellai, Péter Csermely, Katalin Lenti, Andreas Bender, Tamás Korcsmáros, "Neighbours of cancer-related proteins have key influence on pathogenesis and could increase the drug target space for anticancer therapies," *NPJ Systems Biology and Applications* 3, (2017): 2.

72. Heinz Ruffner, Andreas Bauer, Tewis Bouwmeester, "Human protein-protein interaction networks and the value for drug discovery," *Drug Discovery Today* 12, (2007): 709–716.

73. Lyubomir T. Vassilev, Binh T. Vu, Bradford Graves, Daisy Carvajal, Frank Podlaski, Zoran Filipovic, Norman Kong, Ursula Kammlott, Christine Lukacs, Christian Klein, Nader Fotouhi, Emily A. Liu, "In vivo activation of the p53 pathway by small-molecule antagonists of MDM2," *Science* 303, (2004): 844–848.

74. Dezső Módos, Krishna C. Bulusu, Dávid Fazekas, János Kubisch, Johanne Brooks, István Marczell, Péter M. Szabó, Tibor Vellai, Péter Csermely, Katalin Lenti, Andreas Bender, Tamás Korcsmáros, "Neighbours of cancer-related proteins have key influence on pathogenesis and could increase the drug target space for anticancer therapies," *NPJ Systems Biology and Applications* 3, (2017): 2.

75. Peter Csermely, Tamás Korcsmáros, Huba J. M. Kiss, Gábor London, Ruth Nussinov, "Structure and dynamics of biological networks: a novel paradigm of drug discovery. A comprehensive review," *Pharmacology and Therapeutics* 138, (2013): 333–408. Madhukar S. Dasika, Anthony Burgard, Costas D. Maranas, "A computational framework for the topological analysis and targeted disruption of signal transduction networks," *Biophysical Journal* 91, (2006): 382–398. Alexei Vazquez, "Optimal drug combinations and minimal hitting sets," *BMC Systems Biology* 3, (2009): 81. Hee Sook Lee, Taejeong Bae, Ji-Hyun Lee, Dae Gyu Kim, Young Sun Oh, Yeongjun Jang, Ji-Tea Kim, Jong-Jun Lee, Alessio Innocenti, Claudiu T Supuran, Luonan Chen, Kyoohyoung Rho, Sunghoon Kim, "Rational drug repositioning guided by an integrated pharmacological network of protein, disease and drug," *BMC Systems Biology* 6, (2012): 80.

76. Andriy Marusyk, Vanessa Almendro, Kornelia Polyak, "Intra-tumour heterogeneity: a looking glass for cancer?" *Nature Reviews of Cancer* 12, (2012): 323–334.

77. Peter Csermely, János Hódsági, Tamás Korcsmáros, Dezső Módos, Áron Ricardo Perez-Lopez, Kristóf Szalay, Dávid V. Veres, Katalin Lenti, Lin-Yun Wu, Xiang-Sun Zhang, "Cancer stem cells display extremely large evolvability: alternating plastic and rigid networks as a potential mechanism. Network models, novel therapeutic target strategies and the contributions of hypoxia, inflammation and cellular senescence," *Seminars in Cancer Biology* 30, (2015): 42–51. Dávid M. Gyurkó, Dániel V. Veres, Dezső Módos, Katalin Lenti, Tamás Korcsmáros, Peter Csermely, "Adaptation and learning of molecular networks as a description of cancer development at the systems-level: potential use in anti-cancer therapies," *Seminars in Cancer Biology* 23, (2013): 262–269. Wanyin Chen, Jihu Dong, Jacques Haiech, Marie-Claude Kilhoffer, Maria Zeniou, "Cancer stem cell quiescence and plasticity as major challenges in cancer therapy," *Stem Cells International* 2016, (2016): 1740936. Mary R. Doherty, Jacob M. Smigiel, Damian J. Junk, Mark W. Jackson, "Cancer stem cell plasticity drives therapeutic resistance," *Cancers (Basel)* 8, (2016): 8. Marina Carla Cabrera, Robert E. Hollingsworth, Elaine M. Hurt, "Cancer stem cell plasticity and tumor hierarchy," *World Journal of Stem Cells* 7, (2015): 27–36.

78. Mary R. Doherty, Jacob M. Smigiel, Damian J. Junk, Mark W. Jackson, "Cancer stem cell plasticity drives therapeutic resistance," *Cancers (Basel)* 8, (2016): 8. Sui Huang, Stuart Kauffman, "How to escape the cancer attractor: rationale and limitations of multi-target drugs," *Seminars in Cancer Biology* 23, (2013): 270–278. Sui Huang, "Genetic and non-genetic instability in tumor progression: link between the fitness landscape and the epigenetic landscape of cancer cells," *Cancer Metastasis Reviews* 32, (2013): 423–448. Corbin E. Meacham, Sean J. Morrison, "Tumour heterogeneity and cancer cell plasticity," *Nature* 501, (2013): 328–337. Angela Oliveira Pisco, Amy Brock, Joseph Zhou, Andreas Moor, Mitra Mojtahedi, Dean Jackson, Sui Huang, "Non-Darwinian dynamics in therapy-induced cancer drug resistance," *Nature Communications* 4, (2013): 2467.

79. Peter Csermely, "The wisdom of networks: a general adaptation and learning mechanism of complex systems. The network core triggers fast responses to known stimuli; innovations require the slow network periphery and are encoded by core-remodeling," *Bioessays* 40, (2018): 201700150.

80. Peter Csermely, "Plasticity-rigidity cycles: a general adaptation mechanism," (2015): http://arxiv.org/abs/1511.01239.

81. Peter Csermely, János Hódsági, Tamás Korcsmáros, Dezső Módos, Áron Ricardo Perez-Lopez, Kristóf Szalay, Dávid V. Veres, Katalin Lenti, Lin-Yun Wu, Xiang-Sun Zhang, "Cancer stem cells display extremely large evolvability: alternating plastic and rigid networks as a potential mechanism. Network models, novel therapeutic target strategies and the contributions of hypoxia, inflammation and cellular senescence," *Seminars in Cancer Biology* 30, (2015): 42–51. Dávid M. Gyurkó, Dániel V. Veres, Dezső Módos, Katalin Lenti, Tamás Korcsmáros, Peter Csermely, "Adaptation and learning of molecular networks as a description of cancer development at the systems-level: potential use in anti-cancer therapies," *Seminars in Cancer Biology* 23, (2013): 262–269.

82. Peter Csermely, János Hódsági, Tamás Korcsmáros, Dezső Módos, Áron Ricardo Perez-Lopez, Kristóf Szalay, Dávid V. Veres, Katalin Lenti, Lin-Yun Wu, Xiang-Sun Zhang, "Cancer stem cells display extremely large evolvability: alternating plastic and rigid networks as a potential mechanism. Network models, novel therapeutic target strategies and the contributions of hypoxia, inflammation and cellular senescence," *Seminars in Cancer Biology* 30, (2015): 42–51. Dávid M. Gyurkó, Dániel V. Veres, Dezső Módos, Katalin Lenti, Tamás Korcsmáros, Peter Csermely, "Adaptation and learning of molecular networks as a description of cancer development at the systems-level: potential use in anti-cancer therapies," *Seminars in Cancer Biology* 23, (2013): 262–269.

83. Peter Csermely, Tamás Korcsmáros, Huba J. M. Kiss, Gábor London, Ruth Nussinov, "Structure and dynamics of biological networks: a novel paradigm of drug discovery. A comprehensive review," *Pharmacology and Therapeutics* 138, (2013): 333–408. Dávid M. Gyurkó, Dániel V. Veres, Dezső Módos, Katalin Lenti, Tamás Korcsmáros, Peter Csermely, "Adaptation and learning of molecular networks as a description of cancer development at the systems-level: potential use in anti-cancer therapies," *Seminars in Cancer Biology* 23, (2013): 262–269. Áron Ricardo Perez-Lopez, Kristóf Z. Szalay, Dénes Türei, Dezső Módos, Katalin Lenti, Tamás Korcsmáros, Peter Csermely, "Targets of drugs are generally, and targets of drugs having side effects are specifically good spreaders of human interactome perturbations," *Scientific Reports* 5, (2015): 10182. Dezső Módos, Krishna C. Bulusu, Dávid Fazekas, János Kubisch, Johanne Brooks, István Marczell, Péter M. Szabó, Tibor Vellai, Péter Csermely, Katalin Lenti, Andreas Bender, Tamás Korcsmáros, "Neighbours of cancer-related proteins have key influence on pathogenesis and could increase the drug target space for anticancer therapies," *NPJ Systems Biology and Applications* 3, (2017): 2.

84. For some examples, see Romano Demicheli, Dinah Faith T. Quiton, Marco Fornili, William J. M. Hrushesky, "Cancer as a changed tissue's way of life (when to treat, when to watch and when to think)," *Future Oncology* 12, (2016): 647–657. Li Wenbo, Jin Wang, "Uncovering the underlying mechanism of cancer tumorigenesis and development under an immune microenvironment from global quantification of the landscape," *Journal of the Royal Society Interface* 14, (2017): 20170105. Robert J. Seager, Cynthia Hajal, Fabian Spill, Roger D. Kamm, Muhammad H. Zaman, "Dynamic interplay between tumour, stroma and immune system can drive or prevent tumour progression," *Convergent Science Physical Oncology* 3, (2017): 3. Joseph X. Zhou, Roberto Taramelli, Edoardo Pedrini, Theo Knijnenburg, Sui Huang, "Extracting intercellular signaling network of cancer tissues using ligand-receptor expression patterns from whole-tumor and single-cell transcriptomes," *Scientific Reports* 7, (2017): 8815. Frank Winkler, Wolfgang Wick, "Harmful networks in the brain and beyond," *Science* 359, (2018): 1100–1101. Jan P. Böttcher, Eduardo Bonavita, Probir Chakravarty, Hanna Blees, Mar Cabeza-Cabrerizo, Stefano Sammicheli, Neil C. Rogers, Erik Sahai, Santiago Zelenay, Caetano Reis e Sousa, "NK cells stimulate recruitment of cDC1 into the tumor microenvironment promoting cancer immune control," *Cell* 172, (2018): 1022–1037. Shicheng Su, Jianing Chen, Herui Yao, Jiang Liu, Shubin Yu, Liyan Lao, Minghui Wang, Manli Luo, Yue Xing, Fei Chen, Di Huang, Jinghua Zhao, Linbin Yang, Dan Liao, Fengxi Su, Mengfeng Li, Qiang Liu, Erwei Song, "Cancer-associated fibroblasts promote cancer formation and chemoresistance by sustaining cancer stemness," *Cell* 172, (2018): 841–856.

85. For example, see Luca Albergante, J. Julian Blow, Timothy J. Newman, "Buffered qualitative stability explains the robustness and evolvability of transcriptional networks," *eLife* 3, (2014): e02863. Angela Oliveira Pisco, Amy Brock, Joseph Zhou, Andreas Moor, Mitra Mojtahedi, Dean Jackson, Sui Huang, "Non-Darwinian dynamics in therapy-induced cancer drug resistance," *Nature Communications* 4, (2013): 2467. Ranran Zhang, Mithun Vinod Shah, Jun Yang, Susan B. Nyland, Xin Liu, Jong K. Yun, Réka Albert, Thomas P. Loughran Jr., "Network model of survival signaling in large granular lymphocyte leukemia," *Proceedings of the National Academy of Sciences of the USA* 105 (2008): 16308–16313. Stefan R. Maetschke, Mark A. Ragan, "Characterizing cancer subtypes as attractors of Hopfield networks," *Bioinformatics* 30, (2014): 1273–1279. Kristof Z. Szalay, Ruth Nussinov, Peter Csermely, "Attractor structures of signaling networks: consequences of different conformational barcode dynamics and their relations to network-based drug design," *Molecular Informatics* 33, (2014): 463–468. Anthony Szedlak, Giovanni Paternostro, Carlo Piermarocchi, "Control of asymmetric Hopfield networks and application to cancer attractors," *PLoS One* 9, (2014): e105842. Chunhe Li, Jin Wang, "Quantifying the landscape for development and cancer from a core cancer stem cell circuit," *Cancer Research* 75, (2015): 2607–2618. Joseph X. Zhou, Zerrin Isik, Caide Xiao, Irit Rubin, Stuart A. Kauffman, Michael Schroeder, Sui Huang, "Systematic drug perturbations on cancer cells reveal diverse exit paths from proliferative state," *Oncotarget* 7, (2016): 7415–7425. Qin Li, Anders Wennborg, Erik Aurell, Erez Dekel, Jie-Zhi Zou, Yuting Xu, Sui Huang, Ingemar Ernberg, "Dynamics inside the cancer cell attractor reveal cell heterogeneity, limits of stability, and escape," *Proceedings of the National Academy of Sciences of the USA* 113, (2016): 2672–2677. Ruoshi Yuan, Suzhan Zhang, Jiekai Yu, Yanqin Huang, Demin Lu, Runtan Cheng, Sui Huang, Ping Ao, Shu Zheng, Leroy Hood, Xiaomei Zhu, "Beyond cancer genes: colorectal cancer as robust intrinsic states formed by molecular interactions," *Open Biology* 7, (2017): 170169.

86. Stuart Kaufman, "Differentiation of malignant to benign cells," *Journal of Theoretical Biology* 31, (1971): 429–451. Joseph X. Zhou, Zerrin Isik, Caide Xiao, Irit Rubin, Stuart A. Kauffman, Michael Schroeder, Sui Huang, "Systematic drug perturbations on cancer cells reveal diverse exit paths from proliferative state," *Oncotarget* 7, (2016): 7415–7425.

7 The Role of Genomic Dark Matter in Cancer

Using AI to Shine a Light on It

Why Cancer Genes Are Not the Whole Story

Kahn Rhrissorrakrai and Laxmi Parida

Overview

Although the human genome contains more than three billion base pairs (bp), only a small fraction has been studied with the level of depth necessary to adequately characterize its biological function. Until the recent reduction in costs of whole-genome sequencing (WGS) technologies and increased throughput of next-generation sequencers, much of the research focus has been on the protein-coding regions of the genome. These areas representing just 3 percent of the genome can be more easily associated with functional roles of the translated proteins in the cell and the organism than the other 97 percent. The focus on coding regions is even more pronounced in the study of cancer, where most of our genetic/molecular explanations of different cancer etiologies are based on mutations in the coding regions or large chromosomal events whose impact on cancer phenotype is ultimately attributed to their effect on gene expression. Yet, there have been efforts and resources such as Encyclopedia of DNA Elements (ENCODE),[1] The Cancer Genome Atlas (TCGA),[2] International Cancer Genome Consortium (ICGC) (https://icgc.org), and Pan-Cancer Analysis of Whole Genomes (PCAWG)[3] that have not only significantly increased our understanding of coding regions relevant to cancer but have also reignited interest in genomic noncoding regions, as well as increased our awareness of how much of the genome has so far remained "dark matter" and uncharacterized.

In this chapter, we will describe the current state of knowledge regarding the genomic noncoding and "dark-matter" regions, as well as offer a perspective on how bioinformatics methods, such as Big Data analytics and artificial intelligence (AI), can be leveraged to elucidate the role these regions are playing in cancer. The integration of data from throughout these 97 percent of the genome will give us a more complete understanding of the development and progression of an extraordinarily heterogeneous disease such as cancer. Efforts to characterize dark-matter information also provide a context for using the power of "Big Data" and AI approaches to discover new relationships between the genome and the cancer phenotype and thus help to expand our knowledge of cancer biology.

7.1   The Influence of DNA Sequence on Cancer Development

7.1.1   Coding Regions: Driver versus Passenger Paradigm

For much of modern cancer research, initiation, growth, and metastasis of malignant cells, as well as treatment outcomes and patient survival, have been viewed as a direct result of a limited number of somatic mutations in a limited number of genes (often called the somatic mutation theory [SMT] of cancer[4]). These somatic variants that are not inherited from the parent but originate/accumulate during the lifetime of the organism are often found in relatively high abundance inside any given tumor.[5] Yet, heritable germline mutations still play a role in cancer predisposition and progression as they number in the millions for any individual, and many cancers are found to have hereditary and familial components that are transmitted through the germline, such as *MLH1* and *TERT* promoter mutations.[6] While any individual germline mutation would not typically be considered a driving force in carcinogenesis, there are cases, such as *RB1* loss, that support the *two-hit hypothesis*,[7] which posits that where there exists an inherited, recessive loss-of-function mutation in a tumor suppressor gene and a second somatic loss-of-function mutation is acquired during the lifetime of the organism in the other copy of the same gene, effectively a complete loss of function phenotype is generated. Although germline mutations may potentiate the impact of a somatic alteration, the relative rarity of heritable mutations relevant for cancer has guided much of cancer research to be been aimed toward the study of somatic alterations as the oncogenic drivers.

Somatic alterations may include DNA sequence changes such as single-nucleotide variations (SNVs), short indels (less than 50 bp), large structural variants, copy number variations (CNVs), and epigenetic changes that do not involve a sequence change at the DNA level. These somatic alterations, particularly in the coding region of a gene, have long been the focus of cancer study,[8] and hundreds to thousands of somatic variants, most of which are relatively rare in healthy cells, have been characterized in tumor tissues.[9]

New sequencing and bioinformatic methods have improved our ability to more deeply characterize the genomic landscapes of cancers. Furthermore, it has highlighted the complexity of somatic alterations and the limitations of the "classic" concepts of the SMT of cancer, leading to the development of additional concepts, such as "driver" and "passenger" mutations[10] to retain explanatory power. Recent calls for a critical reassessment of the conceptual framework underlying cancer research are in part motivated by the outcomes of large cancer genomics studies and the understanding that possibly a nonreductionist approach to cancer research might be required.[11] Driver mutations are those alterations that are thought to be causally linked to carcinogenesis, confer a growth advantage to cancer cells, and are under positive selection. Passenger mutations are typically under neutral selection and do not confer a growth advantage to cancer cells. If present clonally, such passengers likely existed prior to expansion or were induced by the same carcinogenic stress factors that cause the driver mutations. While the notion of driver and passenger mutations has been a popular mode of understanding the development of cancers, recently this well-established paradigm has been expanded to include so-called mini-drivers,[12] passenger and germline mutations that were previously thought to be benign or as not contributing to the cancer phenotype, particularly those located in noncoding regions. Recent studies have shown these passenger alterations may in fact be mildly deleterious for cancer cells, and the accumulation of these mutations may have a phenotypic impact on the tumor.[13] The effect of these "mild" alterations, which may not in themselves be deleterious for the cell individually and thus escape selective pressures, in aggregate may exert a significant opposing force to the set of driver mutations in the tumor that could ultimately prove lethal to tumor cells. Thus, the patient would experience a positive effect as tumor progression would be slowed down due to increased tumor cell death (based on a mechanism such as Muller's ratchet effect[14]). Likewise to deleterious effects, mild cancer-promoting mutations may also accumulate in a tumor or precancerous cell and confer positive selective advantages that can lead to or support the environment for cancer progression.[15] The accumulation of mildly cancer-promoting or cancer-suppressing mutations is likely occurring simultaneously, and it is possible that for tumors lacking well-known classical driver mutations, the balance of passenger mutations shifted toward having increased numbers of "subtle" tumor-promoting alterations that in aggregate are driving the cancer forward.[16] The advent of sequencing methodologies of greater depth and throughput has made clear that only a small fraction of genes is altered in more than 20 percent of cancers.[17] This demonstrates the need for the development of more complex and quantitative models that can consider the subtler effects of a combination of factors that may span all parts of the genome to explain cancer development and progression.

7.2   Known Noncoding Elements

The overwhelming number of mutations, whether somatic or germline, occurs in noncoding regions of the genome. Although there have been many recent advances and efforts to characterize elements in noncoding regions and their functional relevance to cancer, much of it remains "dark," or without a clear biological understanding of its contribution to disease etiology, even for noncoding variants of established cancer genes.[18] To illustrate the challenges in illuminating genomic dark matter, it is important to appreciate the diversity and functionality of currently known noncoding elements relevant to cancer, as it will inform how such elements are detected in genomic dark regions and which computational methods are best suited. These noncoding elements can broadly be placed into two categories: *regulatory elements* and *noncoding RNAs*, which will be described in more detail later. There exist additional noncoding elements whose disruption may contribute to cancer. For example, alterations residing in introns, particularly at splice junctions, can affect the final transcribed RNA by affecting normal splicing, as in the case of exon 5 skip events in the *BRCA1* gene that have been associated with a predisposition to hereditary breast and ovarian cancer.[19] Recent work has begun to systematically characterize noncoding elements and their role in cancer.[20]

7.2.1 Regulatory Elements

Regulatory elements represent a major source of noncoding alterations that contribute to the progression of cancer. Within the noncoding DNA, there are regions where proteins and RNAs bind to specific target sequences in order to regulate the expression of nearby genes. These sequences are either located proximally or distally to the target gene. Mutations in these areas can lead to dysregulation of gene expression. Furthermore, the majority of epigenetic changes that are indicative of many cancers as well as changes that have been correlated with mutational burden have been found in noncoding regions.[21] Although alterations in the regulation of the expression or stability of a gene product may be subtle, the accumulation of many such changes may both predispose and ultimately drive a cancerous state.[22]

7.2.2 Promoters and 5' and 3' Untranslated Regions

The *cis*-acting elements (acting on the same strand of DNA) proximal to a gene are, for obvious reasons, noncoding regions of intense study because of their more easily associated role in the regulation of their respective gene's expression and function. Functional alterations here will often affect transcription factor (TF) binding sites, RNA binding proteins (RBPs), or microRNAs (miRNAs). Alterations to the promoter region of a gene can lead to dysregulation as the composition or conformation of the bound TFs may change. The quintessential example of how promoter mutations can activate putative oncogenes are the frequent mutations in the promoter of the telomerase reverse transcriptase (*TERT*) gene that lead to[23] increased *TERT* expression due to altered binding of E-twenty-six (ETS) family TFs. The 5' untranslated region (UTR) is important for proper transcription and translation of the gene and contains elements necessary for translation initiation, such as the 5' cap structure, internal ribosomal entry sites (IRES), and translation initiation motifs.[24] A recent report has shown that the microRNA *mir-1254* bound to the *CCAR1* 5' UTR helps to stabilize both molecules and, in so doing, can resensitize tamoxifen-resistant breast cancers to tamoxifen.[25] Similarly, the 3' UTR contains binding sites for different RBPs and short and long noncoding RNA that often affect the stability and translation of the transcript. Disruptions in this region will, like the 5' UTR, affect composition and strength of bound regulatory elements. The 3' UTR of a gene may also contain multiple polyadenylation sites (PAS), and depending on which PAS is polyadenylated, multiple isoforms of the transcript may be produced each with different stability, translational efficacy, and export. These multiple isoforms have been shown to distinguish cancer types.[26]

7.2.3 Enhancers and Silencers

Compared to the study of promoter regions and UTRs, predicting whether an alteration in a distal element, such as an enhancer or silencer, is functional is highly challenging. Enhancers are elements that can be located far or near to their site of action, can act in *cis* or *trans*, and may have different sequence orientations.[27] Often, they encode short motifs that serve as TF binding sites. Similarly, silencers work to repress gene expression or may act as insulators between an enhancer and its target to alter expression.[28] These elements exist throughout the noncoding regions, whether in seemingly empty stretches of open chromatin, introns, or untranslated regions. Enhancers may be differentially used through development,[29] and although many techniques can be applied to identify likely enhancers, they are most reliably detected by following RNA polymerase II binding.[30] Yet drawing the connection between an enhancer and its target requires the use of a variety of analytic techniques,[31] including accounting for the chromatin looping that is often required for bringing an enhancer in proximity to its target promoter. These approaches typically leverage chromatin conformation capture methods based on cross-linking and ligation, and they include HI-C,[32] C5,[33] C4,[34] and C3[35] that reveal three-dimensional interactions between distant regions of nuclear chromatin. For example, a study found that for a set of patients with T-cell acute lymphoblastic leukemia, a specific single-nucleotide variant in an enhancer 4 kb upstream of the oncogene *LMO1* created a TF binding site for MYB.[36] This additional MYB site led to increased expression of *LMO1* for these patients. These enhancer and silencer regions are now being more actively explored for their role in cancer biology.[37]

7.2.4 Chromatin Structure

Chromatin accessibility is a prerequisite for gene regulation and expression. Gene locations must be made accessible to the transcriptional machinery in order for expression to occur. Thus, changes in the epigenetic state that cause chromatin to open or close can have wide-ranging effects. Mutations in the regulation of histone acetylation via histone deacetylases (HDACs) affect the chromatin state and have been associated with many cancers.[38] Regions of open chromatin where active transcription may be occurring have also been found to have increased mutational load as it is believed that the transcriptional machinery may impede DNA repair mechanisms, thus resulting in the accumulation of mutations in highly transcribed regions.[39] Changes in the chromosome conformation through alterations of the three-dimensional structure have also been associated with noncoding variants in cancer.[40] Recent work has highlighted such chromatin-related changes as a rich area of study,[41] and specific examples are well characterized in the literature.[42] For example, *IGFBP3*, a breast cancer–associated gene, has been shown to have long-range interactions with *EGFR* due to changes in chromatin organization.[43] *MITF*-related chromosome loops have been found in both melanoma patient samples and cell lines.[44] If accurate models can be built explaining the link between accessibility, mutational density and frequencies, and epigenetic factors, then the in-depth study of noncoding DNA through whole-genome sequencing may begin to reveal areas of the genome that are affecting chromatin structure in a cancer-specific manner and could therefore serve as potential biomarkers for therapeutic response.

7.3   Noncoding RNAs

The role of noncoding RNAs in cancer is becoming better understood in light of large-scale studies such as TCGA.[45] These transcription products (e.g., short and long noncoding RNAs) can serve to regulate the expression of genes or may serve as part of functional complexes (e.g., small nuclear RNAs that aid in the preprocessing of mRNA and small nucleolar RNAs such as ribosomal RNAs [rRNAs] and transfer RNAs [tRNAs]). To date, the roles of microRNAs (miRNAs) and long noncoding RNAS (lncRNAs) in cancer are the most studied and best understood.

7.3.1   Short Noncoding RNAs

There is an abundance of studies on the role and impact of short noncoding RNAs, including miRNAs and small-interfering RNAs (siRNAs), on organismal development, cellular function, and cancer.[46] These small RNAs typically mediate gene expression through the repression or degradation of mRNA. miRNAs bind the 3′ UTR and will recruit the RNA-induced silencing complex (RISC) to degrade the RNA via the action of Dicer and Argonaute proteins. Rather than initiate degradation, miRNAs can also simply prevent translation by making the bound mRNA no longer accessible to the ribosome.[47] siRNAs operate in a similar fashion to miRNAs by helping to recruit RISC to cleave the mRNA; however, these noncoding RNAs are longer than miRNAs and are able to bind anywhere within the mRNA. The frequent deletion of miR-15a/16-1 in chronic lymphocytic leukemia was among the first studies to demonstrate the role miRNAs can play in cancer,[48] including tumor-suppressive activity. Mouse models have shown that transgenic expression of miR-155 can induce lymphomagenesis,[49] and several miRNAs have been found to be oncogenic or tumor suppressors.[50]

7.3.2   Long Noncoding RNAs

lncRNAs are typically described as transcripts greater than 200 bp that do not encode for a protein.[51] The roles of lncRNAs are more varied than their shorter counterparts and may include protein-DNA scaffolds and competitive transcripts for the binding of miRNAs and RNA-binding proteins, as well as act directly as enhancers or suppressors of gene expression. lncRNA expression has also been found to be highly tissue and cell specific,[52] and in some cases, this specificity suggests their use as biomarkers for specific cancers. Within these elements, germline and somatic alterations can lead to their dysregulation, as is often observed in cancer patients.[53] It may also be that the overexpression of pseudogenes, which give rise to noncoding RNAs that resemble gene products but no longer encode functional proteins, may still compete for the binding of regulatory molecules and thereby affect the regulation of the intended target of these factors. Next-generation sequencing has made possible the discovery of thousands of lncRNAs and implicated their potential role in cancer.[54] It has already been shown in multiple cancers, such as breast, melanoma, or colorectal, that lncRNAs can a have significant role in their progression.[55]

7.3.3 tRNAs

tRNAs, which were historically thought to have a mere passive role in translation by bringing amino acids to actively translating ribosomes, have recently been shown to also have a significant role in cancer progression.[56] Changes in tRNA abundance have been associated with changes in protein expression levels in cells, and this modulation of their levels can provide a selective growth advantage depending on the cellular function of the protein. tRNA^{Arg} CCG and tRNA^{Glu} UUC have been shown to be upregulated in breast cancer cell lines and in metastatic cells compared to cells from the primary tumor. Evidence also suggests that hypoxic stress–induced tRNA-derived fragments can suppress the development of breast cancer metastasis.[57] tRNA and tRNA-derived fragments affecting cancer development and progression have been found in diverse cancer types, including breast, lung, cervical, pancreatic, and skin cancers.[58]

7.4   Genomic Dark Matter and Its Role in Carcinogenesis

Moving into the truly unknown space of the noncoding genome, we find the *dark matter*. In the past, all noncoding DNA would have been deemed dark matter, or junk DNA, because they did not encode the "traditional" functional machinery of the cell (i.e., proteins).[59] Yet, as technology has enabled the cataloguing and functional characterization of the noncoding DNA, what was once "dark" is now becoming increasingly "illuminated." Today dark-matter DNA can be defined as the current space of an unannotated genome (i.e., noncoding DNA excluding areas like the promoter, UTR, introns, enhancers, and silencers). In this region lies an underexplored frontier of genomic science. This category of DNA includes areas such as open chromatin—regions that are available for transcription and expression and whose three-dimensional confirmation can affect the function and organization of many elements of the genome. Noncoding genomic markers are often criticized for being insufficient to explain cellular phenotypes since they do not always clearly contribute to regulatory functions of transcription, posttranscription, translation, or posttranslational modifications. It is here that AI offers a way forward to discover functional elements in dark-matter DNA that have long been overlooked, particularly those elements that may play a role in disease, and how they may interact with each other in novel and subtle ways.

Changes in the DNA dark-matter space are only recently being appreciated for their potential role in disease etiology. The global chromatin remodeling and structural variation that has been found to enable the interaction of long range *cis-* and *trans-*acting elements, such as in the critical region for Williams-Beuren syndrome, exemplifies the type of complex interaction of genomic elements associated with disease that awaits characterization in the dark-matter space.[60] Machine learning (ML) applied to methylation data from a diverse set of cancer samples has found there to be reasonable agreement with the histological classification of these cancers and also novel groupings of histologically similar tumors of different cancer types as well as groupings that differed from the World Health Organization (WHO) classification.[61] This suggests that there are novel molecular cancer subtypes awaiting discovery. The noncoding DNA harboring most of these epigenetic changes in cancer are gaining greater appreciation as potential markers of tumorigenesis.[62] The epigenetic changes resulting from the accumulation of genetic alterations occurring in promoters, enhancers, and silencer regions can have an aggregative effect on the progression of cancer. We have the opportunity to train models on these signals found in annotated regions and use them to find similar patterns in the dark-matter space that have gone overlooked, possibly as a result of their infrequency in a given cohort or cancer type or because the appropriate set of covariates has not yet been discovered. This is where machine learning provides hope for making these types of discoveries, as cancer is the result of a highly complex system of interactions that will not necessarily conform to "single-hit" or somatic-only mechanisms.

Since cancer patients represent the largest source of sequenced normal and disease tissue, it is within this disease that we have our greatest potential for understanding the complex interplay between the myriad genomic elements. There has yet to be a systematic analysis of the dark-matter DNA across different cancer types and populations. As we are now performing increasing numbers of whole-genome studies at scale, these questions can now be addressed, although there must be methods at hand that can handle the orders of magnitude greater number of genomic locations as compared to looking at SNP arrays or even whole-exome analysis. AI is certainly the method of choice to uncover these novel functional elements and their role in cancer development and progression.

7.5   Computational Methods to Illuminate Cancer Genome Dark Matter

Here we will present an overview of some important computational approaches, particularly how "Big Data," ML, and AI can and are being leveraged to expand our understanding of the contribution of noncoding and dark-matter DNA sequences to cancer.

7.5.1   Bioinformatics Tools and Resources

Without the advancements in computational power, bioinformatics, and machine learning, our ability to perform WGS and other large-scale high-throughput -omic analyses to understand the complex features of genomes would not have been possible. Cancer sequencing and omics initiatives like TCGA, ICGC, and PCAWG have not only delivered unprecedented amounts of sequence information of cancer genomes, but importantly, they have also revealed major gaps in our understanding of carcinogenesis, including the high variation in the penetrance of mutations, the lack of clear drivers for many sequenced patients, the high degree of genomic inter- and intrapatient heterogeneity, and the shortcomings of relying entirely on somatic alterations as the sole explanatory mechanism for cancer.

A large segment of the available bioinformatics and computational tools has been focused on the prediction of drivers in somatic coding regions. Most use some form of statistical model to ultimately describe the recurrence of DNA sequence alterations and its correlation with cellular function.[63] The initial step for any process to analyze and annotate a genomic variant begins with its identification.[64] For these statistical approaches, overall organism and tissue-specific mutation rates and transcriptional activity in normal and cancer tissue must be taken into account to get a sense of the expected DNA sequence alteration rate in both diseased and healthy tissues. Only with an accurate understanding of data from healthy samples can new signals specific to disease be confidently identified. Such methods can then prioritize mutations that seem to rise above background levels, whether from the background of healthy individuals, other cancer types, or even other treatment response categories.[65] After identification of these alterations, tools such as SIFT,[66] PolyPhen,[67] and GWAVA[68] predict the functional relevance for specific cellular processes or impact of germline and somatic nonsynonymous mutations.[69] Other tools, such as ExPecto, go further to predict downstream phenotypes from different variant types, predicting the impact of noncoding mutations on gene expression in different tissue types.[70]

These approaches serve to generate hypotheses and direct experimental and in silico research focus. In this regard, much of computational biology is oriented toward hypothesis generation to reduce the possible experimental space to something more tractable for expensive and time-consuming wet lab experiments. Any new experiment performed based upon the hypotheses generated provides valuable feedback to these computational tools so that they can update their underlying models, irrespective of whether the experimental result is positive or negative. This feedback loop is a vital means for iterating our understanding of cancer. Notably, computational tools offer the biologist a broader perspective of the landscape of functional elements that may affect their mechanism of interest in subtle and unexpected ways while synthesizing diverse information types. As the breadth of data increases with each technological advance, these computational tools will be a necessary means by which we understand cancer biology.

7.5.2 Big Data and Statistical Learning

The detection of cancer relevant noncoding variants is made more challenging due to the relative lack of selective pressures resulting in many rare, noncoding variants. While large-scale analyses across thousands of patient samples are yielding numerous noncoding alterations that show cancer specificity,[71] the rarity for the vast majority of variants can stymie both statistical approaches for variant detection and functional prediction as well as ML approaches. Despite this reality, ML methods have been put forth to identify such variants.[72] We will review the fundamentals of these ML methods to better understand their applications in characterizing noncoding DNA and discuss their current impact on cancer research. There are two broad categories of learning methods: supervised and unsupervised.

Unsupervised learning is where ML is applied to find structure or patterns in the data without labeled response or class information. This is often used to identify novel signals and is typically implemented via some form of cluster analysis. Clustering methods can vary widely in their approach, although most make use of some defined distance between samples to find categories or groupings that best describe the data. Common clustering methods include hierarchical, k-means, self-organizing maps, and hidden Markov models. Unsupervised techniques make few assumptions on the data, and along with identifying novel patterns, it can be used to help impute missing information. These have been used extensively throughout cancer research for decades.

Supervised learning methods, on the other hand, use labeled response or class information and attempt to discover a mapping function from the input data to the label. These approaches yield models that can then be applied to new data for the prediction of the sample labels. Supervised approaches require the use of some training data sets and a separate test set on which to evaluate the model. The structure and quantity of training data depend on the supervised method chosen. Furthermore, care must be taken with certain methods to avoid overfitting, that is, inferring a function that is so complex and specific to the training data that it is not sufficiently generalizable to accurately predict the labels of the test data set. These supervised approaches may be offered DNA sequence patterns from the noncoding region that have functional relevance in a cancer and then be used to scan the dark-matter DNA for additional sites that may have similar features, thus suggesting similarly functionality. The identification of such sites would further elucidate these unknown regions and may point to additional factors for cancer development. There exists a variety of traditional machine learning methods that have been successfully applied to biological data to characterize sequence features and functional elements. We will describe a few of the more prevalent supervised methods and their role in the cancer analysis.

Cancer research has benefited from machine learning analysis for decades, beginning with simple regression-based approaches that look for functions that minimize a cost function or models that draw boundaries between labeled data sets (figure 7.1A). Kernel methods, such as support vector machines (SVMs), are more sophisticated learning functions that use a kernel or similarity function. A simple example of such a function is a linear kernel or dot product of two vectors. The advantage of using a kernel function is that they are able to operate on very high-dimensional data without requiring knowledge of the input data, and instead it returns a simple number, irrespective of the feature space. SVMs have been popular instantiations of kernel methods to analyze genomic data, classify disease samples,[73] and identify functional variants.[74] SVMs have been used extensively in cancer research[75] from the early days of using microarray data to classify cancer[76] to the prediction of patient response to a variety of chemotherapies[77] to the estimate of disease recurrence.[78] Decision trees and random forests (RFs) have similarly been applied for unsupervised regression tasks but were primarily used for the development of classification and prediction models. Recently, RFs have been successfully applied to distinguish disease-associated noncoding variants from healthy controls.[79] They have also shown it is possible to predict a cancer cell of origin from mutational load and epigenetic data,[80] and a variation of the RF that considers intertumoral heterogeneity was able to improve drug response prediction in different cancer types.[81] RFs and SVMs are also well suited to identifying informative features of the classification or regression tasks. This makes the discovery of the specific elements playing a role in the phenotype of interest all the more attainable. Bayesian networks (BNs) have been a popular choice for supervised learning in cancer, because these networks can operate on a relatively small amount of the data, and by being probabilistic, they can handle the uncertainty concomitant of any given event. One can integrate multiple modes of data into a single BN model along with prior expert knowledge. For example, it is possible to include clinical patient history information, genomic

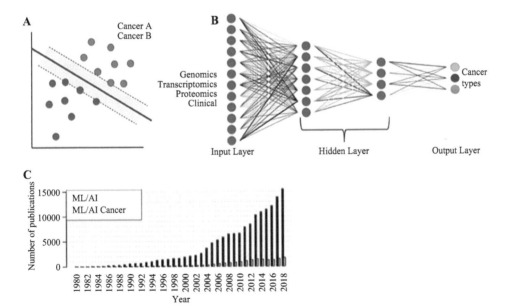

Figure 7.1
(A) Example of discriminating boundary learned between two cancer class labels. (B) Example of neural network structure to classify or predict a sample cancer type that can take a variety of data inputs and will learn the optimal weight functions through the hidden layers until a final output label is given. (C) Number of publications over time in PubMed mentioning ML/AI (with terms "deep learning," "machine learning," "artificial intelligence," and "neural networks") and ML/AI + "cancer" as of January 18, 2019.

profiling data, and image analysis results into a single Bayesian network to give a probability for the phenotype of interest.[82] Bayesian networks were recently used to discover TRIB1 as a novel regulator of cancer progression and survival.[83] Such classic ML methods continue to show their relevance and power in identifying novel signals in complex, often noisy data.

While AI and deep learning algorithms, primarily using neural networks, have been in existence since the 1940s, these approaches have only made a resurgence in the past fifteen years with the advent of sufficiently advanced computing hardware for these methods to be efficiently deployed at scale. AI is becoming more relevant and valuable in cancer research as the scale, diversity, and complexity of data generated expand. Artificial neural networks (ANNs) are a particular embodiment of AI that have seen renewed interest as the state of computational power has grown dramatically in recent years (figure 7.1B). ANN models are ordered as successive layers of nodes, or "neurons," whereby within each layer, the input undergoes some transformation to serve as input to the following layer. With each layer, the data are further abstracted, and each layer may be considered a unique representation of the data. Within a layer, there are often linear and nonlinear transformations that enable these approaches to identify nonlinearities in the data. The output of an ANN is, in the case of backpropagation, then evaluated against the correct labels, and adjustments are made to the neuron weights in the network and the process repeats, ultimately attempting to minimize some error function. In this way, the ANN is learning.

ANNs are widely being applied to a host of biological problems from variant calling to diagnosis to treatment planning. For example, DeepVariant is an AI-driven approach to identify SNPs and small indels that makes use of deep neural networks to learn generalized models that can operate across genome builds and even species.[84] This system outperformed the current state-of-the-art statistical methods in a 2016 Food and Drug Administration (FDA)–administered SNP identification challenge, the Truth Challenge, without any specialized knowledge of genomics or next-generation sequencing. A key technology underpinning deep learning systems like DeepVariant are ANNs and, in this instance, a convolutional neural network (CNN).[85] CNNs are typically applied to image analysis tasks and natural language processing with great success.[86] These neural network approaches are now being leveraged for sequencing and variant analysis.[87] For DeepVariant, all reads relevant for a putative variant are represented in an image and provided to the CNN so that patterns among reads that are associated to variants may be learned. This AI system's accuracy in variant identification against GATK and others suggests that manual statistical models could eventually be replaced by a deep learning system. NNs have been used to predict the impact of mutations, including noncoding variants, on the expression of genes throughout the genome, as is the case with ExPecto.[88] While it prioritizes variants in disease- or trait-associated loci, it is scalable to regulatory promoter mutations where it has already profiled more than 140 million mutations.

Autoencoders are an instance of ANNs that are typically used in unsupervised applications. It learns a representation of the data and, in so doing, is able to reduce the dimensionality of the data into a form that is more tractable for traditional ML methods. This is valuable in the context of cancer genomics where often the feature space far exceeds the sample space. In this undersampled space, denoising stacked autoencoders have been used on high-dimensional gene expression data to find linkages between genes in breast cancer to classify cancer cells and can suggest genes relevant for disease etiology.[89] Autoencoders can also be applied in combination with other ML methods for *semisupervised* learning where the autoencoder infers the labels used for the supervised learner. This is useful for identifying novel signals when the response label information is incomplete or missing. This combination of deep learning and ML methods is an important means of dealing with the high feature space and relatively low sample numbers that often accompany omic profiling. It can do so while not sacrificing the detection of likely subtle effects of many different alterations that may be driving a cancerous cell toward malignancy.

A challenge for many NN approaches is that for the biologist, it can operate as something of a "black box," where it is difficult to extract the understanding of the data that the NN learned in order to achieve its performance. Visible neural networks have been introduced to take advantage of the learning features of the ANN by designing neurons in a more structured way that leverages domain expertise.[90] Such a structure would yield a more transparent view as to how the ANN arrived at its output. Even without a clear understanding of the learned underlying model of most NNs, ANNs and other AI systems have been used for cancer detection,[91] cancer diagnosis,[92] and clinical decision support.[93] Watson for Genomics is a cloud-based system that leverages AI to analyze genomic profiles of cancer patients, identify relevant actionable alterations, and compile relevant molecularly targeted therapeutic options while automatically internalizing the latest scientific literature through natural language processing.

These ML and AI approaches can similarly be applied to noncoding and dark-matter regions to uncover patterns hitherto irretrievable via traditional statistical approaches, whether because of limitations in sequencing technologies or underlying inter- and intratumoral heterogeneity. DeepSEA has been shown to predict the effect of the mutations, even in the noncoding dark-matter genome, on chromatin structure, the corresponding impact on the epigenetic states, and transcription factor binding site availability.[94] It has further been evaluated against the Human Gene Mutation Data (HGMD) to accurately prioritize HGMD indels after being trained against only HGMD SNPs. DeepSEA highlights the potential applications for ML and AI to characterize the noncoding and dark-matter elements of the genome and to eventually connect them to the development of cancer.

7.5.3   Aggregating to Find Emergent Features

Most mutations observed in a given cancer, whether in the coding region and especially in the noncoding regions, are not particularly penetrant, and so looking for individual alterations as a biomarker for a disease in a patient can be problematic. Fewer than 150 genes are considered to be significantly mutated across cancer types.[95] And while a gene within this set may be mutated in upward of 95 percent of a given cancer type, like *TP53* in serous ovarian cancers, it is not to say that a specific variant ever achieves that level of penetrance in any specific disease. Well-known recurrent mutations in colorectal adeno-carcinoma, such as *KRAS* G12 and *NRAS* Q61, may only be found in 30 percent and 5 percent of patients, respectively. Even noncoding mutations that would be considered frequent, such as the *PMS2* promoter mutations in melanoma, are only found in 10 percent of patients.[96] This leaves large numbers of patients without clear cancer drivers and high-lights again the challenges with the cancer driver and SMT paradigms as there may be a host of other factors at work in any given tumor.

These levels of recurrence make the challenge of characterizing whether an alteration or combination of alterations is playing a role in a disease type all the more difficult. Approaches are needed that look beyond repetitive patterns between patients and consider modes of aggregating patients according to a phenotype and identifying emergent patterns in aggregated populations. Early work into a novel method for agglomerating mutation signals across patients shows that even when focusing specifically on mutations that exclusively fall into the noncoding or dark-matter portions of the genome, it is possible to distinguish between different subtypes of hematological cancers that all develop from the same cell of origin.[97] ReVeaL is able to distinguish between blood cancer subtypes with high accuracy, whereas standard approaches using alleles or mutational load often were no better than chance in predicting subtype. It has already been established that there exist clear trends between different cancers when it comes to their tumor mutational burden (TMB),[98] and it is almost trivial to distinguish between diseases like skin squamous cell carcinoma, which has a very high TMB, and bone marrow myelodysplastic syndrome (MDS), which has an extremely low TMB. Yet most cancers are within a highly overlapping range of TMB, which is particularly the case for blood cancers, which have very similar TMB. Thus, ML and AI methods like RF, SVM, and NN tend to fail when looking at dif-ferences in variant and gene-level alterations because they are fundamentally looking for differences in mutational load or frequency differences between disease types. ReVeaL's processing of the data enables detection of the discriminatory signal in the dark-matter DNA, even when excluding regulatory regions and portions that are annotated as noncoding RNA, using a variety of machine learning and AI approaches. This result also highlights the impor-tance of understanding the characteristics and nuances of the data being analyzed to better process them and offer an input for any learning system with as little noise as possible.

The biggest limitation of any Big Data cancer analysis often lies in an inadequate understanding of the confounding technical and biological issues of the data. To address such challenges, broad initiatives are needed to coordinate diverse research strategies to improve data availability and sharing, to understand available computational methodolo-gies, and to establish strong feedback mechanisms between insights made from data science and that of experimental or clinical validation studies. Projects like the Joint Design of Advanced Computing Solutions for Cancer (JDACS4C) is one such project from the

National Cancer Institute and the Department of Energy to advance precision oncology through advanced computing (https://cbiit.cancer.gov/ncip/hpc/jdacs4c) at molecular, cellular, and population scales. It is only through the wide availability of data to multiple research groups that we can discover more of the functional elements of a disease while avoiding the artifacts of too small sample size and biases in study design. Any machine learning method is only as good as the data provided and to a large extent can only learn based upon what it has seen. Crowdsourcing challenges such as DREAM have been successful at coalescing a community of researchers around specific biological questions to determine best practices and discover unseen technical and biological confounding factors. DREAM can even serve as a model of how to develop data collection and analytic strategies[99] and has already found success in the assessment of structural variation detection techniques[100] and predicting essential genes in cancer cell lines.[101]

For the study of noncoding alterations, it is important to recognize that the selective pressures will vary widely across the noncoding areas, whether in the regulatory regions, intergenic regions, or areas of open chromatin. Aggregative methods that can accommodate the widely varying mutational patterns of these regions observed across patients will enable AI systems to make better use of whole-genome data to uncover regions of the genome whose functional relevance to cancer has been underappreciated. Models that look for hotspot regions within the noncoding genome are examples of aggregative methods that bin mutations spatially along the genome. Recently, a study found eleven noncoding hotspots in CTCF binding sites, a DNA-binding protein important for genomic architecture, when analyzing 212 gastric cancer patients.[102] These sites, like those identified via DeepSea, showed a significant association to nearby chromosomal changes and affected gene expression of neighboring genes. Extending these sorts of aggregative approaches to aggregate over the samples of a cancer type may lead to the identification of functional elements in the dark-matter region whose mutation frequencies are low.

7.6   Conclusions

With ever-increasing computational power, the role of Big Data analytics, machine learning, and artificial intelligence will continue to expand and become a vital tool for understanding the many facets of cancers. The algorithms that underpin these approaches are capable of integrating vast amounts of heterogeneous information types and of discovering new patterns in genomic data that may explain the etiology of disease, particularly in cases where there do not appear to be classic "driver" alterations or mechanisms of therapeutic resistance. Only in the past five years has there been incredible interest in the application of AI for biological questions (figure 7.1C). As we recognize the insufficiency of the tenets of the classical somatic mutation theory of cancer and even the driver gene paradigm, we will need to look toward methods that are able to integrate signals from across a variety of data types and at a scale (both in the sample and feature space) not yet seen before in biology. We are now beginning to see studies analyzing patients at multiple scales to identify molecular covariates to polygenic health risks.[103] Price et al. found that over the course of nine months, as they performed longitudinal whole-genome, metabolomic, proteomic, microbiomic, and clinical assays on 108 patients along with daily activity tracking, they were able to detect changes in the molecular state that predicted changes in the health state and in some cases were an early indication of disease state, such as liver disease and diabetes. Additional studies that took a similar multiscale data integration approach have made associations between Alzheimer disease and herpesvirus[104] or have been able to develop predictive models of cancer prognosis and recurrence.[105] CNNs are already being used to make connections between histopathology images and genetic mutations such that from an image of tumor cells, it is possible to predict some of the underlying mutations.[106] These Big Data analytic approaches will enable the discovery of biomarkers that are not only population specific to a highly granular level but will offer much-needed guidance as to the unique mechanisms an individual's cancer may be employing in its development, growth, and response to treatment.

AI creates opportunities for the vast, underexplored regions of the noncoding and dark-matter portions of the genome to be leveraged in combination with our understanding of alterations in the coding region. This will provide much-needed context for those patients who do not fit into the traditional model of cancer development. Such whole-genome integration, in conjunction with clinical patient information, will further shift cancer research from the study of a few important genes to a broader appreciation of the myriad mechanisms that come into unique functional combinations within an individual patient. With such knowledge, we will have the tools to improve outcomes for all patients.

Notes

1. Consortium, E.P., An integrated encyclopedia of DNA elements in the human genome. *Nature*, 2012. 489(7414): p. 57–74.

2. Network, C.G.A.R., et al., The Cancer Genome Atlas Pan-Cancer analysis project. *Nat Genet*, 2013. 45: p. 1113–1120.

3. Campbell, P.J., et al., Pan-cancer analysis of whole genomes. bioRxiv, 2017: p. 162784.

4. Soto, A.M. and C. Sonnenschein, One hundred years of somatic mutation theory of carcinogenesis: Is it time to switch? *BioEssays*, 2014. 36(1): p. 118.

5. Helleday, T., S. Eshtad, and S. Nik-Zainal, Mechanisms underlying mutational signatures in human cancers. *Nat Rev Genet*, 2014. 15(9): p. 585.

6. Horn, S., et al., TERT promoter mutations in familial and sporadic melanoma. *Science*, 2013. 339(6122): p. 959–961. Ward, R.L., et al., Identification of constitutional MLH1 epimutations and promoter variants in colorectal cancer patients from the Colon Cancer Family Registry. *Genet Med*, 2013. 15(1): p. 25–35.

7. Knudson, A.G., Mutation and cancer: statistical study of retinoblastoma. *Proc Natl Acad Sci USA*, 1971. 68: p. 820–823.

8. Araya, C.L., et al., Identification of significantly mutated regions across cancer types highlights a rich landscape of functional molecular alterations. *Nat Genet*, 2016. 48: p. 117–125.

9. Alexandrov, L.B., et al., Signatures of mutational processes in human cancer. *Nature*, 2013. 500(7463): p. 415. Consortium, G.P. and others, An integrated map of genetic variation from 1,092 human genomes. *Nature*, 2012.

491(7422): p. 56. Kandoth, C., et al., Mutational landscape and significance across 12 major cancer types. *Nature*, 2013. 502(7471): p. 333. Vogelstein, B., et al., Cancer genome landscapes. *Science*, 2013. 339(6127): p. 1546–1558.

10. Greenman, C., et al., Patterns of somatic mutation in human cancer genomes. *Nature*, 2007. 446: p. 153–158.

11. Soto, A.M. and C. Sonnenschein, One hundred years of somatic mutation theory of carcinogenesis: Is it time to switch? *BioEssays*, 2014. 36(1): p. 118. Brucher, B.L.D.M. and I.S. Jamall, Somatic mutation theory: why it's wrong for most cancers. *Cell Physiol Biochem*, 2016. 38(5): p. 1663–1680.

12. Castro-Giner, F., P. Ratcliffe, and I. Tomlinson, The mini-driver model of polygenic cancer evolution. *Nat Rev Cancer*, 2015. 15: p. 680–685.

13. McFarland, C.D., et al., Impact of deleterious passenger mutations on cancer progression. *Proc Natl Acad Sci USA*, 2013. 110(8): p. 2910–2915. McFarland, C.D., L.A. Mirny, and K.S. Korolev, Tug-of-war between driver and passenger mutations in cancer and other adaptive processes. *Proc Natl Acad Sci USA*, 2014. 111(42): p. 15138–15143.

14. Muller, H.J., The relation of recombination to mutational advance. *Mut Res*, 1964. 1(1): p. 2–9.

15. Goyal, S., et al., Dynamic mutation—selection balance as an evolutionary attractor. *Genetics*, 2012. 191(4): p. 1309–1319.

16. Castro-Giner, F., P. Ratcliffe, and I. Tomlinson, The mini-driver model of polygenic cancer evolution. *Nat Rev Cancer*, 2015. 15: p. 680–685. Bennett, L., et al., Mutation pattern analysis reveals polygenic mini-drivers associated with relapse after surgery in lung adenocarcinoma. *Sci Rep*, 2018. 8(1): p. 14830.

17. Lawrence, M.S., et al., Discovery and saturation analysis of cancer genes across 21 tumour types. *Nature*, 2014. 505(7484): p. 495–501.

18. Zhang, W., et al., A global transcriptional network connecting noncoding mutations to changes in tumor gene expression. *Nat Genet*, 2018. 50(4): p. 613–620. Khurana, E., et al., Role of non-coding sequence variants in cancer. *Nat Rev Genetics*, 2016. 17: p. 93–108. Santana Dos Santos, E., et al., Non-coding variants in BRCA1 and BRCA2 genes: potential impact on breast and ovarian cancer predisposition. *Cancers*, 2018. 10(11).

19. Claes, K., et al., Pathological splice mutations outside the invariant AG/GT splice sites of BRCA1 exon 5 increase alternative transcript levels in the 5' end of the BRCA1 gene. *Oncogene*, 2002. 21(26): p. 4171–4175.

20. Khurana, E., et al., Role of non-coding sequence variants in cancer. *Nat Rev Genetics*, 2016. 17: p. 93–108. Zhou, S., A.E. Treloar, and M. Lupien, Emergence of the noncoding cancer genome: a target of genetic and epigenetic alterations. *Cancer Discov*, 2016. 6(11): p. 1215–1229. Fredriksson, N.J., et al., Systematic analysis of noncoding somatic mutations and gene expression alterations across 14 tumor types. *Nat Genet*, 2014. 46(12): p. 1258–1263. Weinhold, N., et al., Genome-wide analysis of noncoding regulatory mutations in cancer. *Nat Genet*, 2014. 46: p. 1160–1165.

21. Polak, P., et al., Cell-of-origin chromatin organization shapes the mutational landscape of cancer. *Nature*, 2015. 518: p. 360–364.

22. Castro-Giner, F., P. Ratcliffe, and I. Tomlinson, The mini-driver model of polygenic cancer evolution. *Nat Rev Cancer*, 2015. 15: p. 680–685. Hornshoj, H., et al., Pan-cancer screen for mutations in non-coding elements with conservation and cancer specificity reveals correlations with expression and survival. NPJ Genom Med, 2018. 3: p. 1.

23. Horn, S., et al., TERT promoter mutations in familial and sporadic melanoma. *Science*, 2013. 339: p. 959–961.

24. Dvir, S., et al., Deciphering the rules by which 5'-UTR sequences affect protein expression in yeast. *Proc Natl Acad Sci USA*, 2013. 110(30): p. E2792–E2801.

25. Li, G., et al., CCAR1 5' UTR as a natural miRancer of miR-1254 overrides tamoxifen resistance. *Cell Res*, 2016. 26: p. 655–673.

26. Litt, D.B., et al., Hybrid lithographic and DNA-directed assembly of a configurable plasmonic metamaterial that exhibits electromagnetically induced transparency. *Nano Lett*, 2018. 18: p. 859–864.

27. Banerji, J., S. Rusconi, and W. Schaffner, Expression of a beta-globin gene is enhanced by remote SV40 DNA sequences. *Cell*, 1981. 27(2 Pt 1): p. 299–308. Bulger, M. and M. Groudine, Functional and mechanistic diversity of distal transcription enhancers. *Cell*, 2011. 144: p. 327–339.

28. Yang, J. and V.G. Corces, Chromatin insulators: a role in nuclear organization and gene expression. *Adv Cancer Res*, 2011. 110: p. 43–76.

29. Zentner, G.E., P.J. Tesar, and P.C. Scacheri, Epigenetic signatures distinguish multiple classes of enhancers with distinct cellular functions. *Genome Res*, 2011. 21: p. 1273–1283.

30. De Santa, F., et al., A large fraction of extragenic RNA pol II transcription sites overlap enhancers. *PLoS Biol*, 2010. 8: p. e1000384. Kolovos, P., et al., Enhancers and silencers: an integrated and simple model for their function. *Epigenetics Chromatin*, 2012. 5: p. 1.

31. Fishilevich, S., et al., GeneHancer: genome-wide integration of enhancers and target genes in GeneCards. *Database*, 2017. 2017.

32. Lieberman-Aiden, E., et al., Comprehensive mapping of long-range interactions reveals folding principles of the human genome. *Science*, 2009. 326: p. 289–293.

33. Dostie, J., et al., Chromosome Conformation Capture Carbon Copy (5C): a massively parallel solution for mapping interactions between genomic elements. *Genome Res*, 2006. 16(10): p. 1299–1309.

34. Simonis, M., et al., Nuclear organization of active and inactive chromatin domains uncovered by chromosome conformation capture-on-chip (4C). *Nat Genet*, 2006. 38: p. 1348–1354.

35. Dekker, J., et al., Capturing chromosome conformation. *Science*, 2002. 295(5558): p. 1306–1311.

36. Li, Z., et al., APOBEC signature mutation generates an oncogenic enhancer that drives LMO1 expression in T-ALL. *Leukemia*, 2017. 31: p. 2057–2064.

37. Herz, H.M., D. Hu, and A. Shilatifard, Enhancer malfunction in cancer. *Mol Cell*, 2014. 53(6): p. 859–866. Koche, R.P. and S.A. Armstrong, Genomic dark matter sheds light on EVI1-driven leukemia. *Cancer Cell*, 2014. 25: p. 407–408.

38. Ropero, S. and M. Esteller, The role of histone deacetylases (HDACs) in human cancer. *Molecular Oncology*, 2007. 1: p. 19–25.

39. Sabarinathan, R., et al., Nucleotide excision repair is impaired by binding of transcription factors to DNA. *Nature*, 2016. 532: p. 264–267.

40. Hnisz, D., et al., Activation of proto-oncogenes by disruption of chromosome neighborhoods. *Science*, 2016. 351: p. 1454–1458.

41. Jia, R., et al., Novel insights into chromosomal conformations in cancer. *Mol Cancer*, 2017. 16: p. 173.

42. Zapata, L., et al., Signatures of positive selection reveal a universal role of chromatin modifiers as cancer driver genes. *Sci Rep*, 2017. 7(1): p. 13124. Morgan, M.A. and A. Shilatifard, Chromatin signatures of cancer. *Genes Dev*, 2015. 29(3): p. 238–249. Suzuki, H., et al., Genome-wide profiling of chromatin signatures reveals epigenetic regulation of MicroRNA genes in colorectal cancer. *Cancer Res*, 2011. 71(17): p. 5646–5658.

43. Zeitz, M.J., et al., Genomic interaction profiles in breast cancer reveal altered chromatin architecture. *PLoS One*, 2013. 8(9): p. e73974.

44. Bastonini, E., et al., Chromatin barcodes as biomarkers for melanoma. *Pigment Cell Melanoma Res*, 2014. 27(5): p. 788–800.

45. Anastasiadou, E., L.S. Jacob, and F.J. Slack, Non-coding RNA networks in cancer. *Nat Rev Cancer*, 2018. 18: p. 5–18. Romano, G., et al., Small non-coding RNA and cancer. *Carcinogenesis*, 2017. 38: p. 485–491. Schwarzer, A., et al., The non-coding RNA landscape of human hematopoiesis and leukemia. *Nat Commun*, 2017. 8: p. 218. Vallone, C., et al., Non-coding RNAs and endometrial cancer. *Genes*, 2018. 9. Yan, X., et al., Comprehensive genomic characterization of long non-coding RNAs across human cancers. *Cancer Cell*, 2015. 28: p. 529–540.

46. Carthew, R.W. and E.J. Sontheimer, Origins and Mechanisms of miRNAs and siRNAs. *Cell*, 2009. 136: p. 642–655. Di Leva, G., M. Garofalo, and C.M. Croce, MicroRNAs in cancer. *Annu Rev Pathol*, 2014. 9: p. 287–314.

47. Thermann, R. and M.W. Hentze, Drosophila miR2 induces pseudo-polysomes and inhibits translation initiation. *Nature*, 2007. 447: p. 875–878.

48. Calin, G.A., et al., Frequent deletions and down-regulation of micro- RNA genes miR15 and miR16 at 13q14 in chronic lymphocytic leukemia. *Proc Natl Acad Sci USA*, 2002. 99(24): p. 15524–15529.

49. Costinean, S., et al., Pre-B cell proliferation and lymphoblastic leukemia/high-grade lymphoma in E(mu)-miR155 transgenic mice. *Proc Natl Acad Sci USA*, 2006. 103(18): p. 7024–7029.

50. Di Leva, G., M. Garofalo, and C.M. Croce, MicroRNAs in cancer. *Annu Rev Pathol*, 2014. 9: p. 287–314.

51. Kapranov, P., et al., RNA maps reveal new RNA classes and a possible function for pervasive transcription. *Science*, 2007. 316: p. 1484–1488.

52. Mercer, T.R., M.E. Dinger, and J.S. Mattick, Long non-coding RNAs: insights into functions. *Nat Rev Genetics*, 2009. 10: p. 155–159.

53. Yan, X., et al., Comprehensive genomic characterization of long non-coding RNAs across human cancers. *Cancer Cell*, 2015. 28: p. 529–540.

54. Hu, X., et al., The role of long noncoding RNAs in cancer: the dark matter matters. *Curr Opin Genet Dev*, 2018. 48: p. 8–15.

55. Yan, X., et al., Comprehensive genomic characterization of long non-coding RNAs across human cancers. *Cancer Cell*, 2015. 28: p. 529–540. Leucci, E., et al., Melanoma addiction to the long non-coding RNA SAMMSON. *Nature*, 2016. 531(7595): p. 518–522. Redis, R.S., et al., Allele-specific reprogramming of cancer metabolism by the long non-coding RNA CCAT2. *Mol Cell*, 2016. 61(4): p. 640.

56. Goodarzi, H., et al., Endogenous tRNA-derived fragments suppress breast cancer progression via YBX1 displacement. *Cell*, 2015. 161: p. 790–802. Pavon-Eternod, M., et al., tRNA over-expression in breast cancer and functional consequences. *Nucleic Acids Res*, 2009. 37: p. 7268–7280.

57. Goodarzi, H., et al., Endogenous tRNA-derived fragments suppress breast cancer progression via YBX1 displacement. *Cell*, 2015. 161: p. 790–802.

58. Huang, S.Q., et al., The dysregulation of tRNAs and tRNA derivatives in cancer. *J Exp Clin Cancer Res*, 2018. 37(1): p. 101.

59. Carey, N., *Junk DNA: A Journey through the Dark Matter of the Genome*. 2015, New York: Columbia University Press.

60. Gheldof, N., et al., Structural variation-associated expression changes are paralleled by chromatin architecture modifications. *PLoS One*, 2013. 8(11): p. e79973.

61. Capper, D., et al., DNA methylation-based classification of central nervous system tumours. *Nature*, 2018. 555(7697): p. 469–474.

62. Zhou, S., A.E. Treloar, and M. Lupien, Emergence of the noncoding cancer genome: a target of genetic and epigenetic alterations. *Cancer Discov*, 2016. 6(11): p. 1215–1229.

63. Lawrence, M.S., et al., Mutational heterogeneity in cancer and the search for new cancer-associated genes. *Nature*, 2013. 499: p. 214–218.

64. Cibulskis, K., et al., Sensitive detection of somatic point mutations in impure and heterogeneous cancer samples. *Nat Biotechnol*, 2013. 31: p. 213–219. Koboldt, D.C., et al., VarScan: variant detection in massively parallel sequencing of individual and pooled samples. *Bioinformatics*, 2009. 25: p. 2283–2285. Larson, D.E., et al., SomaticSniper: identification of somatic point mutations in whole genome sequencing data. *Bioinformatics*, 2012. 28: p. 311–317. Saunders, C.T., et al., Strelka: accurate somatic small-variant calling from sequenced tumor-normal sample pairs. *Bioinformatics*, 2012. 28: p. 1811–1817.

65. Lawrence, M.S., et al., Mutational heterogeneity in cancer and the search for new cancer-associated genes. *Nature*, 2013. 499: p. 214–218.

66. Ng, P.C. and S. Henikoff, SIFT: Predicting amino acid changes that affect protein function. *Nucleic Acids Res*, 2003. 31: p. 3812–3814.

67. Adzhubei, I.A., et al., A method and server for predicting damaging missense mutations. *Nat Methods*, 2010. 7(4): p. 248–249.

68. Ritchie, G.R.S., et al., Functional annotation of noncoding sequence variants. *Nat Methods*, 2014. 11(3): p. 294.

69. Gnad, F., et al., Assessment of computational methods for predicting the effects of missense mutations in human cancers. *BMC Genomics*, 2013. 14 Suppl 3: p. S7. Kircher, M., et al., A general framework for estimating the relative pathogenicity of human genetic variants. *Nat Genet*, 2014. 46(3): p. 310.

70. Zhou, J., et al., Deep learning sequence-based ab initio prediction of variant effects on expression and disease risk. *Nat Genet*, 2018. 50(8): p. 1171–1179.

71. Hornshoj, H., et al., Pan-cancer screen for mutations in non-coding elements with conservation and cancer specificity reveals correlations with expression and survival. *NPJ Genom Med*, 2018. 3: p. 1.

72. Ritchie, G.R.S., et al., Functional annotation of noncoding sequence variants. *Nat Methods*, 2014. 11(3): p. 294. Libbrecht, M.W. and W.S. Noble, Machine learning applications in genetics and genomics. *Nat Rev Genetics*, 2015. 16: p. 321–332. Shihab, H.A., et al., An integrative approach to predicting the functional effects of

non-coding and coding sequence variation. *Bioinformatics*, 2015. 31: p. 1536–1543. Zhou, J. and O.G. Troyanskaya, Predicting effects of noncoding variants with deep learning—based sequence model. *Nat Methods*, 2015. 12(10): p. 931.

73. Furey, T.S., et al., Support vector machine classification and validation of cancer tissue samples using microarray expression data. *Bioinformatics*, 2000. 16(10): p. 906–914. Guyon, I., et al., Gene selection for cancer classification using support vector machines. *Machine Learning*, 2002. 46(1–3): p. 389.

74. Kircher, M., et al., A general framework for estimating the relative pathogenicity of human genetic variants. *Nat Genet*, 2014. 46(3): p. 310.

75. Huang, S., et al., Applications of support vector machine (SVM) learning in cancer genomics. *Cancer Genomics Proteomics*, 2018. 15(1): p. 41–51.

76. Golub, T.R., et al., Molecular classification of cancer: class discovery and class prediction by gene expression monitoring. *Science*, 1999. 286(5439): p. 531–537.

77. Huang, C., et al., Machine learning predicts individual cancer patient responses to therapeutic drugs with high accuracy. *Sci Rep*, 2018. 8(1): p. 16444.

78. Kim, W., et al., Development of novel breast cancer recurrence prediction model using support vector machine. *J Breast Cancer*, 2012. 15(2): p. 230–238.

79. Ritchie, G.R.S., et al., Functional annotation of noncoding sequence variants. *Nat Methods*, 2014. 11(3): p. 294.

80. Polak, P., et al., Cell-of-origin chromatin organization shapes the mutational landscape of cancer. *Nature*, 2015. 518: p. 360–364.

81. Rahman, R., et al., Heterogeneity aware random forest for drug sensitivity prediction. *Sci Rep*, 2017. 7(1): p. 11347.

82. Emaminejad, N., et al., Fusion of quantitative image and genomic biomarkers to improve prognosis assessment of early stage lung cancer patients. *IEEE Trans Biomed Eng*, 2016. 63(5): p. 1034–1043. Exarchos, K.P., Y. Goletsis, and D.I. Fotiadis, Multiparametric decision support system for the prediction of oral cancer reoccurrence. *IEEE Trans Inf Technol Biomed*, 2012. 16(6): p. 1127–34. Ni, Y., F.C. Stingo, and V. Baladandayuthapani, Integrative Bayesian network analysis of genomic data. *Cancer Informatics*, 2014. 13: p. 39–48.

83. Gendelman, R., et al., Bayesian network inference modeling identifies TRIB1 as a novel regulator of cell-cycle progression and survival in cancer cells. *Cancer Res*, 2017. 77(7): p. 1575–1585.

84. Poplin, R., et al., Creating a universal SNP and small indel variant caller with deep neural networks. BioRxiv, 2017: p. 092890.

85. LeCun, Y., et al., Gradient-based learning applied to document recognition. *Proc IEEE*, 1998. 86(11): p. 2278–2324.

86. Collobert, R. and J. Weston. A unified architecture for natural language processing: deep neural networks with multitask learning. In *Proceedings of the 25th International Conference on Machine Learning*. New York: ACM, 2008. Dumoulin, V. and F. Visin, A guide to convolution arithmetic for deep learning. arXiv preprint arXiv:1603.07285, 2016. Girshick, R., et al. Deformable part models are convolutional neural networks. In *Proceedings of the IEEE Conference on Computer Vision and Pattern Recognition*. Boston: IEEE, 2015.

87. Zhou, J. and O.G. Troyanskaya, Predicting effects of noncoding variants with deep learning—based sequence model. *Nat Methods*, 2015. 12(10): p. 931. LeCun, Y., Y. Bengio, and G. Hinton, Deep learning. *Nature*, 2015. 521(7553): p. 436. Quang, D. and X. Xie, DanQ: a hybrid convolutional and recurrent deep neural network for quantifying the function of DNA sequences. *Nucleic Acids Res*, 2016. 44(11): p. e107.

88. Zhou, J., et al., Deep learning sequence-based ab initio prediction of variant effects on expression and disease risk. *Nat Genet*, 2018. 50(8): p. 1171–1179.

89. Danaee, P., R. Ghaeini, and D.A. Hendrix, A deep learning approach for cancer detection and relevant gene identification. *Pac Symp Biocomput*, 2017. 22: p. 219–229.

90. Ma, J., et al., Using deep learning to model the hierarchical structure and function of a cell. *Nat Methods*, 2018. 15: p. 290–298.

91. Danaee, P., R. Ghaeini, and D.A. Hendrix, A deep learning approach for cancer detection and relevant gene identification. *Pac Symp Biocomput*, 2017. 22: p. 219–229.

92. Naushad, S.M., et al., Artificial neural network-based exploration of gene-nutrient interactions in folate and xenobiotic metabolic pathways that modulate susceptibility to breast cancer. *Gene*, 2016. 580: p. 159–168. Yuan,

Y., et al., DeepGene: an advanced cancer type classifier based on deep learning and somatic point mutations. *BMC Bioinformatics*, 2016. 17: p. 476.

93. Lisboa, P.J. and A.F.G. Taktak, The use of artificial neural networks in decision support in cancer: a systematic review. *Neural Networks*, 2006. 19: p. 408–415. Patel, N.M., et al., Enhancing next-generation sequencing-guided cancer care through cognitive computing. *Oncologist*, 2018. 23: p. 179–185. Wrzeszczynski, K.O., et al., Comparing sequencing assays and human-machine analyses in actionable genomics for glioblastoma. *Neurol Genet*, 2017. 3(4): p. e164.

94. Zhou, J. and O.G. Troyanskaya, Predicting effects of noncoding variants with deep learning—based sequence model. *Nat Methods*, 2015. 12(10): p. 931.

95. Kandoth, C., et al., Mutational landscape and significance across 12 major cancer types. *Nature*, 2013. 502(7471): p. 333.

96. Chalmers, Z.R., et al., Analysis of 100,000 human cancer genomes reveals the landscape of tumor mutational burden. *Genome Med*, 2017. 9(1): p. 34.

97. Parida, L., et al., Defining subtle cancer subtypes using the darkest DNA. In *Proceedings: AACR Annual Meeting 2019*. Atlanta: AACR, 2019. Parida, L., et al., Dark-matter matters: discriminating subtle blood cancers using the darkest DNA. *PLOS Comput Biol*, 2019. 15(8): p. e1007332.

98. Chalmers, Z.R., et al., Analysis of 100,000 human cancer genomes reveals the landscape of tumor mutational burden. *Genome Med*, 2017. 9(1): p. 34.

99. Azencott, C.A., et al., The inconvenience of data of convenience: computational research beyond post-mortem analyses. *Nat Methods*, 2017. 14(10): p. 937–938.

100. Lee, A.Y., et al., Combining accurate tumor genome simulation with crowdsourcing to benchmark somatic structural variant detection. *Genome Biol*, 2018. 19(1): p. 188.

101. Gonen, M., et al., A Community challenge for inferring genetic predictors of gene essentialities through analysis of a functional screen of cancer cell lines. *Cell Syst*, 2017. 5(5): p. 485–497 e3.

102. Guo, Y.A., et al., Mutation hotspots at CTCF binding sites coupled to chromosomal instability in gastrointestinal cancers. *Nat Commun*, 2018. 9(1): p. 1520.

103. Price, N.D., et al., A wellness study of 108 individuals using personal, dense, dynamic data clouds. *Nat Biotechnol*, 2017. 35(8): p. 747–756.

104. Readhead, B., et al., Multiscale analysis of independent Alzheimer's cohorts finds disruption of molecular, genetic, and clinical networks by human herpesvirus. *Neuron*, 2018. 99(1): p. 64–82 e7.

105. Exarchos, K.P., Y. Goletsis, and D.I. Fotiadis, Multiparametric decision support system for the prediction of oral cancer reoccurrence. *IEEE Trans Inf Technol Biomed*, 2012. 16(6): p. 1127–1134. Chen, Y.-C., et al., Risk classification of cancer survival using ANN with gene expression data from multiple laboratories. *Comput Biol Med*, 2014. 48: p. 1–7. Gevaert, O., et al., Predicting the prognosis of breast cancer by integrating clinical and microarray data with Bayesian networks. *Bioinformatics*, 2006. 22(14): p. e184–e190.

106. Coudray, N., et al., Classification and mutation prediction from non-small cell lung cancer histopathology images using deep learning. *Nat Med*, 2018. 24(10): p. 1559–1567.

III THE TIME DIMENSION OF CANCER

8 Darwinism, Not Mutationalism, for New Cancer Therapies

Jacob Scott, David Basanta, and Andriy Marusyk

Overview

Despite decades of research and drug development efforts, most metastatic cancers remain incurable. Even the most effective targeted therapies, which elicit strong tumor shrinkage without incurring significant side effects, provide only a temporary relief. As long as therapies fail to kill all of the tumor cells, and as long as tumor cells retain epigenetic plasticity and genetic instability, the remaining cells eventually evolve under the therapy-imposed selective pressures, enabling tumors to resume growth, eventually leading to clinical relapse. Even though we know that tumors are changing under therapy, this knowledge is currently not factored in when choosing how we treat patients. The standard of care in targetable cancers is a continuous administration of the front-line drug until tumors stop responding and relapse, at which point the front-line agent is rendered useless, and a second-line drug option has to be chosen. Then, the cycle repeats, until we eventually run out of effective options. Under this scenario, we are following the tumor's lead, choosing the response after the tumor "made its move" and has changed, in most cases irreversibly. This strategy is not necessarily the only choice or the best choice. We posit that explicit consideration of the evolutionary processes that drive resistance can offer new approaches to stop or at least delay the relapse. However, to take evolution into account, we first need to understand its basic principles and capture this understanding in predictive models. We argue that the current dominant dogma—namely, that mutations "drive" cancer evolution—poses a formidable roadblock limiting useful conceptual developments and preventing integration of the impact of cellular contexts, environmental impact, phenotypic plasticity, and nongenetic heritable variability. Removing this roadblock and refocusing on the original Darwinian view of evolution as adaptation, achieved through the interplay of diversification of heritable phenotypes and environmental selection, should provide a more balanced framework to understand cancer evolution. We argue that applying Darwinian concepts in their original form to understand cancer enables new, potentially groundbreaking approaches toward therapies that optimize long-term patient survival, rather than short-term killing of tumor cells.

8.1   Acquired Resistance as a Major Clinical Challenge

Although it is not that difficult to kill tumor cells, achieving the selectivity needed to do so while sparing normal cells and tissues is far more challenging. The first types of drugs that became available to oncologists were cytotoxic compounds covered under the broad term of "chemotherapies." Chemotherapies preferentially target proliferating cells by interfering with cell division, achieving selectivity toward cancer cells due to their higher proliferation rates. Additionally, compromised DNA integrity checkpoints and a high degree of genetic instability within cancer cells exacerbate the challenge of completing DNA replication and cell division without making catastrophic errors. Thus, chemotherapies essentially push tumor cells "over the cliff" more easily than normal cells. This selectivity, however, is relatively limited, as normal cells and tissue suffer as well, which translates into narrow therapeutic windows. Consequently, in many cases, chemotherapies ultimately fail because of the limitations of both the tolerated dose versus the effective dose and the numbers of cycles that the drugs can be administered, before side effects start outweighing antitumor benefits. Nevertheless, tumors typically become less and less responsive with each subsequent cycle of chemotherapy, indicative of evolving resistance, and patients die of chemorefractory disease.

The molecular biology revolution and the dramatic expansion of biomedical research provided critical insights into the biology of cancers, eventually leading to the development of a different class of drugs—one that enabled truly "targeted" therapies. Instead of relying on general cell toxicities, leading to relatively weak differential vulnerabilities of cancer cells, these drugs target well-defined signaling or repair pathways ("addictions") specifically within tumor cells. Targeted therapies represent a rather broad class of drugs, covering small-molecule and antibody-based therapies directed against proteins that generate abnormal signaling, such as those resulting from gene fusion (e.g., *BCR-ABL*), genetic mutations (e.g., *EGFR*), or gene overexpression (e.g., *HER2*). Additionally, the term applies to therapies directed against preexisting dependencies (e.g., antihormonal therapies in prostate and breast cancers) or acquired ones (e.g., PARP inhibitors). Finally, targeted therapies might be directed against certain elements of the tumor microenvironment, such as antiangiogenic agents. Much higher selectivity of targeted therapies creates a widened therapeutic window, which typically leads to better clinical responses. Further, the increased drug tolerability enables long-term, daily administration of the drugs, often resulting in months or even years of tumor control. Thus, when available, they are often used as front-line therapeutic options. Although systemic toxicity-related limitations are less prominent with this class of drugs, the issue of therapy resistance became even more pronounced with such targeted therapies. In advanced, metastatic cancers, targeted therapies ultimately fail with almost no exception, as the initially responsive tumors acquire resistance and relapse. When effective backup, or second-line, options are available, they ultimately face the same fate.

Frustratingly, even though we know that tumors can and do evolve therapy resistance, we currently do not take this knowledge into account when designing therapies, with oncologists only reacting to the development of resistance, rather than being able to preempt it. Therapies are optimized toward achieving maximal short-term responses, through continuous administration of highest tolerated doses of the drugs, for as long as tumors respond. This approach is not the only or the best course of action, and it should be possible to interfere with the evolutionary mechanisms of resistance, to block, redirect, or at least delay it. At present, we lack the knowledge needed: despite the massive research efforts focusing on molecular mechanisms, our understanding of how resistance evolves remains rudimentary, as we lack proper conceptual frameworks and factual knowledge.

8.2   Somatic Evolution: Mutation Driven or a Darwinian Process?

Most cancers start from a single, initially normal cell and end up as large phenotypically and genetically diverse populations of highly abnormal cells that have acquired multiple malignant phenotypic features, broadly referred to as the hallmarks of cancer.[1] Since most of these malignant phenotypes are currently believed to be caused by genetic mutations, acquired through carcinogenesis, the concept that cancers are a result of somatic evolution has been generally accepted in the field.

The idea of cancer as a product of somatic evolution is typically attributed to the seminal 1976 paper by Peter Nowell, "The Clonal Evolution of Tumor Cell Populations," and the 1975 paper by John Cairns, "Mutation Selection and the Natural History of Cancer." Nowell's paper conceptualized somatic evolution in general Darwinian terms, where mutational diversification generated subclones, enabling differential survival of some of the genetic variants. The process was described as a "stepwise sequence of mutations," with the differences in mutational status between different tumors not only reflecting the stochastic nature of mutations but also being "partially determined by the environmental pressures on selection." Cairns's paper, focused on a discussion of how tissue organization might be limiting the selective effect of oncogenic mutations, was even more grounded in the Darwinian paradigm.

However, the currently dominant framework to conceptualize somatic evolution was primarily shaped by a highly influential 1990 paper by Eric Fearon and Bert Vogelstein, who argued that initiation and clinical progression of cancers are the direct result of the occurrence and accumulation of specific mutations.[2] Since the accuracy of DNA replication is not perfect, each cell division leads to the emergence of a few point mutations. Further, since most tumor cells are aneuploid, and since aneuploidy is strongly linked with genomic instability, each cell division is also linked with a high probability of deleting, amplifying, or translocating parts of chromosomes. The majority of the new mutations are nonconsequential "passengers." However, some of these random point or chromosomal mutations are "drivers" of cancer/tumor evolution as they create (e.g., *BCR-ABL* fusion), activate (e.g., *RAS* mutations), or amplify (e.g., *HER2* amplification) an oncogene or inactivate/delete a tumor suppressor gene (e.g., mutation/loss of *TP53*). In this framework, somatic cancer/tumor evolution can be fully accounted for by the "bad luck"[3] of the occurrence of specific "driver" mutations that unlock a "hallmarks of cancer" phenotype.

The highly influential cancer biology textbook by Robert Weinberg[4] and many key opinion-forming reviews on the subject refer to this mutation-centric depiction of somatic evolution as "Darwinian"[5]—but is it? At the first glance, the reference is warranted, as the ideas of fitness and selective advantage are considered at play. However, instead of reflecting a measure of being fit for a specific context, "fitness" becomes context independent in the mutation-centric paradigm, as increased selective advantage is considered an essentially invariable consequence of the occurrence of "driver" mutations. Following the compelling evidence for the existence of substantial genetic heterogeneity in tumors,[6] branching mutational trees describing clonal phylogeny within tumors have been often used to point out parallels to Darwin's depiction of the tree of life (figure 8.1B,C). However, Darwin was not the inventor of the concept of branching evolution, nor did he make such a claim. The concept can be traced back to the ancient idea of the Tree of Life, and branching trees were used to describe phylogenetic relationships in the kingdom of life by many scientists, including Lamarck and Augier, well before Darwin.

Darwin's key contribution to biology was the proposal of a mechanism behind the phylogenetic tree (i.e., natural selection). Referring to mutation-driven evolution as Darwinian has a truly Orwellian twist to it, as the concept of "driver" mutations (when applied to evolution, rather than a specific phenotype) essentially revives the long-rejected arguments expressed by the explicitly anti-Darwinian saltationalist/mutationalist camp.[7] If the mutation-centric take on somatic evolution is correct, then mutationalists were right all along but have unluckily chosen the wrong natural phenomena to apply their ideas to. If this is the case, and if somatic evolution operates under principles that are fundamentally different from those operating in natural populations of species, then cancer biologists should avoid using the term "Darwinian" in the description of somatic evolution. However, as we and others have argued, despite some unique features,[8] Darwinian evolution by natural selection provides a better conceptual framework to understand how and why cancers evolve.[9] Perhaps more important, it also offers new ways to approach cancer therapies, as we will articulate below.

What then, exactly, is Darwin's concept of evolution by natural selection? Despite being arguably the most powerful idea in biology, captured in the famous phrase of Dobzhansky, "Nothing makes sense in biology except in the light of [Darwinian] evolution," the concept itself is fairly simple. Preexisting and ongoing diversification of heritable phenotypes enables populations to change and adapt through the action of natural selection (i.e., differential survival of individuals whose phenotypes are more "fit" for their complex, multifactorial environment). Thus, the whole process is "driven" by the interplay of two forces: (1) diversification of heritable phenotypes and (2) selective survival/reproduction of fitter individuals

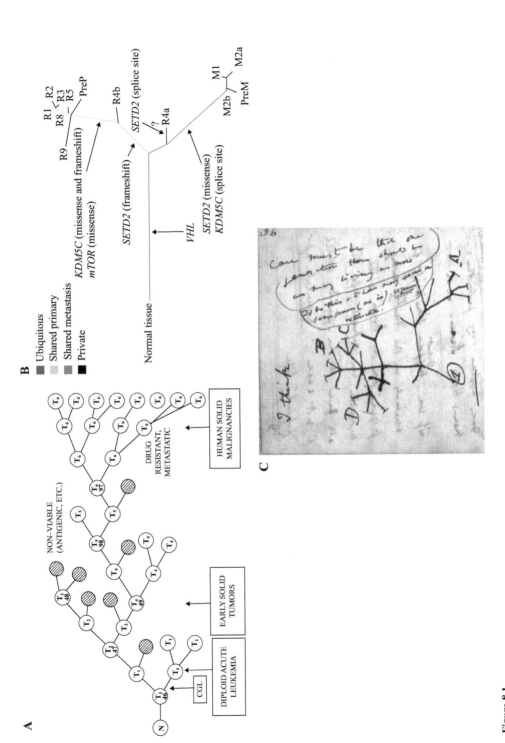

Figure 8.1

(A) Schematic representation of somatic evolution, arising from mutational diversification and shaped by environmental selection. Figure taken from the seminal publication of Nowel.[10] (B) Inferences of clonal evolution from sequencing tumor DNA allow reconstruction of phylogenetic trees of cancer cells (from Gerlinger et al.[11]). (C) Branching evolutionary tree, illustrating speciation from Darwin's 1837 notebook.

(figure 8.2A). Diversification provides a substrate, on which natural selection operates. The direction and the outcome of the process are shaped by "options" provided by *diversification* and continuous optimization through *selection* by the environment for the fittest variants.

Notably, the etymology of "fitness" assumes a coupling to a specific context—thus understanding Darwin's process is impossible without taking this context into consideration. Under stable conditions, this Darwinian process enables populations to reach and maintain an optimal fitness. Once the context changes, the combination of diversification and selection enables the population to adapt to the new context (figure 8.2B). That fitness is a moving target is self-evident for any student of natural populations, but not so much for the majority of cancer biologists. As was the case for views of many prominent geneticists prior to the evolutionary synthesis, lack of appreciation of adaptation to shifting contexts may stem from the artificial nature of experimental laboratory settings, where keeping things constant is essential for reproducibility. But it is not an adequate reflection of the clinical reality, where contextual changes in space and time are omnipresent.

It should be noted that Darwin's original take on evolution was not married to any specific mechanism of diversification. Darwin was not influenced by Mendel's studies, and his speculative mechanism of pangenesis was proven wrong by genetics several decades later. Furthermore, Darwin was quite open to the now-ridiculed (perhaps not entirely justly) ideas of Lamarckian induction. Thus, we find it deeply ironic that, consistent with the mutation-centric misinterpretation of Darwinism in cancer research, the term "Darwinian" is often restricted exclusively to a selection of preexisting or newly emerging genetic mutations, with Darwinian explanations being rejected when encountering nongenetic variability, even when the observed phenomenon clearly involves differential sur-

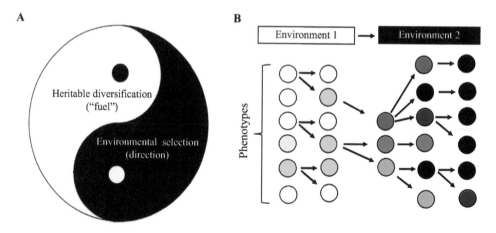

Figure 8.2
(A) Darwinian evolution represents an interplay between diversification of heritable phenotypes and natural selection, which preferentially "picks" the variants that are fitter to a given context. (B) Schematic illustration of a population adapting to environmental change based on Darwinian evolution.

vival of fitter phenotypes.[12] We posit that the Darwinian idea of evolution through natural selection provides a basis for incorporation of all sources of phenotypic variability within somatic evolution, including those defined by nongenetic changes in gene expression, as long as the phenotypes affect the differential survival of tumor cells. Whereas arguments for the need to extend the evolutionary synthesis to provide a more inclusive consideration of influences shaping heritable phenotypes have been made very explicitly in the field of evolutionary theory,[13] this debate has so far produced no noticeable reverberation in the field of cancer research where the rationale for including nongenetic inheritance mechanisms would be even more important as it would have clear therapeutic implications.

8.3   Therapeutic Implications of the Darwinian Take on Somatic Evolution

Whether somatic evolution is viewed through a mutation-centric or a Darwinian lens is not just a matter of theoretical preference. Instead, it has profound implications for how we should treat cancers. Within the currently dominant mutation-centric paradigm, acquired therapy resistance is fully explainable by resistance-driving mutations that either exist before treatment begins or occur during or because of it. Due to the stochastic nature of mutations, the idea of "bad luck" describing cancer initiation/progression[14] is fully applicable to mutations that "drive" resistance. The near inevitability of resistance is fully explainable by the population sizes of tumor cells and mutational probabilities.[15] Consequently, there is nothing that can be done to incorporate considerations of evolution into therapeutic approaches, apart from reducing mutational target size through surgical or therapeutic tumor debulking. Since in this framework, cancer evolution is reduced to the acquisition of "driver" mutations, the focus is on identifying, characterizing, and targeting the oncogenic "drivers"—which is essentially the exclusive focus of current cancer research and, in this view, the only sensible course of action.

From the standpoint of a truly/classical Darwinian paradigm, however, the near-inevitable development of therapy resistance stems from the reciprocal interplay of diversification *and* selection. In most cases, therapies fail to kill all of the tumor cells, due to the combined effects of constraints of therapeutic windows, phenotypic and genetic heterogeneity of tumor cells, and heterogeneity of tumor microenvironments, leading to unequal access of drugs and environmental factors (such as nutrients, oxygen, and growth factors). This heterogeneity of surviving tumor cells and the ability to generate new variants through the ongoing mutational and, importantly, also the nonmutational diversification mechanisms enables populations of tumor cells to evolve (figure 8.2B). As populations of cancer cells adapt to the treatment, they become less and less sensitive to the action of the drugs, eventually leading to a net positive population growth, reflected in clinical relapse. In contrast to the helplessness in dealing with the bad luck of mutation-driven process, embracing a Darwinian explanation enables a much more proactive attitude, as both diversification *and* selection could be subjects of clinical interventions.

8.3.1   Intervening in the Process of Diversification

Mutational diversification. Since mutations resulting from replication errors and chromo-somal aberrations are major sources of diversification of heritable phenotypes, as also stated by the mutation-centric paradigm, debulking tumor size to reduce mutational load is still a sensible approach. However, embracing the Darwinian paradigm extends the range of potential interventions and offers far more nuanced approaches. Since abnormal tumor microenvironments and inflammation have been linked with increased reactive oxygen species (ROS) and genomic instability,[16] therapies directed at restoration (sometimes, normalization) of the microenvironment and reduction of inflammation might reduce the ability of surviving tumor cells to diversify their phenotypes. On the other hand, many of the traditional chemotherapies, such as taxanes, significantly increase chromosomal insta-bility. Whereas the question of the fitness impact of point mutations in somatic cells remains to be resolved, there is undeniable evidence that chromosomal instability, which probably represents the biggest contributor to genetic diversity within tumors, comes at a solid fitness cost, as gains and losses of large fragments of DNA typically involve multiple genes, thus disrupting optimal stoichiometry of multiple protein complexes and taxing pathways responsible for maintaining proper protein folding.[17] As a result, where lack of chromosomal instability constrains diversification of tumor genotypes, too much vari-ability overburdens tumor cells with deleterious mutations and limits the selection of optimal viable phenotypes. Both experimental studies and analyses of clinical data suggest that aggressive cancers exist within a constrained range of genetic instability.[18] Rather than maximizing the short-term tumor cell kill rate, it might be possible to optimize chemothera-peutic treatments to increase the rates of genomic alterations, thus driving down fitness of populations of tumor cells.

In addition to the well-characterized replication errors, microsatellite instability, and chromosomal instability, genetic diversity of cancers might be shaped by additional power-ful mechanisms. Recent studies by the group of Paul Mischel uncovered an unexpected novel source of genomic diversification, which blurs the lines of heritability.[19] Certain segments of tumor DNA, including those with well-characterized oncogenes, such as *EGFR*, can exist outside of chromosomes in a form resembling plasmids in bacteria. Compared to traditional losses and amplifications of large segments of genomic DNA, this plasmid-like extrachromosomal DNA offers tumor cells much greater evolutionary flexibility and dramatically restricts fitness costs. Studies in our lab[20] as well as several published reports suggest that spontaneous fusions between tumor cells or between tumor and nontumor cells, followed by ploidy reduction, could provide populations of tumor cells with essentially a form of parasexual recombination,[21] a mechanism that is a major source of diversification in the majority of asexual species, such as many fungi and unicel-lular organisms. Since both extrachromosomal inheritance and fusion-related diversifica-tion must have a mechanistic underpinning, they should in principle be targetable.

Epigenetic diversification. Nongenetic variability is typically outside of the scope of the mutation-centric view of cancer evolution, although some researchers have viewed stable changes in gene expression as an "epimutation." As we have argued above, embracing the Darwinian paradigm, however, provides a framework for incorporation of nongenetic sources of phenotypic variability, as long as this phenotypic variability affects cell fitness. All of the normal cells within our body share the same "wild-type" genotype, apart from rearrangement in the genes encoding T- and B-cell receptors in lymphocytes and a few cases of polyploidy in specialized differentiated cells, such as in mature hepatocytes or in megakariocytes in the bone marrow. Thus, the very same genotype encodes a dizzying range of "building blocks" required to define the phenotypes of different cell types, which make up blood, skin, connective tissue, and so on. Whereas normal cells restrict gene expression to only a subset of genes, required to provide a well-defined, tissue-specific phenotype, tumorigenesis inevitably involves tapping into the nongenetic source of variability, as all tumors display convergent disruption of epigenetic regulation, enabling access to additional "building blocks," compared to their normal counterparts.[22] While nongenetic phenotypic features have a wide range of heritability, even relatively transient, noise-driven cell-cell variability could provide a basis for differential survival, thus "lubricating the machinery of natural selection"[23] for more stable phenotypes. Even normal, differentiated cells can cope with microenvironmental stress via considerable transient changes in gene expression within the reaction norm of the genome. Higher phenotypic plasticity might enable tumor cells to gradually evolve their genetic reaction norm in response to persistent microenvironmental changes (such as lower pH), in turn leading to the emergence of clinically more aggressive phenotypes.[24] On the other end of the spectrum, some alterations, such as silencing of p16 by DNA methylation, are essentially irreversible, thus paralleling the functional impact of bona fide genetic mutations. Epigenetic regulation of gene expression takes place at multiple levels through choreographed actions of different classes of enzymes, many of which are targetable, with many drugs clinically available. Whereas the initial rationale for the development of these drugs was based on the direct cytotoxic and cytostatic impact on tumor cells, many of them can be potentially repurposed with an explicit consideration of the impact on phenotypic diversity and plasticity. In fact, some of the epigenetic inhibitors are already used for this purpose as "differentiation therapies" and achieve exactly that—restricting phenotypic diversity toward clinically benign phenotypic states.

8.3.2   Interfering with Selection Forces

Given the evolvability of populations of tumor cells, the eventual development of resistance to continuous administration of targeted therapies is inevitable (as long as some of the tumor cells can survive the drug) whether one embraces a mutation-centric or Darwinian view of somatic evolution. The key distinction, however, is that from a mutation-centric perspective, continuous administration of the highest tolerated dose of the drug clearly represents the optimal strategy. Maximizing short-term tumor cell kill rates aims at optimization of shrinkage of mutational target size, and the rest comes down to the "bad luck" of either preexistence or occurrence of resistance-conferring mutations. This approach makes therapies more straightforward, as the drugs can be optimized in short-term assays, and keeping a constant drug schedule simplifies the logistics for both oncologists and patients. However, the superiority of continuous drug administration is not obvious from a Darwinian perspective. While reduction of the population size is still desirable, continuous application of strong selective pressures ensures that resistant phenotypes will eventually get selected and eventually become fixed in the population. However, what are the sensible evolution-informed alternatives to the current therapeutic scheduling?

Adaptive therapy scheduling based on evolutionary tradeoffs. Studies in the field of evolutionary ecology have firmly established the principle of "evolutionary tradeoffs,"[25] that is, an increase in fitness in one context is inevitably linked with lower fitness under a different context. In the specific case of targeted therapies, the obvious realization of this principle is that phenotypes that enable tumor cells to thrive without treatment are incompatible with proliferation and survival once the drug is administered. Conversely, phenotypes associated with acquired resistance are likely to be linked with fitness penalties under a different context. From a therapeutic perspective, the easiest trade-off to exploit is that of lower fitness of the resistant type in the absence of the drug. This phenomenon has been well documented in many cases of acquired resistance. Mechanistically, lower fitness in the absence of treatment could be linked with higher energy expense associated with the expression of multiple drug resistance pumps or with overactivation of oncogenic signaling, which, while enabling incomplete shutdown of the target during therapies, brings target activity well above the optimal range in the absence of the drugs.

These fitness trade-offs could be exploited by creating push-pull evolutionary dynamics between therapy-sensitive and therapy-resistant cells through cycling a targeted therapy with treatment breaks in a way that also keeps control over tumor growth (figure 8.3A). This approach, pioneered by Robert Gatenby at Moffitt Cancer Center and named adaptive therapy, has been validated in several preclinical xenograft models of breast cancer.[26] Importantly, adaptive use of therapy breaks can be useful in more complex scenarios, involving more than two cell types, as evidenced by success of a recent pilot clinical trial of castration-resistant prostate cancer.[27] Alternating abiraterone (a drug that disrupts the ability of tumor cells to produce testosterone from the steroid precursors available in serum) therapy with therapy breaks through scheduling, informed by an understanding of tumor growth dynamics and the evolutionary dynamics of the development of resistance, has enabled a more than twofold extension of progression-free survival while using less drug and thus reducing side effects and cost.[28]

Lower fitness in the absence of therapy is not, however, a mandatory manifestation of the evolutionary trade-off principle. In some cases, resistant phenotypes are fitness neutral or even fitter than their therapy-naïve counterparts.[29] In these cases, it might be possible to consider the principle of an evolutionary double bind.[30] Alluringly, resistance to the

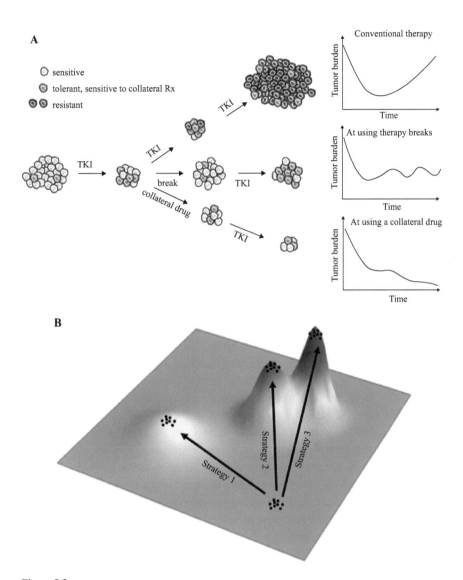

Figure 8.3
(A) Evolutionary trade-offs could be used therapeutically to forestall the expansion of resistant subpopulations, thus preventing or at least delaying tumor relapse. (B) Different drugs and treatment strategies can "push" populations of tumor cells toward different evolutionary trajectories, leading to different fitness peaks. In this schematic reinterpretation of the fitness landscape metaphor, height of peaks indicates the degree of malignancy of phenotypes. Strategy 1 is the most desirable as it leads tumor cells to a well—isolated local fitness peak with lower fitness.

primary drug is often linked to collateral sensitivity to a different pharmacological agent(s). For example, a recent study reported that clinical resistance to BRAF and MEK inhibitors in BRAF mutant melanoma is often linked with increased sensitivity to histone deacetylase inhibitor vorinostat.[31] The authors used this knowledge to design a clinical trial within a conventional approach (i.e., waiting for relapse on BRAF/MEK inhibitors before switching to vorinostat). Perhaps not surprisingly, this approach failed to provide a substantial clinical benefit, as tumors quickly developed resistance to vorinostat and relapsed. We posit that explicit considerations of selection forces might enable a more effective use of evolutionary trade-offs manifested as collateral sensitivities. Instead of cycling the frontline drug with therapy breaks, it should be possible to cycle the primary drug with a collateral agent, which might be capable of achieving better results than the original adaptive therapy, as a pair of collateral drugs enables a better control of population size, thus limiting diversification (figure 8.3A).

Taking selection forces into account is not limited to considerations of creating a push-pull evolutionary dynamic. For example, Carlo Maley and colleagues have proposed the idea of "benign cell boosters"—creating selective pressures to enable benign tumor cells to outcompete more clinically malignant phenotypes, resulting in a clinically benign tumor with a low risk of progression.[32] Although the feasibility of this approach has yet to be tested experimentally, John Nagy has argued about a similar phenomenon, termed "hypertumor," where outgrowth of a highly proliferative but not self-sufficient subpopulation could lead to eventual tumor extinction.[33] While it is still unclear whether clinical observations of the extinction of histologically advanced tumors (such as some cases of childhood neuroblastomas) could be attributable to the "hypertumor" effect, we might have observed an experimental validation of the phenomenon, where an aggressive clone outgrew and outcompeted subpopulations, responsible for the establishment of proper tumor infrastructure, eventually resulting in tumor collapse.[34] A twist on the idea of benign cell boosters is the so-called sucker's gambit[35]—favoring an expansion of a subpopulation that can be effectively wiped out with a therapy. A recent study by Reinshagen and colleagues provided experimental validation of this idea, where tumor cells engineered to produce a cell death receptor, to which they were insensitive, were used to destroy nonengineered cells in both a primary and a metastatic setting and were then eliminated through the induction of a suicide switch.[36]

Considerations of selection forces are not limited to the scenarios involving limited numbers of stable subpopulations. In many cases, therapy-resistant phenotypes can arise through nongenetic mechanisms and lack strict heritability.[37] Our mathematical modeling studies revealed that adaptive therapy strategies based on a pair of collateral drugs might be superior to conventional therapies given a scenario where cells are switching between distinct phenotypic states.[38] Additionally, while clinical drug resistance is often assumed to be caused by a single mechanism, this is likely to be a misleading oversimplification. Multifactorial resistance to antibiotics is very common in bacteria and other animal and plant pests. Similarly, emerging evidence,[39] including our studies,[40] suggests that resistance to targeted therapy might be arising through a gradual acquisition of multiple cooperating mechanisms. Consideration of selection forces and evolutionary dynamics becomes even more important under these scenarios. Using intelligently designed therapeutic schedules, it might be possible to steer evolving populations of tumor cells toward local fitness peaks corresponding to clinically benign phenotypes[41] (figure 8.3B).

Considering the tumor microenvironment and cell-cell interactions. Whereas therapies that directly kill tumor cells generate the most obvious and dramatic selective pressures, one needs to remember that the tissue context experienced by tumor cells is highly structured and complex. Considering this context and developing tools to intelligently manipulate it could represent another promising avenue for evolution-informed therapies. This idea is not entirely untested, as for example with VEGF inhibitors that target tumor vasculature. Initially, these therapies were developed with the rationale of depriving tumors of oxygen and nutrients, that is, causing indirect (microenvironmental) toxicity, ultimately directed at tumor cells.[42] However, severe restriction of blood supply has been shown to backfire in some of the cases, leading to induction/selection of more aggressive tumor cell phenotypes.[43] Antiangiogenic therapies still hold substantial promise, especially when developed with a more nuanced consideration of normalizing vascularization, allowing better drug delivery and reducing selection pressure for the evolution of more aggressive phenotypes.[44]

Paracrine signals between different subpopulations of tumor cells and different cells within the tumor microenvironment represent another essential part of the context.[45] Understanding this ecological dimension and taking this knowledge into consideration could revolutionize cancer therapies.[46] For example, our previous work has demonstrated that tumor growth and metastatic dissemination could result from the action of a minor subpopulation of tumor cells by altering a complex network of interactions involving fibroblasts and neutrophils.[47] Sensitivity of tumor cells to many types of targeted therapies could be dramatically reduced in the proximity of cancer-associated fibroblasts, through mechanisms involving signaling that impacts the extracellular matrix and paracrine factors. Recent dramatic advances in immune therapies also fall into this category. These phenomena are typically viewed as straightforward antitumor effects driven by typical immune mechanisms. However, considering the impact of crosstalk between tumor cells and different elements of the tumor microenvironment and the immune response, as well as the selective pressures imposed by the active immune response, could potentially lead to a more effective and informed use of immune therapies.

8.4   Synopsis and Future Directions

Despite some early attempts at evolution-informed therapies, their full potential remains mostly unrealized. In contrast to the vast and growing number of studies that elucidate proximal, molecular mechanisms underlying malignant and therapy-resistant phenotypes at ever-increasing levels of detail, our understanding of sources of diversification, selection pressures, and the evolutionary dynamics of cancer cells remains rudimentary. Our knowledge is mostly based on assumptions and inferences, and experimental studies in this area are very limited. As we have argued above, a gross misunderstanding of how evolution "works" that pervades most of current cancer research is the major barrier to developments in this area. Another reason is the complexity of the cellular tumor ecosystem, which appears to go well beyond our current capacities to understand or systematically investigate. Not only is every individual patient's cancer unique, but advanced cancers display significant genetic, phenotypic, and microenvironmental heterogeneity. This is further complicated by the fact that tumors are not simply the sums of their parts—as distinct populations of tumor cells interact with each other, as well as with cellular and noncellular components of tumor microenvironments (which itself is very complex), and these interactions change in space and time as the tumor develops and evolves. This staggering complexity cannot be understood within the current focus on specific proximal mechanisms, such as in-depth characterization of specific genetic mutations, copy-number alterations, metabolic changes, and so on. Moreover, as we dig deeper into the underlying molecular mechanisms, the complexity exponentially increases, as more details are discovered about the old players and novel layers of regulation and interaction are uncovered. Since tackling these questions from the molecular oncology perspective is impossible, doing more of the same (i.e., finding new druggable targets, aiming to incrementally prolong patient survival through discovery of new drugs and drug combinations) might appear to be the only sensible course of action, despite being intractably slow and expensive, while a cure remains out of reach.

However, a complete understanding of molecular mechanisms is not a mandatory prerequisite for the ability to influence and manage complex, evolving ecosystems. Identifying key interactions and using the first principles of evolution and ecology, captured in mathematical formalisms, parameterized for a given system, have allowed evidence-based management of complex, evolving ecosystems such as fisheries, parks, and farmlands, even in the absence of detailed knowledge of every individual species and interaction. We postulate that it is not a far stretch to suggest that first principles of ecology and evolution, captured in predictive mathematical models, could be applied to understand and manage cancers. Owing to the sheer amount of knowledge, cancers appear to be extremely complex, but from an ecological perspective, they are simpler than most natural ecosystems.[48] We posit that applying eco-evolutionary principles toward understanding and disrupting the acquisition of therapy resistance would provide opportunities to improve clinical outcomes using the array of clinical tools that are already at our disposal.

For this to happen, the hegemony of reductionist and gene-centric approaches as the only valid direction in cancer research needs to be overturned. Sadly, the prevalent implicit assumption in the field is that we have correctly figured out all of the key underlying principles, and what is left to understand are the details (some of which might be extremely important). Thus, most cancer researchers accept these underlying assumptions as an unquestionable truth, rather than taking a more Bayesian approach of leaning on these assumptions but being ready to challenge and revisit them in light of new knowledge. Consequently, important aspects of biology end up outside of the focus of active scientific inquiry, and given the hypercompetitive funding climate, where success depends on agreed consensus between reviewers, stepping out of the box becomes extremely challenging. Interestingly, this situation resembles the state of physics at the dawn of the twentieth century, when the consensus opinion was that Newtonian physics has explained everything, leaving only to sort out the details. Then, quantum physics and the theory of relativity completely reshaped the field. Whereas similarly transformative developments may still lie ahead in biology, we argue that a lot could be gained by simply taking one keystone idea that has been successfully applied to many biological phenomena, namely, Darwinian evolution, and by applying it correctly to understand cancers—without twisting it out of its original meaning. We believe that, similar to other areas of science, progress is inevitable. However, the sooner cancer research can break the spell of the current, stifling gene-centric dogma, enabling less biased and more creative scientific endeavors, the sooner we can become more efficient in turning billions of research and development spending into saving of lives of cancer patients.

Acknowledgments

We thank Nara Yoon for help preparing illustrations.

Notes

1. Hanahan, D. & Weinberg, R. A. The hallmarks of cancer. *Cell* 100, 57–70 (2000). Hanahan, D. & Weinberg, R. A. Hallmarks of cancer: the next generation. *Cell* 144, 646–674, doi:10.1016/j.cell.2011.02.013 (2011).

2. Fearon, E. R. & Vogelstein, B. A genetic model for colorectal tumorigenesis. *Cell* 61, 759–767 (1990).

3. Tomasetti, C. & Vogelstein, B. Cancer etiology. Variation in cancer risk among tissues can be explained by the number of stem cell divisions. *Science* 347, 78–81, doi:10.1126/science.1260825 (2015).

4. Weinberg, R. *The Biology of Cancer, Second Edition.* (Taylor & Francis Group, 2013).

5. Cahill, D. P., Kinzler, K. W., Vogelstein, B. & Lengauer, C. Genetic instability and Darwinian selection in tumours. *Trends Cell Biol* 9, M57–60 (1999).

6. Gerlinger, M. et al. Intratumor heterogeneity and branched evolution revealed by multiregion sequencing. *N Engl J Med* 366, 883–892, doi:10.1056/NEJMoa1113205 (2012). Navin, N. et al. Tumour evolution inferred by single-cell sequencing. *Nature* 472, 90–94, doi:10.1038/nature09807 (2011).

7. Mayr, E. *The Growth of Biological Thought: Diversity, Evolution, and Inheritance.* (Belknap Press, 1982).

8. Marusyk, A. Obstacles to the Darwinian framework of somatic cancer evolution. In *Ecology and Evolution of Cancer*, 223 (2017).

9. Greaves, M. Darwinian medicine: a case for cancer. *Nat Rev Cancer* 7, 213–221, doi:10.1038/nrc2071 (2007). Gatenby, R. A., Gillies, R. J. & Brown, J. S. Of cancer and cave fish. *Nat Rev Cancer* 11, 237–238 (2011). Scott, J. & Marusyk, A. Somatic clonal evolution: a selection-centric perspective. *Biochim Biophys Acta Rev Cancer* 1867, 139–150, doi:10.1016/j.bbcan.2017.01.006 (2017). DeGregori, J. Challenging the axiom: does the occurrence of oncogenic mutations truly limit cancer development with age? *Oncogene* 32, 1869–1875, doi:10.1038/onc.2012.281 (2013). DeGregori, J. D. G. *Adaptive Oncogenesis: A New Understanding of How Cancer Evolves*

inside Us. (Harvard University Press, 2018). Merlo, L. M., Pepper, J. W., Reid, B. J. & Maley, C. C. Cancer as an evolutionary and ecological process. *Nat Rev Cancer* 6, 924–935, doi:10.1038/nrc2013 (2006). Marusyk, A., Almendro, V. & Polyak, K. Intra-tumour heterogeneity: a looking glass for cancer? *Nat Rev Cancer* 12, 323–334, doi:10.1038/nrc3261 (2012).

10. Nowell, P. C. The clonal evolution of tumor cell populations. *Science* 194, 23–28 (1976).

11. Gerlinger, M. et al. Intratumor heterogeneity and branched evolution revealed by multiregion sequencing. *N Engl J Med* 366, 883–892, doi:10.1056/NEJMoa1113205 (2012).

12. Biddy, B. A. et al. Single-cell mapping of lineage and identity in direct reprogramming. *Nature* 564, 219–224, doi:10.1038/s41586-018-0744-4 (2018).

13. Pigliucci, M. et al. *Evolution, the Extended Synthesis*. (MIT Press, 2010).

14. Tomasetti, C. & Vogelstein, B. Cancer etiology. Variation in cancer risk among tissues can be explained by the number of stem cell divisions. *Science* 347, 78–81, doi:10.1126/science.1260825 (2015).

15. Bozic, I. & Nowak, M. A. Timing and heterogeneity of mutations associated with drug resistance in metastatic cancers. *Proc Natl Acad Sci U S A* 111, 15964–15968, doi:10.1073/pnas.1412075111 (2014).

16. Radisky, D. C. et al. Rac1b and reactive oxygen species mediate MMP-3-induced EMT and genomic instability. *Nature* 436, 123–127, doi:10.1038/nature03688 (2005).

17. Santaguida, S. & Amon, A. Short- and long-term effects of chromosome mis-segregation and aneuploidy. *Nat Rev Mol Cell Biol* 16, 473–485, doi:10.1038/nrm4025 (2015).

18. Andor, N. et al. Pan-cancer analysis of the extent and consequences of intratumor heterogeneity. *Nat Med* 22, 105–113, doi:10.1038/nm.3984 (2016). Chandhok, N. S. & Pellman, D. A little CIN may cost a lot: revisiting aneuploidy and cancer. *Curr Opin Genet Dev* 19, 74–81, doi:10.1016/j.gde.2008.12.004 (2009). Godek, K. M. et al. Chromosomal instability affects the tumorigenicity of glioblastoma tumor-initiating cells. *Cancer Discov* 6, 532–545, doi:10.1158/2159–8290.CD-15–1154 (2016).

19. Turner, K. M. et al. Extrachromosomal oncogene amplification drives tumour evolution and genetic heterogeneity. *Nature* 543, 122–125, doi:10.1038/nature21356 (2017).

20. Myroshnychenko, D. et al. Spontaneous cell fusions as a mechanism of parasexual recombination in tumor cell populations. *bioRxiv*, doi:10.1101/2020.03.09.984419 (2020).

21. Duelli, D. & Lazebnik, Y. Cell-to-cell fusion as a link between viruses and cancer. *Nat Rev Cancer* 7, 968–976, doi:10.1038/nrc2272 (2007). Jacobsen, B. M. et al. Spontaneous fusion with, and transformation of mouse stroma by, malignant human breast cancer epithelium. *Cancer research* 66, 8274–8279, doi:10.1158/0008–5472 .CAN-06–1456 (2006). Su, Y. et al. Somatic cell fusions reveal extensive heterogeneity in basal-like breast cancer. *Cell Rep* 11, 1549–1563, doi:10.1016/j.celrep.2015.05.011 (2015).

22. Marusyk, A. Obstacles to the Darwinian framework of somatic cancer evolution. In *Ecology and Evolution of Cancer*, 223 (2017).

23. Brock, A., Chang, H. & Huang, S. Non-genetic heterogeneity—a mutation-independent driving force for the somatic evolution of tumours. *Nat Rev Genet* 10, 336–342, doi:10.1038/nrg2556 (2009).

24. Robertson-Tessi, M., Gillies, R. J., Gatenby, R. A. & Anderson, A. R. Impact of metabolic heterogeneity on tumor growth, invasion, and treatment outcomes. *Cancer Res* 75, 1567–1579, doi:10.1158/0008–5472.CAN-14 –1428 (2015).

25. Stearns, S. C. Trade-offs in life-history evolution. *Functional Ecol* 3, 259–268, doi:10.2307/2389364 (1989).

26. Enriquez-Navas, P. M. et al. Exploiting evolutionary principles to prolong tumor control in preclinical models of breast cancer. *Sci Transl Med* 8, 327ra324, doi:10.1126/scitranslmed.aad7842 (2016). Gatenby, R. A., Silva, A. S., Gillies, R. J. & Frieden, B. R. Adaptive therapy. *Cancer Res* 69, 4894–4903, doi:10.1158/0008–5472 .CAN-08–3658 (2009).

27. Zhang, J., Cunningham, J. J., Brown, J. S. & Gatenby, R. A. Integrating evolutionary dynamics into treatment of metastatic castrate-resistant prostate cancer. *Nat Commun* 8, 1816, doi:10.1038/s41467-017-01968-5 (2017).

28. Zhang, J., Cunningham, J. J., Brown, J. S. & Gatenby, R. A. Integrating evolutionary dynamics into treatment of metastatic castrate-resistant prostate cancer. *Nat Commun* 8, 1816, doi:10.1038/s41467-017-01968-5 (2017).

29. Kaznatcheev, A., Peacock, J., Basanta, D., Marusyk, A. & Scott, J. G. Fibroblasts and alectinib switch the evolutionary games played by non-small cell lung cancer. *Nat Ecol Evol* 3, 450–456, doi:10.1038/s41559-018 -0768-z (2019).

30. Basanta, D. & Anderson, A. R. Exploiting ecological principles to better understand cancer progression and treatment. *Interface Focus* 3, 20130020, doi:10.1098/rsfs.2013.0020 (2013). Gatenby, R. A., Brown, J. & Vincent, T. Lessons from applied ecology: cancer control using an evolutionary double bind. *Cancer Res* 69, 7499–7502, doi:10.1158/0008-5472.CAN-09-1354 (2009).

31. Wang, L. et al. An acquired vulnerability of drug-resistant melanoma with therapeutic potential. *Cell* 173, 1413–1425 e1414, doi:10.1016/j.cell.2018.04.012 (2018).

32. Maley, C. C., Reid, B. J. & Forrest, S. Cancer prevention strategies that address the evolutionary dynamics of neoplastic cells: simulating benign cell boosters and selection for chemosensitivity. *Cancer Epidemiol Biomarkers Prev* 13, 1375–1384 (2004).

33. Nagy, J. D. Competition and natural selection in a mathematical model of cancer. *Bull Math Biol* 66, 663–687, doi:10.1016/j.bulm.2003.10.001 (2004).

34. Marusyk, A. et al. Non-cell-autonomous driving of tumour growth supports sub-clonal heterogeneity. *Nature* 514, 54–58, doi:10.1038/nature13556 (2014).

35. Maley, C. C., Reid, B. J. & Forrest, S. Cancer prevention strategies that address the evolutionary dynamics of neoplastic cells: simulating benign cell boosters and selection for chemosensitivity. *Cancer Epidemiol Biomarkers Prev* 13, 1375–1384 (2004).

36. Reinshagen, C. et al. CRISPR-enhanced engineering of therapy-sensitive cancer cells for self-targeting of primary and metastatic tumors. *Sci Transl Med* 10, eaao3240, doi:10.1126/scitranslmed.aao3240 (2018).

37. Sharma, S. V. et al. A chromatin-mediated reversible drug-tolerant state in cancer cell subpopulations. *Cell* 141, 69–80, doi:10.1016/j.cell.2010.02.027 (2010).

38. Yoon, N., Vander Velde, R., Marusyk, A. & Scott, J. G. Optimal therapy scheduling based on a pair of collaterally sensitive drugs. *Bull Math Biol*, doi:10.1007/s11538-018-0434-2 (2018).

39. Bakhoum, S. F. et al. Chromosomal instability drives metastasis through a cytosolic DNA response. *Nature* 553, 467–472, doi:10.1038/nature25432 (2018).

40. Vander Velde, R. et al. Resistance to targeted therapies as a multifactorial, gradual adaptation to inhibitor specific selective pressures. *Nat Commun* 11, 2393, doi:10.1038/s41467-020-16212-w (2020).

41. Dhawan, A. et al. Collateral sensitivity networks reveal evolutionary instability and novel treatment strategies in ALK mutated non-small cell lung cancer. *Sci Rep* 7, 1232, doi:10.1038/s41598-017-00791-8 (2017). Nichol, D. et al. Steering evolution with sequential therapy to prevent the emergence of bacterial antibiotic resistance. *PLoS Comput Biol* 11, e1004493, doi:10.1371/journal.pcbi.1004493 (2015). Nichol, D. et al. Antibiotic collateral sensitivity is contingent on the repeatability of evolution. *Nat Commun* 10, 334, doi:10.1038/s41467-018-08098-6 (2019).

42. Folkman, J. Angiogenesis in cancer, vascular, rheumatoid and other disease. *Nat Med* 1, 27–31 (1995).

43. Paez-Ribes, M. et al. Antiangiogenic therapy elicits malignant progression of tumors to increased local invasion and distant metastasis. *Cancer Cell* 15, 220–231, doi:10.1016/j.ccr.2009.01.027 (2009). Ebos, J. M. et al. Accelerated metastasis after short-term treatment with a potent inhibitor of tumor angiogenesis. *Cancer Cell* 15, 232–239, doi:10.1016/j.ccr.2009.01.021 (2009).

44. Jain, R. K. Normalization of tumor vasculature: an emerging concept in antiangiogenic therapy. *Science* 307, 58–62, doi:10.1126/science.1104819 (2005). Jain, R. K. Antiangiogenesis strategies revisited: from starving tumors to alleviating hypoxia. *Cancer Cell* 26, 605–622, doi:10.1016/j.ccell.2014.10.006 (2014).

45. Tabassum, D. P. & Polyak, K. Tumorigenesis: it takes a village. *Nat Rev Cancer* 15, 473–483, doi:10.1038/nrc3971 (2015).

46. Basanta, D. & Anderson, A. R. Exploiting ecological principles to better understand cancer progression and treatment. *Interface Focus* 3, 20130020, doi:10.1098/rsfs.2013.0020 (2013). Gatenby, R. & Brown, J. The evolution and ecology of resistance in cancer therapy. *Cold Spring Harb Perspect Med* 8, a033415, doi:10.1101/cshperspect.a033415 (2018).

47. Marusyk, A. et al. Non-cell-autonomous driving of tumour growth supports sub-clonal heterogeneity. *Nature* 514, 54–58, doi:10.1038/nature13556 (2014). Janiszewska, M. et al. Subclonal cooperation drives metastasis by modulating local and systemic immune microenvironments. *Nat Cell Biol* 21, 879–888, doi:10.1038/s41556-019-0346-x (2019).

48. Maley, C. C. et al. Classifying the evolutionary and ecological features of neoplasms. *Nat Rev Cancer* 17, 605–619, doi:10.1038/nrc.2017.69 (2017).

9 Cancer as a Reversion to an Ancestral Phenotype

Kimberly J. Bussey and Paul C. W. Davies

Overview

The atavism theory postulates that cancer is the reversion to ancestral single-cell behaviors in cells of multicellular organisms, possibly representing a speciation event. This reconceptualization explains why cancers share specific cell behaviors, termed "hallmarks of cancer," independent of the tissue of origin; why cancers are pervasive across the tree of life; and why they are seemingly an inevitable concomitant of multicellularity. It also accounts for the fact that although changes at the genomic level may be a cause of cancer, the cancer phenotype may nevertheless be suppressed by surrounding stroma that is physiologically normal. Moreover, the atavism theory emphasizes the fact that *adaptability* itself is a selected trait. This shift in perspective has important ramifications for clinical practice. If, through the redeployment of unicellular behaviors, tumor cells are actually primed to respond to selective pressures by rapid evolution and the generation of phenotypic heterogeneity, then a "take no prisoners" approach to treatment (i.e., aiming to kill all cancer cells with one drug) will inevitably lead to treatment resistance and simultaneously damage the normal multicellular mechanisms that have evolved to suppress carcinogenesis.

9.1 Introduction

What is cancer? This is a fundamental question that still lacks an adequate answer. Cancers or cancer-like phenomena are found across the tree of life in multicellular organisms,[1] suggesting that it is deeply embedded in the nature of multicellular life and has deep evolutionary roots. The "hallmarks of cancer"[2] describe up to eight specific cellular functions or behaviors that cells express in their transition to a cancerous tumor, including uncontrolled growth, uninhibited mobility, and resistance to cell death. The current paradigm of carcinogenesis views these aberrant cell behaviors as "normal gone wrong" and ascribes the acquisition of such behaviors to the gradual accumulation of genomic alterations/mutations.[3] This is presumed because most cancers have significant evidence of genomic alterations at both large (chromosomal) and small (one to hundreds of base pairs) scales, and models of both skin cancer and colon cancer suggest a stepwise transition from premalignant to malignant growth due to an incremental accumulation of mutations.[4] Tumors with little evidence of genomic alterations at the DNA sequence level often have evidence of disrupted epigenetic modifications, particularly DNA methylation.[5] This has been the rationale for "targeted" therapeutics that are meant to be active only in cells with a specific acquired genomic alteration. However, a common feature in the targeted therapy setting is the inevitable emergence of therapy-resistant cells, usually resulting in relapse and eventual death from the disease.

The problem with the current genome/mutation-centric paradigm is that very few examples exist of both necessary and sufficient genomic changes that would invariably result in tumor formation under multiple contexts. Most oncogenic alterations are neither necessary, sufficient, nor context independent. For example, the first chromosomal translocation documented in human cancers, t(9;22)(q34;q11), also known as the Philadelphia chromosome,[6] is a necessary and sufficient cause of chronic myeloid leukemia *only* when occurring at a certain stage of hematopoietic cell differentiation. This translocation creates the fusion transcript *bcr-abl*. The resulting constitutive expression of abl leads to an alteration in intracellular signaling that causes uncontrolled proliferation and inhibition of normal differentiation. However, there are several documented cases of individuals expressing the fusion protein in leukocytes without evidence of the disease.[7]

More generally, the almost exclusive focus on genomic alteration as the main mechanism of carcinogenesis has led to the underappreciation of the causal roles of the tissue context, such as the tumor microenvironment, physical parameters, and the host immune system, among others. Work by a number of groups has demonstrated that when malignant breast cancer cells carrying considerable genome alterations are transplanted into normal mammary fat pads, tumors fail to grow and cells revert to near-normal breast tissue phenotype, even though they retain their abnormal genomes.[8] Another example of a tissue factor that influences malignancy of cancer cells comes from experiments showing that the melanoma phenotype of skin cancer cells can be suppressed by modulating their resting electric membrane potential, despite the fact that they carry genomic alterations.[9] And the successes and failures of immunotherapy firmly establish immune surveillance as an important force that influences tumor cell behavior. Although it is now clear from a large body of evidence that the current genome/mutation-centered paradigm is not sufficient to explain the well-established behaviors of tumors, as well as tumor microenvironments and overall patient health, novel conceptual frameworks are still lacking.

We propose here to reconceptualize cancer as a kind of atavism, in this case the reversion to ancestral single-cell behaviors in cells of multicellular organisms, perhaps representing a speciation event. In this view, cancer is not the result of the acquisition of hallmarks of cancer cell behaviors via the accumulation of genetic modifications at the DNA level. Rather, the hallmarks of cancer result from redeploying, or reawakening, single-cell biology upon exposure to certain stress factors in cells that have evolved to be part of a multicellular organism. Once a cell shows cancer hallmark behavior, its genome *as a whole* is no longer wired as it would be in normal development and tissue maintenance, but rather in a way that supports single-cell survival. This perspective frames cancer in terms of evolution, both organismal over millions of years and cellular over the course of tumor development. It puts cancer in an ecological context, again both at the organismal level that is exposed to environmental factors and at the cellular level that is exposed to the tissue microenvironment with the tumor using the patient as an environment with resources to exploit and dangers to avoid. It also places adaptability and evolvability at the heart of therapeutics, as they are the basis of tumor heterogeneity, a defining characteristic of the neoplastic phenotype.[10]

9.2   Cancer and Multicellularity

The transition from unicellular to multicellular life is arguably one of the most important evolutionary steps life has ever taken. It has happened multiple times; in animals, the acquisition of simple multicellularity occurred sometime around 1,000 million years ago, with complex multicellularity arising between 900 and 600 million years ago. There are five characteristics of cell-level cooperation displayed in multicellular life: (1) proliferation inhibition, (2) controlled cell death, (3) maintenance of extracellular environments, (4) division of labor, and (5) resource allocation.[11] Comparing these foundations with the hallmarks of cancer,[12] it becomes apparent that cancer represents a fundamental breach of the cooperative contract between cells that permits multicellular life. If cancer is the reversion to single-cell behavior in cells that have evolved for hundreds of millions of years in the context of multicellularity, then we need to identify how the evolution of multicellularity has appropriated programs for single-cell survival and rewired them to respond to either different environmental inputs or developmentally regulated signals. For example, in the unicellular eukaryotic cell cycle of budding yeast (*Saccharomyces cerevisiae*), nutrient abundance is signaled via the MAP-kinase signaling cascade and may initiate a round of division. In multicellular organisms, while cell division remains the result of the MAP-kinase signaling cascade, the input is not nutrient abundance but specific growth factors that are themselves regulated in space and time as a consequence of development. If our hypothesis of cancer as a reversion to unicellular behavior is correct, then we would expect tumor cells to reverse the unicellular-to-multicellular wiring of molecular interactions and, as a result, restore ancestral patterns of cellular information flow.

9.2.1   Boveri's Hypothesis: The First Mechanistic Paradigm of Carcinogenesis

Theodor Boveri was a renowned German biologist of the late nineteenth and early twentieth centuries who made a number of fundamental discoveries in cell biology. He first postulated that chromosomes are individual and the site of unique qualitative phenotypic traits. From this he also concluded that Mendelian linkage of traits is due to colocalization on chromosomes. Furthermore, he suggested that during mitosis, chromosomes represent defined portions of interphase chromatin. He also realized the importance of the centrosome as an organelle involved in cell division and first described the chromosomal details of meiosis. Experiments on multipolar mitoses of sea urchin embryos made him conclude that each somatic cell needs to have two complete sets of chromosomes, one from each parent.[13] From his observations, he also postulated that "the tumour problem is a cell problem" and that cancer would be a "consequence of certain abnormal chromosome constitutions." Boveri summarized his chromosomal hypothesis of carcinogenesis in his 1914 book, *The Origin of Malignant Tumours*.[14] In his monograph, Boveri anticipated many later concepts, such as essential/housekeeping genes and nonessential genes, the "multiple-hit" hypothesis, the cancer stem cell idea, "oncogenes" and "tumor suppressor genes," tumor heterogeneity, and tumor evolution. He put forth his central thesis that malignant tumors emerge as a result of abnormal chromosomal arrangements caused by aberrant mitosis and that these would cause an irreversible defect that results in the loss of "normal reactions of the cell to its environment and the organism as a whole." It was at the time also known that multipolar mitoses were found enriched in tumor tissue. Boveri formulated a number of hypotheses stemming from his main assumption, namely, that malignant tumors would start from a single cell, as a very rare event; that tumor cells would continue to proliferate due to either a random loss of chromosomal factors (now called genes) that suppress cell division or an excess of chromosomal factors that promote division; and that "such unrestrained proliferation is no doubt a very primitive property of cells." This idea of proliferation as an ancient trait of cells, a kind of "default" state, was widely held at the time and based on simple evolutionary arguments. However, Boveri explicitly stressed the fact that metazoan cells *revert* to "unrestrained proliferation" simply when the "normal relationship of the cell with its surroundings is permanently disturbed."[15]

9.2.2   The Other Half of Boveri's Hypothesis: Cancer as an Atavism

Among historically informed cancer scientists, Boveri's hypothesis is generally considered the first definition of cancer as a disease caused by genes and DNA alterations. However, Boveri himself was acutely aware of the importance of the cellular/tissue context, or as he writes in several places in his monograph: the relationship of the cell with its surroundings, including the organism as a whole. Moreover, the idea of unrestrained proliferation as an evolutionary ancient trait was for him almost a given, as he wrote,

The main thesis is admittedly hypothetical, namely whether an abnormal chromosome constitution can be produced such that the cells that harbour it are driven to unrestrained proliferation. This assumption must be made *ad hoc*, but there is much to be said for it. *Above all, I regard it as beyond doubt that the tendency to multiply indefinitely is a primaeval property of cells and that the inhibition of multiplication in metazoan cells occurs secondarily under the influence of the environment.* Cells normally submit to this inhibition and reassert their original proliferative drive only when there are certain changes in their environment. If this is so, one must presuppose that there exists a specific cellular apparatus that is sensitive to conditions in the environment. Given such an apparatus, one must assume that it is susceptible to aberrations that incur the loss of sensitivity to the conditions of the environment. Then, the inherent proliferative drive of the cell is released and proceeds without taking any notice of the requirements of the rest of the body.

If now we regard the malignant cell as one that has lost certain properties and hence its normal reactivity to the rest of the body, then this change may well be enough to induce an altruistic cell to revert to its egoistical mode and thus release its multiplication from restraint. (Relapse of "organo-typic growth" to "cytotypic growth," to use the picturesque terminology of R. Hertwig.) But it is also possible that, in the tissue cells of metazoa, special inhibitory mechanisms have developed that have to be eradicated before unrestrained multiplication can take place.[16] (emphasis added)

We thus like to think that Boveri might have agreed with our description of cancer as an atavism: malignant tumors behave like parasitic single-cell collectives "without taking any notice of the requirements of the rest of the body."[17] If cancer is the "reexpression" of unicellular behavior pathways that are still available *within the reaction norm of the current genome*, then the risk of cancer is a concomitant feature of multicellularity, innate to the multicellular system, an accident waiting to happen. We use the term "atavism" here to describe the fact that metazoan cells still have the capacity to behave in a coherent way, "out of context," in a unicellular fashion to optimize their individual survival if stressful (micro) environmental conditions make this a necessary evolutionary strategy. We propose that these unicellular-like ("cancer hallmark") cell behaviors correspond to a complete rewiring of the most highly conserved integrated genomic pathways that have coevolved for hundreds of millions of years with the newer parts of the genome. This rewiring would correspond to a fallback to highly conserved adaptive shortcuts in a complex network of molecular interactions, which give the cell a phenotypic survival advantage (see also chapter 6, this volume, and Henry Heng's book, *Debating Cancer: The Paradox in Cancer Research*[18]).

How does the view of cancer as an atavism move us forward in understanding its place in biological systems and human health? For a start, it shifts our focus away from single mutations or genes to the broader functional and evolutionary context in which cancer occurs. Moreover, it allows us to make predictions, in this case about genomic and phenotypic behaviors that should be universal features of cancers, irrespective of tissue of origin. If cancer is the atavistic reexpression of unicellular programs, then several predictions follow:

1. Genes that are causally implicated in cancer should be older than the emergence of complex multicellularity around 600 million years ago.

2. Cancer cells should show a transcriptional shift toward unicellular behavior pathways.

3. Cancer cells should employ such unicellular behavior in response to cellular and environmental stresses.

4. The cancer phenotype should be suppressed when a tumor is placed into a physiologically normal multicellular environment. A corollary to this is that while the immediate microenvironment of a malignant tumor may appear morphologically normal, it is not

physiologically normal (i.e., there should always be a gradient or field effect with respect to "tissue normality" in the vicinity of a tumor, even if it is subtle).

5. Population diversity within a tumor should be an independent prognostic factor for patient outcome.

9.2.3   Gene Age and Cancer

Phylostratigraphy is a method to trace any given gene's lineage through its homology across species,[19] and it can be used to test the prediction that genes causally implicated in cancer are, as a group, older than the emergence of multicellularity. (Henceforth, we define "causally implicated" cancer genes as those listed in COSMIC, a highly curated database of genes for which there is reliable evidence of a link with cancer.[20]) Once a gene's orthologous grouping has been established, the species that represents the closest evolutionary link to the last common ancestor assigns a "date" to the gene. This is done for each gene, generating a distribution of either phylostratigraphic groups or gene ages. One can then take the subset of genes of interest, in our case COSMIC genes, and ask whether their age or phylostratigraphic distribution differs substantially from the distribution of all genes. Whether looking by phylostrata[21] or gene age,[22] COSMIC genes are indeed enriched in genes older than the emergence of complex multicellularity, with most having evolved just prior to the advent of simple metazoan life around 1,000 million years ago. Interestingly, COSMIC genes with recessive phenotypes (i.e., when all copies of the gene in a cancer cell need to be nonfunctional for the phenotype to occur) are enriched for genes with ages that correspond to the emergence of cellular life.[23]

Gene age also provides insights into mutational patterns in cancer. The likelihood for mutations to happen is not uniform across the genome. It is determined by the interplay of several levels of genomic organization, ranging from where in the nucleus a locus resides—and therefore how easily accessible it is for both damage and repair to occur—to how much transcription takes place at this location and whether or not the cell can survive a mutation at that genomic position. This difference in mutation probability across different regions of the genome is itself the result of adaptive evolution and has led to the current probability structure of mutation events. This also means that in the course of evolution, some genomic regions show little change over time and others can even be "preferred" regions of genomic change. For example, many genomes contain regions of repeated chromosomal rearrangement; these are evolutionarily reused breakpoints (EBRs), where double-strand breaks leading to structural alterations of chromosomes have repeatedly occurred and have been selected for, presumably because changes involving genes surrounding the break point had a selective advantage for the organism. Genomes also contain regions that show significant homology across species in sequence, gene order, and gene orientation. These are known as homologous synteny blocks (HSBs). In humans, genes younger than 1,000 million years, coinciding with the evolution of Metazoa, are enriched in EBRs and depleted in HSBs, while genes that are premetazoan are enriched in HSBs and excluded from EBRs.[24] Further complicating matters, mutations can be scattered as single events across the genome, but they can also occur in clusters of events that are located more closely together than would be predicted by the overall number of mutations in the genome. When looking at non-inherited single-nucleotide variants in individuals without cancer, there is a depletion of clustered variants in HSBs. Conversely, EBRs are enriched for clustered variants.[25] In cancer, the pattern for clusters enriched in EBRs and depleted in HSBs remains genome-wide. However, at the level of individual genes, clustered mutations are no longer excluded from those genes localized in HSBs. This means that genes normally protected from mutation by virtue of their genomic locations are no longer protected. Furthermore, COSMIC genes are enriched almost twofold in clustered mutations, even though they are older and more likely to reside in HSBs than other genes.[26] These changes in the mutation pattern within EBRs and HSBs highlight an important point: somatic variants in cancer and normal tissues *are found in different genomic regions.* Why cancer somatic variants are relocated into genes in HSBs is still an open question, but two likely explanations come to mind. One is a reversion of genome maintenance programs to unicellular wiring in response to tumor microenvironmental stress factors. The other is the existence of a recovery bias due to selection moving to the cellular level from the organismal level (i.e., the mutation is tolerated *only* at a cellular level and is never detected when it must be tolerated during embryogenesis and development).

9.2.4   Cancer Transcriptomes

The detection of the dysregulation of transcriptional programs in cancer has progressed from observational studies in the early 2000s to clinical translation for molecular subtyping, risk stratification, and therapy choice. As an example, tests such as Mammaprint[27] and Oncotype DX® DCIS score[28] are now employed for risk stratification and therapeutic choice among patients with breast cancer. The transcriptomes of tumor cells usually retain enough features of their tissue of origin to identify them but often show signs of aberrant signal transduction, elevated proliferation, and resistance to cell death. The atavism hypothesis/theory predicts this pattern of gene expression, because it represents a switch toward features of unicellular behavior.

Support for interpreting genomic changes seen in cancers as a switch toward unicellular gene expression patterns comes from recent work on transcriptional responses to rapid acquisition of doxorubicin resistance[29] and phylostratigraphy applied to gene expression data in cancer.[30] Using an in vitro system that generated a concentration gradient of doxorubicin to select for the emergence of resistance, Wu and colleagues identified a set of genes that showed no mutations from RNA sequencing but nevertheless displayed significant differences in gene expression levels compared to wild-type cells. Using estimates of gene age, they showed that genes that are older than 1,000 million years, predating the emergence of multicellularity, were the source of these expression differences.[31] The observed differences in gene expression without evidence of mutation are indicative of the structural rewiring of the genome that the tumor cells have undergone, which leads to an increased capacity to deal with environmental challenges like DNA-damaging agents.

Trigos and colleagues performed a systematic analysis of gene expression and gene phylostratigraphy. Using data from seven different cancer types, they found that the transcriptomes of tumor cells have a larger contribution from genes from the two most ancient phylostrata in their analysis, corresponding to genes found already in unicellular life (see figure 9.1). Furthermore, in prostate cancer, the increase in the proportion of the transcriptome coming from ancient genes corresponded to a loss of cellular differentiation as defined by increasing Gleason score.[32] In both analyses, upregulation and downregulation of genes of different origins (unicellular [UC] or multicellular [MC]) depend upon the cellular pathway. This suggests that reversion is not haphazard and random but a coordinated and systemic reestablishment of unicellular gene expression patterns. Exploring this further, Trigos et al. found a compartmentalization of correlated gene expression in both normal tissue and tumors, where pairs of either UC or of MC genes were generally positively correlated. This points to a modularity of expression based on gene evolutionary history. In contrast, UC-MC correlations are more negative in tumors than they are in normal tissues. Trigos et al. hypothesize that loss of the crosstalk between UC and MC genes is a key factor in tumorigenesis.[33] This is further supported by their work showing recurrent point mutations in cancer are enriched in the regulator genes linking UC and MC gene subnetworks, while copy-number alterations affected downstream targets in regions of the gene regulatory network that are distinctly unicellular or multicellular.[34]

9.2.5   Reengaging Unicellular Programs

If cancer is detrimental to multicellular organisms, why hasn't there been a strong selective pressure to eliminate it? A likely answer is because the unicellular processes that underlie the cancer phenotype are also essential to multicellular development and somatic tissue maintenance (e.g., in wound healing). The discovery of a compartmentalization of gene expression by gene evolutionary age pertaining to UC and MC modules is intriguing.[35] So too is the observation that in cancer, there are 15 percent more somatic mutations outside of genes compared to those within them and that recurrent mutations are enriched in early metazoan genes that regulate the interaction between UC and MC modules.[36] These facts invite us to consider the regulation of, and cellular responses to, the extracellular environment. What is targeted in the rewiring of the genome during tumorigenesis is the regulatory circuit(s) that repurpose unicellular cell behavior to benefit multicellular existence. Cancer is often described in terms of normal cells "going wrong" and "running amok" (i.e., being out of control). But the characterization of cancer in terms of "rogue cells" misses potentially a key property of multicellular life, which is that cells of a metazoan organism can revert to unicellular behavior in a coherent manner under certain stressful conditions. That is, the normal-to-cancer transition is a systematic and, in many cases, predictable trans-

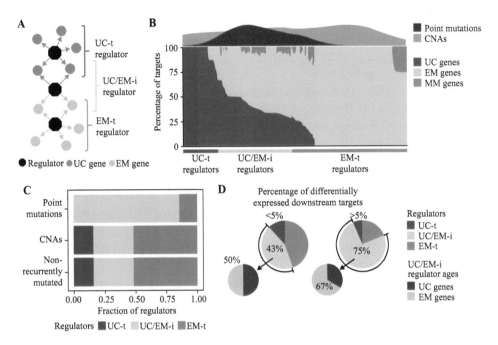

Figure 9.1
Point mutations in regulators affect UC-EM gene regulation. (A) Classification of regulators by the age of their downstream targets. UC-t regulators mostly regulate UC genes, EM-t regulators EM genes, and UC/EM-i regulators are at the interface of UC and EM genes. (B) (Lower panel) Percentage of UC, EM, and MM target genes in regulators. Upper panel: Distribution of recurrent point mutations (dark gray) and copy number abnormalities (CNAs; light gray) across regulators. UC/EM-i regulators are enriched in point mutations. (C) Fraction of regulators with point mutations, CNAs, and those nonrecurrently altered. More than 85 percent of regulators affected by point mutations are UC/EM-i regulators. The fraction of regulators of each class affected by CNAs is similar to those not affected by recurrent mutations, indicating a lack of preferential alteration of a particular regulator class by CNAs. (D) Effect of point mutations in regulators on the expression of downstream targets. Point mutations with a high downstream effect (>5 percent differentially expressed targets) are more likely to be UC/EM-i regulators of EM origin. Low-impact mutations (<5 percent differentially expressed targets) affect a higher proportion of regulators of a UC origin. UC, unicellular; EM, early metazoan; MM, mammal specific (Figure 3, reproduced from Trigos et al., 2019, with permission).

formation of cellular functions, a reversion to ancestral functionality. If this hypothesis is correct, it leads to further predictions. The onset of multicellularity involved the repurposing of certain unicellular functions to adapt somatic cells to their extracellular environment. For example, the directed motility of unicellular organisms in response to nutrient sensing evolved into the critical ability of somatic cells to migrate to designated locations during development, under the influence of regulated morphogenetic gradients. The onset of cancer triggered by regulatory breakdown in multicellular organisms implies that individual somatic cells will tend to revert to their original unicellular behavior, in this case becoming motile through loss of cell polarity or cell-cell contacts, facilitating the search for a more favorable environment (metastasis).

Another example, which we have investigated in detail, involves mutagenic DNA repair mechanisms. A very ancient unicellular survival mechanism is known as stress-induced mutation (SIM). It is a process by which bacteria increase their mutation rate in the face of environmental stress that becomes sufficiently threatening to induce the SOS response at the same time as they are facing double-strand DNA damage.[37] During this process, the repair mechanism of double-strand DNA breaks (DSBs) switches from high-fidelity homologous recombination repair to an alternative pathway that employs the error-prone DNA polymerase DinB. The result of this switch is a trail of self-inflicted damage either side of the double-strand break out for many kilobases, displaying a spectrum of both single-nucleotide variants (SNVs) and amplification events. The upshot of such a dramatic increase in focused, localized mutations is the possibility of adaptation to the challenging environment through random generation of adaptive genotypes.[38] This propensity of bacteria to evolve their way out of trouble is associated with a distinctive signature, namely, a clustering of SNVs around the site of the DSB with a decreasing probability of encountering an SNV with increasing distance from the break.[39]

In humans, there are orthologous genes to DinB that include Pol κ and Pol η, which are responsible for translesion synthesis in order to bypass damaged bases during normal replication and are also employed during microhomology-mediated break repair.[40] One of the hallmarks of most cancers is genomic instability, in particular a plethora of DSBs that result in ongoing structural rearrangements detected by cytogenetics.[41] It has been reported that SNVs in cancer display clustering patterns too, attributed to the activity of Pol η but also of the AID/APOBEC family of deaminases.[42] These observations, combined with our finding that ancient COSMIC genes were enriched for DNA repair pathways, particularly double-strand break repair,[43] led us to ask whether the clusters observed in cancer and normal somatic tissues featured the telltale pattern displayed by bacteria (i.e., a decrease in mutation density from the center of a cluster indicative of SIM, which can be thought of as a cluster "shape" if one were to plot a histogram of mutations as a function of distance). We did indeed find that clusters in both cancer and healthy somatic tissue show the characteristic pattern, with normal cells showing evidence for a highly regulated process with similarly "shaped" clusters, while cancers exhibited a large degree of heterogeneity in cluster "shape." We also determined that the action of translesion synthesis polymerases resulted in more SNV clusters than either AID or APOBEC. Significantly, diversity in cluster "shape" (i.e., the distribution of mutation density within the cluster) was a predictor of shorter overall survival in patients with cancer.[44]

The retention over evolutionary time scales of a DNA repair mechanism that *increases* the mutation rate in response to cellular stress may seem counterintuitive. To be sure, its components are useful for translesion synthesis in replication and certain types of homology-related repair. But why keep such a system intact in the context of multicellularity when the original inputs now result not in cellular evolution but cellular death via apoptosis or necroptosis? One reason is that the generation of somatic genome diversity is critical to the functioning of the immune system via somatic hypermutation of V regions that generate antibody diversity.[45] Somatic genome diversity is also implicated in polyploidy in megakaryocyte development[46] and the appearance of tetraploidy in normally functioning liver and cardiac tissues.[47] On general grounds, it makes sense that organs have evolved some level of intercellular genomic diversity in order to deal with organism homeostasis and organ-specific demands and stresses that go with tissue maintenance, particularly in the face of injury or illness. Transcriptional regulation can only go so far, and if all the genomic eggs are put in one basket—that is, if every cell has exactly the same response potential in the face of challenges—then the system is much less robust. Intercellular heterogeneity deployed as an evolutionary adaptive mechanism in response to stress may also explain the probabilistic nature of tumor formation. If a population of cells is genetically diverse, then the effect of carcinogens will vary from cell to cell as well as from individual to individual, which may explain why some people who smoke for years never develop lung cancer while others with the same exposure do. It may also explain why estimates of the contribution of environment, genetics, and "bad luck" to carcinogenesis vary so much between studies. The underlying assumption, that every cell in a body or organ has the same genomic makeup, is simply wrong. Finally, we note that the distinctive mutational clusters around DSBs are found not only in normal somatic tissues but also in the germline,[48] suggesting that multicellular signals for inducing the mutational response have their source in development. The idea of tightly regulated diversity-generating bursts of mutation and structural rearrangement during development, both over generations within the population as well as within the body, receives support from the discovery of a significant somatic diversity among neurons.[49]

9.2.6 Cancer and the Extracellular Environment

One of the most perplexing observations of cancer biology is that the cancer phenotype, particularly the properties of unregulated growth and unrestricted movement, can be suppressed by placing tumor cells in a physiologically normal environment.[50] This has led some to question the paramount importance attached to somatic mutations in tumor formation and cell behavior in general,[51] which is favored still by most oncologists. This dominant view has also diverted attention (and research funding) from other cellular mechanisms that play an important role in determining phenotype at the tissue level.

For example, recent work in bioelectricity may explain the aforementioned observation in the context of the atavism theory of cancer. In their normal resting state, cells possess a potential difference, V_{mem}, across the cell membrane, which facilitates electrical signaling, interaction, and coordinated responses to changing environmental conditions, both within and between cells. Signals are propagated via ion channels, a fundamental feature of cells that predates the emergence of multicellularity (figure 9.2[52]). Bacteria use bioelectric signaling for quorum sensing and maintaining microbial community diversity by using potassium pulses that transiently alter resting membrane potential.[53] In multicellular organisms,

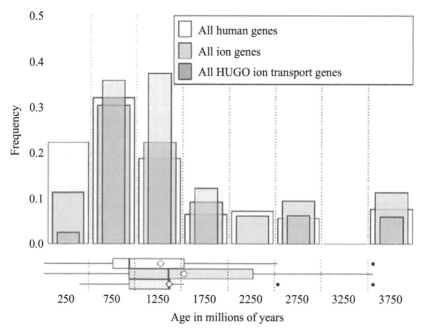

Figure 9.2
Age distribution of genes involved in ion transport. The background distribution of all human genes is in light gray. The list of genes involved in ion transport as annotated by HUGO is in dark gray. Genes annotated by the GO term "ion transport" are in medium gray.

patterns of bioelectric polarization due to differences in bioelectric potentials across membranes and between cells contribute to morphogenetic signaling by establishing orientation in three-dimensional space.[54] Importantly, however, cancer cells display altered membrane potential V_{mem}, in some cases becoming completely depolarized. In species with regenerative capacities, such as salamanders, bioelectric control creates a permissive environment during regeneration that is capable of normalizing cancer behavior. Recent work in frogs demonstrates that depolarization of instructional cell resting membrane potential is sufficient to induce hyperproliferation, invasion, and motility in distant melanocytes via metalloprotease expression in response to serotonin signaling.[55]

In contrast to other cellular processes where the unicellular gene expression program has been subjugated to multicellular control (a regulation that breaks down in cancer), bioelectrical control in multicellular organisms remains fundamentally unicellular in nature. Thus, cancer cells appear to maintain the ability to respond appropriately, at least in terms of proliferation, motility, and invasion, to environments where the resting membrane potential of cancer cells reverts to a more polarized state (i.e., a higher V_{mem}). This implies that the tumor microenvironment has lost the ability to enforce an elevated V_{mem}. A test of this hypothesis would be to repeat the experiments of Ana Soto and colleagues[56] involving the transplanting of cancer cells or normal mammary epithelial cells into cleared mammary fat pads that are normal or had been exposed to carcinogen. In these experiments, it was the mammary fat pad, not the tumor or epithelial cells, that determined whether near-normal mammary morphology resulted from a particular combination.[57] Furthermore, the age of the rat determined the strength of this "normalization." Among virgin rats, the ability of the fat pad to enforce normal mammary gland growth increased as the animals aged.[58] In repeating the experiments, the V_{mem} of both the stroma and the tumor over time should be measured, as well as the stroma's ability to resist changes in membrane potential. According to our hypothesis, stroma from normal mammary glands should be found to reverse changes in the V_{mem} of the transplanted cancer cells. Conversely, stroma from tumorous glands should fail to repolarize transplanted tumor cells and would depolarize transplanted normal epithelial cells. In both cases, the genomic complement of the tumor or epithelial cells would remain the same. Thus, we would predict that if tumor cells that had repolarized were removed from that environment, they would revert to a depolarized state and reexpress the phenotypes of hyperproliferation and motility they had originally. Why? Because formation of a malignant tumor requires both alteration of the genome at massive scale and a concomitant alteration of surrounding healthy tissue that disrupts the normal regulation of V_{mem}. The atavism theory predicts that since bioelectrical signaling has remained at its core unicellular, responding to the same inputs with the same outputs, cancer cells will not evolve away from it. Why? Cancer cells might not actually perceive a change of the external control mechanisms of bioelectric signaling as a stress that requires an adaptive response. This is in contrast to something like growth factor signaling, where the outputs of cell division and motility triggered by the unicellular inputs of nutrients have been replaced by multicellular inputs of developmentally regulated growth factor concentrations. As has been seen clinically, inhibiting such signaling almost always results in the development of therapeutic resistance.[59] Therefore, manipulation of the bioelectrical control of the tumor should not induce the same levels of therapeutic resistance seen with current modalities, opening up the promise of long-term disease control.

9.2.7   Cancer as a Speciation Event

The atavism theory treats cancer as the integrated reexpression of cell behaviors characteristic of unicellular organisms and lays no claim to how and/or why this happens, although we have laid out arguments and evidence for how it might. Tumors evolve over time, creating a moving target for therapeutic treatment. The study of cancer genomes reveals that the observed genomic instability comprises both large-scale changes in the form of chromosomal instability (CIN) and mutations on a much smaller scale, such as SNVs and small insertion and deletion events (indels). These distinct patterns of instability point to two types of tumor evolution: punctuated macroevolution and stepwise microevolution. They take place because tumor cells are subject to strong selective forces, for example, hypoxia produced by a tumor outgrowing local blood supply or the many effects of chemotherapy. Punctuated evolution is driven by CIN, while stepwise evolution is facilitated by Darwinian mutations.[60] The upshot of CIN is large-scale karyotype evolution that leads in the case of a viable genotype to a complete rewiring of gene regulatory networks and can therefore be legitimately described as a speciation event.

The association of karyotype evolution with speciation events is well established. The mechanisms, both in terms of dissemination of new chromosomes throughout a population as well as the selective advantages of rearranged chromosomes leading to this association, are still a subject of study and debate. Explanations for a failure of hybrid karyotypes leading to speciation over time include increased nuclear-mitochondrial genome mismatch due to reduction in recombination frequency, generation of duplication-deletion events during recombination involving inversions, and missegregation of both normal and rearranged chromosomes during meiosis.[61] On the face of it, such mechanisms should be subject to strong selective pressures *against* karyotype evolution. And yet, there are multiple examples, both over evolutionary time and in the somatic context of cancer, where karyotype evolution has indeed occurred. While such accounts may explain why divergent karyotypes remain divergent, they do not explain how the divergence came about in the first place. What is apparent is that either through programmed or random events, the genome went through a period of instability.

Ploidy is the number of chromosomes that is specific for an organism and passed on to its offspring. In sexually reproducing species, gametes are haploid (having only one copy of each chromosome), while somatic tissues are usually diploid (having two copies of each chromosome, one from each parent). Changes in ploidy, either due to problems with cell division or in response to developmental or extracellular cues, are common. Rounds of endoreduplication—genome replication without cell division—result in greater than sixteen copies of the genome in human megakaryocytes and in fruit fly salivary glands and nurse cells.[62] Other human tissues, such as liver, cardiac muscle, smooth muscle of the aorta and uterus, and adrenal glands, respond to cellular stress with low levels (four to sixteen copies) of polyploidy.[63] Changes in ploidy lead to changes in gene expression that combine features of both proliferation and differentiation.[64] Alterations of ploidy have been linked to the transcriptional programs regulated by MYC, a well-known oncogene.[65]

Extra copies of entire haploid sets of chromosomes are not themselves tumorigenic. However, aneuploidy, alterations in copy number of one or more chromosomes, can be,[66] particularly in the context of stress-induced polyploidy.[67] The result is a highly rearranged and new karyotype that, if found in any context other than cancer, would be regarded as a new species. Thus, cancer can be viewed as a speciation event.[68] Support for this interpretation can be seen in the examples of transmissible cancers found in the wild. Canine venereal cancer is transmissible in dogs and has been circulating for at least 11,000 years. Its chromosome count is 59 with thirteen to seventeen rearranged chromosomes that are meta- or submetacentric.[69] By comparison, the constitutional karyotype of dogs has a diploid number of seventy-eight subtelomeric chromosomes. Similarly, two different transmissible cancers, devil facial tumor disease 1 and 2, are circulating in the Tasmanian Devil population. Both show different rearranged near-diploid karyotypes with several structurally abnormal chromosomes.[70]

Large-scale genomic rearrangement characterized by CIN is a stochastic event in the context of stress-induced polyploidy.[71] It represents a punctuated macroevolutionary step that results in a systems-level rearrangement of the genome and therefore information flow.[72] The new information flow is reflected in alteration of the packaging of the genome in aneuploid cells compared to diploid cells.[73] It is also reflected in the observed reduction in crosstalk between cellular networks of unicellular versus multicellular origin.[74] This information flow is robust to perturbation and optimized for adaptation, likely through the reliance on "fuzzy" inheritance.[75]

If we regard cancer as a speciation event, defined by redeployment of conserved unicellular behavior and driven by large-scale chromosome rearrangement, we need to incorporate the role of stromal-tumor interactions as described in the studies above. We suggest a variation on the familiar "two-hit" hypothesis originally proposed by Nordling[76] and formalized by Knudson.[77] In our variation, the two required hits are (1) a speciation event that leads to the reexpression of unicellular programs that (2) takes place in the environmental context of a limited stromal ability to enforce a highly polarized tumor cell V_{mem} and other tissue-level mechanisms that normally suppress unicellular cell behavior. Such a model could potentially explain both tumor dormancy as well as metastatic potential. Tumors take hold only when the stroma fails to enforce the bioelectric control of cell division and motility that has remained fundamentally unicellular throughout evolutionary time. Such failure might arise from a wound-healing type of response in the stroma, which the cancer then perpetuates. Another possibility is a response to infection, limiting the ability of the stroma to remain as polarized as it would be under normal physiological conditions. Alterations of stromal bioelectrical control might also influence the ability of the immune system to function as a watchdog against tumor formation. As such, it might explain the association of chronic inflammation with cancer incidence and mortality.[78]

9.3 Conclusions

The atavism theory of cancer provides an overarching conceptual framework to explain why cancer is pervasive across the tree of life and why it is an inevitable concomitant of multicellularity. It provides several testable predictions. In addition, it accounts for the fact that although genomic alterations are a cause of cancer, the cancer phenotype may nevertheless be suppressed by surrounding stroma that is physiologically normal. The atavism theory emphasizes the role of *adaptability* as a selected trait. This shift in perspective has important ramifications for clinical practice. If, through the redeployment of unicellular behavior, tumor cells are primed to respond to selective pressures through evolution and the generation of phenotypic heterogeneity, then a "take no prisoners" approach to treatment risks creating a monster beyond our control. Furthermore, in the process of treatment, the normal multicellular mechanisms that have evolved to quell such an uprising are damaged and so no longer function as they should.

How, then, do we move forward with applying our hypothesis/theory to yield results that can be tested ultimately in the clinic? Several areas of further study are suggested by our analysis.

1. We need to develop a model of "information flow" in the gene regulatory networks of cancer cells (see also chapters 4, 5, and 6, this volume). What exactly are the network nodes that link unicellular and multicellular gene expression modules? Do they represent a control network in themselves, and if so, what are their inputs under normal physiological conditions and how does that change during tumorigenesis? Can we develop therapeutic strategies that reinstate or mimic multicellular control of the unicellular network, either directly or through manipulating the extracellular environment?

2. More attention needs to be paid to the phenotypic outcomes of large-scale rearrangements of cancer genomes. How do clonal sweeps and the induction of periods of heightened mutational rates play out over time in cancer? Are they sequential or simultaneous? How does the evolutionary history of a given species' genome constrain the outcome of genomic instability, both large scale and small scale? Can we define a general "cancer genome" that is characterized by physical properties such as chromatin packing and chromosome architecture that produces a unicellular phenotype in a cell of a multicellular organism? How does the physical alteration of the genome relate to the information flow through the gene regulatory networks of cancer cells?

3. We need to address stromal bioelectric control capabilities. How do various tissues and organ systems maintain control of V_{mem} throughout development and in response to perturbations of homeostasis such as injury? How has the evolution of multicellularity interfaced with bioelectrical control? Does cancer cell V_{mem} play a role in the possible control network defined by the links between unicellular and multicellular genes, and if so, how? More generally, we ask the fundamental question of whether and how bioelectric information networks couple to and coevolve with biochemically wired control networks.

4. While not strictly an argument of the atavism theory/hypothesis, as it has been made before based on evolutionary arguments, there is a strong need to reevaluate the push for complete tumor eradication as a treatment goal. There are lessons to be learned from agricultural pest management. In our paradigm, tumors are a heterogeneous mixture of cells that have various degrees of sensitivity to treatment and are characterized by highly adaptive unicellular responses to extreme selective forces. Furthermore, there is a fitness cost to drug resistance that allows sensitive cells to keep resistant cells in check. Thus, the goal of treatment is to keep tumors stable or slowly reduce their growth by maintaining enough sensitive cells in the tumor population to keep resistant cell numbers low. This is one area where research is quickly translating into clinical trials. So-called adaptive therapy trials

are currently being implemented that seek to balance tumor size control with drug dose minimization, to avoid annihilating all sensitive cells. Mathematical models based on this balancing act suggest that, for most tumors, treatment failure and recurrence are reduced (see also chapter 10, this volume). The first clinical implementations of these paradigms are under study for both prostate and thyroid cancer (ClinicalTrials.gov study numbers NCT03511196 and NCT03630120, respectively).

In this chapter, we have outlined an attempt to reconceptualize cancer as a systemic reversion to an ancestral phenotype (an "atavism") in which ancient unicellular modalities, retained in all extant genomes, are reactivated as a response to stress. This theory has the ability to explain several otherwise baffling properties of cancer and makes a number of specific predictions. The therapeutic implications of the theory are far-reaching, and we advocate several follow-up studies to test our atavism hypothesis in experimental and clinical practice.

Acknowledgments

This work was supported in part by National Institutes of Health grant U54 CA217376. We thank our colleagues, particularly Charles Lineweaver, Mark Vincent, Robert Austin, Susan Rosenberg, Luis Cisneros, and David Goode, for their astute observations and work in understanding the relationship between unicellularity and cancer. Thanks also to Michael Levin and Danny Adams for insight into how bioelectrical control affects vertebrate physiology beyond neurons. Special thanks to Carlo Maley and Athena Aktipis for their perceptive discussions about the role of multicellularity, cooperation, and the evolution of those traits throughout the tree of life.

Notes

1. Aktipis CA, et al. (2015) Cancer across the tree of life: cooperation and cheating in multicellularity. *Phil Trans R Soc B* 370(1673):20140219.

2. Hanahan D, Weinberg RA (2011) Hallmarks of cancer: the next generation. *Cell* 144(5):646–674.

3. Weinberg RA (1989) Oncogenes, antioncogenes, and the molecular bases of multistep carcinogenesis. *Cancer Res* 49(14):3713–3721.

4. Fearon ER, Vogelstein B (1990) A genetic model for colorectal tumorigenesis. *Cell* 61(5):759–767. Schulz WA ed. (2005) Cancers of the skin. *Molecular Biology of Human Cancers: An Advanced Student's Textbook* (Springer Netherlands), pp 255–270.

5. Schwartzentruber J, et al. (2012) Driver mutations in histone H3.3 and chromatin remodelling genes in paediatric glioblastoma. Nature 482(7384):226–231. The Cancer Genome Atlas Network (2012) Comprehensive molecular portraits of human breast tumours. Nature 490(7418):61–70. Bender S, et al. (2013) Reduced H3K27me3 and DNA hypomethylation are major drivers of gene expression in K27M mutant pediatric high-grade gliomas. Cancer Cell 24(5):660–672. Buczkowicz P, et al. (2014) Genomic analysis of diffuse intrinsic pontine gliomas identifies three molecular subgroups and recurrent activating ACVR1 mutations. Nat Genet 46(5):451–456. Fontebasso AM, et al. (2014) Recurrent somatic mutations in ACVR1 in pediatric midline high-grade astrocytoma. Nat Genet 46:462–466. The St Jude Children's Research Hospital–Washington University

Pediatric Cancer Genome Project, et al. (2014) The genomic landscape of diffuse intrinsic pontine glioma and pediatric non-brainstem high-grade glioma. Nat Genet 46(5):444–450. Roy DM, Walsh LA, Chan TA (2014) Driver mutations of cancer epigenomes. Protein Cell 5(4):265–296. Taylor KR, et al. (2014) Recurrent activating ACVR1 mutations in diffuse intrinsic pontine glioma. Nat Genet 46(5):457–461. Castel D, et al. (2015) Histone H3F3A and HIST1H3B K27M mutations define two subgroups of diffuse intrinsic pontine gliomas with different prognosis and phenotypes. Acta Neuropathol (Berl) 130(6):815–827. Mackay A, et al. (2017) Integrated molecular meta-analysis of 1,000 pediatric high-grade and diffuse intrinsic pontine glioma. Cancer Cell 32(4):520–537.

6. Nowell PC, Hungerford DA (1960) A minute chromosome in human chronic granulocytic leukemia. *Science* 132(3438):1497. Rowley JD (1973) A new consistent chromosomal abnormality in chronic myelogenous leukaemia identified by quinacrine fluorescence and Giemsa staining. *Nature* 243(5405):290–293.

7. Biernaux C, Loos M, Sels A, Huez G, Stryckmans P (1995) Detection of major bcr-abl gene expression at a very low level in blood cells of some healthy individuals. Blood 86(8):3118–3122. Bose S, Deininger M, Gora-Tybor J, Goldman JM, Melo JV (1998) The presence of typical and atypical BCR-ABL fusion genes in leukocytes of normal individuals: biologic significance and implications for the assessment of minimal residual disease. Blood 92(9):3362–3367. Ismail SI, Naffa RG, Yousef A-MF, Ghanim MT (2014) Incidence of bcr-abl fusion transcripts in healthy individuals. Mol Med Rep 9(4):1271–1276.

8. Maffini MV, Calabro JM, Soto AM, Sonnenschein C (2005) Stromal regulation of neoplastic development: age-dependent normalization of neoplastic mammary cells by mammary stroma. *Am J Pathol* 167(5):1405–1410. Maffini MV, Soto AM, Calabro JM, Ucci AA, Sonnenschein C (2004) The stroma as a crucial target in rat mammary gland carcinogenesis. *J Cell Sci* 117(8):1495–1502. Weaver VM, et al. (1997) Reversion of the malignant phenotype of human breast cells in three-dimensional culture and in vivo by integrin blocking antibodies. *J Cell Biol* 137(1):231–245. Sternlicht MD, et al. (1999) The stromal proteinase MMP3/stromelysin-1 promotes mammary carcinogenesis. *Cell* 98(2):137–146. Ricca BL, et al. (2018) Transient external force induces phenotypic reversion of malignant epithelial structures via nitric oxide signaling. *eLife* 7:e26161.

9. Blackiston D, Adams DS, Lemire JM, Lobikin M, Levin M (2011) Transmembrane potential of GlyCl-expressing instructor cells induces a neoplastic-like conversion of melanocytes via a serotonergic pathway. *Dis Model Mech* 4(1):67–85. Levin M, Martyniuk CJ (2018) The bioelectric code: an ancient computational medium for dynamic control of growth and form. *Biosystems* 164:76–93.

10. Fidler IJ (1978) Tumor heterogeneity and the biology of cancer invasion and metastasis. Cancer Res 38(9):2651–2660. Heppner GH, Miller BE (1983) Tumor heterogeneity: biological implications and therapeutic consequences. Cancer Metastasis Rev 2(1):5–23. Dexter DL, Leith JT (1986) Tumor heterogeneity and drug resistance. J Clin Oncol 4(2):244–257. Heim S, Mitelman F (1989) Primary chromosome abnormalities in human neoplasia. Adv Cancer Res 52:1–43. Shackney SE, Shankey TV (1995) Genetic and phenotypic heterogeneity of human malignancies: finding order in chaos. Cytometry 21(1):2–5. Marusyk A, Polyak K (2010) Tumor heterogeneity: causes and consequences. Biochim Biophys Acta 1805(1):105–117. Greaves M (2015) Evolutionary determinants of cancer. Cancer Discov 5(8):806–820. Greaves M (2018) Nothing in cancer makes sense except…. BMC Biol 16(1):22.

11. Aktipis CA, et al. (2015) Cancer across the tree of life: cooperation and cheating in pmulticellularity. *Phil Trans R Soc B* 370(1673):20140219.

12. Hanahan D, Weinberg RA (2011) Hallmarks of cancer: the next generation. *Cell* 144(5):646–674.

13. Manchester KL (1995) Theodor Boveri and the origin of malignant tumours. Trends Cell Biol 5(10): 384–387. Opitz JM (2016) Annals of morphology THEODOR BOVERI (1862–1915) To commemorate the centenary of his death and contributions to the Sutton–Boveri hypothesis. Am J Med Genet A 170(11):2803–2829.

14. Boveri T (1929) *The Origins of Malignant Tumors* (The Williams and Wilkins Company). Boveri T (2008) Concerning the origin of malignant tumours by Theodor Boveri, translated and annotated by Henry Harris. J Cell Sci 121(S1):1–84. Boveri T (2008) *Concerning the Origin of Malignant Tumours*, translated and annotated by Henry Harris (Cold Spring Harbor Press and The Company of Biologists).

15. Boveri T (2008) Concerning the origin of malignant tumours by Theodor Boveri, translated and annotated by Henry Harris. J Cell Sci 121(S1):1–84.

16. Boveri T (2008) Concerning the origin of malignant tumours by Theodor Boveri, translated and annotated by Henry Harris. J Cell Sci 121(S1):1–84.

17. Boveri T (2008) Concerning the origin of malignant tumours by Theodor Boveri, translated and annotated by Henry Harris. J Cell Sci 121(S1):1–84.

18. Heng HHQ (2015) Debating Cancer: The Paradox in Cancer Research (World Scientific Publishing Company).

19. Domazet-Lošo T, Brajković J, Tautz D (2007) A phylostratigraphy approach to uncover the genomic history of major adaptations in metazoan lineages. *Trends Genet* 23(11):533–539.

20. Forbes SA, et al. (2015) COSMIC: exploring the world's knowledge of somatic mutations in human cancer. *Nucleic Acids Res* 43(D1):D805–D811.

21. Domazet-Lošo T, Tautz D (2010) Phylostratigraphic tracking of cancer genes suggests a link to the emergence of multicellularity in metazoa. *BMC Biol* 8(1):66.

22. Cisneros L, et al. (2017) Ancient genes establish stress-induced mutation as a hallmark of cancer. *PLoS ONE* 12(4):e0176258.

23. Cisneros L, et al. (2017) Ancient genes establish stress-induced mutation as a hallmark of cancer. *PLoS ONE* 12(4):e0176258.

24. Cisneros L, et al. (2017) Ancient genes establish stress-induced mutation as a hallmark of cancer. *PLoS ONE* 12(4):e0176258.

25. Cisneros L, et al. (2017) Ancient genes establish stress-induced mutation as a hallmark of cancer. *PLoS ONE* 12(4):e0176258.

26. Cisneros L, et al. (2017) Ancient genes establish stress-induced mutation as a hallmark of cancer. *PLoS ONE* 12(4):e0176258.

27. Cardoso F, et al. (2016) 70-Gene signature as an aid to treatment decisions in early-stage breast cancer. *N Engl J Med* 375(8):717–729.

28. Rakovitch E, et al. (2015) A population-based validation study of the DCIS score predicting recurrence risk in individuals treated by breast-conserving surgery alone. *Breast Cancer Res Treat* 152(2):389–398.

29. Wu A, et al. (2015) Ancient hot and cold genes and chemotherapy resistance emergence. *Proc Natl Acad Sci USA* 112(33):10467–10472.

30. Trigos AS, Pearson RB, Papenfuss AT, Goode DL (2017) Altered interactions between unicellular and multicellular genes drive hallmarks of transformation in a diverse range of solid tumors. *Proc Natl Acad Sci USA* 114(24):6406–6411.

31. Wu A, et al. (2015) Ancient hot and cold genes and chemotherapy resistance emergence. *Proc Natl Acad Sci USA* 112(33):10467–10472.

32. Trigos AS, Pearson RB, Papenfuss AT, Goode DL (2017) Altered interactions between unicellular and multicellular genes drive hallmarks of transformation in a diverse range of solid tumors. *Proc Natl Acad Sci USA* 114(24):6406–6411.

33. Trigos AS, Pearson RB, Papenfuss AT, Goode DL (2017) Altered interactions between unicellular and multicellular genes drive hallmarks of transformation in a diverse range of solid tumors. *Proc Natl Acad Sci USA* 114(24):6406–6411.

34. Trigos AS, Pearson RB, Papenfuss AT, Goode DL (2019) Somatic mutations in early metazoan genes disrupt regulatory links between unicellular and multicellular genes in cancer. *eLife* 8:e40947.

35. Trigos AS, Pearson RB, Papenfuss AT, Goode DL (2017) Altered interactions between unicellular and multicellular genes drive hallmarks of transformation in a diverse range of solid tumors. *Proc Natl Acad Sci USA* 114(24):6406–6411.

36. Cisneros L, et al. (2017) Ancient genes establish stress-induced mutation as a hallmark of cancer. *PLoS ONE* 12(4):e0176258. Trigos AS, Pearson RB, Papenfuss AT, Goode DL (2019) Somatic mutations in early metazoan genes disrupt regulatory links between unicellular and multicellular genes in cancer. *eLife* 8:e40947.

37. Shee C, Gibson JL, Rosenberg SM (2012) Two mechanisms produce mutation hotspots at DNA breaks in Escherichia coli. *Cell Rep* 2(4):714–721. Rosenberg SM, Shee C, Frisch RL, Hastings PJ (2012) Stress-induced mutation via DNA breaks in Escherichia coli: a molecular mechanism with implications for evolution and medicine. *BioEssays* 34(10):885–892. McKenzie GJ, Harris RS, Lee PL, Rosenberg SM (2000) The SOS response regulates adaptive mutation. *Proc Natl Acad Sci USA* 97(12):6646–6651.

38. Rosenberg SM, Shee C, Frisch RL, Hastings PJ (2012) Stress-induced mutation via DNA breaks in Escherichia coli: a molecular mechanism with implications for evolution and medicine. *BioEssays* 34(10):885–892.

39. Shee C, Gibson JL, Rosenberg SM (2012) Two mechanisms produce mutation hotspots at DNA breaks in Escherichia coli. *Cell Rep* 2(4):714–721.

40. Sakofsky CJ, et al. (2015) Translesion polymerases drive microhomology-mediated break-induced replication leading to complex chromosomal rearrangements. *Mol Cell* 60(6):860–872.

41. Roschke AV, Stover K, Tonon G, Schäffer AA, Kirsch IR (2002) Stable karyotypes in epithelial cancer cell lines despite high rates of ongoing structural and numerical chromosomal instability. *Neoplasia N Y N* 4(1):19–31. Heim S, Mitelman F (2015) Nonrandom chromosome abnormalities in cancer. *Cancer Cytogenetics* (Wiley-Blackwell), pp 26–41. Mitelman F, Heim S (2015) How it all began. *Cancer Cytogenetics* (Wiley-Blackwell), pp 1–10.

42. Roberts SA, et al. (2012) Clustered mutations in yeast and in human cancers can arise from damaged long single-strand DNA regions. *Mol Cell* 46(4):424–435. Lada AG, et al. (2012) AID/APOBEC cytosine deaminase induces genome-wide kataegis. *Biol Direct* 7:47. Burns MB, Temiz NA, Harris RS (2013) Evidence for APOBEC3B mutagenesis in multiple human cancers. *Nat Genet* 45(9):977–983. Taylor BJ, et al. (2013) DNA deaminases induce break-associated mutation showers with implication of APOBEC3B and 3A in breast cancer kataegis. *eLife* 2:e00534. Roberts SA, et al. (2013) An APOBEC cytidine deaminase mutagenesis pattern is widespread in human cancers. *Nat Genet* 45(9):970–976. Supek F, Lehner B (2017) Clustered mutation signatures reveal that error-prone DNA repair targets mutations to active genes. Cell 170(3):534–547.e23.

43. Cisneros L, et al. (2017) Ancient genes establish stress-induced mutation as a hallmark of cancer. *PLoS ONE* 12(4):e0176258.

44. Cisneros L, Vaske C, Bussey KJ (2018) Determining the relationship between a measure of stress-induced mutagenesis and patient survival in cancer. In *Proceedings of the 109th Annual Meeting of the American Association for Cancer Research* (AACR, Chicago, IL), p 3381.

45. Murphy K, Travers P, Walport M, Janeway C (2012) *Janeway's Immunobiology—NLM Catalog—NCBI*, 8th ed. (Garland Science).

46. Ravid K, Lu J, Zimmet JM, Jones MR (2002) Roads to polyploidy: the megakaryocyte example. *J Cell Physiol* 190:7–20. https://doi.org/10.1002/jcp.10035

47. Anatskaya OV, Vinogradov AE (2007) Genome multiplication as adaptation to tissue survival: Evidence from gene expression in mammalian heart and liver. *Genomics* 89:70–80. https://doi.org/10.1016/j.ygeno.2006.08.014.

48. Jónsson H, et al. (2017) Parental influence on human germline de novo mutations in 1,548 trios from Iceland. *Nature* 549:519–522. https://doi.org/10.1038/nature24018.

49. McConnell MJ, et al. (2017) Intersection of diverse neuronal genomes and neuropsychiatric disease: the Brain Somatic Mosaicism Network. *Science* 356:eaal1641. https://doi.org/10.1126/science.aal1641.

50. Maffini MV, Calabro JM, Soto AM, Sonnenschein C (2005) Stromal regulation of neoplastic development: age-dependent normalization of neoplastic mammary cells by mammary stroma. *Am J Pathol* 167(5):1405–1410. Maffini MV, Soto AM, Calabro JM, Ucci AA, Sonnenschein C (2004) The stroma as a crucial target in rat mammary gland carcinogenesis. *J Cell Sci* 117(8):1495–1502. Weaver VM, et al. (1997) Reversion of the malignant phenotype of human breast cells in three-dimensional culture and in vivo by integrin blocking antibodies. *J Cell Biol* 137(1):231–245. Sternlicht MD, et al. (1999) The stromal proteinase MMP3/stromelysin-1 promotes mammary carcinogenesis. *Cell* 98(2):137–146. Ricca BL, et al. (2018) Transient external force induces phenotypic reversion of malignant epithelial structures via nitric oxide signaling. *eLife* 7:doi:10.7554/eLife.e26161.

51. Soto AM, Sonnenschein C (2014) One hundred years of somatic mutation theory of carcinogenesis: is it time to switch? Cause to reflect. *BioEssays* 36:118–120. https://doi.org/10.1002/bies.201300160.

52. Cisneros L, et al. (2017) Ancient genes establish stress-induced mutation as a hallmark of cancer. *PLoS ONE* 12(4):e0176258.

53. Prindle A, Liu J, Asally M, Ly S, Garcia-Ojalvo J, Süel GM (2015) Ion channels enable electrical communication in bacterial communities. *Nature* 527:59–63. https://doi.org/10.1038/nature15709. Humphries J, et al. (2017) Species-independent attraction to biofilms through electrical signaling. *Cell* 168:200–209.e12. https://doi.org/10.1016/j.cell.2016.12.014.

54. Levin M, Martyniuk CJ (2018) The bioelectric code: an ancient computational medium for dynamic control of growth and form. *Biosystems* 164:76–93.

55. Blackiston D, Adams DS, Lemire JM, Lobikin M, Levin M (2011) Transmembrane potential of GlyCl-expressing instructor cells induces a neoplastic-like conversion of melanocytes via a serotonergic pathway. *Dis Model Mech* 4(1):67–85.

56. Maffini MV, Calabro JM, Soto AM, Sonnenschein C (2005) Stromal regulation of neoplastic development: age-dependent normalization of neoplastic mammary cells by mammary stroma. *Am J Pathol* 167(5):1405–1410. Maffini MV, Soto AM, Calabro JM, Ucci AA, Sonnenschein C (2004) The stroma as a crucial target in rat mammary gland carcinogenesis. *J Cell Sci* 117(8):1495–1502.

57. Maffini MV, Soto AM, Calabro JM, Ucci AA, Sonnenschein C (2004) The stroma as a crucial target in rat mammary gland carcinogenesis. *J Cell Sci* 117(8):1495–1502.

58. Maffini MV, Calabro JM, Soto AM, Sonnenschein C (2005) Stromal regulation of neoplastic development: age-dependent normalization of neoplastic mammary cells by mammary stroma. *Am J Pathol* 167(5):1405–1410.

59. Heng HH, et al. (2013) Chromosomal instability (CIN): what it is and why it is crucial to cancer evolution. *Cancer Metastasis Rev* 32:325–340. https://doi.org/10.1007/s10555-013-9427-7.

60. Dion-Côté A-M, Barbash DA (2017) Beyond speciation genes: an overview of genome stability in evolution and speciation. *Curr Opin Genet Dev Evol Genet* 47:17–23. https://doi.org/10.1016/j.gde.2017.07.014.

61. Anatskaya OV, Vinogradov AE (2010) Somatic polyploidy promotes cell function under stress and energy depletion: evidence from tissue-specific mammal transcriptome. *Funct Integr Genomics* 10:433–446. https://doi.org/10.1007/s10142-010-0180-5.

62. Ravid K, Lu J, Zimmet JM, Jones MR (2002) Roads to polyploidy: the megakaryocyte example. *J Cell Physiol* 190:7–20. https://doi.org/10.1002/jcp.10035.

63. Ravid K, Lu J, Zimmet JM, Jones MR (2002) Roads to polyploidy: the megakaryocyte example. *J Cell Physiol* 190:7–20. https://doi.org/10.1002/jcp.10035.

64. Anatskaya OV, Vinogradov AE (2007) Genome multiplication as adaptation to tissue survival: evidence from gene expression in mammalian heart and liver. *Genomics* 89:70–80. https://doi.org/10.1016/j.ygeno.2006.08.014. Vazquez-Martin A, et al. (2016) Somatic polyploidy is associated with the upregulation of c-MYC interacting genes and EMT-like signature. *Oncotarget* 7:75235–75260. https://doi.org/10.18632/oncotarget.12118.

65. Duesberg P, et al. (2000) Aneuploidy precedes and segregates with chemical carcinogenesis. *Cancer Genet Cytogenet* 119:83–93. https://doi.org/10.1016/S0165-4608(99)00236-8.

66. Duesberg P, Rasnick D (2000) Aneuploidy, the somatic mutation that makes cancer a species of its own. *Cell Motil* 47:81–107. https://doi.org/10.1002/1097-0169(200010)47:2<81::AID-CM1>3.0.CO;2-#.

67. Dion-Côté A-M, Barbash DA (2017) Beyond speciation genes: an overview of genome stability in evolution and speciation. *Curr Opin Genet Dev Evol Genet* 47:17–23. https://doi.org/10.1016/j.gde.2017.07.014.

68. Ye CJ, Regan S, Liu G, Alemara S, Heng HH (2018) Understanding aneuploidy in cancer through the lens of system inheritance, fuzzy inheritance and emergence of new genome systems. *Mol Cytogenet* 11:31. https://doi.org/10.1186/s13039-018-0376-2. Barski G, Cornefert-Jensen F (1966) Cytogenetic study of sticker venereal sarcoma in European dogs. *J Natl Cancer Inst* 37:787–797. https://doi.org/10.1093/jnci/37.6.787.

69. Murchison EP, et al. (2014) Transmissible dog cancer genome reveals the origin and history of an ancient cell lineage. *Science* 343:437–440. https://doi.org/10.1126/science.1247167. Murchison EP, et al. (2012) Genome sequencing and analysis of the Tasmanian Devil and its transmissible cancer. *Cell* 148:780–791. https://doi.org/10.1016/j.cell.2011.11.065.

70. Pye RJ, et al. (2016) A second transmissible cancer in Tasmanian Devils. *Proc Natl Acad Sci USA* 113:374–379. https://doi.org/10.1073/pnas.1519691113. Liu G, et al. 2014. Genome chaos: survival strategy during crisis. Cell Cycle Georget Tex 13:528–537. https://doi.org/10.4161/cc.27378.

71. Barski G, Cornefert-Jensen F (1966) Cytogenetic study of sticker venereal sarcoma in European dogs. *J Natl Cancer Inst* 37:787–797. https://doi.org/10.1093/jnci/37.6.787. Nandakumar V, et al. (2012) Isotropic 3D nuclear morphometry of normal, fibrocystic and malignant breast epithelial cells reveals new structural alterations. *PLoS One* 7:e29230. https://doi.org/10.1371/journal.pone.0029230.

72. Barski G, Cornefert-Jensen F (1966) Cytogenetic study of sticker venereal sarcoma in European dogs. *J Natl Cancer Inst* 37:787–797. https://doi.org/10.1093/jnci/37.6.787.

73. Nandakumar V, et al. (2016) Vorinostat differentially alters 3D nuclear structure of cancer and non-cancerous esophageal cells. *Sci Rep* 6:30593. https://doi.org/10.1038/srep30593. Nordling CO (1953) A new theory on the cancer-inducing mechanism. *Br J Cancer* 7:68–72. https://doi.org/10.1038/bjc.1953.8.

74. Trigos AS, Pearson RB, Papenfuss AT, Goode DL (2017) Altered interactions between unicellular and multicellular genes drive hallmarks of transformation in a diverse range of solid tumors. *Proc Natl Acad Sci USA* 114(24):6406–6411.

75. Barski G, Cornefert-Jensen F (1966) Cytogenetic study of sticker venereal sarcoma in European dogs. *J Natl Cancer Inst* 37:787–797. https://doi.org/10.1093/jnci/37.6.787.

76. Nordling CO (1953) A new theory on the cancer-inducing mechanism. *Br J Cancer* 7(1):68–72.

77. Knudson AG (1971) Mutation and cancer: statistical study of retinoblastoma. *Proc Natl Acad Sci USA* 68:820–823. https://doi.org/10.1073/pnas.68.4.820.

78. Deng FE, Shivappa N, Tang Y, Mann JR, Hebert JR (2017) Association between diet-related inflammation, all-cause, all-cancer, and cardiovascular disease mortality, with special focus on prediabetics: findings from NHANES III. *Eur J Nutr* 56(3):1085–1093. Taniguchi K, Karin M (2018) NF-κB, inflammation, immunity and cancer: coming of age. *Nat Rev Immunol* 18:309–324. https://doi.org/10.1038/nri.2017.142. Gallaher JA, Enriquez-Navas PM, Luddy KA, Gatenby RA, Anderson ARA (2018) Spatial heterogeneity and evolutionary dynamics modulate time to recurrence in continuous and adaptive cancer therapies. *Cancer Res* 78:2127–2139. https://doi.org/10.1158/0008–5472.CAN-17–2649.

10 Time and Timing in Oncology

What Therapy Scheduling Can Teach Us about Cancer Biology

Larry Norton

Overview

When the notion that drugs could be useful in the treatment of cancer was young, we thought it would be simple: find cancer's biological Achilles' heel and hit it with a medicinal arrow and the disease would be quickly vanquished.[1] So we looked for that vulnerable spot and the answer seemed obvious. We knew that one of the key abnormalities in cancer was increased proliferation, so it was logical to attack the molecules key to mitosis: enzymes, metabolites, and DNA itself to induce apoptosis.[2]

Fast-forward to our current times, when our knowledge of the biochemical deviations that differentiate cancer cells from normal cells has deepened so profoundly that we can design treatments that attack—precisely, as we intend—just these molecules.[3] And this possibility encourages us to pursue the original paradigm: "hit the heel and the enemy will die." Furthermore, it sometimes works! But when it does not, which is most often, we do not question the underlying metaphor but merely search for new targets, or newer, better arrows to hit old targets, or group arrows into new combinations.

Yet there is a clear flaw in the metaphor: Achilles does not stand still. A major difficulty with the concept of a static Achilles' heel is that a crucial variable—time—is ignored or, at the very least, deemphasized. This is problematic, because cancer is a highly dynamic process. Morphology, genomics, volume, metabolism, microenvironmental relationships, and countless other critical factors are continuously changing in a time-dependent fashion.[4] Having a good quiver of arrows is essential, but they are useless unless one knows not only how but also, even more important, *when* to shoot them at a target in motion.

So *time* is an essential variable in understanding and intervening optimally in malignancy. Yet scheduling—the application of drugs as a function of time—is often the last thing considered in the design of treatments. Yet if scheduling is not done properly, it could be a very common source of therapeutic failure. Furthermore, reliance on trial and error is a grossly inefficient way to construct an optimal schedule. One needs engineering: the application of mathematical concepts to design.[5]

It is for this reason that the science of drug scheduling needs to progress in parallel with the science of molecularly targeted drug design. Failure to do so would tragically relegate active compounds to the discard pile. Not only must we select the best drugs for the individual patient but also the best dose levels and schedules and durations of therapy. In other words: to be successful, precision medicine must respect time.

Moreover, in our quest for the best schedules, we might well gain biological insights that could inform future advances in mechanism-based therapeutics. This is because there are certainly biological reasons why one schedule is better than others. These reasons may involve not only the kinetics of drug circulation, tissue penetration, metabolism, and elimination but also the intercellular and intracellular molecular processes that determine time-dependent events. An entity, whether a mythological hero or a molecule or a cell, that is in motion is not the same as that entity at rest, irrespective of whether that motion occurs in space or in time. Indeed, it is possible that the very motion itself could be a therapeutic target.

We will consider this topic in several sections: What are the kinetic laws of growth of cancers? Why do they grow that way? What is the phenomenology of response to anticancer therapy? How has this been exploited for improving chemotherapy? Is there a central principle to be articulated? Can these ideas be extrapolated to the new generation of mechanism-based therapies, including immunotherapy? Does our study of this topic help us understand the nature of neoplasia in new and productive ways?

For the sake of brevity, we will focus largely on breast cancer in this chapter, but the fundamental concepts are applicable more widely.

10.1 Historical Background

The earliest studies of the relationships between tumor growth kinetics and therapy were founded on the assumption that cancers were largely a manifestation of aberrantly increased cell division.[6] This focus on proliferation is evident in the very language we use to describe cancer: *neoplasia, dysplasia, hyperproliferation*, and so on. Language often guides thought, and here it has had a substantial impact on the design and testing of anticancer agents, even to the present day. Specifically, in most cases, we design drugs to perturb molecular pathways known to be involved in mitosis: we screen candidate drugs for their ability to kill cancer cells in culture and in highly artificial, rapidly proliferating experimental animal models, and we evaluate them clinically on the basis of their power to kill dividing cells and cause shrinkage of tumors. In other words, the view that "cancer is a disease of cell proliferation," which we may call *mitosis centricity*, still dominates therapeutics despite ample confirmation that a myriad other defects are intrinsic to malignancy.[7] We will reexamine this hegemony below, but for now let us consider the historical context and the implications of this mitosis centricity for drug dosing and scheduling.

Were cancer simply a disease of unregulated mitosis, then the growth kinetics of tumors would be expected to be exponential: one cell begets two over one cell-cycle time, two begets four, four begets eight, and so on. Indeed, this pattern of growth was observed in the experimental model most dominant in its day (the third quarter of the twentieth century): murine leukemia. This observation not only lent support to mitosis centricity but also provided a convenient laboratory model for anticancer (i.e., antimitotic) drug development.[8] This work, carried out at the Southern Research Institute in Birmingham, Alabama, under contract with the U.S. National Cancer Institute, was vital to early anticancer drug development not only because of its specific discoveries regarding active agents but also, or mainly, because it eschewed empiricism in favor of a comprehensive mathematical theory called the *log-kill hypothesis*.

It is difficult to overstate the influence of the log-kill hypothesis. Contemporary scientists and physicians who are generations removed from the age in which it was invented, and may well be ignorant of its champions or even of its existence, still adhere to its precepts.

The hypothesis states that a given dose of an effective drug kills a given percentage of the cancer cells present, regardless of the absolute number of cells at the time of treatment.[9] That is, the killing of 90 percent of cells equates to a log-kill of one, and this applies to a larger tumor as much as to a smaller tumor of the same biology, in both cases leaving 10 percent of the original number of cells to survive. The killing of 99 percent of cells means a log-kill of two and so on. And there were corollary principles:

• When using more than one drug, their log-kills are additive. That is, if drug A kills 90 percent and B kills 90 percent, then A plus B kill 99 percent.

• Higher dose levels of drugs always result in higher log-kills.

• Because a certain percent cell-kill (for example, one log) always means a certain percent cell-survival (10 percent in that case), large cancers are harder to cure than small cancers, quite simply because there are more cells present in the larger tumor, so more of these cells will escape treatment even if all the cells individually were sensitive to the treatment.

• Conversely, application of effective high log-kill therapy at the time of minimal disease should produce high rates of cure.

The subsequent application of these concepts effectively describes much of modern cancer medicine. This includes combination therapy, using drugs at their maximum tolerated dose levels and pre- and postoperative adjuvant therapy for operable disease. Indeed, we owe most of the advances we have made so far in the drug treatment of cancer to our adherence to these principles, which is probably why they are so venerated that they are rarely even noticed.

Nevertheless, we must recognize that in most cases, our victories based on these principles have only been partial. Furthermore, using the example of breast cancer, there are common clinical observations that should make us question the infallibility of the log-kill idea:

• Treating small volumes of metastatic disease is no more curative than treating larger bulks of metastatic disease. This is evidenced by the failure of postoperative surveillance to improve cure rates.[10]

• The perioperative use of adjuvant drug therapy is widely regarded as one of our major advances. Yet cures are not universal and might even be explained entirely by lead-time bias.[11]

• Preoperative shrinkage of cancers to complete disappearance is associated with better outcome for the individuals experiencing such outcomes. However, the causal link is still unclear.[12] That is, the susceptibility of a cancer to be induced into complete remission might be a *marker* of good prognosis rather than the obtaining of complete remission the actual *cause* of the good prognosis.

• Extremely high dose levels of chemotherapy do not improve cure rates even though they do increase response rates.[13]

• Combination chemotherapy is not always better than the sequential use of single agents.[14]

So here is a paradox: (1) the statement "a given dose of drug kills a fixed proportion of the cancer cells" is clearly observed in murine leukemia, and hence it is true in that system, and (2) the clinical translation of this concept suggests that it is not true, at least in many cases. Moreover, it has long been recognized that the log-kill hypothesis is counterintuitive: if therapeutic molecules kill cells by interacting with cellular molecules, including receptor sites, then a fixed amount of drug should kill a fixed number of cells, not a fixed proportion.

A solution to this paradox may be found in the observation that murine leukemias grow exponentially, but human cancers and many other animal models of cancer do not. In exponential growth, the *proportional* growth rate is fixed. In the simple example above, the doubling time, a proportion of two, is the length of one cell cycle. So an alternative way of stating the log-kill hypothesis, which is entirely consistent with the observations in murine leukemia on which it was based, is that a given dose of drug kills a number of cells proportional to that tumor's growth rate.[15] Restating this, the phrase "a given dose of drug kills a fixed proportion of the cancer cells" and the new phrase "a given dose of drug kills a variable proportion of the cancer cells depending on the rate of growth of the population" are equivalent but only in the case of exponential growth.

But what is critical about this second phrase is that in addition to being true in murine leukemia, we will see below that it is as applicable to clinical human cancers that do not follow exponential growth kinetics.[16] Moreover, the restated log-kill hypothesis might reach deeply into cancer biology once we mechanistically understand why anticancer drugs really target rates of change, not absolute cell numbers.

10.2   What Is the Growth Kinetics of Clinical Cancers?

Human cancers of epithelial origin usually follow *sigmoid* growth curves. We know this not only from direct and indirect evidence but also by logical inference.[17] For example, take the case of a breast cancer of 1 centimeter in diameter that is found by a screening mammogram. Say we look at a screening mammogram taken one year earlier and we find at the site where the cancer grew a tiny calcification, but no mass, meaning that the tumor

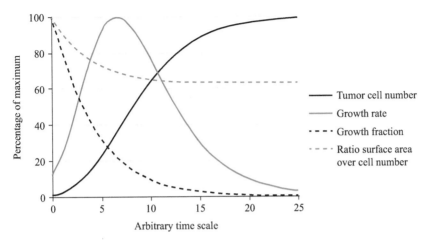

Figure 10.1
The characteristics of Gompertzian-like sigmoid growth: The number (N) of cells as a function of time follows an S-shaped curve that approaches an upper maximum N_∞. The point of maximum growth rate occurs at about one-third of that maximum tumor size. The growth rate N' is proportional to the growth fraction G times the number of cells N. That growth faction is at its maximum at one cell and falls as the number of cells increases. The ratio S/N of the surface area S of the mass of cells to the number of cells N also falls as the mass increases. Indeed, the G in classical Gompertzian growth and the ratio S/N are highly correlated, as is evident in the figure by observing their parallel behavior over time. Indeed, they are correlated with R^2 in excess of 0.99 (not shown). Mathematical details: $G(N) = (0.06)\log_e(N_\infty/N)$; $N' = N \cdot G(N)$; $S/N = N^\wedge((2.7/3) - 1)$. N' is maximum at $(N_\infty) \cdot \exp(-1)$.

at that point in its growth rate could have been no larger than 1 millimeter in diameter. Now let us project the course of growth of a 1-millimeter *exponential* tumor growing to 1 centimeter in one year. By exponential growth, were that mass untreated, it would grow to over 3 centimeters in diameter in three months from diagnosis and 10 centimeters in diameter in three months after that. This kind of explosive growth is extraordinarily rare in human breast cancer. So, some slowing of growth over time must be the rule.

Indeed, the pattern of growth most commonly seen in human epithelial cancers is similar to that described by Benjamin Gompertz in 1825[18] (figure 10.1). The technical definition of a Gompertz-type curve is that as the mass of cells gets larger, it increases in size by a constantly *decreasing* proportion of itself over any chosen time interval rather than a *fixed* proportion as in exponential growth.[19] That yields a sigmoid (S-shaped) growth curve, starting off somewhat like exponential growth in that the growth rate increases with increasing volume but eventually reaching a phase where the growth rate decreases with increasing volume, approaching a plateau phase of hardly any perceptible growth at all.

10.3   Why Do Tumors Grow That Way?

Many possible explanations for sigmoid growth have been proposed over the decades since it was first observed.[20] The most familiar hypothesis, that cancers outgrow their blood supply, was never quite plausible to surgeons or pathologists who observed abundant vascularization when examining these cancers grossly and microscopically. And certainly the vast body of evidence that neoangiogenesis is a hallmark of cancer has rendered this theory outmoded.[21]

A perhaps useful new explanation came from observations of a clinically relevant animal model of breast cancer. In this model, the expression of genes known to mediate lung metastases also made the cancers grow faster at the site of implantation, the prepared mammary fat pad.[22] Moreover, this faster growth was not due to a higher percentage of dividing cells than that observed in slower-growing cell lines that had different gene expression profiles and a lower incidence of lung metastases. So here is the puzzle: how can a mass grow faster without having a higher percentage of dividing cells? To explain this, we offered and later proved the hypothesis of *self-seeding*: in addition to the cells being able to seed the lung, they were also capable of being "metastatic" back to the fat pad of origin.[23] The faster growth rate is observed, because the mass in the fat pad is not one homogeneous cellular unit but a conglomerate of individual masses, each attracting new bone marrow–derived endothelial blood vessel precursors and leukocytes with growth-promoting properties.

This hypothesis explains many mysteries in clinical oncology, which are discussed elsewhere.[24] But, relevant to this chapter, it also provides a mechanism for sigmoidal growth: seeding must happen at the surface of a mass, and the ratio of the surface of a mass to its volume decreases as that object grows larger.[25] Growth primarily at the surface—the place where the tumor mass meets its microenvironment—means that the proportional growth rate must also decrease as the mass grows larger, which is the definition we gave above of sigmoid growth (figure 10.1).

In this regard, the reason why the term "sigmoid" is a better term than "Gompertzian" is that while the general shapes of sigmoid and Gompertzian growth curves are similar (Gompertzian being one particular kind of sigmoid curve), the equations describing them are different.[26] In Gompertzian growth, the rate of growth is the product of the number of cells and a "growth fraction," proportional to the logarithm of (N_∞/N), where N_∞ is the ultimate "plateau size" of the growing mass of N cells. In the self-seeding equation, the growth fraction is proportional to the ratio of the surface area S of the mass of N cells and the volume, which is N. That ratio is dependent on the geometric regularity of the mass, with a very regular mass having a lower S/N ratio than a very irregular, convoluted mass. Indeed, at an ultimate, a totally disorganized mass would have a S/N ratio close to 1. The growth rate is therefore highly dependent on the geometry of the cancerous mass as well as the cell-cycle time. More disorganized/irregular tumors have higher surface-to-volume ratios and therefore should (theoretically) and do (actually) tend to have faster growth rates. But as we just said, the growth fraction is *proportional* to the S/N ratio. That constant of proportionality must depend on host factors, including the relative robustness of the growth-promoting microenvironment and especially the ability of certain immune cells to modulate growth. Consequently, the host modulators of the growth fraction might be one of the most important topics in contemporary cancer biology, as underscored by the recent surge of interest in immuno-oncology.

10.4   The Phenomenology of Therapeutic Cell Kill

Why is this important to cancer therapeutics? When one examines sigmoid tumors experimentally, it is clear that the proportion of cells killed by an effective therapy is not fixed but, rather, varies with tumor size. That is, for sigmoid tumors, the phrase "a given dose of drug kills a fixed proportion of the cancer cells" is not true, but the new phrase "a given dose of drug kills a variable proportion of the cancer cells, depending on the rate of growth of the population" does indeed apply. Furthermore, the cell kill rate follows a simple rule in that it is indeed proportional to the growth rate as predicted by the restated log-kill hypothesis.

Explicitly, a given drug at a given dose kills a higher fraction of cancer cells in faster-growing, small tumors than in larger, more slowly growing masses of similar biology.[27] This is the so-called *Norton-Simon hypothesis*: it is all about *rates*. In exponential growth, the growth rate is proportional to tumor size, and therefore a cell-kill proportional to tumor size is also proportional to the tumor's unperturbed growth rate. But in sigmoid growth, the growth rate changes with tumor size, so the proportion of cell-kill changes as well. Furthermore, at least for many agents, the cell-kill rate is not a strictly linear function of dose level, but the relationship between cell-kill and absolute dose level has its own sigmoid curve. Higher dose levels are certain to be more toxic, but they are not certain to be more effective in terms of proportional cell killing.[28]

It must be emphasized that the Norton-Simon hypothesis is not a de novo finding but rather an extension, a rewording, of the log-kill hypothesis so that it is applicable to tumors that grow other than exponentially. Hence, it owes its origin as much to the pioneering work of Howard Skipper and Frank Schabel as to the subsequent observations that motivated its formulation.[29]

As with the original log-kill hypothesis, the Norton-Simon hypothesis generates clinically testable applications. One is that the "density" of therapy is important. *Density* is defined as the frequency of drug administration, with denser treatments having shorter intervals between administrations. The rationale for this manipulation is that shorter inter-treatment intervals reduce the time for the cancer to recover from cell-kill.[30] This is especially important when one is trying, as in the setting of perioperative chemotherapy, to eradicate subclones of tumors that are small. Small tumors, by sigmoid kinetics, regrow quickly after their numbers are reduced by effective drug therapy since their surface-to-volume ratios are relatively high. Denser therapies not only minimize such regrowth by reducing the time of regrowth, but they take advantage of the more rapid regrowth by virtue of the relationship between growth rate and cell-kill rate: faster regrowth equals more vulnerability.

But there is yin *and* yang at work here: in an effort to minimize intertreatment intervals safely, one could erroneously choose a dose level of drug that is so low as to not be very effective. So the ideal regimen would balance dose level against intertreatment interval, which cannot be done simply empirically but requires mathematical engineering. This becomes especially apparent when one designs combinations of more than one drug. If, to be able to give the combination safely, one has to reduce the dose level of each component too far, no degree of reduction in intertreatment interval can provide an optimal cell-kill. In that case, it would be preferable to give the drugs as single agents or combinations of a reduced number of components (say doublets rather than triplets) so as to restore adequate dose levels of all the drugs employed. Many active drugs can be delivered this way in sequence.

10.5   Optimizing Cell Kill with Cytotoxic Chemotherapy

The above paragraph presents logical conclusions that may be drawn from the Norton-Simon hypothesis. However, logical conclusions do not always correspond to real-life observations, so it is essential to subject them to experimental verification. When this has been done, the results have been quite positive. For example, sequential therapy has been tested in several major clinical trials for adjuvant chemotherapy of operable breast cancer and, as predicted, has proven superior to simultaneous combinations.[31] Moreover, studies have shown that increasing the density of the therapy is advantageous. This was found not only in a pivotal trial conducted by the Cancer and Leukemia Group B but also in meta-analyses of subsequent studies.[32] A definitive, comprehensive meta-analysis by the Early Breast Cancer Trialists' Collaborative Group has recently been published that is reassuringly confirmatory.[33]

The major pharmaceutical advance that helped tip the yin-yang balance in favor of efficacy over toxicity, indeed that makes dose density practical at all, is the use of granulocyte colony-stimulating factor, as this enables compensation for the loss of white blood cells due to cytotoxic anticancer agents. Filgrastim was the first drug of this class used to facilitate dose densification, but this was soon switched to pegfilgrastim for convenience of administration.[34] The pivotal CALGB study that clearly established the efficacy and lack of incremental toxicity of dose density used doxorubicin, cyclophosphamide, and paclitaxel, all at optimal dose levels as determined by prior clinical trials.[35] Dose level refers to the amount of drug in milligrams per meter squared. The drugs were used sequentially, doxorubicin for four cycles followed by paclitaxel for four cycles followed by cyclophosphamide for four cycles, or as a doxorubicin-cyclophosphamide combination for four cycles, followed in sequence by four cycles of paclitaxel. The doxorubicin-cyclophosphamide combination did not violate the principle of dose density because the dose levels of each of the drugs were not compromised in forming the combination. Finding the right dose level for each agent was a necessary precursor to this pivotal trial. This required several stepwise preliminary studies that were conducted by cooperative groups.[36]

Once proper dose levels were identified, the use of granulocyte colony-stimulating factor allowed the investigators to ask the key question: in the experimental arm, the patients received their treatment each two weeks rather than the conventional three weeks. Paclitaxel, used sequentially after the doxorubicin-cyclophosphamide, was also given each two weeks (as permitted by the use of filgrastim) in the experimental arm. Dose levels were identical in all arms. Subsequent work has shown that paclitaxel could also be given weekly, which, we must recognize, is another dose-dense regimen compared with three-weekly administrations, with equal efficacy in some subgroups and somewhat lower toxicity.[37]

Trastuzumab and now pertuzumab for the treatment of HER2 overexpressing breast cancer can be safely integrated into such a regimen (dose-dense Adriamycin cyclophosphamide–Taxol Herceptin [AC-TH] or Adriamycin cyclophosphamide–Taxol Herceptin pertuzumab [AC-THP]) with good effect.[38] Hence, valid comparisons of other regimens with a regimen like AC-TH or AC-THP are not possible if the AC-TH or AC-THP is not given in a dose-dense fashion.[39] In this regard, it is important to point out that trastuzumab may be given simultaneously with paclitaxel without diminishing the efficacy of the chemotherapy. It was a prediction of the Norton-Simon hypothesis that the combination AC-TH would be more effective than AC-T followed by trastuzumab, which has been confirmed by clinical trial.[40]

The primary objective of these studies was the application of the theory to developing improved treatments for patients. Moreover, the fact that sequential dose-dense chemotherapy works is evidence in favor of its motivating mathematical theory. In the long run, the establishment of the validity of mathematical modeling in general, and even this model in particular, may prove more important than specific applications. This is because it might lead to novel, less toxic treatments, different from those employing cytotoxic chemotherapy drugs that were developed during an era dominated by mitotic-centricity.

10.6   Approaching a Molecular Understanding of Drug Responses

In essence, the validity of dose density reveals two fundamentals of oncology. The first is the Norton-Simon hypothesis: therapy primarily affects rates of change of numbers of cancer cells, causing declining cancer cell numbers as a consequence. As we have seen, this applies to exponential growth, in which the rates of change are proportional to the number of cells present, as well as sigmoid growth in which the relationship between rate of growth and cell number is more complex. The second is that there is no positive linear relationship between dose level and biological effect. Although there are as yet no clear mechanistic explanations for these principles, there are some thought-stimulating hints. Time, as before, is the core variable.

If one has an effective drug that can be given at an active dose level every day, is continuous treatment the best way to use that agent? In some ways, this question is similar to the issue of the efficacy/dose-level ratio, which we have seen is not constant in that efficacy does not always rise in lock step with increasing dose level. So we must ask, does efficacy always rise with increasing duration of continuous treatment?

This question was addressed by examining the anticancer effects of the 5-fluorouracil prodrug capecitabine in animal models.[41] We looked at its impact on growth curves. This is a more sophisticated way of looking at total anticancer influence than just evaluating response rates or tumor volumes at a fixed time point after treatment. We found that two weeks of daily capecitabine are not twice as effective as one week of exposure even though the longer duration of exposure was more toxic. Indeed, the antigrowth pressure peaked at eight to ten days of a fourteen-day regimen. It was for this reason that a seven-day on, seven-day off regimen was tested in the clinic in preparation for a randomized trial, now under way, of the new schedule versus the conventional fourteen-day on, seven-day off schedule. The pilot study did indicate the preservation of activity with a strong suggestion of reduced toxicity, which of course is the desired result, awaiting further confirmation.[42]

But this raises an important question: why would antigrowth activity peak at a certain time of exposure and then eventually decline? A suggestion may be forthcoming from work in other systems in which it is clear that exposure to some agents induces, over time, resistance mechanisms that are also time dependent. For example, the viability of cancers of the prostate frequently depends on signaling from their androgen receptors. This is decreased in cancers with PTEN deletions, but inhibition of PI3K activates androgen receptor signaling indirectly by relieving feedback inhibition of kinases in the epidermal growth factor family.[43] This relationship is mutual in that inhibiting the androgen receptor, a standard manipulation in the treatment of prostate cancer, activates AKT signaling. The presence of cross-regulation is a survival mechanism in the same way that prey-predator models in ecology allow for the stability of cross-regulating populations.[44] The inhibition of one oncogenic pathway over time promotes the emergence of another survival-promoting pathway. A similar example, also involving PI3K-AKT, implicates mTORC1-mediated inhibition of upstream signaling and FOXO-modulated effects.[45] By this mechanism, inhibiting AKT relieves positive feedback loops such that it induces the expression and activation by phosphorylation of receptor tyrosine kinases such as HER3, IGF-1R, and the insulin receptor.[46] This is a time-dependent process and is rationally hypothesized to account for the ability of continuous exposure to AKT inhibitors to induce their own resistance. Since constitutive feedback inhibition of upstream signaling pathways is such a fundamental constituent of cancer biology, it is likely that many such examples of therapeutic relief of inhibition as a cause of drug resistance will be discovered. Analysis of the kinetics of these processes might allow the use of agents sequentially, in which a second agent applied at the proper time addresses the induced activity consequent to the application of a first agent. This could provide effective therapy while avoiding the toxicity that might be associated with the simultaneous combination of signaling inhibitors.

Resistance to ionizing radiation is another impediment to optimal treatment and one that might be addressed by kinetic modeling. For example, the malignant cells in glioblastomas of the brain have been shown to be heterogeneous and unstable with regard to differentiation states and that these different states have different sensitivities to radiation therapy.[47] By theoretical modeling as informed by laboratory experiments, it has been determined that unconventional scheduling can improve therapeutic results. The extension of this work to diseases other than glioblastoma is in progress. Since one of the earliest observations that led to the development of the Norton-Simon hypothesis used the response of an experimental tumor to radiation therapy, there is a possible convergence of these lines of investigation.[48]

10.7 Beyond Mitosis Centricity

As described above, a tenable explanation for sigmoid growth could be the geometry of solid masses, with the ratio of the surface area of the tumor mass to its volume being a major determinant of the growth rate. The surface area reflects the magnitude of the "outside" of the tumor, the part of the tumor mass in contact with its microenvironment, including blood and lymphatic vessels and infiltrating leukocytes. The volume is its "inside," the cancer cells not in direct contact with the microenvironment. The mathematics of this is illustrated in figure 10.1. As objects get larger, that ratio decreases, so the relative growth rate, defined as the absolute growth rate divided by the size of the object, decreases as well. That the efficacy of therapy is proportional to the rate of growth raises the possibility that therapy acts at that interface between the "outside" and the microenvironment. Hence, self-seeding and the Norton-Simon hypothesis could be mechanistically linked, shifting interest from mitosis itself as a target to a new target: the communication between cancer cells and the cellular and acellular components of the microenvironments.

Biological experiments have tested this possibility. One of the most highly expressed genes in lung-seeding and self-seeding breast cancers is CXCL1.[49] Cancer cells that over-express CXCL1 and 2 by transcriptional activation or amplification attract certain myeloid cells that produce cytokines, particularly S100A8/9, that enhance cell survival.[50] The same myeloid cells, CD11b+Gr1+, also suppress anticancer T cells, and hence may be promoters of cancer cell survival.[51] Many chemotherapy drugs, in addition to their presumed direct effects on cancer cells, induce TNFα production by endothelial cells, which via NF-κB amplifies the CXCL1/2-CXCR2-S100A8/9 loop and thereby rescues cells from apoptosis. One of the more attractive aspects of focusing on such time-dependent interactions in cancer therapeutics is that the number of such interactions may be smaller than the number of genomic alterations underlying carcinogenesis, which are in the thousands and may vary from cell to cell even within one patient's cancer.[52] This applies to immunotherapeutics as well as medicinal therapeutics in that timed manipulations of the tumor itself in conjunction with immune-checkpoint inhibition have shown considerable activity in the laboratory and are amenable to study in the clinic.[53]

While it is commonly assumed that the stromal cells found within cancers are normal, co-opted to support the growth of the genomically abnormal cancer cells, this assumption has recently been challenged by DNA sequencing of breast cancer–infiltrating leukocytes.[54] In about half of cases, these leukocytes were found to harbor known oncogenic mutations not found in the epithelial cells of the cancer. What role these leukocytes are playing in carcinogenesis or drug resistance is under investigation, but it is important to note that their anatomic location implicates an outside/inside ratio relationship in keeping with the discussions above. Hence, they may be part of the narrative linking geometry, growth rates, and therapeutic sensitivity/resistance and therefore might be novel targets for timed interventions.

Yet another mechanism by which seeding, growth, and drug sensitivity/resistance may be linked is the process of cell mobility itself. For example, a published case of metastatic lung cancer manifested tumor regrowth in many sites after an initial response to crizotinib. The drug resistance was due to an acquired mutation in CD74-ROS1 that was identical in all the progressing sites and not present in normal tissues.[55] It is statistically close to impossible for an identical mutation to arise in disparate sites by chance, leaving the likelihood that a cell with this mutation arose in one site and then seeded the rest.[56] This same phenomenon might have been observed in a metastatic breast cancer mutant in PIK3CA that was treated with a PI3Kα inhibitor.[57] Upon the occurrence after initial response of tumor regrowth in metastatic sites, it was found that convergent genetic evolution toward different molecular mechanisms of PTEN-null status was etiological. But, curiously, four of fourteen metastatic sites may have demonstrated the same molecular mechanism of biallelic PTEN loss, suggesting a seeding of three sites from one in which that alteration occurred. Hence, time-dependent movement of cells between metastatic sites might be important clinically, even as a therapeutic target. Therefore, efforts to classify tumor types and organs of involvement as "sponges" that attract cancer cells and "seeders" that release these cells are merited.[58]

10.8   Conclusion

Cancers of the breast and other organs grow by sigmoid kinetics. Furthermore, it has been shown, at least in breast cancer, that this phenomenon can be exploited to improve cancer chemotherapy scheduling. The implications of this observation are conceptual as well as practical. That is, both the etiology of sigmoid growth and the way anticancer therapy affects rates of change are fundamentals of cancer biology that might meritoriously be explored both biochemically and mathematically. Such efforts are in progress but should be expanded to keep pace with advances in molecular biology, intra- and extracellular signaling, and immunobiology as they relate to cancer. It is therefore imperative for students of cancer biology, medicinal therapeutics, radiation therapeutics, and other cancer scientists to include time and timing in their investigations. As the here presented examples have demonstrated, it is certainly productive to apply often simple mathematical models to cancer, as these can lead to very successful treatment approaches—even without the need for detail-rich information on presumed causal mechanisms at the genetic and molecular level. This would provide the data needed to develop an engineering framework for therapy scheduling. The regrettable alternative might be that good therapies may be abandoned erroneously, valuable treatments and prevention strategies targeting novel mechanisms may remain unexplored, and our basic as well as applied oncologic sciences may not meet their societal obligation—to eradicate cancer.

Notes

1. Karnofsky DA, Burchenal JH, Escher GC. Chemotherapy of neoplastic diseases. *Med Clin North Am.* 1950 Mar;34(2):439–58.

2. Shapiro DM, Gellhorn A. Combinations of chemical compounds in experimental cancer therapy. *Cancer Res.* 1951 Jan;11(1):35–41.

3. Chabner BA, Ellisen LW, Iafrate AJ. Personalized medicine: hype or reality. *Oncologist.* 2013 Jun;18(6):640–43.

4. Norton L. Cancer stem cells, self-seeding, and decremented exponential growth: theoretical and clinical implications. *Breast Dis.* 2008;29:27–36.

5. Norton L. Conceptual and practical implications of breast tissue geometry: toward a more effective, less toxic therapy. *Oncologist.* 2005 Jun–Jul;10(6):370–81.

6. Karnofsky DA. The bases for cancer chemotherapy. *Stanford Med Bull.* 1948 Feb;6(1):257–69.

7. Hanahan D, Weinberg RA. Hallmarks of cancer: the next generation. *Cell.* 2011 Mar 4;144(5):646–74.

8. Skipper HE. Laboratory models: some historical perspective. *Cancer Treat Rep.* 1986 Jan;70(1):3–7.

9. Schabel FM Jr, Skipper HE, Trader MW, Laster WR Jr, Cheeks JB. Combination chemotherapy for spontaneous AKR lymphoma. *Cancer Chemother Rep 2.* 1974 Mar;4(1):53–72.

10. Henry LN, Hayes DF, Ramsey SD, Hortobagyi GN, Barlow WE, Gralow JR. Promoting quality and evidence-based care in early-stage breast cancer follow-up. *J Natl Cancer Inst.* 2014 Mar 13;106(4):dju034.

11. Herdman R, Norton L, editors. *Saving Women's Lives: Strategies for Improving Breast Cancer Detection and Diagnosis: A Breast Cancer Research Foundation and Institute of Medicine Symposium.* Institute of Medicine (US) Committee on New Approaches to Early Detection and Diagnosis of Breast Cancer; Washington (DC): National Academies Press (US); 2005. McArthur HL, Hudis CA. Advances in adjuvant chemotherapy of early stage breast cancer. *Cancer Treat Res.* 2008;141:37–53.

12. Rastogi P, Anderson SJ, Bear HD, Geyer CE, Kahlenberg MS, Robidoux A, Margolese RG, Hoehn JL, Vogel VG, Dakhil SR, Tamkus D, King KM, Pajon ER, Wright MJ, Robert J, Paik S, Mamounas EP, Wolmark N. Preoperative chemotherapy: updates of National Surgical Adjuvant Breast and Bowel Project Protocols B-18 and B-27. *J Clin Oncol.* 2008 Feb 10;26(5):778–85. Cortazar P, Geyer CE Jr. Pathological complete response in neoadjuvant treatment of breast cancer. *Ann Surg Oncol.* 2015 May;22(5):1441–46. Epub 2015 Mar 2. Cortazar P, Zhang L, Untch M, Mehta K, Costantino JP, Wolmark N, Bonnefoi H, Cameron D, Gianni L, Valagussa P, Swain SM, Prowell T, Loibl S, Wickerham DL, Bogaerts J, Baselga J, Perou C, Blumenthal G, Blohmer J, Mamounas EP, Bergh J, Semiglazov V, Justice R, Eidtmann H, Paik S, Piccart M, Sridhara R, Fasching PA, Slaets L, Tang S, Gerber B, Geyer CE Jr, Pazdur R, Ditsch N, Rastogi P, Eiermann W, von Minckwitz G. Pathological complete response and long-term clinical benefit in breast cancer: the CTNeoBC pooled analysis. *Lancet.* 2014 Feb 13;384(9938):164–72. pii: S0140–6736(13)62422–8. Von Minckwitz G, Fontanella C. Comprehensive review on the surrogate endpoints of efficacy proposed or hypothesized in the scientific community today. *J Natl Cancer Inst Monogr.* 2015 May;2015(51):29–31.

13. Peters WP1, Rosner GL, Vredenburgh JJ, Shpall EJ, Crump M, Richardson PG, Schuster MW, Marks LB, Cirrincione C, Norton L, Henderson IC, Schilsky RL, Hurd DD. Prospective, randomized comparison of high-dose chemotherapy with stem-cell support versus intermediate-dose chemotherapy after surgery and adjuvant

chemotherapy in women with high-risk primary breast cancer: a report of CALGB 9082, SWOG 9114, and NCIC MA-13. *J Clin Oncol.* 2005 Apr 1;23(10):2191–200. Epub 2005 Mar 14. Lake DE, Hudis CA. High-dose chemotherapy in breast cancer. *Drugs.* 2004;64(17):1851–60.

14. DeVita VT Jr, Young RC, Canellos GP Combination versus single agent chemotherapy: a review of the basis for selection of drug treatment of cancer. *Cancer.* 1975 Jan;35(1):98–110. Linden HM, Haskell CM, Green SJ, Osborne CK, Sledge GW Jr, Shapiro CL, Ingle JN, Lew D, Hutchins LF, Livingston RB, Martino S. Sequenced compared with simultaneous anthracycline and cyclophosphamide in high-risk stage I and II breast cancer: final analysis from INT-0137 (S9313). *J Clin Oncol.* 2007 Feb 20;25(6):656–61. Oakman C, Francis PA, Crown J, Quinaux E, Buyse M, De Azambuja E, Margeli Vila M, Andersson M, Nordenskjöld B, Jakesz R, Thürlimann B, Gutiérrez J, Harvey V, Punzalan L, Dell'orto P, Larsimont D, Steinberg I, Gelber RD, Piccart-Gebhart M, Viale G, Di Leo A. Overall survival benefit for sequential doxorubicin-docetaxel compared with concurrent doxorubicin and docetaxel in node-positive breast cancer—8-year results of the Breast International Group 02–98 phase III trial. *Ann Oncol.* 2013 May;24(5):1203–11. Dear RF, McGeechan K, Jenkins MC, Barratt A, Tattersall MH, Wilcken N. Combination versus sequential single agent chemotherapy for metastatic breast cancer. *Cochrane Database Syst Rev.* 2013 Dec 18;12:CD008792.

15. Norton L, Simon R. Tumor size, sensitivity to therapy, and design of treatment schedules. *Cancer Treat Rep.* 1977 Oct;61(7):1307–17. Norton L, Simon R. The Norton-Simon hypothesis revisited. *Cancer Treat Rep.* 1986 Jan;70(1):163–69.

16. Simon R, Norton L. The Norton-Simon hypothesis: designing more effective and less toxic chemotherapeutic regimens. *Nat Clin Pract Oncol.* 2006 Aug;3(8):406–7. Gompertz B. On the nature of the function expressive of the law of human mortality, and on a new mode of determining the value of life contingencies. *Phil Trans R Soc London* 1825:115:513–83.

17. Norton L, Simon R, Brereton HD, Bogden AE. Predicting the course of Gompertzian growth. *Nature.* 1976 Dec 9;264(5586):542–45. Norton L. A Gompertzian model of human breast cancer growth. *Cancer Res.* 1988 Dec 15;48(24 Pt 1):7067–71. Norton L. Cancer stem cells, self-seeding, and decremented exponential growth: theoretical and clinical implications. *Breast Dis.* 2008;29:27–36.

18. Gompertz B. On the nature of the function expressive of the law of human mortality, and on a new mode of determining the value of life contingencies. *Phil Trans R Soc London* 1825:115:513–83.

19. Norton L. Cancer stem cells, self-seeding, and decremented exponential growth: theoretical and clinical implications. *Breast Dis.* 2008;29:27–36. Norton L, Simon R. Tumor size, sensitivity to therapy, and design of treatment schedules. *Cancer Treat Rep.* 1977 Oct;61(7):1307–17. Norton L, Simon R. The Norton-Simon hypothesis revisited. *Cancer Treat Rep.* 1986 Jan;70(1):163–69. Simon R, Norton L. The Norton-Simon hypothesis: designing more effective and less toxic chemotherapeutic regimens. *Nat Clin Pract Oncol.* 2006 Aug;3(8):406–7. Gompertz B. On the nature of the function expressive of the law of human mortality, and on a new mode of determining the value of life contingencies. *Phil Trans R Soc London* 1825:115:513–83. Norton L, Simon R, Brereton HD, Bogden AE. Predicting the course of Gompertzian growth. *Nature.* 1976 Dec 9;264(5586):542–45. Norton L. A Gompertzian model of human breast cancer growth. *Cancer Res.* 1988 Dec 15;48(24 Pt 1):7067–71. Norton, L. Cell kinetics in normal tissues and in tumors of the young. In *Cancer in the Young*, Levine, AS (Ed.). Masson, New York (1982): 53–82.

20. Norton, L. Cell kinetics in normal tissues and in tumors of the young. In *Cancer in the Young*, Levine, AS (Ed.). Masson, New York (1982): 53–82.

21. Hanahan D, Weinberg RA. Hallmarks of cancer: the next generation. *Cell.* 2011 Mar 4;144(5):646–74.

22. Minn AJ, Gupta GP, Padua D, Bos P, Nguyen DX, Nuyten D, Kreike B, Zhang Y, Wang Y, Ishwaran H, Foekens JA, van de Vijver M, Massagué J. Lung metastasis genes couple breast tumor size and metastatic spread. *Proc Natl Acad Sci USA.* 2007 Apr 17;104(16):6740–45.

23. Norton L, Massagué J. Is cancer a disease of self-seeding? *Nat Med.* 2006 Aug;12(8):875–78. Kim MY, Oskarsson T, Acharyya S, Nguyen DX, Zhang XH, Norton L, Massagué J. Tumor self-seeding by circulating cancer cells. *Cell.* 2009 Dec 24;139(7):1315–26.

24. Norton L. Cancer stem cells, EMT, and seeding: a rose is a rose is a rose? *Oncology (Williston Park).* 2011 Jan;25(1):30, 32. Comen E, Norton L, Massagué J. Clinical implications of cancer self-seeding. *Nat Rev Clin Oncol.* 2011 Jun;8(6):369–77. Comen EA, Norton L, Massagué J. Breast cancer tumor size, nodal status, and prognosis: biology trumps anatomy. *J Clin Oncol.* 2011 Jul 1;29(19):2610–12.

25. Norton L. Cancer stem cells, self-seeding, and decremented exponential growth: theoretical and clinical implications. *Breast Dis.* 2008;29:27–36. Norton L. Conceptual and practical implications of breast tissue geometry: toward a more effective, less toxic therapy. *Oncologist.* 2005 Jun–Jul;10(6):370–81.

26. Norton L, Massagué J. Is cancer a disease of self-seeding? *Nat Med.* 2006 Aug;12(8):875–78.

27. Norton L, Simon R. Tumor size, sensitivity to therapy, and design of treatment schedules. *Cancer Treat Rep.* 1977 Oct;61(7):1307–17.

28. Peters WP1, Rosner GL, Vredenburgh JJ, Shpall EJ, Crump M, Richardson PG, Schuster MW, Marks LB, Cirrincione C, Norton L, Henderson IC, Schilsky RL, Hurd DD. Prospective, randomized comparison of high-dose chemotherapy with stem-cell support versus intermediate-dose chemotherapy after surgery and adjuvant chemotherapy in women with high-risk primary breast cancer: a report of CALGB 9082, SWOG 9114, and NCIC MA-13. *J Clin Oncol.* 2005 Apr 1;23(10):2191–200. Epub 2005 Mar 14. Henderson IC, Berry DA, Demetri GD, Cirrincione CT, Goldstein LJ, Martino S, Ingle JN, Cooper MR, Hayes DF, Tkaczuk KH, Fleming G, Holland JF, Duggan DB, Carpenter JT, Frei E III, Schilsky RL, Wood WC, Muss HB, Norton L. Improved outcomes from adding sequential paclitaxel but not from escalating doxorubicin dose in an adjuvant chemotherapy regimen for patients with node-positive primary breast cancer. *J Clin Oncol.* 2003 Mar 15;21(6):976–83. Budman DR, Berry DA, Cirrincione CT, Henderson IC, Wood WC, Weiss RB, Ferree CR, Muss HB, Green MR, Norton L, Frei E III. Dose and dose intensity as determinants of outcome in the adjuvant treatment of breast cancer. The Cancer and Leukemia Group B. *J Natl Cancer Inst.* 1998 Aug 19;90(16):1205–11. Fisher B, Anderson S, DeCillis A, Dimitrov N, Atkins JN, Fehrenbacher L, Henry PH, Romond EH, Lanier KS, Davila E, Kardinal CG, Laufman L, Pierce NH, Abramson N, Keller AM, Hamm JT, Wickerham DL, Begovic M, Tan-Chiu E, Tian W, Wolmark N. Further evaluation of intensified and increased total dose of cyclophosphamide for the treatment of primary breast cancer: findings from National Surgical Adjuvant Breast and Bowel Project B-25. *J Clin Oncol.* 1999 Nov;17(11):3374–88. Winer EP, Berry DA, Woolf S, Duggan D, Kornblith A, Harris LN, Michaelson RA, Kirshner JA, Fleming GF, Perry MC, Graham ML, Sharp SA, Keresztes R, Henderson IC, Hudis C, Muss H, Norton L. Failure of higher-dose paclitaxel to improve outcome in patients with metastatic breast cancer: cancer and leukemia group B trial 9342. *J Clin Oncol.* 2004 Jun 1;22(11):2061–68.

29. Skipper HE. Laboratory models: some historical perspective. *Cancer Treat Rep.* 1986 Jan;70(1):3–7.

30. Norton L, Simon R. Tumor size, sensitivity to therapy, and design of treatment schedules. *Cancer Treat Rep.* 1977 Oct;61(7):1307–17. Norton L, Simon R. The Norton-Simon hypothesis revisited. *Cancer Treat Rep.* 1986 Jan;70(1):163–69. Simon R, Norton L. The Norton-Simon hypothesis: designing more effective and less toxic chemotherapeutic regimens. *Nat Clin Pract Oncol.* 2006 Aug;3(8):406–7. Morris PG, McArthur HL, Hudis C, Norton L. Dose-dense chemotherapy for breast cancer: what does the future hold? *Future Oncol.* 2010 Jun;6(6):951–65.

31. Bonadonna G, Zambetti M, Moliterni A, Gianni L, Valagussa P. Clinical relevance of different sequencing of doxorubicin and cyclophosphamide, methotrexate, and fluorouracil in operable breast cancer. *J Clin Oncol.* 2004 May 1;22(9):1614–20. Linden HM, Haskell CM, Green SJ, Osborne CK, Sledge GW Jr, Shapiro CL, Ingle JN, Lew D, Hutchins LF, Livingston RB, Martino S. Sequenced compared with simultaneous anthracycline and cyclophosphamide in high-risk stage I and II breast cancer: final analysis from INT-0137 (S9313). *J Clin Oncol.* 2007 Feb 20;25(6):656–61. Oakman C, Francis PA, Crown J, Quinaux E, Buyse M, De Azambuja E, Margeli Vila M, Andersson M, Nordenskjöld B, Jakesz R, Thürlimann B, Gutiérrez J, Harvey V, Punzalan L, Dell'orto P, Larsimont D, Steinberg I, Gelber RD, Piccart-Gebhart M, Viale G, Di Leo A. Overall survival benefit for sequential doxorubicin-docetaxel compared with concurrent doxorubicin and docetaxel in node-positive breast cancer—8-year results of the Breast International Group 02–98 phase III trial. *Ann Oncol.* 2013 May;24(5):1203–11.

32. Hudis C, Seidman A, Baselga J, Raptis G, Lebwohl D, Gilewski T, Moynahan M, Sklarin N, Fennelly D, Crown JP, Surbone A, Uhlenhopp M, Riedel E, Yao TJ, Norton L. Sequential dose-dense doxorubicin, paclitaxel, and cyclophosphamide for resectable high-risk breast cancer: feasibility and efficacy. *J Clin Oncol.* 1999 Jan;17(1):93–100. Citron ML, Berry DA, Cirrincione C, Hudis C, Winer EP, Gradishar WJ, Davidson NE, Martino S, Livingston R, Ingle JN, Perez EA, Carpenter J, Hurd D, Holland JF, Smith BL, Sartor CI, Leung EH, Abrams J, Schilsky RL, Muss HB, Norton L. Randomized trial of dose-dense versus conventionally scheduled and sequential versus concurrent combination chemotherapy as postoperative adjuvant treatment of node-positive primary breast cancer: first report of Intergroup Trial C9741/Cancer and Leukemia Group B Trial 9741. *J Clin Oncol.* 2003 Apr 15;21(8):1431–39. Cognetti F, Bruzzi P, De Placido S, De Laurentiis M, Boni C, Aitini E, Durando A, Turletti A, Valle E, Garrone O, Puglisi F, Montemurro F, Barni S, Di Blasio B, Gamucci T, Colantuoni G, Olmeo N, Tondini C, Parisi AM, Bighin C, Pastorino S, Lambertini M, Del Mastro L. [S5–06] Epirubicin and cyclophosphamide (EC) followed by paclitaxel (T) versus fluorouracil, epirubicin and cyclophosphamide (FEC) followed by T, all given every 3 weeks or 2 weeks, in node-positive early breast cancer (BC) patients (pts). Final results of the gruppo Italiano mammella (GIM)-2 randomized phase III study. San Antonio Breast Cancer Symposium, 2013. Del Mastro L, De placido S, Bruzzi P, De Laurentiis M, Boni C, Cavazzini G, Durando A, Turletti A, Nistico C, Valle E, Garrone O, Puglisi F, Montemurro F, Barni S, Ardizzoni A, Gamucci T, Colantuoni G, Giuliano M, Gravina A, Papaldo P, Bighin C, Bisagni G, Gorestieri V, Cognetti F, Gruppo Italiano Mammella (GIM) investigators. Fluorouracil and dose-dense chemotherapy in adjuvant treatment of patients

with early-stage breast cancer: an open-label, 2×2 factorial, randomized phase 3 trial. *Lancet.* 2015 May 9;385(9980):1863–72. McArthur HL, Hudis CA. Dose-dense therapy in the treatment of early-stage breast cancer: an overview of the data. *Clin Breast Cancer.* 2007 Dec;8 Suppl 1:S6–S10. Lyman GH, Barron RL, Natoli JL, Miller RM. Systematic review of efficacy of dose-dense versus non-dose-dense chemotherapy in breast cancer, non-Hodgkin lymphoma, and non-small cell lung cancer. *Crit Rev Oncol Hematol.* 2012 Mar;81(3):296–308. Lyman GH, Dale DC, Culakova E, Poniewierski MS, Wolff DA, Kuderer NM, Huang M, Crawford J. The impact of the granulocyte colony-stimulating factor on chemotherapy dose intensity and cancer survival: a systematic review and meta-analysis of randomized controlled trials. *Ann Oncol.* 2013 Oct;24(10):2475–84. Bonilla L, Ben-Aharon I, Vidal L, Gafter-Gvili A, Leibovici L, Stemmer SM. Dose-dense chemotherapy in nonmetastatic breast cancer: a systematic review and meta-analysis of reandomized controlled trials. *J Natl Cancer Inst.* 2010 Dec 15:102(24):1845–54. Lemos Duarte, da Silveira Nogueira Lima JP, Passos Lima CS, Deeke Sasse A. Dose-dense chemotherapy versus conventional chemotherapy for early breast cancer: a systematic review with meta-analysis. *Breast.* 2012 Jun;21(3):343–49. Petrelli F, Cabiddu M, Coinu A, Borgonovo K, Ghilardi M, Lonati V, Barni S. Adjuvant dose-dense chemotherapy in breast cancer: a systematic review and meta-analysis of random-ized trials. *Breast Cancer Res Treat.* 2015 Jun;151(2):251–59.

33. Increasing the dose intensity of chemotherapy by more frequent administration or sequential scheduling: a patient-level meta-analysis of 37,298 women with early breast cancer in 26 randomized trials. Early Breast Cancer Trialists' Collaborative Group (EBCTCG). *Lancet.* 2019 Apr 6;393(10179):1440–52. Epub 2019 Feb 8.

34. Citron ML, Berry DA, Cirrincione C, Hudis C, Winer EP, Gradishar WJ, Davidson NE, Martino S, Livings-ton R, Ingle JN, Perez EA, Carpenter J, Hurd D, Holland JF, Smith BL, Sartor CI, Leung EH, Abrams J, Schilsky RL, Muss HB, Norton L. Randomized trial of dose-dense versus conventionally scheduled and sequential versus concurrent combination chemotherapy as postoperative adjuvant treatment of node-positive primary breast cancer: first report of Intergroup Trial C9741/Cancer and Leukemia Group B Trial 9741. *J Clin Oncol.* 2003 Apr 15;21(8):1431–39. Burstein HJ. Myeloid growth factor support for dose-dense adjuvant chemotherapy for breast cancer. *Oncology (Williston Park).* 2006 Dec;20(14 Suppl 9):13–15. Liu MC, Demetri GD, Berry DA, Norton L, Broadwater G, Robert NJ, Duggan D, Hayes DF, Henderson IC, Lyss A, Hopkins J, Kaufman PA, Marcom PK, Younger J, Lin N, Tkaczuk K, Winer EP, Hudis CA, Cancer and Leukemia Group B. Dose-escalation of filgrastim does not improve efficacy: clinical tolerability and long-term follow-up on CALGB study 9141 adju-vant chemotherapy for node-positive breast cancer patients using dose-intensified doxorubicin plus cyclophos-phamide followed by paclitaxel. *Cancer Treat Rev.* 2008 May;34(3):223–30.

35. Citron ML, Berry DA, Cirrincione C, Hudis C, Winer EP, Gradishar WJ, Davidson NE, Martino S, Livings-ton R, Ingle JN, Perez EA, Carpenter J, Hurd D, Holland JF, Smith BL, Sartor CI, Leung EH, Abrams J, Schilsky RL, Muss HB, Norton L. Randomized trial of dose-dense versus conventionally scheduled and sequential versus concurrent combination chemotherapy as postoperative adjuvant treatment of node-positive primary breast cancer: first report of Intergroup Trial C9741/Cancer and Leukemia Group B Trial 9741. *J Clin Oncol.* 2003 Apr 15;21(8):1431–39. Henderson IC, Berry DA, Demetri GD, Cirrincione CT, Goldstein LJ, Martino S, Ingle JN, Cooper MR, Hayes DF, Tkaczuk KH, Fleming G, Holland JF, Duggan DB, Carpenter JT, Frei E III, Schilsky RL, Wood WC, Muss HB, Norton L. Improved outcomes from adding sequential paclitaxel but not from escalat-ing doxorubicin dose in an adjuvant chemotherapy regimen for patients with node-positive primary breast cancer. *J Clin Oncol.* 2003 Mar 15;21(6):976–83. Budman DR, Berry DA, Cirrincione CT, Henderson IC, Wood WC, Weiss RB, Ferree CR, Muss HB, Green MR, Norton L, Frei E III. Dose and dose intensity as determinants of outcome in the adjuvant treatment of breast cancer. The Cancer and Leukemia Group B. *J Natl Cancer Inst.* 1998 Aug 19;90(16):1205–11. Fisher B, Anderson S, DeCillis A, Dimitrov N, Atkins JN, Fehrenbacher L, Henry PH, Romond EH, Lanier KS, Davila E, Kardinal CG, Laufman L, Pierce HI, Abramson N, Keller AM, Hamm JT, Wickerham DL, Begovic M, Tan-Chiu E, Tian W, Wolmark N. Further evaluation of intensified and increased total dose of cyclophosphamide for the treatment of primary breast cancer: findings from National Surgical Adjuvant Breast and Bowel Project B-25. *J Clin Oncol.* 1999 Nov;17(11):3374–88. Winer EP, Berry DA, Woolf S, Duggan D, Kornblith A, Harris LN, Michaelson RA, Kirshner JA, Fleming GF, Perry MC, Graham ML, Sharp SA, Keresztes R, Henderson IC, Hudis C, Muss H, Norton L. Failure of higher-dose paclitaxel to improve outcome in patients with metastatic breast cancer: cancer and leukemia group B trial 9342. *J Clin Oncol.* 2004 Jun 1;22(11):2061–68.

36. Henderson IC, Berry DA, Demetri GD, Cirrincione CT, Goldstein LJ, Martino S, Ingle JN, Cooper MR, Hayes DF, Tkaczuk KH, Fleming G, Holland JF, Duggan DB, Carpenter JT, Frei E 3rd, Schilsky RL, Wood WC, Muss HB, Norton L. Improved outcomes from adding sequential paclitaxel but not from escalating doxorubicin dose in an adjuvant chemotherapy regimen for patients with node-positive primary breast cancer. *J Clin Oncol.* 2003 Mar 15;21(6):976–83. Budman DR, Berry DA, Cirrincione CT, Henderson IC, Wood WC, Weiss RB, Ferree

CR, Muss HB, Green MR, Norton L, Frei E III. Dose and dose intensity as determinants of outcome in the adjuvant treatment of breast cancer. The Cancer and Leukemia Group B. *J Natl Cancer Inst.* 1998 Aug 19;90(16):1205–11. Fisher B, Anderson S, DeCillis A, Dimitrov N, Atkins JN, Fehrenbacher L, Henry PH, Romond EH, Lanier KS, Davila E, Kardinal CG, Laufman L, Pierce HI, Abramson N, Keller AM, Hamm JT, Wickerham DL, Begovic M, Tan-Chiu E, Tian W, Wolmark N. Further evaluation of intensified and increased total dose of cyclophosphamide for the treatment of primary breast cancer: findings from National Surgical Adjuvant Breast and Bowel Project B-25. *J Clin Oncol.* 1999 Nov;17(11):3374–88. Winer EP, Berry DA, Woolf S, Duggan D, Kornblith A, Harris LN, Michaelson RA, Kirshner JA, Fleming GF, Perry MC, Graham ML, Sharp SA, Keresztes R, Henderson IC, Hudis C, Muss H, Norton L. Failure of higher-dose paclitaxel to improve outcome in patients with metastatic breast cancer: cancer and leukemia group B trial 9342. *J Clin Oncol.* 2004 Jun 1;22(11):2061–68. Liu MC, Demetri GD, Berry DA, Norton L, Broadwater G, Robert NJ, Duggan D, Hayes DF, Henderson IC, Lyss A, Hopkins J, Kaufman PA, Marcom PK, Younger J, Lin N, Tkaczuk K, Winer EP, Hudis CA, Cancer and Leukemia Group B. Dose-escalation of filgrastim does not improve efficacy: clinical tolerability and long-term follow-up on CALGB study 9141 adjuvant chemotherapy for node-positive breast cancer patients using dose-intensified doxorubicin plus cyclophosphamide followed by paclitaxel. *Cancer Treat Rev.* 2008 May;34(3):223–30.

37. Fennelly D, Aghajanian C, Shapiro F, O'Flaherty C, McKenzie M, O'Connor C, Tong W, Norton L, Spriggs D. Phase I and pharmacologic study of paclitaxel administered weekly in patients with relapsed ovarian cancer. *J Clin Oncol.* 1997 Jan;15(1):187–92. Budd GT, Barlow WE, Moore HCF, Hobday TJ, Stewart JA, Isaacs C, Salim M, Cho JK, Rinn K, Albain KS, Chew HK, Von Burton G, Moore TD, Srkalovic G, McGregor BA, Flaherty LE, Livingston RB, Lew D, Gralow J, Hortobagyi GN. Comparison of two schedules of paclitaxel as adjuvant therapy for breast cancer. *J Clin Oncol.* 2013; 31(suppl):abstr CRA1008.

38. Morris PG, Dickler M, McArthur HL, Traina T, Sugarman S, Lin N, Moy B, Come S, Godfrey L, Nulsen B, Chen C, Steingart R, Rugo H, Norton L, Winer E, Hudis CA, Dang CT. Dose-dense adjuvant Doxorubicin and cyclophosphamide is not associated with frequent short-term changes in left ventricular ejection fraction. *J Clin Oncol.* 2009 Dec 20;27(36):6117–23. Morris PG, Iyengar NM, Patil S, Chen C, Abbruzzi A, Lehman R, Steingart R, Oeffinger KC, Lin N, Moy B, Come SE, Winer EP, Norton L, Hudis CA, Dang CT. Long-term cardiac safety and outcomes of dose-dense doxorubicin and cyclophosphamide followed by paclitaxel and trastuzumab with and without lapatinib in patients with early breast cancer. *Cancer.* 2013 Nov 15;119(22):3943–51.

39. Slamon D, Eiermann W, Robert N, Pienkowski T, Martin M, Press M, Mackey J, Glaspy J, Chan A, Pawlicki M, Pinter T, Valero V, Liu MC, Sauter G, von Minckwitz G, Visco F, Bee V, Buyse M, Bendahmane B, Tabah-Fisch I, Lindsay MA, Riva A, Crown J; Breast Cancer International Research Group. Adjuvant trastuzumab in HER2-positive breast cancer. *N Engl J Med.* 2011 Oct 6;365(14):1273–83.

40. Perez EA, Suman VJ, Davidson NE, Gralow JR, Kaufman PA, Visscher DW, Chen B, Ingle JN, Dakhil SR, Zujewski J, Moreno-Aspitia A, Pisansky TM, Jenkins RB. Sequential versus concurrent trastuzumab in adjuvant chemotherapy for breast cancer. *J Clin Oncol.* 2011 Dec 1;29(34):4491–97. Shulman LN, Cirrincione CT, Berry DA, Becker HP, Perez EZ, O'Regan R, Martino S, Atkins JN, Mayer E, Schneider CJ, Kimmick G, Norton L, Muss H, Winer EP, Hudis C. Six cycles of doxorubicin and cyclophosphamide or paclitaxel are not superior to four cycles as adjuvant chemotherapy for breast cancer in women with zero to three positive axillary nodes: Cancer and Leukemia Group B 40101. *J Clin Oncol.* 2012 Nov 20;30(33):4071–76.

41. Traina TA, Dugan U, Higgins B, Kolinsky K, Theodoulou M, Hudis CA, Norton L.Optimizing chemotherapy dose and schedule by Norton-Simon mathematical modeling. *Breast Dis.* 2010;31(1):7–18.

42. Traina TA, Theodoulou M, Feigin K, Patil S, Tan KL, Edwards C, Dugan U, Norton L, Hudis C. Phase I study of a novel capecitabine schedule based on the Norton-Simon mathematical model in patients with metastatic breast cancer. *J Clin Oncol.* 2008 Apr 10;26(11):1797–802. Comen E, Morris PG, Norton L. Translating mathematical modeling of tumor growth patterns into novel therapeutic approaches for breast cancer. *J Mammary Gland Biol Neoplasia.* 2012 Dec;17(3–4):241–49. Fournier C, Tisman G, Kleinman R, Park Y, Macdonald WD. Clinical evidence for overcoming capecitabine resistance in a woman with breast cancer terminating in radiologically occult micronodular pseudo-cirrhosis with portal hypertension: a case report. *J Med Case Rep.* 2010 Apr 21;4:112.

43. Carver BS, Chapinski C, Wongvipat J, Hieronymus H, Chen Y, Chandarlapaty S, Arora VA, Le C, Koutcher J, Scher H, Scardino PT, Rosen N, Sawyers CL. Reciprocal feedback regulation of PI3K and androgen receptor signaling in PTEN-deficient prostate cancer. *Cancer Cell.* 2011 May 17;19(5):575–86.

44. Inchausti P, Ginzburg LR. Maternal effects mechanism of population cycling: a formidable competitor to the traditional predator-prey view. *Philos Trans R Soc Lond B Biol Sci.* 2009 Apr 27;364(1520):1117–24.

45. Chandarlapaty S, Sawai A, Scaltriti M, Rodrik-Outmezguine V, Grbovic-Huezo O, Serra V, Majumder PK, Baselga J, Rosen N. AKT inhibition relieves feedback suppression of receptor tyrosine kinase expression and activity. *Cancer Cell.* 2011 Jan 18;19(1):58–71.

46. Geretti E, Leonard SC, Dumont N, Lee H, Zheng J, De Souza R, Gaddy DF, Espelin CW, Jaffray DA, Moyo Nielsen UB, Wickham TJ, Hendriks BS. Cyclophosphamide-mediated tumor priming for enhanced delivery and anti-tumor activity of HER2-targeted liposomal doxorubicin (MM-302). *Mol Cancer Ther.* 2015 Sep;14(9):2060–71.

47. Leder K, Pitter K, Laplant Q, Hambardzumyan D, Ross BD, Chan TA, Holland EC, Michor F. Mathematical modeling of PDGF-driven glioblastoma reveals optimized radiation dosing schedules. *Cell.* 2014 Jan 30;156(3):603–16.

48. Norton L, Simon R. Growth curve of an experimental solid tumor following radiotherapy. *J Natl Cancer Inst.* 1977 Jun;58(6):1735–41.

49. Minn AJ, Gupta GP, Padua D, Bos P, Nguyen DX, Nuyten D, Kreike B, Zhang Y, Wang Y, Ishwaran H, Foekens JA, van de Vijver M, Massagué J. Lung metastasis genes couple breast tumor size and metastatic spread. *Proc Natl Acad Sci USA.* 2007 Apr 17;104(16):6740–45.

50. Acharyya S, Oskarsson T, Vanharanta S, Malladi S, Kim J, Morris PG, Manova-Todorova K, Leversha M, Hogg N, Seshan VE, Norton L, Brogi E, Massagué J. A CXCL1 paracrine network links cancer chemoresistance and metastasis. *Cell.* 2012 Jul 6;150(1):165–78.

51. Kao J, Ko EC, Eisenstein S, Sikora AG, Fu S, Chen SH. Targeting immune suppressing myeloid-derived suppressor cells in oncology. Crit Rev *Oncol Hematol.* 2011 Jan;77(1):12–19.

52. Parker JS, Perou CM. Tumor heterogeneity: focus on the leaves, the trees, or the forest? *Cancer Cell.* 2015 Aug 10;28(2):149–50.

53. Waitz R, Solomon SB, Petre EN, Trumble AE, Fassò M, Norton L, Allison JP. Potent induction of tumor immunity by combining tumor cryoablation with anti-CTLA-4 therapy. *Cancer Res.* 2012 Jan 15;72(2):430–39. Epub 2011 Nov 22. Diab A, Solomon SB, Comstock C, Maybody M, Sacchini V, Durack JC, Blum B, Yuan J, Patil S, Neville DA, Sung JS, Kotin A, Morris EA, Brogi E, Morrow M, Wolchok JD, Allison J, Hudis C, Norton L, McArthur HL. A pilot study of preoperative, single-dose ipilimumab and/or cryoablation in women with early-stage, resectable breast cancer. *J Clin Oncol.* 2013;31(26, suppl):67.

54. Kleppe M, Comen E, Wen HY, Bastian L, Blum B, Rapaport FT, Keller M, Granot Z, Socci N, Viale A, You D, Benezra R, Weigelt B, Brogi E, Berger MF, Reis-Filho JS, Levine RL, Norton L. Somatic mutations in leukocytes infiltrating primary breast cancers. *npj Breast Cancer.* 2015 Jun 10;1:15005.

55. Awad MM, Katayama R, McTigue M, Liu W, Deng YL, Brooun A, Friboulet L, Huang D, Falk MD, Timofeevski S, Wilner KD, Lockerman EL, Khan TM, Mahmood S, Gainor JF, Digumarthy SR, Stone JR, Mino-Kenudson M, Christensen JG, Iafrate AJ, Engelman JA, Shaw AT. Acquired resistance to crizotinib from a mutation in CD74-ROS1. *N Engl J Med.* 2013 Jun 20;368(25):2395–401.

56. Gerlinger M, Norton L, Swanton C. Acquired resistance to crizotinib from a mutation in CD74-ROS1. *N Engl J Med.* 2013 Sep 19;369(12):1172–73.

57. Juric D, Castel P, Griffith M, Griffith OL, Won HH, Ellis H, Ebbesen SH, Ainscough BJ, Ramu A, Iyer G, Shah RH, Huynh T, Mino-Kenudson M, Sgroi D, Isakoff S, Thabet A, Elamine L, Solit DB, Lowe SW, Quadt C, Peters M, Derti A, Schegel R, Huang A, Mardis ER, Berger MF, Baselga J, Scaltriti M. Convergent loss of PTEN leads to clinical resistance to a PI(3)Kα inhibitor. *Nature.* 2015 Feb 12;518(7538):240–44. Epub 2014 Nov 17.

58. Newton PK, Mason J, Bethel K, Bazhenova L, Nieva J, Norton L, Kuhn P. Spreaders and sponges define metastasis in lung cancer: a Markov chain Monte Carlo mathematical model. *Cancer Res.* 2013 May 1;73(9):2760–69. Epub 2013 Feb 27.

IV THE MICRO-/ENVIRONMENT DIMENSION OF CANCER

11 Tissue Tension Modulates Metabolism and Chromatin Organization to Promote Malignancy

Roger Oria, Dhruv Thakar, and Valerie M. Weaver

Overview

The extracellular matrix (ECM) modulates embryogenesis, tissue development, and homeostasis.[1] Cell-ECM interactions direct the growth, survival, and migration of cells within a tissue to drive processes such as gastrulation and branching morphogenesis and orchestrate tissue organization and differentiation. Importantly, ECM composition, posttranslational modifications, and organization influence its biochemical and biophysical properties or stiffness. ECM stiffness-induced tissue tension and mechanical force also influence tissue development, and ECM stiffness directly regulates stem cell behavior and cell differentiation. Moreover, development and differentiation are accompanied by striking changes in cellular metabolism, and cell-ECM adhesion and ECM stiffness modify cell metabolism to regulate cell behavior and tissue morphogenesis.[2] Indeed, embryonic development and tissue-specific differentiation are associated with profound changes in gene expression that is sustained by epigenetic alterations that are in turn influenced by the metabolic state of the cell that modifies chromatin organization and gene expression. Accordingly, a functional dialogue between ECM composition and stiffness modulates tissue development and homeostasis by influencing cell metabolism to alter chromatin organization and gene expression. Not surprisingly, conditions that stiffen the ECM, such as chronic inflammation, also modify cell metabolism and perturb tissue organization to foster diseases such as cancer.

Malignancy associates with alterations in ECM deposition, remodeling, and crosslinking that change its composition and organization and elevates its stiffness. Indeed, tissue fibrosis, which is characterized by elevated levels of remodeled, crosslinked, and stiffened ECM, is a major feature of many solid cancers. The stiffened tumor ECM promotes integrin focal adhesion assembly and cellular signaling that stimulate tumor cell proliferation, survival, and migration to promote malignant transformation, as well as induces a mesenchymal transition that fosters tumor metastasis. The stiff ECM additionally reduces nutrient diffusion and compromises the vasculature-inducing tissue hypoxia that further restricts nutrient accessibility. Tumor cells adapt to this "hostile and mechanically challenged" microenvironment by rewiring their metabolism and reorganizing their epigenome through posttranslational modifications of their chromatin, which ultimately reprograms their gene expression and fosters expression of a malignant and aggressive phenotype.

In this chapter, we summarize what is currently known about cancer-associated ECM remodeling and stiffening. We describe how an abnormal stiffened tumor ECM can drive cell proliferation, survival, invasion, and migration and how this could mediate changes in cellular metabolism. We discuss how a stiffened ECM can directly influence chromatin organization and gene transcription, even in the absence of any DNA sequence change, and how this could be indirectly linked to changes in metabolism that reprogram cells toward malignancy and tumor aggression. We further discuss the impact these findings have on causal explanatory models of carcinogenesis and the transition from premalignancy to malignancy.

11.1   The Desmosplastic Response and Malignancy

The interstitial ECM within epithelial tissues is a highly organized polymeric meshwork of macro-molecules consisting of fibrous collagenous proteins, as well as proteoglycans and glycoproteins that sequester various growth factors, hormones, cytokines, and metabolites. The molecular composition of the interstitial ECM determines its biochemical properties while its three-dimensional (3D) organization and posttranslational modifications, such as its crosslinking, dictate its mechanical properties, including its tensile properties and stiffness. Stromal fibroblasts are the major stromal cell type that synthesizes, secretes, and reorganizes the interstitial ECM and tunes its biochemical composition and mechanical properties to maintain the normal function of a healthy tissue.[3]

Malignant transformation is accompanied by a desmoplastic response that is similar to the tissue response to injury or another traumatic insult. Accordingly, tumors resemble a chronically wounded tissue.[4] Tumor desmoplasia is marked by ECM deposition, metalloproteinase-mediated remodeling, and lysyl oxidase (LOX)– and lysyl hydroxylase (LH)–mediated collagen crosslinking, which collaboratively alter the composition and organization of the ECM. Carcinoma-associated fibroblasts (CAFs) are the primary cells responsible for increasing the levels of and directing the aberrant remodeling and crosslinking of the tumor stromal ECM.[5] CAFs are a heterogeneous population of mesenchymal PDGF, FAP, FSP1, and/or alpha smooth muscle actin (αSMA)–positive cells localized within the interstitial stroma. Some of these CAFs are highly contractile and thus are competent to remodel the interstitial ECM.[6] Contractile CAFs alter the composition, crosslinking, and organization of the stromal ECM and thereby change its topography,[7] porosity,[8] and stiffness, and these ECM features in turn promote the tumorigenic behavior of the tissue (see figure 11.1).[9]

Collagen is the most abundant protein in mammals and a major component of the interstitial ECM.[10] Collagen I in particular is a fibrillar collagen that is more abundantly expressed in the tumor stroma by the tumor-associated CAFs.[11] Intracellular lysyl hydroxylases expressed by the CAFs modify the newly synthesized collagen, and subsequently, the N- and C-terminal propeptides in the secreted procollagen[12] are cleaved by fibroblast-secreted peptidases followed by further collagen modification by extracellular LOX enzymes. Thereafter, the newly synthesized and secreted collagen molecules spontaneously crosslink to assemble into triple helical collagen fibers.[13] The level of LH and LOX-mediated crosslinking in the interstitial collagen dictates the stiffness of the stromal ECM. Some CAFs also exhibit high levels of fibronectin fibrillogenesis, and fibronectin fibrils are assembled into a scaffold with fibrillar collagen to increase the load-bearing capacity of the collagen.[14] Not surprisingly, LOX and LH and fibronectin are frequently overexpressed in the stroma of many tumors, including colorectal,[15] breast, pancreatic, and lung,[16] and this phenotype could explain why many tumors, including those of the pancreas and breast, are often twice as stiff as their normal counterparts.[17] Elasticity measurements in mouse mammary tumors and human specimens have revealed that the stiffness of the stroma increases as a function of tumor evolution and aggression, and they show that it associates positively with collagen accumulation, particularly at the tumor-stromal fibronectin-enriched interface.[18] Importantly, elevated expression of LOX family enzymes and higher fibronectin levels both correlate with poor overall cancer patient survival.[19] The increased collagen levels and crosslinking that stiffen the tissue stroma could also account for the higher solid stress found in many solid tumors, a phenotype that has been shown to contribute significantly to tumor aggression and poor patient outcome because it impedes the vasculature, induces hypoxia, and compromises drug delivery.[20]

CAFs can also collaborate with invading tumor cells to remodel the interstitial tumor collagens to increase their alignment, thickness, and stiffness, thereby promoting tumor cell migration and invasion (see figure 11.1).[21] For instance, collagen I fibers exhibit a strain-stiffening behavior such that increased density of cellular forces applied by the migrating cells will further align the fibers and increase their density and mechanical properties to reinforce the collagenous migratory tracks.[22] Consistently, experimental murine and in vitro three-dimensional tissue culture models have demonstrated that invasive mammary carcinomas develop progressively thickened, linearized collagenous fibers that are oriented perpendicularly into the primary tumor mass and extend into and throughout the interstitial stroma.[23] Intravital imaging revealed that experimental mammary tumor cells migrate more efficiently along thick, linear collagenous tracks, and furthermore, these tracks can enhance vascular intravasation.[24] Consistently, the presence of perpendicular thick collagen fibers adjacent to breast tumor lesions[25] and thicker collagen fibers surrounding

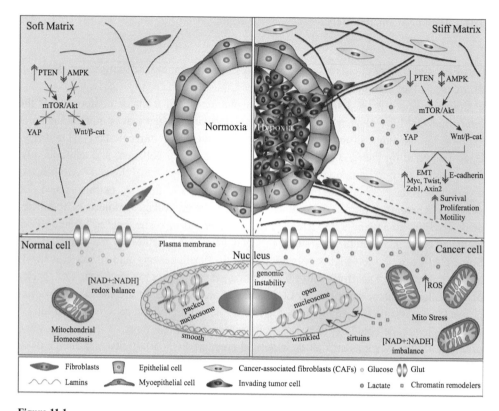

Figure 11.1

Alterations in tissue stiffness and metabolic state affect mechano-regulated and glycolytic pathways during breast cancer progression. In healthy tissue, stromal cells maintain intact homeostatic mechanical properties of the extracellular matrix through an exquisite balance of protein secretion, extracellular matrix remodeling, and protein degradation. Mammary epithelial cells are organized into growth-arrested polarized ductal structures with adherens junctions (E-cadherin mediated) and central lumens. Moreover, the extracellular environment maintains a physiological pH, ranging between 7.2 and 7.5 (Gatenby, R.A., Gawlinski, E.T., Gmitro, A.F., Kaylor, B. & Gillies, R.J. Acid-mediated tumor invasion: a multidisciplinary study. *Cancer Res* 66, 5216–5223 [2006]), and optimal provision of metabolites. Glucose (light gray circles) translocates to the cell interior through glucose transport proteins (GLUTs) to fuel cellular energetic demands via oxidative phosphorylation (OXPHOS) within the mitochondria. Acetyl-CoA and other intermediate metabolites may induce posttranslational modifications to DNA and histones, but the condensed state of chromatin prevents the activation of oncogenes that drive transformation and tumor development. On the other hand, during cancer progression, fibroblasts alter their phenotype to become more contractile cancer-associated fibroblasts (CAFs), and they begin to generate excessive extracellular matrix (ECM) molecules, particularly collagen I. This collagen production in conjunction with lysyl oxidase—enzymes responsible for collagen crosslinking—act to stiffen the stromal tissue. Moreover, contractile CAFs remodel the ECM and align collagen fibers, which further stimulates directed cell migration. CAFs present dysfunctional mitochondria and aerobic glycolysis, therefore rewiring their metabolic state to one suitable for increased glucose uptake. This increase in glucose uptake is favored by an overexpression of GLUTs. Stiffer matrix promotes the loss of PTEN, a master tumor suppressor gene, and activates oncogenic pathways such as the Hippo and Wnt signaling axes. Moreover, increased ATP production activates AMPK leading to the inhibition of mTOR and stimulation of the PI3K/Akt survival signaling pathway. Similarly, epithelial cells undergo malignant transformation that may be accompanied by an EMT transition, which is associated with enhanced aerobic glycolysis and downregulation of E-cadherin. During aerobic glycolysis, lactate (dark gray circles) produced by cells is secreted into the surrounding environment and matrix, thereby contributing to extracellular acidification. Acidification sustains hypoxia and stimulates cellular expression of glycolytic genes via hypoxia inducible factors (HIFs). Cell nuclear morphology is also altered by a reduction in nuclear elasticity that may result from a combination of downregulated lamin A/C expression and higher forces transmitted to the nucleus due to gastric cancer tumors and pancreatic ductal adenocarcinoma (PDAC) lesions predicts poorer five-year overall patient survival.[26] These data provide evidence of a compelling link between a remodeled, stiffened ECM and tumor progression.

11.2 Feeling the Extracellular Matrix: Adhesion-Mediated Mechanotransduction

Malignant transformation is thought to be initiated by genetic mutations and amplification of key oncogenic pathways and the loss of tumor suppressor activity that collaborate to promote the aberrant growth, survival, and migration of the cancer cells within the transformed tissue.[27] However, many oncogenes also simultaneously increase tumor cell tension by activating ROCK and myosins, and this increased contractility permits the tumors to remodel and stiffen their adjacent ECM.[28] Thus, the elevated oncogene-induced tumor cell tension collaborates with the contractile CAFs to accelerate the remodeling and stiffening of the tumor's ECM. Interestingly, recent evidence suggests that elevated glucose metabolism increases cell tension, suggesting metabolic dysfunction in a tissue might also promote ECM remodeling and stiffening. If true, given that tumors are characterized by perturbed glucose metabolism, this could enhance their contractile phenotype and would contribute to ECM remodeling and stiffening and tissue dysfunction. Regardless, reducing tumor cell tension in culture and in vivo or preventing CAF-mediated ECM remodeling impedes malignant transformation and tumor progression in experimental models, suggesting oncogenic transformation and metabolic dysfunction may promote a malignant phenotype, at least in part, by altering tissue tension.

Cells respond to the stiffness of their ECM microenvironment by applying intracellular tension via actomyosin contractile forces at a magnitude that corresponds to the resistance or stiffness of the ECM. Cells accomplish this by engaging transmembrane cell-matrix adhesion structures of which integrin heterodimers are the best studied.[29] Mature cell-matrix adhesions, also known as focal adhesions (henceforth called focal adhesions [FAs]) are large protein hubs that provide mechanical linkage between ECM ligands and the intracellular actomyosin cytoskeleton through transmembrane integrins and adaptor proteins. FAs are the main cellular structures that transmit actomyosin contractile forces from the cell to the ECM.[30] The integrin-generated focal contacts (nascent adhesions) that cells initially make following ECM ligation are quite weak and transient. However, if the ECM

Figure 11.1 (continued)

increased matrix stiffness. Morphological changes to the nucleus may favor the relaxation of condensed chromatin, therefore providing more accessible regions for gene modification. These newly accessible regions can then be further remodeled with intermediate metabolites or sirtuins to promote posttranslational modifications of DNA and histones. These posttranslational modifications can alter the expression of DNA repair and hypoxic or proliferation-related genes to exacerbate tumor initiation, progression, and metastasis.

is sufficiently stiff, these adhesions will mature to assemble into larger, more stable FAs. Thus, cells interacting with soft ECM form transient nascent adhesions, whereas those that interact with a stiff ECM assemble stable FAs. Cells interacting with a stiff ECM assemble FAs because they are able to stimulate contractile forces through the actomyosin machinery that orchestrate the recruitment and favor the unfolding of the integrin-associated adhesion plaque proteins talin and vinculin to assemble into larger adhesion complexes. Indeed, the size of the FAs assembled in cells scales with the level of rigidity.[31] Once stabilized, integrin FAs recruit focal adhesion kinase (FAK) and other plaque proteins, including paxillin, that nucleate additional signaling molecules and exclude inhibitors, such as p190RhoGAP, to favor the activation of Rho-ROCK (Rho Kinase) GTPases.[32] Rho is a member of a family of small GTPases that activate ROCK. Cells with activated ROCK show increased myosin light chain phosphorylation that promotes actin stress fiber assembly and myosin dipoles that stimulate contractile forces within the actomyosin cytoskeleton. By these means, a stiffened tumor stroma tunes the tension of the malignant cell to reinforce integrin adhesion assembly, and this in turn potentiates ECM remodeling, which ultimately further increases intracellular tension via a positive feedback loop that has been termed "mechano-reciprocity."[33]

In a transformed tissue irrespective of whether the elevated tumor cell tension is initiated by oncogenic transformation, a fibrotic stiffened ECM, or aberrant glucose metabolism, tumor cell tension promotes cell proliferation, survival, and migration, thus enabling invasion and malignant transformation. Elevated tumor cell tension can also induce tumor cell transdifferentiation into a mesenchymal-like phenotype that then promotes tumor progression and aggression.[34] For example, when nonmalignant mammary epithelial cells (MECs) are cultured on or within a soft extracellular matrix (with a Young's modulus of ~400 Pa, a physical, mechanical measure of material stiffness) that exhibits a rigidity that is similar to that of a normal breast stroma, the cells form normal, growth-arrested polarized acini (milk ducts) with a central lumen. These "differentiated" acini are characterized by stable adherens cell-cell junctions and an assembled endogenous laminin-rich basement membrane.[35] By contrast, if the nonmalignant MECs are cultured either on or within a stiffer "tumor-like" ECM (with a Young's modulus of ~5 kPa), they form continuously growing, disorganized cellular aggregates with loose cell-cell adherens junctions and prominent FAs characterized by activated FAK (^{p397}FAK), vinculin, and high extracellular signal-regulated kinase (ERK) activity—similar to a cancerous tissue (see figure 11.1). However, consistent with integrin-mediated stimulation of Rho-dependent cytoskeletal contractility, acini morphogenesis and growth arrest can also be induced in MECs grown within a stiffer ECM if their Rho/Rho-associated protein kinase (ROCK) activity is experimentally inhibited. Importantly, tumorigenic derivatives of these nonmalignant MECs show amplified epidermal growth factor receptor–mediated activation of ERK and elevated Rho-ROCK activity and consequently high actomyosin contractility. When these transformed MECs are embedded within a soft ECM, they form continuously growing, disorganized, and invasive colonies and display prominent [p397]FAK and vinculin-positive FAs. Yet, the malignant phenotype of these tumor colonies can be reverted toward "normal-like acini" if the tumor cell actomyosin contractility is reduced by treatment with a ROCK inhibitor. By these molecular mechanisms, either oncogene signaling *or* a stiffened tissue stroma, induced by activated contractile CAFs, chronic inflammation, or metabolically dysfunctional cells in the tissue, increases actomyosin-generated tumor cell tension to promote the malignant behavior of the tissue.[36] Accordingly, the malignant phenotype of a tissue is functionally linked to an altered cellular tensional homeostasis. Thus, the perturbed tensional homeostasis found in tumors is mediated through cell-ECM interactions and altered actomyosin contractility that is triggered not only by oncogenic transformation but also by modifications in cellular metabolism or a pathologically stiffened ECM.

11.3   Tissue Tension Induces Dysregulation of Cellular Metabolism in Cancer

In response to the increased metabolic demands associated with proliferation and migration, and as an adaptation to their stiffened stroma, cancer cells switch from oxidative phosphorylation (OXPHOS) to aerobic glycolysis (fermentation of sugars in the presence of oxygen, although fermentation is a process normally only used by cells in the absence of oxygen). The switching of cancer cells to this less efficient metabolic pathway is termed the Warburg effect.[37] For instance, Madin-Darby canine kidney cells increase their aerobic glycolysis when seeded on a rigid ECM but prefer OXPHOS when grown within a compliant, soft collagen gel.[38] Consistently, both nonmalignant and malignant cells can, with some exceptions, reduce their oxygen consumption and OXPHOS-mediated glucose metabolism and revert to glutamine-fueled tricarboxylic acid cycle activity when interacting with a stiff ECM.[39] Interestingly, the switch from OXPHOS respiration to aerobic glycolysis scales with substrate stiffness and depends upon elevated actomyosin activity, because OXPHOS respiration can be recovered in cells interacting with a stiff ECM by inhibiting their contractility.[40]

Tumor cells that have undergone the Warburg switch alter their metabolism so that they ferment glucose in the presence of oxygen to produce adenosine triphosphate (ATP) but do so less efficiently than their healthy counterparts, likely due to mitochondrial dysfunction.[41] Indeed, an abnormally elevated mitochondrial membrane potential correlates with a higher glycolytic rate and increased reactive oxygen species (ROS) production in some cancers[42] and thus may enhance tumor cell survival under stress conditions. The mitochondrial dysfunction exhibited by tumor cells, mediated in part by a hostile mechanically challenged microenvironment, drives changes in cellular respiration and energy biosynthesis, leading to excessive production of ROS and the accumulation of mitochondrial metabolites, such as fumarate and succinate (see figure 11.1). In addition, modifications in the permeability of the outer and inner membranes of the mitochondria influence the epigenetic landscape regulating gene expression by allosterically altering the function of Jumonji demethylases.[43]

Cells exhibiting sustained aerobic glycolysis show an altered NAD/NADH ratio since the lactate dehydrogenase–mediated conversion of pyruvate to lactate requires NAD as a cofactor. NAD is a critical cofactor for polyadenosine diphosphate (ADP-ribose) polymerases (PARPs) and sirtuins, which are important components of DNA damage repair pathways and regulate proteome-wide acetylation patterns that influence the epigenome (via histone modifications) and cytoskeletal stability (via microtubules). Consistently, cancers show dramatic aberrations in their NAD to NADH ratio,[44] which potentially modulate the rate of nucleotide biosynthesis to favor proliferation (see figure 11.1).[45]

The NAD-dependent function of SIRT1 is linked to the induction of the epithelial-mesenchymal transition (EMT),[46] which not only is a feature of many cancers but also is promoted by a stiff, hypoxic microenvironment. Moreover, both SIRT1 and SIRT6 have been implicated in the NAD-fueled maintenance of genomic stability,[47] implying that aerobic glycolysis may promote the accrual of genetic mutations. Indeed, during adipogenesis, the pool of available NAD+ is depleted, either because it is used as a metabolic substrate or donor NAD+ accessibility is limited because it has been diverted to modulate gene expression. NAD+ usage depends upon whether it is in the nucleus, where it can be acted on by the activity of nicotinamide mononucleotide adenylyl transferase (NMNAT) 1, or in the cytosol, where it is modified by NMNAT2. Experiments with human neuro-blastoma cells suggest that cancer cells preferentially use one NMNAT enzyme over the other to meet their metabolic needs, particularly given that glucose starvation decreases expression of the NMNAT2 protein and increases the activity of PARP-1.[48] The ratio of NAD+/NADH can also modulate tumor behavior as revealed by studies in metastatic breast cancer cells where higher levels of NAD+ that tipped the balance of NAD+ to NADH were able to inhibit their growth and dissemination.[49] Importantly, although metabolic rewiring of tumors has been attributed primarily to oncogenic transformation (e.g., Ras, Myc), elevated cell tension induced by a stiffened ECM or aberrant glucose metabolism similarly elicits many of these metabolic changes (e.g., an altered ratio of NADH+/NADH), underscoring the importance of stromal-epithelial interactions in tumorigenesis.

11.4   Cellular Nutrient Sensing Regulates Expression of the Malignant Phenotype

Cells sense extracellular and intracellular metabolites through specialized molecules that respond to nutrient fluctuation. The best understood and most studied metabolic sensors are 5′-adenosine monophosphate (AMP)–activated protein kinase (AMPK) and sirtuins.

AMPK is a metabolic sensor that detects changes in the ratio of ADP (adenosine diphosphate)/ATP and AMP/ATP.[50] Cancer cells under metabolic stress, such as those subjected to glucose deprivation or those exposed to hypoxia or oxidative stress, increase the activity of AMPK. Activated AMPK inhibits anabolic pathways that consume ATP and stimulate glucose uptake to increase ATP and activate mitochondrial biogenesis, anaerobic glycolysis, and fatty acid oxidation.[51] AMPK activity also increases the expression of glycolytic enzymes and substrates required for the Warburg effect[52] by inhibiting mTORC1 and stabilizing HIF1α. Consistent with evidence that a fibrotic tumor alters metabolism to promote malignancy, a stiff ECM and elevated cellular tension acutely enhance glucose uptake and promote glycolysis. Interestingly, however, cancer cells express both low and high levels of AMPK, and its activity can also impair cell proliferation. These findings suggest AMPK has a dual role in cancer. On the one hand, AMPK promotes a malignant phenotype, and on the other hand, it can act as a tumor suppressor and counteracts the Warburg effect to help recover energetic balance in the cell.[53] Indeed, in ovarian, lung, and colorectal cancer, low levels of AMPK activity correlate with poor patient prognosis and reduced overall survival.[54] Similarly, in liver cancer, low levels of kinase B1, a key activator of AMPK, correlate positively with tumor aggression and reduced overall patient survival. Reduced AMPK activity also associates with an aggressive clinical diagnosis and poor prognosis in cancer patients. Consistently, AMPK activity can counteract the metastatic potential of melanoma cells. Whether a chronic elevation of tissue tension ultimately compromises AMPK activity to drive the malignant behavior of a tissue remains unclear.

Sirtuins, also termed class III histone deacetylases, are enzymes that integrate metabolic cues from the environment and employ NAD+ as a cofactor. Sirtuins comprise a family of SIRT 1–7 proteins distinguished according to their specificity and catalytic activity.[55] Sirtuins regulate many cellular processes but are best known for their effects on cellular metabolism and gene transcription.[56] SIRT1 is the most studied and best understood sirtuin. SIRT1 is localized primarily in the nucleus, where it deacetylates lysine residues in histone and nonhistone targets, such as p53, PGC-1α, or FOXO.[57] SIRT1 also stabilizes HIF2α in response to hypoxia to inhibit OXPHOS and promote aerobic glycolysis. SIRT1 can also bind and deacetylate HIF1α to stabilize HIF1α and enhance the expression of HIF1α target genes such as GLUT that sustain tumor glycolysis.[58]

Not surprisingly given their central role in cellular metabolism and gene expression, sirtuins are often overexpressed in tumors where they are implicated in cancer aggression (see figure 11.1). For instance, SIRT1 is implicated in EMT in melanoma cells. By contrast, the loss of SIRT1 has also been linked to oral squamous cell and breast carcinoma metastasis.[59] Indeed, SIRT3 can function as a tumor suppressor by reducing ROS to maintain genomic integrity.[60] Consistently, many breast cancers show significant reductions in SIRT3, and loss of SIRT3 enhances the transcription of and stabilizes HIF1α protein.[61] Loss of SIRT6 can also stabilize HIF1α and promote aerobic glycolysis by fostering expression of GLUT1.[62] Interestingly, inhibiting SIRT1 and SIRT3 in lung and breast cancer cells arrested their growth, and disrupting SIRT1 activity in leukemia cells blocked their proliferation and promoted apoptosis by activating the tumor suppressor p53 axis.[63] Mechanistically, the aberrant tumor-associated expression of sirtuins, which are the key sensors of cellular NAD, compromises the normal, physiological histone acetylation and deacetylation balance in cells altering their gene expression. Thus, chronically aberrant levels of histone acetylation/deacetylation ultimately favor epigenetic chromatin modifications that drive malignancy and tumor aggression because they regulate the expression of key genes implicated in cell growth, survival, migration, and transdifferentiation. For instance, the increased mitochondrial NAD+ levels associated with elevated SIRT3 activity alter the tumorigenic potential of glioma tumor-initiating cells.[64] Increased expression and activity of the master oncogene c-Myc also trigger a positive feedback loop between nicotinamide phosphoribosyltransferase (which controls NAD levels), DBC1 (inhibitor of SIRT1), and SIRT1, which promotes and sustains tumor development.[65] These studies and others have stimulated the development of therapeutic anticancer treatments that target sirtuins and normalize NAD levels. For instance, building on observations that an increase in the pool of NAD+ favors the survival of healthy cells but compromises the viability of premalignant tumors, selective inhibition of NAD+ synthesis promoted apoptosis in human liver carcinoma cells.[66] Likewise, combined SIRT1 and HDAC inhibitors promote apoptosis in leukemia cells.[67] Although the clinical efficacy of these compounds is still being optimized, the sirtuin inhibitor nicotinamide blocked cell proliferation and induced apoptosis in leukemia and human prostate cancers, even at micromolar concentrations.[68] These findings suggest that new anticancer therapeutic opportunities that are aimed at overriding the metabolic impact of a mechanically challenged microenvironment reside in altering both NAD+ levels and targeting sirtuin activity.

11.5 Cell Tension and Metabolism-Sensing Pathways Interact with Each Other and Modulate Chromatin to Regulate Gene Expression

In tumors, the rigid ECM and restricted nutrient availability increase cellular mechanical tension and induce metabolic stress promoting a malignant phenotype. Tissue tension promotes the malignant behavior of cells by activating signaling pathways that enhance cell growth, survival, and migration and by altering the expression of key oncogenes and tumor suppressors. Importantly, tissue tension also modulates cellular metabolism to regulate chromatin organization and gene expression that also promote cell growth, survival, and motility.

Phosphatase and tensin homolog (PTEN) is the master negative regulator of the phosphatidylinositol 3-kinase (PI3K) pathway that is the main pathway relaying nutritional input to drive either proliferation or quiescence. Not surprisingly, loss of PTEN levels and/ or function promotes the malignant transformation of many tissues. PTEN function can be perturbed on multiple levels, including through germline and somatic mutations, genomic deletions, epigenetic and transcriptional silencing, posttranscriptional or posttranslational regulation, and changes in protein-protein interaction.[69] However, while PTEN somatic mutations occur in a large percentage of human cancers, with the highest frequency in the endometrium, breast, central nervous system, skin, and prostate, high tissue tension can also increase PI3K activity either by potentiating growth factor receptor signaling or by reducing PTEN expression through elevated myc and β-catenin activity and altered microRNA expression.[70]

PTEN is a major tumor suppressor because it regulates the activity of mechanistic target of rapamycin (mTOR). mTOR is a serine/threonine kinase within the PI3K family and acts as a metabolic rheostat to regulate cell growth, survival, and proliferation via the activity of mTORC1 and mTORC2.[71] mTOR participates in the PI3K-Akt-mTOR axis. Activated PI3K phosphorylates and primes Akt and mTORC2, and other kinases thereafter complete its activation. Not surprisingly, mTOR activity is stimulated by a myriad of intra- and extracellular metabolites, hormones, growth factors, and cytokines.[72] Moreover, given the central role mTOR plays in regulating cell growth and survival, many cancers exhibit hyperactivated mTOR. Indeed, elevated mTOR signaling increases expression of genes implicated in glycolysis, as well as oncogenesis, including c-Myc, and its loss destabilizes HIF1α to reduce glucose transport and glycolysis.[73]

mTORC1 also regulates the transcription of genes directly involved in glycolysis and chromatin remodeling.[74] For example, Akt has a large number of downstream effectors, such as CREB or FOXO. CREB targets PGC-1α, which is a transcriptional coactivator that interacts with PPAR-γ. CREB is implicated in breast tumorigenesis via the PKC/CREB/PGC-1α complex because it regulates lipid metabolism.[75] Interestingly, mTOR-dependent signaling can also regulate ECM synthesis, as was observed in mesangial cells with high levels of Akt that they also expressed and deposited fibronectin at higher levels.[76] Consistently, compromising AMPK-mediated mitochondrial levels and reducing ATP levels suppressed mTOR activity and reduced the expression of Col1α1 and fibronectin in hepatic stellate cells. In systemic sclerosis, patients exhibit high mTOR activity and produce high quantities of type I collagen.[77] Exposure to high glucose can also influence the posttranslational modification of ECM proteins by altering DNA methylation and the expression of genes that encode proteins involved in fibrillar collagen biosynthesis, folding, and assembly.[78]

A stiff ECM can regulate PTEN and mTOR-Akt signaling by modulating the activity of oncogenes, including myc and β-catenin, as well as the levels of key microRNAs (miRNAs). MicroRNAs are posttranscriptional regulators of gene expression and can modulate levels of critical oncogenes and tumor suppressors.[79] A stiff ECM increases Akt signaling by reducing PTEN. A stiffened ECM and/or elevated integrin signaling increases c-Myc and activates Wnt, which elevates levels of oncogenic miR-18a, leading to a reduction of PTEN. Breast cancer patients expressing high levels of miR-18a also showed significantly lower PTEN that correlated significantly with higher ECM stiffness.[80] In addition, PTEN translation can be suppressed by Yes-associated protein (YAP),[81] a key downstream effector of the Hippo pathway. Interestingly, a stiff ECM induces YAP transcription, suggesting tension-mediated PTEN loss could also be mediated through YAP, a finding consistent with its increased nuclear localization and established role in malignancy.[82] Not surprisingly, recent data implicate cellular metabolism in YAP-mediated transcription. For instance, aerobic glycolysis promotes nuclear YAP localization, whereas AMPK-mediated phosphorylation of YAP at Ser94 impairs YAP-TEAD interactions, which inhibits its transcriptional activity.[83] However, YAP can also regulate cellular metabolism as indicated by its ability to promote the transcription of GLUT3 via interacting with TEAD.[84] Interestingly, during liver fibrosis, YAP/TAZ regulates glutaminolysis and modulates the shift to aerobic glycolysis in activated hepatic stellate cells.[85]

Recent evidence suggests that PTEN and Wnt/β-catenin signaling interact and promote malignancy by modulating cellular metabolism. For instance, reducing β-catenin levels in PTEN-positive melanoma cells expressing Wnt3a ligand increased lactate secretion, implying this shifted the cells from OXPHOS to aerobic glycolysis. PTEN-positive cells in which Wnt3a was hyperactivated also showed an increase in the mitochondrial profusion proteins, mitofusin 1–2 and optic atrophy 1, and decreased levels of the profission protein dynamin 1–like-protein, which tracked with changes in mitochondrial morphology consistent with metabolic dysfunction. Nuclear β-catenin also interacts with transcription factors that regulate autophagy to alter cellular metabolism, including p62.[86] Consistently, Wnt5a expression in melanoma cells enhanced Akt/mTOR signaling and led to a post-transcriptional increase in lactate dehydrogenase, reflecting higher anaerobic glycolysis. Conversely, Wnt5a increased OXPHOS in metastatic breast cancer cells.[87]

Twist1 overexpression is frequently observed in many cancers, including those of the breast, bladder, and stomach, or in glioblastoma. Twist1 is a transcription factor implicated in EMT and has been found associated with tumor initiation, aggression, and metastasis.[88] In experimental models, a stiff ECM induced an EMT by enhancing the phosphorylation and nuclear translocation of Twist at Tyr107. The ability of a stiff ECM to induce Twist-dependent EMT depended upon integrin engagement and was accompanied by elevated matrix metallopeptidases (MMP), collagen, and LOX levels and tumor metastasis.[89] Importantly, Twist can also induce metabolic reprogramming by activating the β1-integrin/FAK/PI3K/AKT/mTOR axis and by suppressing p53. For instance, malignant and non-malignant MECs engineered to overexpress Twist showed increased glucose uptake, higher lactate secretion, and reduced mitochondrial mass that was accompanied by an upregulation of PKM2, LDH-A, and G6PD.[90] These findings underscore the interplay between cell-ECM adhesions, cellular metabolism, and gene transcription and raise the intriguing possibility that a stiffened ECM and elevated tension per se may be sufficient to drive many of the aberrant behaviors that characterize tumors.

11.6   Tension and Metabolism Modify Chromatin Accessibility through Epigenetic Alterations

Accumulating evidence suggest that the nucleus may function as a mechano-transducer (see figure 11.1).[91] For instance, intracellular/cytoskeletal molecules and cell-cell adhesions can directly transmit force to the nucleus through linker of nucleoskeleton and cytoskeleton proteins (LINCs), including nesprins and inner nuclear membrane proteins, such as SUNs.[92] Interestingly, it was recently shown that in response to rigidity, extracellular forces are transmitted to the nucleus through the continuous chain of molecular links between ECM-integrins-talin and the LINC complex.[93] Nesprins bind actin filaments, as well as kinesin and dynein proteins,[94] while SUN proteins directly interact with nuclear lamin proteins, chromatin binding proteins, and proteins that constitute the nuclear pore complex.[95]

The main structural components of the nuclear lamina are the lamins, which belong to the type V family of intermediate filament proteins. Lamins interact with chromatin. In the nucleus, the chromatin is tightly packed into nucleosomes, which consist of histones and DNA that become organized into nucleosomes by forming a histone octamer core around which the DNA is wrapped. Chromatin is classified as heterochromatin or euchromatin, according to its degree of histone-dependent compaction. Heterochromatin is the condensed state, which is associated with lower chromatin accessibility as compared to its decondensed state, which is called euchromatin, and is characterized by a greater degree of accessibility.[96] Indeed, decondensation of chromatin promotes DNA repair.[97]

Chromatin and/or histones are subjected to a myriad of posttranslational modifications,[98] including DNA methylation and histone acetylation, methylation, phosphorylation, and ubiquitination. Growth factors, hormones, and cytokines also stimulate pathways that activate transcription factors that in turn recruit chromatin-modifying enzymes to induce chromatin remodeling.[99] The nature and level of chromatin posttranslational modification and its association with structural elements, including the nuclear lamins and chromatin binding proteins, influence the ratio of heterochromatin versus euchromatin, and this in turn dictates the recruitment of cofactors and transcription factors that regulate gene transcription to modulate cell behavior.

DNA methylation and histone lysine methylation, which are mediated by DNA methyltransferases (DNMTs), promote transcriptional repression,[100] whereas histone acetylation permits gene expression.[101] DNA methylation is implicated in tumor initiation, as was shown for the tumor suppressor p16, whose methylation was increased very early during breast cancer.[102] Intracellular acetyl-CoA is the main molecule that is responsible for histone acetylation, so that elevated glycolysis that increases acetyl-CoA levels leads to an increase in acetylated histones. For example, pancreatic cancer cells exposed to high levels of glucose showed increased levels of DNA methylation.[103] Similarly, colorectal cancer cells showed decreased global acetylated histone H3K27 that positively correlated with a dose-dependent glucose uptake and that was normalized by glucose starvation. Glycolytic metabolites can also alter histone modifications such that the pyruvate and lactate generated during aerobic glycolysis inhibit histone deacetylases (HDACs) to favor histone acetylation. Consistently, a reduced rate of glycolysis correlated with a decrease in acetyl-CoA and NAD+ and an increase in the size and amount of condensed histone-H3 structures at the nuclear periphery in lung carcinoma cells. By this means, an altered tumor-associated cellular metabolism can exert a profound effect on chromatin organization and gene transcription.

The nucleus of cancer cells typically demonstrates characteristic alterations in morphology that are indicative of altered elasticity. For instance, prostate cancer cells, melanoma cells, and breast tumor cells all have lower levels of lamin A/C and consequently show reduced nuclear stiffness that likely accounts for their variable nuclear size and morphology. The increased compliance of the nucleus in tumor cells also facilitates their migration through confined ECMs, including the stiffened, fibrotic interstitial stroma, but concomitantly compromises their ability to withstand mechanical stress.[104] For instance, chromatin structure regulates small, intranuclear deformations/forces, whereas lamins A and C maintain the overall integrity of the nucleus, particularly in response to larger forces.[105] Alterations in nuclear stiffness likely reflect differences in chromatin condensation and chromatin accessibility that enhance gene transcription, whereas larger morphological changes of the nucleus regulate the transport of key transcription factors across the nuclear envelope.[106]

Given links between ECM stiffness and altered gene expression, it is not surprising that cells also modulate their nuclear morphology, area and volume, chromatin organization, and mechanical properties in response to the stiffness of their ECM. For instance, a stiff ECM (>50 kPa) can enlarge nuclear area, decondense chromatin, and increase the overall level of histone H3 acetylation.[107] Chromosome territories and gene expression are also affected by the rigidity of the ECM, as illustrated by the downregulation and diffuse redistribution of histone H3K4me3-modified chromatin that has been associated with transcriptional activity. These changes in chromosomal territories, which are accompanied by a redistribution of H3K27me3 chromatin to the nuclear interior, are consistent with transcriptional repression. In experimental studies in isolated cells, a stiff ECM induced the mislocalization of chromosomes 18 and 19, which in turn affected gene expression. Importantly, in these cells, chromosome 18 was restored to its normal localization at the nuclear periphery by overexpressing lamin B2, but this effect was prevented by the tyr99 phosphorylation of emerin, thereby implicating the LINC complex in this regulation.[108] Consistent with this possibility, lamin B expression is also downregulated in mouse embryonic fibroblasts plated on a soft substrate but only when the LINC complex is compromised.[109]

Clearly, there is compelling evidence to argue that cells can sense and respond to the stiffness of their ECM and that they transmit these cues to the nucleus to alter their shape and modify their chromatin accessibility either directly or indirectly by modulating cellular metabolism. What has yet to be determined is whether these tension-induced changes in chromatin modification and chromosomal localization are accompanied by specific differences in gene expression and whether altering the metabolic state of the cell could modify these phenotypes. Moreover, whether similar effects occur in fibrotic, stiffened tumor tissues and if this influences gene expression to drive a tumor-like behavior in the cells and tissue remain unclear.

11.7   Conclusions

Over the past several decades, the somatic mutation theory of cancer, which posits that cancer initiation and progression depend upon the cumulative acquisition of critical genetic modifications, has dominated the field of cancer research. The somatic mutation theory is based on genomic analyses of primary cancers that revealed that tumors harbor multiple genetic mutations, deletions, and amplifications and on epidemiological data that showed that patients with hereditary mutations have a higher incidence of cancer.[110] The credibility of this perspective has been bolstered by experimental models that have causally implicated genetic alterations in malignant transformation and tumor progression.[111] Nevertheless, and importantly, cancer develops within the context of a tissue that is composed of a cellular (stromal fibroblasts, vascular and lymphatic endothelial cells, resident and infiltrating immune cells, adipocytes) and a noncellular stroma (ECM, soluble factors, extracellular vesicles).[112] The stroma is progressively modified during malignancy, and experimental models have demonstrated that these stromal changes collaborate with the genetically modified cells within the tissue to promote its transformation and metastasis. For instance, malignant transformation is fostered by tissue injury and fibrosis, and metastasis is linked to compromised antitumor immunity.[113] Similarly, tumorigenesis is accompanied by ECM remodeling and stiffening, and experimental data have causally implicated ECM stiffness and collagen crosslinking in malignant transformation and tumor metastasis.[114] Accordingly, the somatic mutation theory of cancer has been expanded to incorporate the stroma as a complicit participant in cancer, which operates by acting as a critical tumor promoter.

Accumulating evidence suggests the stroma likely also initiates cancer by inducing protumorigenic genetic modifications. For instance, chronically inflamed tissues are predisposed to transformation,[115] and anti-inflammatory treatments can significantly reduce the incidence of these cancers.[116] Chronic inflammation induces fibrosis, and fibrotic tissues, including fibrotic livers, have an increased overall risk of malignancy.[117] A fibrotic tissue is stiffer and has a compromised tissue phenotype and elevated cell proliferation reminiscent of transformed tissue. Moreover, conditions such as idiopathic pulmonary fibrosis and hereditary systemic scleroderma are characterized by a stiffer ECM that not only compromises their tissue structure function and induces the aberrant proliferation of cells in these tissues but also increases the risk of developing small cell lung cancer and melanoma, respectively.[118] Importantly, in addition to being stiffer, fibrotic tissues are chronically inflamed.[119] Chronic inflammation is linked to elevated tissue ROS and altered cellular metabolism, both of which have been implicated in generating genetic mutations that promote malignancy.[120] Consistently, transgenic mice engineered to express myeloid cells with elevated ROS acquire genetic mutations that drive colon transformation. This malignant progression could be prevented by reducing mitochondrial stress and by normalizing tissue metabolism.[121] Thus, the ECM stiffening and metabolic alterations stimulated by chronic inflammation can also induce genomic alterations that can initiate cancer, thereby significantly expanding the role of the stroma in malignancy from a promoter to an initiator.

High tissue tension or a stiff ECM can also stimulate epigenetic alterations in chromatin that induce sustained changes in gene expression that modify stem cell fate or that promote tumor aggression by driving an EMT.[122] Thus, high tissue tension enhances glucose transport and decreases OXPHOS respiration,[123] which, in turn, can alter the ratio of NAD/NADH to influence chromatin acetylation. Elevated cellular glucose can also stimulate protein glycosylation[124] that increases the cellular glycocalyx and that can lock the cell into a mechanically activated state characterized by a mesenchymal-like phenotype with its associated enhanced migration and tumor-initiating potential, as was observed in aggressive human glioblastomas subjected to chronically elevated mechanical stress.[125] Accordingly, if tissue tension and metabolic rewiring can change chromatin organization to induce sustained changes in gene expression, then a pathologically stiffened ECM could also drive expression of the malignant phenotype and malignant progression, even in the absence of an initiating genetic modification. This newly proposed paradigm would offer a unique perspective of malignancy that would challenge the current somatic mutation theory of cancer. Clearly, further studies aimed at investigating the triumvirate of ECM stiffness, metabolic rewiring, and the consequent genetic and epigenetic reprogramming induced by these states will provide a deeper insight into the complexities underlying carcinogenesis.

Acknowledgments

We thank K. Tharp and J. Northey for constructive scientific discussions and feedback. We apologize to all colleagues whose work was not cited owing to space limitations. This work was supported by funds from the Department of Defense grant BCRP BC122990 and U.S. National Institutes of Health NCI R01 grants CA222508–01, CA192914, CA174929, CA08592, U01 grant CA202241, and U54 grant CA163155.

Notes

1. Gorfinkiel, N., Blanchard, G.B., Adams, R.J. & Martinez Arias, A. Mechanical control of global cell behaviour during dorsal closure in Drosophila. *Development* 136, 1889–1898 (2009). Barriga, E.H., Franze, K., Charras, G. & Mayor, R. Tissue stiffening coordinates morphogenesis by triggering collective cell migration in vivo. *Nature* 554, 523–527 (2018). Engler, A.J., Sen, S., Sweeney, H.L. & Discher, D.E. Matrix elasticity directs stem cell lineage specification. *Cell* 126, 677–689 (2006). Park, J.S. et al. The effect of matrix stiffness on the differentiation of mesenchymal stem cells in response to TGF-beta. *Biomaterials* 32, 3921–3930 (2011). Fenteany, G., Janmey, P.A. & Stossel, T.P. Signaling pathways and cell mechanics involved in wound closure by epithelial cell sheets. *Curr Biol* 10, 831–838 (2000). Paszek, M.J. et al. Tensional homeostasis and the malignant phenotype. *Cancer Cell* 8, 241–254 (2005).

2. Nieborak, A. & Schneider, R. Metabolic intermediates—cellular messengers talking to chromatin modifiers. *Mol Metab* 14, 39–52 (2018).

3. Lukashev, M.E. & Werb, Z. ECM signalling: orchestrating cell behaviour and misbehaviour. *Trends Cell Biol* 8, 437–441 (1998). Bonnans, C., Chou, J. & Werb, Z. Remodelling the extracellular matrix in development and disease. *Nat Rev Mol Cell Biol* 15, 786–801 (2014). Lu, P., Weaver, V.M. & Werb, Z. The extracellular matrix: a dynamic niche in cancer progression. *J Cell Biol* 196, 395–406 (2012).

4. Dvorak, H.F. Tumors: wounds that do not heal-redux. *Cancer Immunol Res* 3, 1–11 (2015).

5. Tao, L., Huang, G., Song, H., Chen, Y. & Chen, L. Cancer associated fibroblasts: an essential role in the tumor microenvironment. *Oncol Lett* 14, 2611–2620 (2017). Mishra, P.J. et al. Carcinoma-associated fibroblast-like differentiation of human mesenchymal stem cells. *Cancer Res* 68, 4331–4339 (2008).

6. Santi, A., Kugeratski, F.G. & Zanivan, S. Cancer associated fibroblasts: the architects of stroma remodeling. *Proteomics* 18, e1700167 (2018).

7. Provenzano, P.P. et al. Collagen reorganization at the tumor-stromal interface facilitates local invasion. *BMC Med* 4, 38 (2006).

8. Wolf, K. & Friedl, P. Extracellular matrix determinants of proteolytic and non-proteolytic cell migration. *Trends Cell Biol* 21, 736–744 (2011). Wolf, K. et al. Collagen-based cell migration models in vitro and in vivo. *Semin Cell Dev Biol* 20, 931–941 (2009).

9. Acerbi, I. et al. Human breast cancer invasion and aggression correlates with ECM stiffening and immune cell infiltration. *Integr Biol (Camb)* 7, 1120–1134 (2015). Laklai, H. et al. Genotype tunes pancreatic ductal adenocarcinoma tissue tension to induce matricellular fibrosis and tumor progression. *Nat Med* 22, 497–505 (2016).

10. Alberts, B. et al. *Molecular Biology of the Cell, Sixth Edition*. (W.W. Norton & Company, 2014).

11. Nia, H.T. et al. Solid stress and elastic energy as measures of tumour mechanopathology. *Nat Biomed Eng* 1, 0004 (2016). Kharaishvili, G. et al. The role of cancer-associated fibroblasts, solid stress and other microenvironmental factors in tumor progression and therapy resistance. *Cancer Cell Int* 14, 41 (2014).

12. Bellamy, G. & Bornstein, P. Evidence for procollagen, a biosynthetic precursors of collagen. *Proc Natl Acad Sci USA* 68, 1138–1142 (1971).

13. Yamauchi, M. & Sricholpech, M. Lysine post-translational modifications of collagen. *Essays Biochem* 52, 113–133 (2012).

14. Ioachim, E. et al. Immunohistochemical expression of extracellular matrix components tenascin, fibronectin, collagen type IV and laminin in breast cancer: their prognostic value and role in tumour invasion and progression. *Eur J Cancer* 38, 2362–2370 (2002). Yu, M.K., Park, J. & Jon, S. Targeting strategies for multifunctional nanoparticles in cancer imaging and therapy. *Theranostics* 2, 3–44 (2012). Kubow, K.E. et al. Mechanical forces regulate the interactions of fibronectin and collagen I in extracellular matrix. *Nat Commun* 6, 8026 (2015). Harburger, D.S. & Calderwood, D.A. Integrin signalling at a glance. *J Cell Sci* 122, 159–163 (2009). Desgrosellier, J.S. & Cheresh, D.A. Integrins in cancer: biological implications and therapeutic opportunities. *Nat Rev Cancer* 10, 9–22 (2010).

15. Wei, B. et al. Human colorectal cancer progression correlates with LOX-induced ECM stiffening. *Int J Biol Sci* 13, 1450–1457 (2017).

16. Siddikuzzaman, Grace, V.M. & Guruvayoorappan, C. Lysyl oxidase: a potential target for cancer therapy. *Inflammopharmacology* 19, 117–129 (2011).

17. Acerbi, I. et al. Human breast cancer invasion and aggression correlates with ECM stiffening and immune cell infiltration. *Integr Biol (Camb)* 7, 1120–1134 (2015). Laklai, H. et al. Genotype tunes pancreatic ductal adenocarcinoma tissue tension to induce matricellular fibrosis and tumor progression. *Nat Med* 22, 497–505 (2016). Levental, K.R. et al. Matrix crosslinking forces tumor progression by enhancing integrin signaling. *Cell* 139, 891–906 (2009). Plodinec, M. et al. The nanomechanical signature of breast cancer. *Nat Nanotechnol* 7, 757–765 (2012). Bordeleau, F. et al. Tissue stiffness regulates serine/arginine-rich protein-mediated splicing of the extra domain B-fibronectin isoform in tumors. *Proc Natl Acad Sci USA* 112, 8314–8319 (2015). Wang, K. et al. Breast cancer cells alter the dynamics of stromal fibronectin-collagen interactions. *Matrix Biol* 60–61, 86–95 (2017).

18. Paszek, M.J. et al. Tensional homeostasis and the malignant phenotype. *Cancer Cell* 8, 241–254 (2005). Provenzano, P.P. et al. Collagen reorganization at the tumor-stromal interface facilitates local invasion. *BMC Med* 4, 38 (2006). Laklai, H. et al. Genotype tunes pancreatic ductal adenocarcinoma tissue tension to induce matricellular fibrosis and tumor progression. *Nat Med* 22, 497–505 (2016). Nia, H.T. et al. Solid stress and elastic energy as measures of tumour mechanopathology. *Nat Biomed Eng* 1, 0004 (2016). Levental, K.R. et al. Matrix crosslinking forces tumor progression by enhancing integrin signaling. *Cell* 139, 891–906 (2009).

19. Ioachim, E. et al. Immunohistochemical expression of extracellular matrix components tenascin, fibronectin, collagen type IV and laminin in breast cancer: their prognostic value and role in tumour invasion and progression. *Eur J Cancer* 38, 2362–2370 (2002). Huang, S.P. et al. Over-expression of lysyl oxidase is associated with poor prognosis and response to therapy of patients with lower grade gliomas. *Biochem Biophys Res Commun* 501, 619–627 (2018).

20. Plodinec, M. et al. The nanomechanical signature of breast cancer. *Nat Nanotechnol* 7, 757–765 (2012). Kalli, M. & Stylianopoulos, T. Defining the role of solid stress and matrix stiffness in cancer cell proliferation and metastasis. *Front Oncol* 8, 55 (2018). Stylianopoulos, T., Munn, L.L. & Jain, R.K. Reengineering the physical microenvironment of tumors to improve drug delivery and efficacy: from mathematical modeling to bench to bedside. *Trends Cancer* 4, 292–319 (2018). Provenzano, P.P. & Hingorani, S.R. Hyaluronan, fluid pressure, and stromal resistance in pancreas cancer. *Br J Cancer* 108, 1–8 (2013).

21. Conklin, M.W. et al. Collagen alignment as a predictor of recurrence after ductal carcinoma in situ. *Cancer Epidemiol Biomarkers Prev* 27, 138–145 (2018). Riching, K.M. et al. 3D collagen alignment limits protrusions

to enhance breast cancer cell persistence. *Biophys J* 107, 2546–2558 (2014). Burke, K., Tang, P. & Brown, E. Second harmonic generation reveals matrix alterations during breast tumor progression. *J Biomed Opt* 18, 31106 (2013). Grossman, M. et al. Tumor cell invasion can be blocked by modulators of collagen fibril alignment that control assembly of the extracellular matrix. *Cancer Res* 76, 4249–4258 (2016).

22. Motte, S. & Kaufman, L.J. Strain stiffening in collagen I networks. *Biopolymers* 99, 35–46 (2013). Han, W. et al. Oriented collagen fibers direct tumor cell intravasation. *Proc Natl Acad Sci USA* 113, 11208–11213 (2016). van Helvert, S. & Friedl, P. Strain stiffening of fibrillar collagen during individual and collective cell migration identified by AFM nanoindentation. *ACS Appl Mater Interfaces* 8, 21946–21955 (2016).

23. Provenzano, P.P. et al. Collagen reorganization at the tumor-stromal interface facilitates local invasion. *BMC Med* 4, 38 (2006). Levental, K.R. et al. Matrix crosslinking forces tumor progression by enhancing integrin signaling. *Cell* 139, 891–906 (2009). Mekhdjian, A.H. et al. Integrin-mediated traction force enhances paxillin molecular associations and adhesion dynamics that increase the invasiveness of tumor cells into a three-dimensional extracellular matrix. *Mol Biol Cell* 28, 1467–1488 (2017). Rubashkin, M.G. et al. Force engages vinculin and promotes tumor progression by enhancing PI3K activation of phosphatidylinositol (3,4,5)-triphosphate. *Cancer Res* 74, 4597–4611 (2014).

24. Harney, A.S. et al. Real-time imaging reveals local, transient vascular permeability, and tumor cell intravasation stimulated by TIE2hi macrophage-derived VEGFA. *Cancer Discov* 5, 932–943 (2015).

25. Conklin, M.W. et al. Aligned collagen is a prognostic signature for survival in human breast carcinoma. *Am J Pathol* 178, 1221–1232 (2011).

26. Zhou, Z.H. et al. Reorganized collagen in the tumor microenvironment of gastric cancer and its association with prognosis. *J Cancer* 8, 1466–1476 (2017).

27. Laklai, H. et al. Genotype tunes pancreatic ductal adenocarcinoma tissue tension to induce matricellular fibrosis and tumor progression. *Nat Med* 22, 497–505 (2016). Palumbo, A., Jr., Da Costa Nde, O., Bonamino, M.H., Pinto, L.F. & Nasciutti, L.E. Genetic instability in the tumor microenvironment: a new look at an old neighbor. *Mol Cancer* 14, 145 (2015). Polyak, K., Haviv, I. & Campbell, I.G. Co-evolution of tumor cells and their microenvironment. *Trends Genet* 25, 30–38 (2009). Tsai, J.H. & Yang, J. Epithelial-mesenchymal plasticity in carcinoma metastasis. *Genes Dev* 27, 2192–2206 (2013). Singh, A. & Settleman, J. EMT, cancer stem cells and drug resistance: an emerging axis of evil in the war on cancer. *Oncogene* 29, 4741–4751 (2010). Samuel, M.S. et al. Actomyosin-mediated cellular tension drives increased tissue stiffness and beta-catenin activation to induce epidermal hyperplasia and tumor growth. *Cancer Cell* 19, 776–791 (2011).

28. Laklai, H. et al. Genotype tunes pancreatic ductal adenocarcinoma tissue tension to induce matricellular fibrosis and tumor progression. *Nat Med* 22, 497–505 (2016). Samuel, M.S. et al. Actomyosin-mediated cellular tension drives increased tissue stiffness and beta-catenin activation to induce epidermal hyperplasia and tumor growth. *Cancer Cell* 19, 776–791 (2011).

29. Gaspar, P. & Tapon, N. Sensing the local environment: actin architecture and Hippo signalling. *Curr Opin Cell Biol* 31, 74–83 (2014). Humphrey, J.D., Dufresne, E.R. & Schwartz, M.A. Mechanotransduction and extra-cellular matrix homeostasis. *Nat Rev Mol Cell Biol* 15, 802–812 (2014).

30. Geiger, B., Bershadsky, A., Pankov, R. & Yamada, K.M. Transmembrane crosstalk between the extracellular matrix—cytoskeleton crosstalk. *Nat Rev Mol Cell Biol* 2, 793–805 (2001).

31. Trichet, L. et al. Evidence of a large-scale mechanosensing mechanism for cellular adaptation to substrate stiffness. *Proc Natl Acad Sci USA* 109, 6933–6938 (2012). Elosegui-Artola, A. et al. Mechanical regulation of a molecular clutch defines force transmission and transduction in response to matrix rigidity. *Nat Cell Biol* 18, 540–548 (2016). Changede, R., Xu, X., Margadant, F. & Sheetz, M.P. Nascent integrin adhesions form on all matrix rigidities after integrin activation. *Dev Cell* 35, 614–621 (2015). Chrzanowska-Wodnicka, M. & Burridge, K. Rho-stimulated contractility drives the formation of stress fibers and focal adhesions. *J Cell Biol* 133, 1403–1415 (1996). Gardel, M.L., Schneider, I.C., Aratyn-Schaus, Y. & Waterman, C.M. Mechanical integration of actin and adhesion dynamics in cell migration. *Annu Rev Cell Dev Biol* 26, 315–333 (2010).

32. Mierke, C.T. et al. Focal adhesion kinase activity is required for actomyosin contractility-based invasion of cells into dense 3D matrices. *Sci Rep* 7, 42780 (2017).

33. Paszek, M.J. & Weaver, V.M. The tension mounts: mechanics meets morphogenesis and malignancy. *J Mammary Gland Biol Neoplasia* 9, 325–342 (2004).

34. Laklai, H. et al. Genotype tunes pancreatic ductal adenocarcinoma tissue tension to induce matricellular fibrosis and tumor progression. *Nat Med* 22, 497–505 (2016). Mouw, J.K. et al. Tissue mechanics modulate

microRNA-dependent PTEN expression to regulate malignant progression. *Nat Med* 20, 360–367 (2014). Barnes, J.M. et al. A tension-mediated glycocalyx-integrin feedback loop promotes mesenchymal-like glioblastoma. *Nat Cell Biol* 20, 1203–1214 (2018). Pickup, M.W. et al. Development of aggressive pancreatic ductal adenocarcinomas depends on granulocyte colony stimulating factor secretion in carcinoma cells. *Cancer Immunol Res* 5, 718–729 (2017). Chang, T.T., Thakar, D. & Weaver, V.M. Force-dependent breaching of the basement membrane. *Matrix Biol* 57–58, 178–189 (2017).

35. Paszek, M.J. et al. Tensional homeostasis and the malignant phenotype. *Cancer Cell* 8, 241–254 (2005).

36. Paszek, M.J. et al. Tensional homeostasis and the malignant phenotype. *Cancer Cell* 8, 241–254 (2005). Levental, K.R. et al. Matrix crosslinking forces tumor progression by enhancing integrin signaling. *Cell* 139, 891–906 (2009). Solon, J., Levental, I., Sengupta, K., Georges, P.C. & Janmey, P.A. Fibroblast adaptation and stiffness matching to soft elastic substrates. *Biophys J* 93, 4453–4461 (2007).

37. Warburg, O., Wind, F. & Negelein, E. The metabolism of tumors in the body. *J Gen Physiol* 8, 519–530 (1927).

38. Pampaloni, F., Stelzer, E.H., Leicht, S. & Marcello, M. Madin-Darby canine kidney cells are increased in aerobic glycolysis when cultured on flat and stiff collagen-coated surfaces rather than in physiological 3-D cultures. *Proteomics* 10, 3394–3413 (2010).

39. Morris, B.A. et al. Collagen matrix density drives the metabolic shift in breast cancer cells. *EBioMedicine* 13, 146–156 (2016).

40. Mah, E.J., Lefebvre, A., McGahey, G.E., Yee, A.F. & Digman, M.A. Collagen density modulates triple-negative breast cancer cell metabolism through adhesion-mediated contractility. *Sci Rep* 8, 17094 (2018).

41. Warburg, O., Wind, F. & Negelein, E. The metabolism of tumors in the body. *J Gen Physiol* 8, 519–530 (1927). Vazquez, A., Liu, J., Zhou, Y. & Oltvai, Z.N. Catabolic efficiency of aerobic glycolysis: the Warburg effect revisited. *BMC Syst Biol* 4, 58 (2010).

42. Bonnet, S. et al. A mitochondria-K+ channel axis is suppressed in cancer and its normalization promotes apoptosis and inhibits cancer growth. *Cancer Cell* 11, 37–51 (2007).

43. Porporato, P.E., Filigheddu, N., Pedro, J.M.B., Kroemer, G. & Galluzzi, L. Mitochondrial metabolism and cancer. *Cell Res* 28, 265–280 (2018). Xiao, M. et al. Inhibition of alpha-KG-dependent histone and DNA demethylases by fumarate and succinate that are accumulated in mutations of FH and SDH tumor suppressors. *Genes Dev* 26, 1326–1338 (2012).

44. Moreira, J.D. et al. The redox status of cancer cells supports mechanisms behind the warburg effect. *Metabolites* 6, 33 (2016).

45. Vander Heiden, M.G. & DeBerardinis, R.J. Understanding the intersections between metabolism and cancer biology. *Cell* 168, 657–669 (2017).

46. Hao, C. et al. Overexpression of SIRT1 promotes metastasis through epithelial-mesenchymal transition in hepatocellular carcinoma. *BMC Cancer* 14, 978 (2014).

47. Chalkiadaki, A. & Guarente, L. The multifaceted functions of sirtuins in cancer. *Nat Rev Cancer* 15, 608–624 (2015).

48. Ryu, K.W. et al. Metabolic regulation of transcription through compartmentalized NAD(+) biosynthesis. *Science* 360, 6389 (2018).

49. Santidrian, A.F. et al. Mitochondrial complex I activity and NAD+/NADH balance regulate breast cancer progression. *J Clin Invest* 123, 1068–1081 (2013).

50. Hardie, D.G. AMP-activated protein kinase: an energy sensor that regulates all aspects of cell function. *Genes Dev* 25, 1895–1908 (2011).

51. Grahame Hardie, D. Regulation of AMP-activated protein kinase by natural and synthetic activators. *Acta Pharm Sin B* 6, 1–19 (2016).

52. Agarwal, S., Bell, C.M., Rothbart, S.B. & Moran, R.G. AMP-activated protein kinase (AMPK) control of mTORC1 is p53- and TSC2-independent in pemetrexed-treated carcinoma cells. *J Biol Chem* 290, 27473–27486 (2015).

53. Faubert, B. et al. AMPK is a negative regulator of the Warburg effect and suppresses tumor growth in vivo. *Cell Metab* 17, 113–124 (2013).

54. Zulato, E. et al. Prognostic significance of AMPK activation in advanced stage colorectal cancer treated with chemotherapy plus bevacizumab. *Br J Cancer* 111, 25–32 (2014). Cheng, J. et al. Prognostic significance of AMPK in human malignancies: a meta-analysis. *Oncotarget* 7, 75739–75748 (2016).

55. Vassilopoulos, A., Fritz, K.S., Petersen, D.R. & Gius, D. The human sirtuin family: evolutionary divergences and functions. *Hum Genomics* 5, 485–496 (2011).

56. Imai, S., Armstrong, C.M., Kaeberlein, M. & Guarente, L. Transcriptional silencing and longevity protein Sir2 is an NAD-dependent histone deacetylase. *Nature* 403, 795–800 (2000). Knight, J.R. & Milner, J. SIRT1, metabolism and cancer. *Curr Opin Oncol* 24, 68–75 (2012). Satoh, A. et al. Sirt1 extends life span and delays aging in mice through the regulation of Nk2 homeobox 1 in the DMH and LH. *Cell Metab* 18, 416–430 (2013). Sun, T., Jiao, L., Wang, Y., Yu, Y. & Ming, L. SIRT1 induces epithelial-mesenchymal transition by promoting autophagic degradation of E-cadherin in melanoma cells. *Cell Death Dis* 9, 136 (2018).

57. Bai, W. & Zhang, X. Nucleus or cytoplasm? The mysterious case of SIRT1's subcellular localization. *Cell Cycle* 15, 3337–3338 (2016). Hsu, W.W., Wu, B. & Liu, W.R. Sirtuins 1 and 2 are universal histone deacetylases. *ACS Chem Biol* 11, 792–799 (2016). Martinez-Redondo, P. & Vaquero, A. The diversity of histone versus nonhistone sirtuin substrates. *Genes Cancer* 4, 148–163 (2013).

58. Finley, L.W. et al. SIRT3 opposes reprogramming of cancer cell metabolism through HIF1alpha destabilization. *Cancer Cell* 19, 416–428 (2011).

59. Sun, T., Jiao, L., Wang, Y., Yu, Y. & Ming, L. SIRT1 induces epithelial-mesenchymal transition by promoting autophagic degradation of E-cadherin in melanoma cells. *Cell Death Dis* 9, 136 (2018). Joo, H.Y. et al. SIRT1 deacetylates and stabilizes hypoxia-inducible factor-1alpha (HIF-1alpha) via direct interactions during hypoxia. *Biochem Biophys Res Commun* 462, 294–300 (2015).

60. Alhazzazi, T.Y., Kamarajan, P., Verdin, E. & Kapila, Y.L. SIRT3 and cancer: tumor promoter or suppressor? *Biochim Biophys Acta* 1816, 80–88 (2011).

61. Kim, H.S. et al. SIRT3 is a mitochondria-localized tumor suppressor required for maintenance of mitochondrial integrity and metabolism during stress. *Cancer Cell* 17, 41–52 (2010).

62. Zhong, L. et al. The histone deacetylase Sirt6 regulates glucose homeostasis via Hif1alpha. *Cell* 140, 280–293 (2010).

63. Fiorino, E. et al. The sirtuin class of histone deacetylases: regulation and roles in lipid metabolism. *IUBMB Life* 66, 89–99 (2014).

64. Son, M.J. et al. Upregulation of mitochondrial NAD(+) levels impairs the clonogenicity of SSEA1(+) glioblastoma tumor-initiating cells. *Exp Mol Med* 49, e344 (2017).

65. Canto, C. & Auwerx, J. Targeting sirtuin 1 to improve metabolism: all you need is NAD(+)? *Pharmacol Rev* 64, 166–187 (2012). Yaku, K., Okabe, K., Hikosaka, K. & Nakagawa, T. NAD metabolism in cancer therapeutics. *Front Oncol* 8, 622 (2018).

66. Yaku, K., Okabe, K., Hikosaka, K. & Nakagawa, T. NAD metabolism in cancer therapeutics. *Front Oncol* 8, 622 (2018). Hasmann, M. & Schemainda, I. FK866, a highly specific noncompetitive inhibitor of nicotinamide phosphoribosyltransferase, represents a novel mechanism for induction of tumor cell apoptosis. *Cancer Res* 63, 7436–7442 (2003).

67. Son, M.J. et al. Upregulation of mitochondrial NAD(+) levels impairs the clonogenicity of SSEA1(+) glioblastoma tumor-initiating cells. *Exp Mol Med* 49, e344 (2017).

68. Hu, J., Jing, H. & Lin, H. Sirtuin inhibitors as anticancer agents. *Future Med Chem* 6, 945–966 (2014).

69. Song, M.S., Salmena, L. & Pandolfi, P.P. The functions and regulation of the PTEN tumour suppressor. *Nat Rev Mol Cell Biol* 13, 283–296 (2012).

70. Levental, K.R. et al. Matrix crosslinking forces tumor progression by enhancing integrin signaling. *Cell* 139, 891–906 (2009). Mouw, J.K. et al. Tissue mechanics modulate microRNA-dependent PTEN expression to regulate malignant progression. *Nat Med* 20, 360–367 (2014). Chalhoub, N. & Baker, S.J. PTEN and the PI3-kinase pathway in cancer. *Annu Rev Pathol* 4, 127–150 (2009).

71. Laplante, M. & Sabatini, D.M. mTOR signaling in growth control and disease. *Cell* 149, 274–293 (2012). Saxton, R.A. & Sabatini, D.M. mTOR signaling in growth, metabolism, and disease. *Cell* 169, 361–371 (2017).

72. Populo, H., Lopes, J.M. & Soares, P. The mTOR signalling pathway in human cancer. *Int J Mol Sci* 13, 1886–1918 (2012).

73. Fukuda, S. et al. Pyruvate kinase M2 modulates esophageal squamous cell carcinoma chemotherapy response by regulating the pentose phosphate pathway. *Ann Surg Oncol* 22 Suppl 3, S1461–1468 (2015). Sun, Q. et al. Mammalian target of rapamycin up-regulation of pyruvate kinase isoenzyme type M2 is critical for aerobic gly-

colysis and tumor growth. *Proc Natl Acad Sci USA* 108, 4129–4134 (2011). Dengler, V.L., Galbraith, M. & Espinosa, J.M. Transcriptional regulation by hypoxia inducible factors. *Crit Rev Biochem Mol Biol* 49, 1–15 (2014).

74. Peng, T., Golub, T.R. & Sabatini, D.M. The immunosuppressant rapamycin mimics a starvation-like signal distinct from amino acid and glucose deprivation. *Mol Cell Biol* 22, 5575–5584 (2002). Sun, Y. et al. Estradiol promotes pentose phosphate pathway addiction and cell survival via reactivation of Akt in mTORC1 hyperactive cells. *Cell Death Dis* 5, e1231 (2014). Acharya, S. et al. Downregulation of GLUT4 contributes to effective intervention of estrogen receptor-negative/HER2-overexpressing early stage breast disease progression by lapatinib. *Am J Cancer Res* 6, 981–995 (2016).

75. Babbar, M., Huang, Y., An, J., Landas, S.K. & Sheikh, M.S. CHTM1, a novel metabolic marker deregulated in human malignancies. *Oncogene* 37, 2052–2066 (2018).

76. Das, F. et al. High glucose forces a positive feedback loop connecting Akt kinase and FoxO1 transcription factor to activate mTORC1 kinase for mesangial cell hypertrophy and matrix protein expression. *J Biol Chem* 289, 32703–32716 (2014).

77. Su, H.Y. et al. The unfolded protein response plays a predominant homeostatic role in response to mitochondrial stress in pancreatic stellate cells. *PLoS One* 11, e0148999 (2016). Perl, A. Activation of mTOR (mechanistic target of rapamycin) in rheumatic diseases. *Nat Rev Rheumatol* 12, 169–182 (2016).

78. Hall, E. et al. The effects of high glucose exposure on global gene expression and DNA methylation in human pancreatic islets. *Mol Cell Endocrinol* 472, 57–67 (2018).

79. Zhang, B., Pan, X., Cobb, G.P. & Anderson, T.A. microRNAs as oncogenes and tumor suppressors. *Dev Biol* 302, 1–12 (2007).

80. Mouw, J.K. et al. Tissue mechanics modulate microRNA-dependent PTEN expression to regulate malignant progression. *Nat Med* 20, 360–367 (2014).

81. Tumaneng, K. et al. YAP mediates crosstalk between the Hippo and PI(3)K-TOR pathways by suppressing PTEN via miR-29. *Nat Cell Biol* 14, 1322–1329 (2012).

82. Yuan, Y., Zhong, W., Ma, G., Zhang, B. & Tian, H. Yes-associated protein regulates the growth of human non-small cell lung cancer in response to matrix stiffness. *Mol Med Rep* 11, 4267–4272 (2015).

83. Enzo, E. et al. Aerobic glycolysis tunes YAP/TAZ transcriptional activity. *EMBO J* 34, 1349–1370 (2015). Koo, J.H. & Guan, K.L. Interplay between YAP/TAZ and metabolism. *Cell Metab* 28, 196–206 (2018).

84. Wang, W. et al. AMPK modulates Hippo pathway activity to regulate energy homeostasis. *Nat Cell Biol* 17, 490–499 (2015).

85. Du, K. et al. Hedgehog-YAP signaling pathway regulates glutaminolysis to control activation of hepatic stellate cells. *Gastroenterology* 154, 1465–1479 e1413 (2018).

86. Brown, K. et al. WNT/beta-catenin signaling regulates mitochondrial activity to alter the oncogenic potential of melanoma in a PTEN-dependent manner. *Oncogene* 36, 3119–3136 (2017).

87. Sherwood, V. et al. WNT5A-mediated beta-catenin-independent signalling is a novel regulator of cancer cell metabolism. *Carcinogenesis* 35, 784–794 (2014).

88. Yang, J., Mani, S.A. & Weinberg, R.A. Exploring a new twist on tumor metastasis. *Cancer Res* 66, 4549–4552 (2006). Matsuo, N. et al. Twist expression promotes migration and invasion in hepatocellular carcinoma. *BMC Cancer* 9, 240 (2009).

89. Broders-Bondon, F., Nguyen Ho-Bouldoires, T.H., Fernandez-Sanchez, M.E. & Farge, E. Mechanotransduction in tumor progression: the dark side of the force. *J Cell Biol* 217, 1571–1587 (2018). Wei, S.C. et al. Matrix stiffness drives epithelial-mesenchymal transition and tumour metastasis through a TWIST1-G3BP2 mechanotransduction pathway. *Nat Cell Biol* 17, 678–688 (2015).

90. Yang, L. et al. Twist promotes reprogramming of glucose metabolism in breast cancer cells through PI3K/AKT and p53 signaling pathways. *Oncotarget* 6, 25755–25769 (2015).

91. Kirby, T.J. & Lammerding, J. Emerging views of the nucleus as a cellular mechanosensor. *Nat Cell Biol* 20, 373–381 (2018). Cho, S., Irianto, J. & Discher, D.E. Mechanosensing by the nucleus: from pathways to scaling relationships. *J Cell Biol* 216, 305–315 (2017). Guilluy, C. & Burridge, K. Nuclear mechanotransduction: forcing the nucleus to respond. *Nucleus* 6, 19–22 (2015). Boukouris, A.E., Zervopoulos, S.D. & Michelakis, E.D. Metabolic enzymes moonlighting in the nucleus: metabolic regulation of gene transcription. *Trends Biochem Sci* 41, 712–730 (2016).

92. Lombardi, M.L. et al. The interaction between nesprins and sun proteins at the nuclear envelope is critical for force transmission between the nucleus and cytoskeleton. *J Biol Chem* 286, 26743–26753 (2011). Elosegui-Artola, A. et al. Force triggers YAP nuclear entry by regulating transport across nuclear pores. *Cell* 171, 1397–1410 e1314 (2017).

93. Elosegui-Artola, A. et al. Force triggers YAP nuclear entry by regulating transport across nuclear pores. *Cell* 171, 1397–1410 e1314 (2017).

94. Rajgor, D. & Shanahan, C.M. Nesprins: from the nuclear envelope and beyond. *Expert Rev Mol Med* 15, e5 (2013). Wilson, M.H. & Holzbaur, E.L. Nesprins anchor kinesin-1 motors to the nucleus to drive nuclear distribution in muscle cells. *Development* 142, 218–228 (2015).

95. Tapley, E.C. & Starr, D.A. Connecting the nucleus to the cytoskeleton by SUN-KASH bridges across the nuclear envelope. *Curr Opin Cell Biol* 25, 57–62 (2013). Chang, W., Worman, H.J. & Gundersen, G.G. Accessorizing and anchoring the LINC complex for multifunctionality. *J Cell Biol* 208, 11–22 (2015).

96. Luger, K., Dechassa, M.L. & Tremethick, D.J. New insights into nucleosome and chromatin structure: an ordered state or a disordered affair? *Nat Rev Mol Cell Biol* 13, 436–447 (2012). Tessarz, P. & Kouzarides, T. Histone core modifications regulating nucleosome structure and dynamics. *Nat Rev Mol Cell Biol* 15, 703–708 (2014).

97. Liu, X.S., Little, J.B. & Yuan, Z.M. Glycolytic metabolism influences global chromatin structure. *Oncotarget* 6, 4214–4225 (2015).

98. For review, see Bannister, A.J. & Kouzarides, T. Regulation of chromatin by histone modifications. *Cell Res* 21, 381–395 (2011). Portela, A. & Esteller, M. Epigenetic modifications and human disease. *Nat Biotechnol* 28, 1057–1068 (2010). Seto, E. & Yoshida, M. Erasers of histone acetylation: the histone deacetylase enzymes. *Cold Spring Harb Perspect Biol* 6, a018713 (2014).

99. Zentner, G.E. & Henikoff, S. Regulation of nucleosome dynamics by histone modifications. *Nat Struct Mol Biol* 20, 259–266 (2013). Schvartzman, J.M., Thompson, C.B. & Finley, L.W.S. Metabolic regulation of chromatin modifications and gene expression. *J Cell Biol* 217, 2247–2259 (2018).

100. Lim D.H.K. & Maher, E.R. DNA methylation: a form of epigenetic control of gene expression. *Obstetrician Gynaecologist* 12, 37–42 (2010).

101. Yan, C. & Boyd, D.D. Histone H3 acetylation and H3 K4 methylation define distinct chromatin regions permissive for transgene expression. *Mol Cell Biol* 26, 6357–6371 (2006).

102. Merlo, A. et al. 5' CpG island methylation is associated with transcriptional silencing of the tumour suppressor p16/CDKN2/MTS1 in human cancers. *Nat Med* 1, 686–692 (1995). Lee, J.J. et al. Methylation and immunoexpression of p16(INK4a) tumor suppressor gene in primary breast cancer tissue and their quantitative p16(INK4a) hypermethylation in plasma by real-time PCR. *Korean J Pathol* 46, 554–561 (2012).

103. Hall, E. et al. The effects of high glucose exposure on global gene expression and DNA methylation in human pancreatic islets. *Mol Cell Endocrinol* 472, 57–67 (2018).

104. Khan, Z.S., Santos, J.M. & Hussain, F. Aggressive prostate cancer cell nuclei have reduced stiffness. *Biomicrofluidics* 12, 014102 (2018).

105. Stephens, A.D., Banigan, E.J., Adam, S.A., Goldman, R.D. & Marko, J.F. Chromatin and lamin A determine two different mechanical response regimes of the cell nucleus. *Mol Biol Cell* 28, 1984–1996 (2017).

106. Elosegui-Artola, A. et al. Force triggers YAP nuclear entry by regulating transport across nuclear pores. *Cell* 171, 1397–1410 e1314 (2017). Wang, N., Tytell, J.D. & Ingber, D.E. Mechanotransduction at a distance: mechanically coupling the extracellular matrix with the nucleus. *Nat Rev Mol Cell Biol* 10, 75–82 (2009).

107. Kocgozlu, L. et al. Selective and uncoupled role of substrate elasticity in the regulation of replication and transcription in epithelial cells. *J Cell Sci* 123, 29–39 (2010).

108. Pradhan, R., Ranade, D. & Sengupta, K. Emerin modulates spatial organization of chromosome territories in cells on softer matrices. *Nucleic Acids Res* 46, 5561–5586 (2018).

109. Alam, S.G. et al. The mammalian LINC complex regulates genome transcriptional responses to substrate rigidity. *Sci Rep* 6, 38063 (2016).

110. Apostolou, P. & Fostira, F. Hereditary breast cancer: the era of new susceptibility genes. *Biomed Res Int* 2013, 747318 (2013). Obuch, J.C. & Ahnen, D.J. Colorectal cancer: genetics is changing everything. *Gastroenterol Clin North Am* 45, 459–476 (2016). Kobayashi, H., Ohno, S., Sasaki, Y. & Matsuura, M. Hereditary breast and ovarian cancer susceptibility genes (review). *Oncol Rep* 30, 1019–1029 (2013).

111. Pfeifer, C.R., Alvey, C.M., Irianto, J. & Discher, D.E. Genome variation across cancers scales with tissue stiffness—an invasion-mutation mechanism and implications for immune cell infiltration. *Curr Opin Syst Biol* 2, 103–114 (2017).

112. Theocharis, A.D., Skandalis, S.S., Gialeli, C. & Karamanos, N.K. Extracellular matrix structure. *Adv Drug Deliv Rev* 97, 4–27 (2016). Rackov, G. et al. Vesicle-mediated control of cell function: the role of extracellular matrix and microenvironment. *Front Physiol* 9, 651 (2018).

113. Janssen, L.M.E., Ramsay, E.E., Logsdon, C.D. & Overwijk, W.W. The immune system in cancer metastasis: friend or foe? *J Immunother Cancer* 5, 79 (2017).

114. Acerbi, I. et al. Human breast cancer invasion and aggression correlates with ECM stiffening and immune cell infiltration. *Integr Biol (Camb)* 7, 1120–1134 (2015). Laklai, H. et al. Genotype tunes pancreatic ductal adenocarcinoma tissue tension to induce matricellular fibrosis and tumor progression. *Nat Med* 22, 497–505 (2016). Levental, K.R. et al. Matrix crosslinking forces tumor progression by enhancing integrin signaling. *Cell* 139, 891–906 (2009). Mouw, J.K. et al. Tissue mechanics modulate microRNA-dependent PTEN expression to regulate malignant progression. *Nat Med* 20, 360–367 (2014).

115. Coussens, L.M. & Werb, Z. Inflammation and cancer. *Nature* 420, 860–867 (2002). Wu, Y. & Zhou, B.P. Inflammation: a driving force speeds cancer metastasis. *Cell Cycle* 8, 3267–3273 (2009).

116. Ridker, P.M. et al. Effect of interleukin-1beta inhibition with canakinumab on incident lung cancer in patients with atherosclerosis: exploratory results from a randomised, double-blind, placebo-controlled trial. *Lancet* 390, 1833–1842 (2017).

117. Fattovich, G., Stroffolini, T., Zagni, I. & Donato, F. Hepatocellular carcinoma in cirrhosis: incidence and risk factors. *Gastroenterology* 127, S35–50 (2004). Llovet, J.M. et al. Hepatocellular carcinoma. *Nat Rev Dis Primers* 2, 16018 (2016).

118. Boozalis, E., Shah, A.A., Wigley, F., Kang, S. & Kwatra, S.G. Morphea and systemic sclerosis are associated with an increased risk for melanoma and nonmelanoma skin cancer. *J Am Acad Dermatol* 80, 1449–1451 (2019). Maria, A.T.J. et al. Fibrosis development in HOCl-induced systemic sclerosis: a multistage process hampered by mesenchymal stem cells. *Front Immunol* 9, 2571 (2018). Watad, A. et al. Autoantibody status in systemic sclerosis patients defines both cancer risk and survival with ANA negativity in cases with concomitant cancer having a worse survival. *Oncoimmunology* 8, e1588084 (2019).

119. Wynn, T.A. & Ramalingam, T.R. Mechanisms of fibrosis: therapeutic translation for fibrotic disease. *Nat Med* 18, 1028–1040 (2012).

120. Forrester, S.J., Kikuchi, D.S., Hernandes, M.S., Xu, Q. & Griendling, K.K. Reactive oxygen species in metabolic and inflammatory signaling. *Circ Res* 122, 877–902 (2018).

121. Canli, O. et al. Myeloid cell-derived reactive oxygen species induce epithelial mutagenesis. *Cancer Cell* 32, 869–883 e865 (2017).

122. Engler, A.J., Sen, S., Sweeney, H.L. & Discher, D.E. Matrix elasticity directs stem cell lineage specification. *Cell* 126, 677–689 (2006). Walsh, S.M. et al. Novel differences in gene expression and functional capabilities of myofibroblast populations in idiopathic pulmonary fibrosis. *Am J Physiol Lung Cell Mol Physiol* 315, L697–L710 (2018). Stowers, R.S. et al. Matrix stiffness induces a tumorigenic phenotype in mammary epithelium through changes in chromatin accessibility. *Nat Biomed Eng* 3, 1009–1019 (2019). Hinz, B., Mastrangelo, D., Iselin, C.E., Chaponnier, C. & Gabbiani, G. Mechanical tension controls granulation tissue contractile activity and myofibroblast differentiation. *Am J Pathol* 159, 1009–1020 (2001).

123. Mah, E.J., Lefebvre, A., McGahey, G.E., Yee, A.F. & Digman, M.A. Collagen density modulates triple-negative breast cancer cell metabolism through adhesion-mediated contractility. *Sci Rep* 8, 17094 (2018).

124. Polet, F., Martherus, R., Corbet, C., Pinto, A. & Feron, O. Inhibition of glucose metabolism prevents glycosylation of the glutamine transporter ASCT2 and promotes compensatory LAT1 upregulation in leukemia cells. *Oncotarget* 7, 46371–46383 (2016).

125. Barnes, J.M. et al. A tension-mediated glycocalyx-integrin feedback loop promotes mesenchymal-like glioblastoma. *Nat Cell Biol* 20, 1203–1214 (2018).

12 Cancer Metabolism and Therapeutic Perspectives

Exploiting Acidic, Nutritional, and Oxidative Stresses

Maša Ždralević and Jacques Pouysségur

Overview

From anoxia to progressively oxygen-enriched environments, life has evolved highly successful and robust metabolic, enzymatic, and bioenergetics pathways through which the known diversity of organisms from bacteria to humans has emerged, by surviving and adapting to the most extreme environments. Nutrients, their capture, storage, and metabolism, are central for all forms of life, providing energy and biomass. In this chapter, we present how the evolutionary most ancient metabolic pathway is exploited by rapidly growing tissues and tumors: the Myc/hypoxia-induced pathway, also known as fermentative glycolysis, termed in cancer cells also the Warburg effect. We then discuss the potential therapeutic benefit of disrupting this metabolic pathway in models of aggressive cancers at three central molecular pathway nodes, namely, glucose-6-phosphate isomerase, lactate dehydrogenases (LDHA/B), and lactic acid transporters (MCT1/4). Furthermore, we illuminate how cancer cells exploit their metabolic plasticity to survive the metabolic and energetic blockade, arrest their growth, or die. We also present evidence that challenges the widely held view that the Warburg effect is essential for tumor growth. Finally, we express our optimism with respect to future therapeutic approaches: one aimed at reducing tumor lactic acid to induce a robust antitumor immune response and the other by targeting the cystine transporter xCT (SLC7A11), a key component of redox homeostasis, to trigger precise tumor "ferroptosis" cell death.

12.1 Introduction

The uptake and metabolism of nutrients is central to life, providing energy and biomass. In metazoans, hormones and growth factors are central to the control of nutrient uptake and subsequent metabolism as a prerequisite to engage either in cell-cycle entry, cell division, proliferation, or cell maintenance and survival.[1] These tasks are performed by a dual set of universal mitogenic signaling pathways: Ras-Raf-ERK and PI3K-AKT, activated by growth factor receptor tyrosine kinases and some G protein–coupled receptors. The key feature of these two signaling pathways is that they synergistically activate the evo-lutionarily conserved master protein kinase mTORC1 that controls not only protein syn-thesis but also other anabolic processes, including lipid and nucleotide synthesis.[2]

It is now well established that cancer cells of different tissues, once they have emerged through multiple scenarios, enter ultimately a state of progressively acquired growth factor and cellular matrix independence. It was therefore not surprising that among the most representative oncogenes/cancer drivers, constitutively activating mutations in these two signaling pathways have been found frequently: upstream receptor tyrosine kinases (RTKs), Ras-Raf-ERK and PI3K/AKT or Myc, a downstream transcription factor of these pathways conferring partial or full cell intrinsic autonomy.[3] From nutrient-deprived to lethal environments such as anoxia/hyperoxia, cells have developed multiple survival strategies through expression of transcription factors such as ATF4, Myc, HIF1, and NRF2. Cancer cells largely exploit these factors to thrive and resist therapies.[4]

After the outstanding work of Otto Warburg almost a century ago,[5] renewed interest in the fundamental principles of energetic metabolism has reemerged after its importance in cancer progression has been established beyond doubt with modern molecular methods, and our knowledge in this area has increased considerably over the past two decades.[6] There is now a general consensus that reprogrammed metabolism provides a selective advantage during tumorigenesis due to its flexibility to support cell growth and prolifera-tion during periods of enhanced nutrient uptake but also through supporting cell survival by activating autophagy, a type of metabolic dormancy when nutrients are scarce.[7] Addi-tionally, maintenance of redox homeostasis to counteract the increased production of reactive oxygen species (ROS) is an important mechanism that can promote tumorigenesis and metastasis, partially by activating oncogenic signaling pathways and transcription factors.[8] Remarkable genetic and phenotypic tumor heterogeneity is now widely recog-nized,[9] and the intense glycolytic flux and altered glutamine metabolism remain key meta-bolic pathways in most cancers (figure 12.1). However, as developed in section 12.2, this phenotypic metabolic feature is not a specific oncogenic trait driven by mutations but more a consequence dictated by the physiological status and microenvironment of the tumor cells. The expression of metabolism-related genes, controlled by hormones, growth factor signaling cascades, and epigenetic mechanisms, promotes metabolic reprogramming and growth survival advantages before emergence of oncogenic mutation.[10] Thus, epigenetic modifications can support increasing cell proliferation, migration, and pluripotency prior to when detectable mutations in these processes occur.

Accelerated aerobic glycolysis distinguishes cancer cells from normal differentiated cells that derive their energy from respiration/oxidative phosphorylation (OXPHOS), and this key feature of tumor cells has been exploited to detect and image tumors in vivo for decades.[11] Most rapidly developing tumors take up glucose and secrete lactic acid, regard-less of the presence of oxygen and without losing respiratory capacity, a phenomenon referred to as the Warburg effect, discovered by the German physiologist Otto Warburg

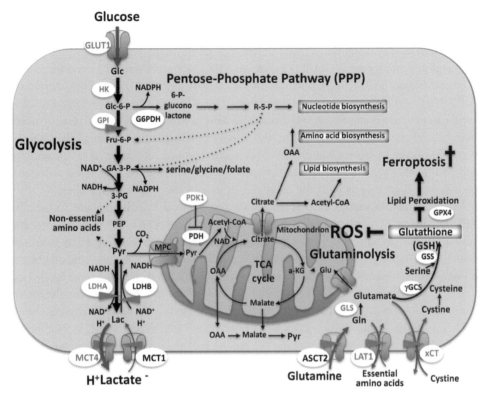

Figure 12.1
Schematic metabolic chart presenting some of the key metabolic pathways and their interconnection in proliferating cells. Most of the increased nutrient uptake is used to support nucleotide, amino acid, and lipid biosynthetic reactions. Considering the fermentative glycolytic pathway (Warburg effect) discussed in the chapter, note from the top—*glucose to lactic acid*—the three enzymatic steps that have been genetically disrupted in our studies (horizontal long arrowheads).

during the 1920s and 1930s.[12] This highly glycolytic phenotype results from the combination of uncontrolled growth signaling, deregulated c-Myc, and hypoxia-induced factor 1 (HIF1) activity, leading to the induction of the glycolytic enzymes and transporters[13] and limiting glycolytic-derived pyruvate to enter the citric acid cycle, also known as the tricarboxylic acid (TCA) or Krebs cycle.[14]

Acceleration of the glycolytic flux provides cancer cells with rapidly produced adenosine triphosphate (ATP), metabolic intermediates, and precursors to sustain biosynthesis and reducing equivalents necessary for cell division (figure 12.1).[15] It is associated with the release of lactic acid, which is then diffused along its concentration gradient from glycolytic toward some oxidative cancer cells[16] or recycled in many physiological situations as glycogen.[17]

It is not surprising that such a fairly common glycolytic cancer phenotype, essential for promoting tumor growth, has attracted a lot of attention as a potential target for treatment approaches. Indeed, many drugs that interfere with glycolysis and transporters of nutrients, such as glucose transporter 1 (GLUT1), hexokinase (HK), 6-phosphofructo 2-kinase-fructose-2, 6-bisphosphatase 3 (PFKFB3), pyruvate kinase isozyme 2 (PKM2), lactate dehydrogenase A (LDHA), and monocarboxylate transporter 1 (MCT1), have been investigated as anti-cancer agents.[18] For example, methyl jasmonate, 2-deoxyglucose (2-DG) and 3-bromopyruvate have been shown to effectively inhibit the activity of HK and the Warburg effect phenotype.[19] However, in clinical studies, all these agents showed substantial systemic toxicity and insufficient target-cell activity and/or specificity, to be exploited. Therefore, at present, no inhibitor of glycolysis has been approved as an anticancer agent in the clinic.[20]

Instead of the pharmacological approaches explored so far, we discuss here the benefits and limitations of disrupting the Warburg effect by CRISPR-Cas9-mediated genetic editing. In fact, we challenge the widely accepted view that the Warburg effect is indeed essential for tumor growth. We have analyzed the Warburg effect, metabolism, and tumor growth following targeted genetic disruption of the glycolytic pathway at three levels (see figure 12.1) by introducing (1) an upstream pathway block at the level of the glucose-6-phosphate isomerase (GPI), (2) a downstream block at the level of lactate dehydrogenases (LDHs, isoforms A and B), and (3) a block of the lactic acid export via H$^+$/lactate$^-$ transporters (MCT1 and MCT4). Exploiting the precision of the genetic approach, we discuss the responses of two aggressive cancer cell lines to the aforementioned pathway blocks in terms of metabolic rewiring and tumor growth and compare our results with the broader literature.

Finally, in the last section of the chapter, we briefly discuss the emerging and exciting possibility of tumor lactic acid as a potent immune suppressor affecting the activity of immune cells and their capacity to infiltrate tumors[21] (figure 12.2). We end by highlighting the key role of the cystine transporter xCT (SLC7A11) in nutrient and redox homeostasis as a promising novel anticancer target through exploitation of "ferroptosis"-induced cell death[22] (figure 12.3).

12.2   The Warburg Effect and Cancer, an Apparent Paradox

The preference that rapidly growing tumors have for glucose fermentation, a low ATP-producing pathway, in contrast to respiration, has long been viewed as paradoxical, and yet this metabolic choice is almost universal for rapid proliferation as long as nutrients and glucose are provided.[23] Glucose fermentation also appears to be the main pathway used for exponential growth of microbes such as bacteria and yeast. Instead of exporting lactate, yeast convert pyruvate into ethanol by a family of well-conserved glycolytic enzymes within the *Saccharomyces* genus.[24] Glycolysis is as well a feature of embryonic

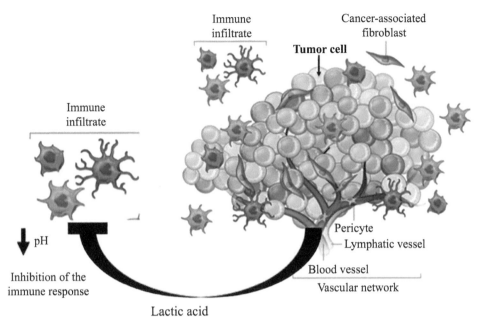

Figure 12.2
Schematic illustration of immunoevasion facilitated by tumor lactic acid. Acidity reduces tumor T-cell infiltration and interferon γ production by natural killer cells (NK),[25] and it induces reprogramming of tumor-associated macrophages into a noninflammatory phenotype.[26]

stem cells (ESCs), which develop under low-oxygen conditions before and immediately after implantation and during reprogramming of somatic into induced pluripotent stem cells (iPSCs).[27] High glycolytic flux is directly responsible for their potential for unlimited proliferation[28] and maintenance of pluripotency.[29] Similar to cancer cells, these metabolic changes prepare cells for upcoming varying cellular energetic demands. In fact, an interesting idea that alteration in metabolic control of cancer cells could be a result of reverting to an embryonic program has already been proposed.[30] Indeed, a shift from oxidative to fermentative metabolism is also involved in differentiated tissues like muscle,[31] neurons,[32] and muscle and adipose tissue of starved mice.[33] This is reminiscent of oxygen depletion in which HIF1-like FOXK1/2 (starvation) induces glycolytic flux into lactate being actively fueled into glycogen stores via gluconeogenesis.[34] The beauty here is that the transcription factors, HIF1 (hypoxia) and FOXK1/2 (starvation), maximize fermentation to lactate by inhibiting the mitochondrial pyruvate dehydrogenase complex (PDH).

In conclusion, fermentative glycolysis, an ancient evolved metabolic pathway, is exploited by rapidly growing tissues and tumors but also occurs in response to nutritional and energetic demands of differentiated tissues. Lactate is recycled in organs as a major metabolic precursor of gluconeogenesis and an energy source.[35]

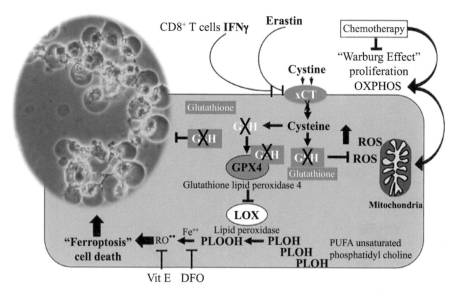

Figure 12.3
Schematic illustration of the cystine transporter xCT in the control of redox homeostasis and induction of "ferroptosis" cell death. *xCT*-KO, like low concentration of erastin, induces a collapse of glutathione (GSH) and induces cell death.[36]

12.3   Is the Warburg Effect Dispensable for Cancer?

We addressed this question by using two aggressive cancer cell lines, derived from human colon adenocarcinoma (LS174T) and mouse B16-F10 melanoma, in which we fully disrupted the "Warburg effect" at either one of the three key glycolytic steps: glucose-6-phosphate-isomerase; lactate dehydrogenase A, B, or A and B; and lactate/H$^+$ symporters MCT1 and MCT4 (figure 12.1, horizontal long arrowheads).

12.3.1   Disruption of Glucose-6-Phosphate Isomerase

GPI (D-glucose-6-phosphate aldose-ketose-isomerase; EC 5.3.1.9) is a ubiquitously expressed housekeeping cytosolic enzyme that catalyzes the reversible interconversion between glucose-6-phosphate (G6P) and fructose-6-phosphate (F6P) in glycolytic and gluconeogenic pathways. It is competitively inhibited by 2-DG, and like most glycolytic genes, its expression is directly induced by c-Myc[37] and HIF1.[38] *GPI* expression is elevated in many cancers, and in lung adenocarcinoma patients, its overexpression correlates with poor prognosis and reduced overall survival.[39] Additionally, GPI overexpression was shown to be involved in carcinogenesis and was associated with poor prognosis in gastric cancer,[40] and *GPI* gene polymorphisms were shown to predict overall survival in hepatocellular carcinoma, therefore serving as potential predictive biomarkers.[41]

GPI-defective mutants were first isolated from Chinese hamster fibroblasts almost forty years ago by using 2-deoxy[^{3}H]-glucose radiation suicide as a selection method.[42] These mutants had impaired hexose transport and very low lactic acid production, which made them entirely dependent on respiration for energy production and growth. However, despite the inhibition of aerobic glycolysis, these mutants retained the in vitro transformed phenotype of the parental line (low serum dependence and loss of anchorage dependence for growth) and tumor-forming capability.[43] These findings suggested that (1) glycolysis is dispensable for cell growth and (2) high aerobic glycolysis can be uncoupled from the in vitro malignant phenotype. Furthermore, *GPI*-defective mutants retained normal control of DNA synthesis and tumor-forming capability, even though these tumors developed more slowly, which might be explained by their higher sensitivity to hypoxia.[44] In other words, changes in the rates of transport and glucose metabolism cannot restore original normal growth regulatory properties of the cells.

We have pursued these studies further by creating complete *GPI-knockout* (KO) cells in two fast-growing cancer cell lines, human colon adenocarcinoma (LS174T) and mouse melanoma (B16-F10), using the CRISPR/Cas9 technique.[45] Both *GPI*-KO cell lines had no measurable GPI enzymatic activity and no detectable secretion of lactic acid, which decreased their growth rate twofold in normoxia.[46] These findings are in agreement with our earlier genetic studies[47] but in contrast with in vitro and in vivo studies exploring pharmacological inhibition of the glycolytic pathway,[48] which was reported to result in inhibition of tumor growth and increased cell death. These discrepancies might be due to the fact that most of the drugs reported so far had off-target effects. In order to compensate for the full ablation of the Warburg phenotype and meet their energetic needs, *GPI*-KO cells rewired their metabolism toward the pentose phosphate pathway (PPP) and OXPHOS, as evidenced by their increased respiratory capacity[49] (figure 12.1). Particularly striking is the example of LS174T, a highly glycolytic cancer cell line with very low respiration on glucose under normal conditions, capable of strong reactivation of OXPHOS when challenged by *GPI* ablation. This created an increased reliance on oxygen; indeed, *GPI*-KO cells, although fully viable, were unable to grow in hypoxia (1 percent O$_2$) and became extremely sensitive to OXPHOS inhibition with phenformin and oligomycin.[50] This vulnerability created by metabolic rewiring is in line with the findings reported for ovarian and hepatocellular cancer cell lines, where *GPI* silencing completely inhibited cancer cell growth in combination with OXPHOS inhibition.[51] Disruption of the glycolytic flow at the level of *GPI* leads to accumulation of its substrate, intracellular G6P, which in turn was proposed to cause a short-term inhibition of hexokinase and a long-term inhibition of glucose transport activity.[52] Indeed, we found decreased GLUT1 expression in both *GPI*-KO cell lines, as well as induction of thioredoxin interacting protein (TXNIP), which is known to impose a strong negative regulation on glucose uptake.[53] Consequently, glucose consumption was severely reduced in *GPI*-KO cells when compared to the wild type (WT).[54]

Finally, complete suppression of aerobic glycolysis caused by genetic ablation of *GPI* in our two aggressive cancer cell lines did not prevent in vivo tumor growth.[55] In line with our earlier and more recent findings of others,[56] tumor growth was impacted only moderately by *GPI* disruption. It was only in combination with mTORC1 or OXPHOS inhibition that *GPI* silencing resulted in significant tumor growth inhibition.[57] Therefore, pharmacological inhibition of glycolysis and tumor growth with 2-DG or 3-bromopyruvate reported previously was effective because their multiple targets affected both glycolysis and OXPHOS. Nevertheless, given its increased expression in different cancers and association with poor prognosis, GPI remains a valid therapeutic target, in combination with OXPHOS inhibition.

In the following section, we have used the same genetic approach to investigate whether in the same two cancer cell lines, the complete disruption of the most downstream and essential glycolytic gene, *LDHA*, is dispensable for in vivo tumor growth like *GPI*.

12.3.2 Disruption of Lactate Dehydrogenase Isoforms A and B

LDH ((*S*)-lactate:NAD$^+$ oxidoreductase; EC 1.1.1.27) is a family of NAD$^+$-dependent enzymes that catalyze the reversible conversion between pyruvate and lactate, with simultaneous oxidation/reduction of the cofactor (NAD$^+$/NADH). Active LDH is a homo- or heterotetramer assembled from two different subunits: M (muscle) and H (heart), encoded by two separate genes, *LDHA* (M) and *LDHB* (H), respectively. There is also a third subunit, LDHC, encoded by a separate *LDHC* gene, but it is expressed only in testes and is probably an evolutionary duplication of the *LDHA* gene.[58] The combinations of the two subunits result in five major LDH isoenzymes, LDH1 (H4), LDH2 (M1H3), LDH3 (M2H2), LDH4 (M3H1), and LDH5 (M4), that are expressed in a tissue-specific manner.[59] LDH1, commonly named LDHB, is prevalent in tissues with high aerobic metabolism, such as the heart and the brain, whereas LDH5 or LDHA is mainly expressed in tissues with anaerobic metabolism, such as skeletal muscle and liver.[60] This difference in tissue distribution is mainly due to the different affinity of each isoenzyme to different substrates, as LDH isoenzymes differ in electrophoretic mobility, Km for pyruvate and lactate, immunological characteristics, thermal stability, and inhibition by coenzyme analogues or excess pyruvate.[61] There are also tissue-specific differences in intracellular localization of isoenzymes; for example, in the heart, LDHA is found both in the cytosol and in mitochondria, whereas LDHB is found only in mitochondria.[62] The existence of both cytosolic and mitochondrial isoforms of LDH is linked to their fundamental role in the intracellular lactate shuttling mechanism, especially in liver and muscle tissues.[63] This concept postulates the role of lactate as an oxidizing substrate used by mitochondrial LDH isoform (i.e., that lactate to pyruvate conversion takes place in mitochondria, and the pyruvate is then metabolized by the TCA cycle).[64]

i) the LDHA isoform: LDHA (LDH5, M4) is a key player of the Warburg effect because it is responsible for the formation of lactate, the final step of the glycolysis. LDHA has the lowest Km for pyruvate of all LDH isoforms and hence catalyzes pyruvate reduction to lactate, accompanied by NAD^+ regeneration, which is mandatory for glycolysis to proceed. LDHA is mainly a cytosolic enzyme, but it has also been found to bind single-stranded DNA in the nucleus, although its involvement in DNA metabolism is not well understood.[65] Nuclear localization of LDHA has been observed in many tumors, including colorectal, breast, prostate, lung, and liver cancer, but with an unidentified underlying cause and function.[66] More recently, it was reported that LDHA translocates into the nucleus in HPV16-positive cervical tumor cells in a ROS-dependent manner and activates the antioxidant response, acting therefore as a pro-oncogenic epigenetic control switch between redox balance and cell proliferation.[67]

LDHA gene expression is regulated by epidermal growth factor (EGF) and tumor necrosis factor alpha (TNFα) levels through protein kinase A (PKA) and protein kinase C (PKC) signaling pathways.[68] Furthermore, transcription factor c-Myc,[69] cAMP,[70] and HIF1[71] are also involved in *LDHA* gene expression regulation.

Posttranslational mechanisms, such as direct phosphorylation[72] and lysine acetylation,[73] were found to alter LDHA levels. Tyrosine phosphorylation by the oncogenic receptor FGFR1 was found to enhance LDHA enzymatic activity, thus promoting the Warburg effect and tumor growth.[74] On the other hand, acetylation had an inhibitory effect on LDHA activity and reduced LDHA protein levels by tagging them for lysosomal degradation.[75]

Clinical significance of serum LDH levels and activity has been shown in many different types of carcinoma.[76] Increased serum LDH levels have been frequently associated with hematological malignancies, such as non-Hodgkin's lymphoma (NHL), and related decreased survival rate and duration of survival[77]; it was shown to be a factor of poor prognosis in Hodgkin's disease[78] and multiple myeloma.[79] Besides the total LDH level, elevated levels of LDH isoenzymes 2 (MH3) and 3 (M2H2) were shown to be indicators of poor overall survival in patients with NHL.[80] However, even if some studies reported that an estimation of LDH isoenzyme proportions was useful in the differential diagnosis of malignant samples,[81] the clinical implications of this approach are still under discussion. Nevertheless, an increased LDHA/LDHB ratio in transformed cells and malignant tissues[82] further supports the pro-oncogenic function of LDHA, the main enzyme responsible for lactic acid production.

LDH5 or LDHA has been acknowledged as a predictive and prognostic biomarker for various malignancies, such as gastric,[83] endometrial,[84] non–small cell lung cancer,[85] colorectal adenocarcinomas,[86] and NHL.[87] Accordingly, tumor lactate levels correlate with formation of metastases and survival of patients; when released into the extracellular matrix, lactate contributes to the immune escape and resistance to radiotherapy, due to its antioxidant properties.[88] Furthermore, a positive correlation between LDHA and vascular endothelial growth factor (VEGF) expression was found in gastric cancer, and high expression of both proteins correlated with lower survival.[89] In general, upregulation of LDHA levels was found in many cancers, such as esophageal squamous cell carcinoma[90] and oral squamous cell carcinoma,[91] where it correlated with formation of metastases, advanced tumor stages, and reduced tumor-specific survival.

The key role of LDHA in maintaining a Warburg phenotype in cancer cells, promoting tumorigenic potential of malignancies, was confirmed by numerous findings showing that LDHA inhibition, gene silencing, or knockdown severely diminished cancer cell tumorigenicity in different cancer types, including breast, lung, liver, lymphoma, and pancreas.[92] LDHA inhibition resulted in increased apoptosis via ROS production (i.e., increased intracellular oxidative stress).[93] It stimulated OXPHOS and mitochondrial oxygen consumption and decreased mitochondrial membrane potential,[94] as well as reduced ATP levels,[95] thus explaining reduced ability of cancer cells to proliferate and form metastasis. These data, together with the fact that LDHA deficiency has no serious consequences under normal conditions, made LDHA a very attractive target for anticancer therapy, especially for invasive cancers. In fact, significant progress in the discovery of small-molecule inhibitors of LDHA has been made, so that there are now a large number of proposed inhibitors available, based on gossypol derivatives, chimeric (bifunctional) molecules, polyphenolic scaffolds, 2,3-dihydroxynaphthoic acid, polyphenolic flavone, N-hydroxyindole scaffolds, dihydropyrimidine and pyrazine, dihydropyrone and cyclohex-2-enone, and quinoline scaffolds.[96] However, due to the highly polar nature and small size of its natural substrate (pyruvate), together with the relatively large and open active site, LDHA is considered a difficult target, and very few LDHA inhibitors have progressed into clinical trials.[97] Recent pharmacological advances described a novel LDHA inhibitor, GNE-140, with the inhibitory activity on both LDHA and LDHB isoforms in the nanomolar range but with poor pharmacokinetic profile in vivo.[98] Resistance to LDHA inhibition by GNE-140 was found to be mediated by an increased reliance on OXPHOS, and thus combined inhibition of both LDHA and OXPHOS was synthetically lethal.[99] Therefore, improvement of the pharmacokinetic profile of GNE-140, combined with simultaneous OXPHOS inhibition, may significantly increase the clinical utility of LDHA inhibition.

Our recent work with *LDHA*-KO cells, created in LS174T and B16 backgrounds by CRISPR-Cas9 gene editing, showed that complete abrogation of LDHA gene and protein expression resulted in only about a 30 percent decrease in secreted lactate levels.[100] This rather unexpected discovery was further corroborated by the finding that lack of *LDHA* did not significantly influence in vitro cell growth under normoxic conditions.[101] This could at least partially be explained by the retained ability of *LDHA*-KO cells to catalyze pyruvate conversion to lactate, although at a reduced rate compared to WT cells, but at levels sufficient to drive glycolysis and lactate production. We argue that this activity is due to the presence of a LDHB isoform in both cell lines, capable of catalyzing the reverse reaction when *LDHA* is no longer present. *LDHA*-KO cells had increased maximal OXPHOS and electron transport system (ETS) capacities, pointing to an increased reliance on OXPHOS for energy production. This metabolic peculiarity made them susceptible to respiratory chain inhibitors, such as phenformin, which significantly reduced their clonal growth.[102] This finding is in agreement with the pharmacological inhibition of LDHA,[103] further arguing in favor of combined targeting of both glycolysis and OXPHOS in treatment approaches.

ii) The LDHB isoform: LDHB (LDH1, H4) is composed of four H subunits, has a high Km for pyruvate, and hence catalyzes the reverse reaction, lactate oxidation to pyruvate, coupled with NAD^+ conversion to NADH. Expression of LDHB was found to be positively regulated, like the A isoform, by the RTK-PI3K-AKT-mTOR pathway both in immortalized mouse cell lines as well as in human cancer cells.[104] In particular, mammalian target of rapamycin (mTOR) and a signal transducer and activator of transcription 3 (STAT3), as a downstream mTOR effector, were found to be positive regulators of LDHB expression.[105] An increasing number of studies are showing a vital role of LDHB in tumorigenesis of several types of cancer,[106] even if *LDHB* expression is silenced by promoter hypermethylation in prostate,[107] breast,[108] and pancreatic cancers.[109] *LDHB* overexpression was found in many different types of cancers, but there are conflicting results regarding its prognostic significance. In triple-negative breast cancer,[110] KRAS-dependent lung adenocarcinoma,[111] maxillary sinus squamous cell cancer,[112] osteosarcoma,[113] and pancreatic ductal adenocarcinoma[114] high LDHB expression correlated with poor patient outcome, whereas in non–small cell lung cancer[115] and hepatocellular carcinoma[116] it correlated with better prognosis, and accordingly, loss of *LDHB* expression was associated with metastatic progression.[117] In addition, high LDHB expression predicted a positive response to neoadjuvant chemotherapy in breast cancer[118] and in oral squamous cell carcinoma.[119] The role of LDHB in cancer still remains to be further elucidated, but it certainly has a key role in the "reverse Warburg effect," where captured lactic acid serves as an energetic fuel for symbiotic cells in tissue/tumor microenvironments.[120]

Complete knockout of the *LDHB* gene by CRISPR-Cas9 in LS174T and B16 cell lines did not significantly alter cell growth and viability in normoxia or hypoxia (1 percent O_2).[121] *LDHB*-KO cells behaved essentially like WT cells in terms of lactate secretion, glycolytic and OXPHOS activity, and sensitivity to OXPHOS inhibitors.[122] It was only when we disrupted both isoforms, *LDHA* and *LDHB*, that we obtained a distinct phenotype. *LDHA/B*–double KO (DKO) cells had no detectable lactate secretion, caused by the complete abolishment of LDH enzymatic activity in both directions.[123] As a consequence, these cells redirected completely their metabolism to OXPHOS, showing an increase in oxidative metabolism of glucose and increased OXPHOS and ETS capacities, which made them extremely sensitive to mitochondrial respiratory chain inhibitors, such as phenformin. Reduced glucose consumption by *LDHA/B*-DKO cells caused growth reduction in normoxia and a complete growth arrest in hypoxia.[124] Comparing glucose and glutamine fluxes between WT and *LDHA/B*-DKO cells revealed only moderate changes, such as more glutamine-derived pyruvate in *LDHA/B*-DKO cells, but *LDHA/B* deletion caused a significant delay of in vivo tumor growth, especially in LS174T cells. Importantly, by making use of the aforementioned specific LDHA and LDHB inhibitor, GNE-140, we showed that a short-term treatment of WT cells with GNE-140 was sufficient to phenocopy the effect of the *LDHA/B*-DKO cells in terms of suppression of glycolysis and reactivation of OXPHOS,[125] thus proving that the *LDHA/B*-DKO phenotype does not result from adaptation during long-term growth selection following the genetic disruption.

These results, based on a genetic approach in two cancer cell lines, showed that both LDHA and LDHB contribute to fermentative glycolysis, and only complete disruption of both isoforms fully suppresses the Warburg phenotype, like *GPI*-KO did in the same cancer cell lines without suppressing tumor growth. These findings demonstrate the remarkable metabolic plasticity of these tumor cells, able to rapidly switch their metabolism to OXPHOS and survive. Therefore, inhibiting both glycolysis and mitochondrial metabolism could be a strategy to eradicate aggressive cancers. Whether this will be clinically manageable remains to be investigated further, as we already suggested.[126]

12.3.3 Sequestration of Lactic Acid in Cancer Cells via Inhibition of Monocarboxylate Transporters

The end product of fermentative glycolysis, lactic acid, has long been considered a metabolic waste product, but today we know this metabolite is one of the most important energy fuels, gluconeogenic precursors, antioxidants, and immune suppressors, as we will discuss in the last section.[127]

Transport of lactic acid into and out of cells is facilitated by a family of four monocarboxylate transporter isoforms (MCTs 1–4), which are extensively studied and characterized.[128] MCTs facilitate the net cotransport or diffusion of one monocarboxylate anion with a single proton, or exchange of one intracellular with one extracellular monocarboxylate, the direction of which depends on the concentration gradients of protons and monocarboxylates across the plasma membrane.[129] MCT1 is expressed in most tissues; its expression is induced by c-Myc and is usually responsible for lactic acid efflux. However, more invasive tumors upregulate another isoform, MCT4, induced by HIF1α.[130] Both MCT1 and MCT4 require the ancillary protein CD147 or basigin (BSG) to facilitate proper expression of active transporters at the plasma membrane.[131] Experimental evidence supports the lactate shuttling hypothesis within some tumors, which posits that in glycolytic hypoxic tumor regions, lactic acid is secreted via MCT4, which is then taken up by oxidative tumor cells via MCT1 and used as respiratory fuel that supports further tumor growth.[132]

In order to transport the large quantities of secreted lactic acid and maintain an appropriate intracellular pH (pHi > 6.8) for tumor growth, MCT1 and MCT4 expressions are upregulated in many tumors.[133] Increased MCT1 and/or MCT4 expression is associated with a poor prognosis in several types of human cancer, such as neuroblastoma, colorectal carcinoma, non–small cell lung cancer, breast cancer, pancreatic ductal adenocarcinoma, melanoma, and gastrointestinal stromal tumors.[134] Growing evidence pointing toward the importance of lactic acid transport for tumor cell survival and proliferation has led to its recognition as an important drug target in cancer[135] and prompted many studies to investigate the role of MCT1 and MCT4 inhibition, gene silencing, or disruption.[136]

Through the export of high concentrations of lactic acid secreted by tumor cells, MCTs play a major role in regulating intracellular pH (pHi). Our group has developed a strong interest in studying different pHi-regulating systems as putative anticancer targets in hypoxic tumors,[137] and therefore we investigated pharmacological and genetic inhibition of lactic acid export as a possible novel anticancer strategy (indicated by horizontal triangles in figure 12.1). Pharmacological inhibition with the specific AstraZeneca MCT1/2 inhibitor (AR-C155858) has been found to induce a rapid decrease in pHi, suppressing tumor growth of Ras-transformed fibroblasts, expressing only MCT1.[138] Ectopic expression of MCT4 in these cells conferred resistance to MCT1/2 inhibition and restored full tumorigenic potential.[139] Similarly, expression of MCT4 in respiration-deficient *ras*-transformed fibroblasts increased the gradient between pHi and extracellular pH (pHe) and led to increased in vivo tumor growth.[140] In non–small cell lung cancer cell lines, genetic disruption of the MCTs chaperone CD174 with zinc finger nucleases reduced the glycolytic rate 2.0- to 3.5-fold and stimulated respiration, making *BSG*-null cells sensitive to inhibitors of mitochondrial respiration in vitro and in vivo.[141] In addition, *BSG*-KO in LS174T and U87 cells led to a decrease in MCT1 and MCT4 transport activity, resulting in intracellular accumulation of lactic and pyruvic acid, as well as consequent pHi decrease and glycolysis inhibition.[142] Similar to the *GPI*-KO and *LDHA/B*-DKO cells, *BSG*-KO cells were able to redirect their metabolism to OXPHOS and resume growth. However, increased reliance on the OXPHOS for survival made *BSG*-KO cells extremely sensitive to phenformin inhibition, which caused a rapid drop in cellular ATP, inducing cell death by "metabolic catastrophe" or energy crisis.[143]

However, most aggressive cancers like colon adenocarcinoma, glioblastoma, and non–small cell lung cancer express both MCT1 and MCT4 isoforms[144]; consequently, homozygous loss of *MCT4* sensitized them to the MCT1 inhibitor.[145] Accordingly, we showed that combined silencing of both MCT1 and MCT4 by short hairpin RNA, or silencing of their chaperone CD174, led to reduced tumor growth in LS174T cells.[146] We have confirmed this finding in LS174T and B16 cell lines, where either knockout of both *MCT1* and *MCT4* (*MCT1/4*-DKO) or double pharmacological inhibition with isoform-specific inhibitors AZ3965 (MCT1i) and AZ39 (MCT4i) abolished cell growth.[147] It is interesting to note that MCT1i or MCT4i alone did not have any effect on cell growth and that upon removal of both drugs (in combination) after a week of treatment cells were able to form colonies again, indicating that complete inhibition of MCT1 and MCT4 induced a cytostatic but not a cytotoxic effect.

In conclusion, the three examples of genetic disruption of glycolysis in different cancer cell lines discussed in this chapter point to a common feature—an extraordinary metabolic plasticity of cancer cells, which is the basis for their ability to adapt and survive in a dynamically changing and, with tumor progression, increasingly hostile tumor microenvironment. Exceptionally, inhibition of both MCT1 and MCT4 had cytostatic effects, due to the mTORC1 inhibition caused by intracellular accumulation of lactic acid.[148] In all three cases, the interruption of glycolytic flux resulted in abolishment of lactic acid secretion and increased reliance on OXPHOS for energetic needs and survival. This metabolic plasticity of cancer cells poses a big challenge for anticancer therapies targeting metabolism, as also reported by other groups,[149] but metabolic vulnerability and sensitivity to mitochondrial respiration inhibition could be explored as an effective anticancer approach.[150]

12.4   Metabolism, Immunoevasion, and Ferroptosis: Therapeutic Perspectives

We reported in the previous section a marked difference in tumor growth when blocking either the production (*LDHA/B*-DKO) or the secretion of lactic acid (*MCT1/4*-DKO). The lactate dehydrogenases block allowed the tumor to escape via OXPHOS at a twofold reduced growth rate.[151] In contrast, the lactic acid export block arrested tumor growth despite OXPHOS reactivation while maintaining cell viability.

The first reason for therapeutic optimism is that this cytostatic effect can be transformed into cell death (energy crisis) when MCTs inhibition is combined with a short exposure to the mitochondrial complex I inhibitor phenformin.[152]

The second reason for optimism with respect to therapeutic approaches aiming to control lactic acid production/export stems from insights into the strategies that evolve in glycolytic tumor cells to evade the immune system (figure 12.2). Tumor acidity appears central in reducing T-cell and natural killer–cell activation, tumor infiltration, interferon γ secretion, and reprogramming of tumor-associated macrophages into a noninflammatory phenotype.[153] Thus, a reduction in tumor acidity via inhibition of lactic acid export could prove valuable in future efforts to improve immune therapy strategies in the clinic.[154]

The third reason for optimism concerns the future exploitation of "ferroptosis"-induced cell death (figure 12.3). Tumor cells relying on OXPHOS following chemotherapeutic treatments are extremely dependent on glutathione (GSH) for survival. Erastin treatment, which phenocopies a cystine transporter knockout (*xCT*-KO), ablates GSH, inducing ferroptotic cell death[155] (figure 12.3). Of great interest is the recent work of Wang et al.,[156] showing that immunotherapy-activated CD8+ T cells increase tumor cell death via interferon γ–mediated inhibition of the cystine transporter xCT. Let's hope that these potential novel therapeutic approaches will meet their promises and gain synergy with therapeutic developments in the field of immune-checkpoint inhibitors.[157]

Acknowledgments

This research was supported by the University Côte d'Azur, IRCAN, CNRS, Centre A. Lacassagne in Nice, France, and by the Centre Scientifique de Monaco (CSM), Monaco, supported by a grant from GEMLUC.

Maša Ždralević was supported by a postdoctoral fellowship from the Fondation ARC on Cancer Research (grant PDF20151203643).

Notes

1. Palm, W., and Thompson, C. B. (2017) Nutrient acquisition strategies of mammalian cells. *Nature*. 546, 234–242. DeBerardinis, R. J., and Chandel, N. S. (2016) Fundamentals of cancer metabolism. *Sci. Adv.* 2, e1600200–e1600200.

2. Saxton, R. A., and Sabatini, D. M. (2017) mTOR signaling in growth, metabolism, and disease. *Cell*. 168, 960–976.

3. Palm, W., and Thompson, C. B. (2017) Nutrient acquisition strategies of mammalian cells. *Nature*. 546, 234–242. Dang, C. Van, and Kim, J. W. (2018) Convergence of cancer metabolism and immunity: an overview. *Biomol. Ther.* 26, 4–9.

4. Semenza, G. L. (2017) Hypoxia-inducible factors: coupling glucose metabolism and redox regulation with induction of the breast cancer stem cell phenotype. *EMBO J.* 36, 252–259.

5. Warburg, O. (1956) On the origin of cancer cells. *Science*. 123, 309–314.

6. Parks, S. K., Mueller-Klieser, W., and Pouyssegur, J. (2020) Lactate in the cancer microenvironment. *Annu. Rev. Cancer Biol.* 4, 141–158.

7. Palm, W., and Thompson, C. B. (2017) Nutrient acquisition strategies of mammalian cells. *Nature*. 546, 234–242. Mazure, N. M., and Pouysségur, J. (2010) Hypoxia-induced autophagy: cell death or cell survival? *Curr. Opin. Cell Biol.* 22, 177–180.

8. DeBerardinis, R. J., and Chandel, N. S. (2016) Fundamentals of cancer metabolism. *Sci. Adv.* 2, e1600200–e1600200.

9. Hensley, C. T., Faubert, B., Yuan, Q., Lev-Cohain, N., Jin, E., Kim, J., Jiang, L., Ko, B., Skelton, R., Loudat, L., Wodzak, M., Klimko, C., Mcmillan, E., Butt, Y., Ni, M., Oliver, D., Torrealba, J., Malloy, C. R., Kernstine, K., Lenkinski, R. E., and Deberardinis, R. J. (2016) Metabolic heterogeneity in human lung tumors HHS Public Access. *Cell*. 11, 681–694.

10. Miranda-Gonçalves, V., Lameirinhas, A., Henrique, R., and Jerónimo, C. (2018) Metabolism and epigenetic interplay in cancer: regulation and putative therapeutic targets. *Front. Genet.* 9, 1–21. Cavalli, G., and Heard, E. (2019) Advances in epigenetics link genetics to the environment and disease. *Nature*. 571, 489–499.

11. Hay, N. (2016) Reprogramming glucose metabolism in cancer: can it be exploited for cancer therapy? *Nat. Rev. Cancer*. 16, 635–649.

12. Warburg, O. (1956) On the origin of cancer cells. *Science*. 123, 309–314. Parks, S. K., Mueller-Klieser, W., and Pouyssegur, J. (2020) Lactate in the cancer microenvironment. *Annu. Rev. Cancer Biol.* 4, 148–151. Kroemer, G., and Pouysségur, J. (2008) Tumor cell metabolism: cancer's Achilles' heel. *Cancer Cell*. 13, 472–482. Vander Heiden, M. G., and DeBerardinis, R. J. (2017) Understanding the intersections between metabolism and cancer biology. *Cell*. 168, 657–669.

13. Brahimi-Horn, M. C., Bellot, G., and Pouysségur, J. (2011) Hypoxia and energetic tumour metabolism. *Curr. Opin. Genet. Dev.* 21, 67–72. Hsieh, A. L., Walton, Z. E., Altman, B. J., Stine, Z. E., and Dang, C. V. (2015) MYC and metabolism on the path to cancer. *Semin. Cell Dev. Biol.* 43, 11–21. Hubbi, M. E., and Semenza, G. L. (2015) Regulation of cell proliferation by hypoxia-inducible factors. *Am. J. Physiol. Cell Physiol.* 309, C775–C782.

14. Kim, J. W., Tchernyshyov, I., Semenza, G. L., and Dang, C. V. (2006) HIF-1-mediated expression of pyruvate dehydrogenase kinase: a metabolic switch required for cellular adaptation to hypoxia. *Cell Metab.* 3, 177–185. Chae, Y. C., Vaira, V., Caino, M. C., Tang, H., Seo, J. H., Kossenkov, A. V, Ottobrini, L., Martelli, C., Lucignani, G., Bertolini, I., Locatelli, M., Bryant, K. G., Ghosh, J. C., Lisanti, S., Ku, B., Bosari, S., Languino, L. R., Speicher, D. W., and Altieri, D. C. (2016) Mitochondrial Akt regulation of hypoxic tumor reprogramming. *Cancer Cell*. 30, 257–272.

15. Vander Heiden, M. G., Cantley, L. C., and Thompson, C. B. (2009) Understanding the Warburg effect: the metabolic requirements of cell proliferation. *Science (80-.)*. 324, 1029–1033.

16. Sonveaux, P., Végran, F., Schroeder, T., Wergin, M. C., Verrax, J., Rabbani, Z. N., De Saedeleer, C. J., Kennedy, K. M., Diepart, C., Jordan, B. F., Kelley, M. J., Gallez, B., Wahl, M. L., Feron, O., and Dewhirst, M.

W. (2008) Targeting lactate-fueled respiration selectively kills hypoxic tumor cells in mice. *J. Clin. Invest.* 118, 3930–3942.

17. Parks, S. K., Mueller-Klieser, W., and Pouyssegur, J. (2020) Lactate in the cancer microenvironment. *Annu. Rev. Cancer Biol.* 4, 141–158. San-Millán, I., and Brooks, G. A. (2017) Reexamining cancer metabolism: lactate production for carcinogenesis could be the purpose and explanation of the Warburg Effect. *Carcinogenesis.* 38, 119–133.

18. Luengo, A., Gui, D. Y., and Vander Heiden, M. G. (2017) Targeting metabolism for cancer therapy. *Cell Chem. Biol.* 24, 1161–1180. Martinez-Outschoorn, U. E., Peiris-Pagés, M., Pestell, R. G., Sotgia, F., and Lisanti, M. P. (2017) Cancer metabolism: a therapeutic perspective. *Nat. Rev. Clin. Oncol.* 14, 11–31.

19. Pusapati, R. V., Daemen, A., Wilson, C., Sandoval, W., Gao, M., Haley, B., Baudy, A. R., Hatzivassiliou, G., Evangelista, M., and Settleman, J. (2016) MTORC1-dependent metabolic reprogramming underlies escape from glycolysis addiction in cancer cells. *Cancer Cell.* 29, 548–562. Goldin, N., Arzoine, L., Heyfets, A., Israelson, A., Zaslavsky, Z., Bravman, T., Bronner, V., Notcovich, A., Shoshan-Barmatz, V., and Flescher, E. (2008) Methyl jasmonate binds to and detaches mitochondria-bound hexokinase. *Oncogene.* 27, 4636–4643. Pedersen, P. L. (2007) Warburg, me and Hexokinase 2: multiple discoveries of key molecular events underlying one of cancers' most common phenotypes, the "Warburg Effect," i.e., elevated glycolysis in the presence of oxygen. *J. Bioenerg. Biomembr.* 39, 211–222.

20. Martinez-Outschoorn, U. E., Peiris-Pagés, M., Pestell, R. G., Sotgia, F., and Lisanti, M. P. (2017) Cancer metabolism: a therapeutic perspective. *Nat. Rev. Clin. Oncol.* 14, 11–31.

21. Colegio, O. R., Chu, N.-Q., Szabo, A. L., Chu, T., Rhebergen, A. M., Jairam, V., Cyrus, N., Brokowski, C. E., Eisenbarth, S. C., Phillips, G. M., Cline, G. W., Phillips, A. J., and Medzhitov, R. (2014) Functional polarization of tumour-associated macrophages by tumour-derived lactic acid. *Nature.* 513, 559–563. Brand, A., Singer, K., Koehl, G. E., Kolitzus, M., Schoenhammer, G., Thiel, A., Matos, C., Bruss, C., Klobuch, S., Peter, K., Kastenberger, M., Bogdan, C., Schleicher, U., Mackensen, A., Ullrich, E., Fichtner-Feigl, S., Kesselring, R., Mack, M., Ritter, U., Schmid, M., Blank, C., Dettmer, K., Oefner, P. J., Hoffmann, P., Walenta, S., Geissler, E. K., Pouyssegur, J., Villunger, A., Steven, A., Seliger, B., Schreml, S., Haferkamp, S., Kohl, E., Karrer, S., Berneburg, M., Herr, W., Mueller-Klieser, W., Renner, K., and Kreutz, M. (2016) LDHA-associated lactic acid production blunts tumor immunosurveillance by T and NK cells. *Cell Metab.* 8, 657–671. Bohn, T., Rapp, S., Luther, N., Klein, M., Bruehl, T. J., Kojima, N., Aranda Lopez, P., Hahlbrock, J., Muth, S., Endo, S., Pektor, S., Brand, A., Renner, K., Popp, V., Gerlach, K., Vogel, D., Lueckel, C., Arnold-Schild, D., Pouyssegur, J., Kreutz, M., Huber, M., Koenig, J., Weigmann, B., Probst, H. C., von Stebut, E., Becker, C., Schild, H., Schmitt, E., and Bopp, T. (2018) Tumor immunoevasion via acidosis-dependent induction of regulatory tumor-associated macrophages. *Nat. Immunol.* 19, 1319–1329.

22. Vučetić, M., Cormerais, Y., Parks, S. K., and Pouysségur, J. (2017) The central role of amino acids in cancer redox homeostasis: vulnerability points of the cancer redox code. *Front. Oncol.* 7, 319. Daher, B., Parks, S. K., Durivault, J., Cormerais, Y., Baidarjad, H., Tambutté, E., Pouyssegur, J., and Vucetic, M. (2019) Genetic ablation of the cystine transporter xCT in PDAC cells inhibits mTORC1, growth, survival and tumor formation via nutrient and oxidative stresses. *Cancer Res.* 79, 3877–3890. Conrad, M., Angeli, J. P. F., Vandenabeele, P., and Stockwell, B. R. (2016) Regulated necrosis: disease relevance and therapeutic opportunities. *Nat. Rev. Drug Discov.* 15, 348–366. Dixon, S. J., Lemberg, K. M., Lamprecht, M. R., Skouta, R., Zaitsev, E. M., Gleason, C. E., Patel, D. N., Bauer, A. J., Cantley, A. M., Yang, W. S., Morrison, B., and Stockwell, B. R. (2012) Ferroptosis: an iron-dependent form of nonapoptotic cell death. *Cell.* 149, 1060–1072. Wang, W., Green, M., Choi, J. E., Gijón, M., Kennedy, P. D., Johnson, J. K., Liao, P., Lang, X., Kryczek, I., Sell, A., Xia, H., Zhou, J., Li, G., Li, J., Li, W., Wei, S., Vatan, L., Zhang, H., Szeliga, W., Gu, W., Liu, R., Lawrence, T. S., Lamb, C., Tanno, Y., Cieslik, M., Stone, E., Georgiou, G., Chan, T. A., Chinnaiyan, A., and Zou, W. (2019) CD8 + T cells regulate tumour ferroptosis during cancer immunotherapy. *Nature.* 569, 270–274. Friedmann Angeli, J. P., Krysko, D. V., and Conrad, M. (2019) Ferroptosis at the crossroads of cancer-acquired drug resistance and immune evasion. *Nat. Rev. Cancer.* 19, 405–414.

23. Kroemer, G., and Pouysségur, J. (2008) Tumor cell metabolism: cancer's Achilles' heel. *Cancer Cell.* 13, 472–482. Vander Heiden, M. G., Cantley, L. C., and Thompson, C. B. (2009) Understanding the Warburg effect: the metabolic requirements of cell proliferation. *Science (80-.).* 324, 1029–1033. Parks, S. K., Mueller-Klieser, W., and Pouyssegur, J. (2020) Lactate in the cancer microenvironment. *Annu. Rev. Cancer Biol.* 4, 141–158.

24. Boonekamp, F. J., Dashko, S., van den Broek, M., Gehrmann, T., Daran, J.-M., and Daran-Lapujade, P. (2018) The genetic makeup and expression of the glycolytic and fermentative pathways are highly conserved

within the Saccharomyces genus. *Front. Genet.* 9, 1–17. Parks, S. K., Mueller-Klieser, W., and Pouyssegur, J. (2020) Lactate in the cancer microenvironment. *Annu. Rev. Cancer Biol.* 4, 141–158.

25. Brand, A., Singer, K., Koehl, G. E., Kolitzus, M., Schoenhammer, G., Thiel, A., Matos, C., Bruss, C., Klobuch, S., Peter, K., Kastenberger, M., Bogdan, C., Schleicher, U., Mackensen, A., Ullrich, E., Fichtner-Feigl, S., Kesselring, R., Mack, M., Ritter, U., Schmid, M., Blank, C., Dettmer, K., Oefner, P. J., Hoffmann, P., Walenta, S., Geissler, E. K., Pouyssegur, J., Villunger, A., Steven, A., Seliger, B., Schreml, S., Haferkamp, S., Kohl, E., Karrer, S., Berneburg, M., Herr, W., Mueller-Klieser, W., Renner, K., and Kreutz, M. (2016) LDHA-associated lactic acid production blunts tumor immunosurveillance by T and NK cells. *Cell Metab.* 24, 657–671.

26. Bohn, T., Rapp, S., Luther, N., Klein, M., Bruehl, T. J., Kojima, N., Aranda Lopez, P., Hahlbrock, J., Muth, S., Endo, S., Pektor, S., Brand, A., Renner, K., Popp, V., Gerlach, K., Vogel, D., Lueckel, C., Arnold-Schild, D., Pouyssegur, J., Kreutz, M., Huber, M., Koenig, J., Weigmann, B., Probst, H. C., von Stebut, E., Becker, C., Schild, H., Schmitt, E., and Bopp, T. (2018) Tumor immunoevasion via acidosis-dependent induction of regulatory tumor-associated macrophages. *Nat. Immunol.* 19, 1319–1329.

27. Nishimura, K., Fukuda, A., and Hisatake, K. (2019) Mechanisms of the metabolic shift during somatic cell reprogramming. *Int. J. Mol. Sci.* 20, 2254.

28. Folmes, C. D. L., Nelson, T. J., Martinez-Fernandez, A., Arrell, D. K., Lindor, J. Z., Dzeja, P. P., Ikeda, Y., Perez-Terzic, C., and Terzic, A. (2011) Somatic oxidative bioenergetics transitions into pluripotency-dependent glycolysis to facilitate nuclear reprogramming. *Cell Metab.* 14, 264–271. Varum, S., Rodrigues, A. S., Moura, M. B., Momcilovic, O., Easley, C. A., Ramalho-Santos, J., Van Houten, B., and Schatten, G. (2011) Energy metabolism in human pluripotent stem cells and their differentiated counterparts. *PLoS One.* 6, e20914.

29. Rodrigues, A. S., Pereira, S. L., Correia, M., Gomes, A., Perestrelo, T., and Ramalho-Santos, J. (2015) Differentiate or die: 3-bromopyruvate and pluripotency in mouse embryonic stem cells. *PLoS One.* 10, e0135617.

30. Vander Heiden, M. G., Cantley, L. C., and Thompson, C. B. (2009) Understanding the Warburg effect: the metabolic requirements of cell proliferation. *Science (80-.).* 324, 1029–1033.

31. Brooks, G. A. (2009) Cell-cell and intracellular lactate shuttles. *J. Physiol.* 587, 5591–5600.

32. Fornazari, M., Nascimento, I. C., Nery, A. A., Caldeira da Silva, C. C., Kowaltowski, A. J., and Ulrich, H. (2011) Neuronal differentiation involves a shift from glucose oxidation to fermentation. *J. Bioenerg. Biomembr.* 43, 531–539.

33. Sukonina, V., Ma, H., Zhang, W., Bartesaghi, S., Subhash, S., Heglind, M., Foyn, H., Betz, M. J., Nilsson, D., Lidell, M. E., Naumann, J., Haufs-Brusberg, S., Palmgren, H., Mondal, T., Beg, M., Jedrychowski, M. P., Taskén, K., Pfeifer, A., Peng, X.-R., Kanduri, C., and Enerbäck, S. (2019) FOXK1 and FOXK2 regulate aerobic glycolysis. *Nature.* 566, 279–283.

34. Sukonina, V., Ma, H., Zhang, W., Bartesaghi, S., Subhash, S., Heglind, M., Foyn, H., Betz, M. J., Nilsson, D., Lidell, M. E., Naumann, J., Haufs-Brusberg, S., Palmgren, H., Mondal, T., Beg, M., Jedrychowski, M. P., Taskén, K., Pfeifer, A., Peng, X.-R., Kanduri, C., and Enerbäck, S. (2019) FOXK1 and FOXK2 regulate aerobic glycolysis. *Nature.* 566, 279–283. Pelletier, J., Bellot, G., Gounon, P., Lacas-Gervais, S., Pouysségur, J., and Mazure, N. M. (2012) Glycogen synthesis is induced in hypoxia by the hypoxia-inducible factor and promotes cancer cell survival. *Front. Oncol.* 2, 1–9.

35. Parks, S. K., Mueller-Klieser, W., and Pouyssegur, J. (2020) Lactate in the cancer microenvironment. *Annu. Rev. Cancer Biol.* 4, 141–158.

36. Daher, B., Parks, S. K., Durivault, J., Cormerais, Y., Baidarjad, H., Tambutté, E., Pouyssegur, J., and Vucetic, M. (2019) Genetic ablation of the cystine transporter xCT in PDAC cells inhibits mTORC1, growth, survival and tumor formation via nutrient and oxidative stresses. *Cancer Res.* 79, 3877–3890. Dixon, S. J., Lemberg, K. M., Lamprecht, M. R., Skouta, R., Zaitsev, E. M., Gleason, C. E., Patel, D. N., Bauer, A. J., Cantley, A. M., Yang, W. S., Morrison, B., and Stockwell, B. R. (2012) Ferroptosis: an iron-dependent form of nonapoptotic cell death. *Cell.* 149, 1060–1072. Wang, W., Green, M., Choi, J. E., Gijón, M., Kennedy, P. D., Johnson, J. K., Liao, P., Lang, X., Kryczek, I., Sell, A., Xia, H., Zhou, J., Li, G., Li, J., Li, W., Wei, S., Vatan, L., Zhang, H., Szeliga, W., Gu, W., Liu, R., Lawrence, T. S., Lamb, C., Tanno, Y., Cieslik, M., Stone, E., Georgiou, G., Chan, T. A., Chinnaiyan, A., and Zou, W. (2019) CD8 + T cells regulate tumour ferroptosis during cancer immunotherapy. *Nature.* 569, 270–274.

37. Kim, J.-W., Zeller, K. I., Wang, Y., Jegga, A. G., Aronow, B. J., O'Donnell, K. A., and Dang, C. V. (2004) Evaluation of Myc E-box phylogenetic footprints in glycolytic genes by chromatin immunoprecipitation assays. *Mol. Cell. Biol.* 24, 5923–5936.

38. Funasaka, T., Yanagawa, T., Hogan, V., and Raz, A. (2005) Regulation of phosphoglucose isomerase/autocrine motility factor expression by hypoxia. *FASEB J.* 19, 1422–30.

39. Pusapati, R. V., Daemen, A., Wilson, C., Sandoval, W., Gao, M., Haley, B., Baudy, A. R., Hatzivassiliou, G., Evangelista, M., and Settleman, J. (2016) MTORC1-dependent metabolic reprogramming underlies escape from glycolysis addiction in cancer cells. *Cancer Cell.* 29, 548–562.

40. Ma, Y.-T., Xing, X., Dong, B., Cheng, X.-J., Guo, T., Du, H., Wen, X.-Z., and Ji, J.-F. (2018) Higher autocrine motility factor/glucose-6-phosphate isomerase expression is associated with tumorigenesis and poorer prognosis in gastric cancer. *Cancer Manag. Res.* 10, 4969–4980.

41. Lyu, Z., Chen, Y., Guo, X., Zhou, F., Yan, Z., Xing, J., An, J., and Zhang, H. (2016) Genetic variants in glucose-6-phosphate isomerase gene as prognosis predictors in hepatocellular carcinoma. *Clin. Res. Hepatol. Gastroenterol.* 40, 698–704.

42. Pouysségur, J., Franchi, A., Salomon, J. C., and Silvestre, P. (1980) Isolation of a Chinese hamster fibroblast mutant defective in hexose transport and aerobic glycolysis: its use to dissect the malignant phenotype. *Proc. Natl. Acad. Sci. USA.* 77, 2698–2701.

43. Pouysségur, J., Franchi, A., Salomon, J. C., and Silvestre, P. (1980) Isolation of a Chinese hamster fibroblast mutant defective in hexose transport and aerobic glycolysis: its use to dissect the malignant phenotype. *Proc. Natl. Acad. Sci. USA.* 77, 2698–2701.

44. Pouysségur, J., Franchi, A., Salomon, J. C., and Silvestre, P. (1980) Isolation of a Chinese hamster fibroblast mutant defective in hexose transport and aerobic glycolysis: its use to dissect the malignant phenotype. *Proc. Natl. Acad. Sci. USA.* 77, 2698–2701. Pouysségur, J., Franchi, A., and Silvestre, P. (1980) Relationship between increased aerobic glycolysis and DNA synthesis initiation studied using glycolytic mutant fibroblasts. *Nature.* 287, 445–447.

45. de Padua, M. C., Delodi, G., Vučetić, M., Durivault, J., Vial, V., Bayer, P., Noleto, G. R., Mazure, N. M., Ždralević, M., and Pouysségur, J. (2017) Disrupting glucose-6-phosphate isomerase fully suppresses the "Warburg effect" and activates OXPHOS with minimal impact on tumor growth except in hypoxia. *Oncotarget.* 8, 87623–87637.

46. de Padua, M. C., Delodi, G., Vučetić, M., Durivault, J., Vial, V., Bayer, P., Noleto, G. R., Mazure, N. M., Ždralević, M., and Pouysségur, J. (2017) Disrupting glucose-6-phosphate isomerase fully suppresses the "Warburg effect" and activates OXPHOS with minimal impact on tumor growth except in hypoxia. *Oncotarget.* 8, 87623–87637.

47. Pouysségur, J., Franchi, A., Salomon, J. C., and Silvestre, P. (1980) Isolation of a Chinese hamster fibroblast mutant defective in hexose transport and aerobic glycolysis: its use to dissect the malignant phenotype. *Proc. Natl. Acad. Sci. USA.* 77, 2698–701.

48. Deep, G., and Agarwal, R. (2013) Targeting tumor microenvironment with silibinin: promise and potential for a translational cancer chemopreventive strategy. *Curr. Cancer Drug Targets.* 13, 486–499. Polanski, R., Hodgkinson, C. L., Fusi, A., Nonaka, D., Priest, L., Kelly, P., Trapani, F., Bishop, P. W., White, A., Critchlow, S. E., Smith, P. D., Blackhall, F., Dive, C., and Morrow, C. J. (2014) Activity of the monocarboxylate transporter 1 inhibitor AZD3965 in small cell lung cancer. *Clin. Cancer Res.* 20, 926–937. Vander Heiden, M. G., Christofk, H. R., Schuman, E., Subtelny, A. O., Sharfi, H., Harlow, E. E., Xian, J., and Cantley, L. C. (2010) Identification of small molecule inhibitors of pyruvate kinase M2. *Biochem. Pharmacol.* 79, 1118–1124.

49. de Padua, M. C., Delodi, G., Vučetić, M., Durivault, J., Vial, V., Bayer, P., Noleto, G. R., Mazure, N. M., Ždralević, M., and Pouysségur, J. (2017) Disrupting glucose-6-phosphate isomerase fully suppresses the "Warburg effect" and activates OXPHOS with minimal impact on tumor growth except in hypoxia. *Oncotarget.* 8, 87623–87637.

50. de Padua, M. C., Delodi, G., Vučetić, M., Durivault, J., Vial, V., Bayer, P., Noleto, G. R., Mazure, N. M., Ždralević, M., and Pouysségur, J. (2017) Disrupting glucose-6-phosphate isomerase fully suppresses the "Warburg effect" and activates OXPHOS with minimal impact on tumor growth except in hypoxia. *Oncotarget.* 8, 87623–87637.

51. Pusapati, R. V., Daemen, A., Wilson, C., Sandoval, W., Gao, M., Haley, B., Baudy, A. R., Hatzivassiliou, G., Evangelista, M., and Settleman, J. (2016) MTORC1-dependent metabolic reprogramming underlies escape from glycolysis addiction in cancer cells. *Cancer Cell.* 29, 548–562.

52. Pouysségur, J., Franchi, A., Salomon, J. C., and Silvestre, P. (1980) Isolation of a Chinese hamster fibroblast mutant defective in hexose transport and aerobic glycolysis: its use to dissect the malignant phenotype. *Proc.*

Natl. Acad. Sci. USA. 77, 2698–2701. Ullrey, D. B., Franchi, A., Pouyssegur, J., and Kalckar, H. M. (1982) Down-regulation of the hexose transport system: metabolic basis studied with a fibroblast mutant lacking phosphoglucose isomerase. *Proc. Natl. Acad. Sci. USA.* 79, 3777–3779.

53. Stoltzman, C. A., Kaadige, M. R., Peterson, C. W., and Ayer, D. E. (2011) MondoA senses non-glucose sugars: regulation of thioredoxin-interacting protein (TXNIP) and the hexose transport curb. *J. Biol. Chem.* 286, 38027–38034. Wu, N., Zheng, B., Shaywitz, A., Dagon, Y., Tower, C., Bellinger, G., Shen, C.-H., Wen, J., Asara, J., McGraw, T. E., Kahn, B. B., and Cantley, L. C. (2013) AMPK-dependent degradation of TXNIP upon energy stress leads to enhanced glucose uptake via GLUT1. *Mol. Cell.* 49, 1167–1175.

54. de Padua, M. C., Delodi, G., Vučetić, M., Durivault, J., Vial, V., Bayer, P., Noleto, G. R., Mazure, N. M., Ždralević, M., and Pouysségur, J. (2017) Disrupting glucose-6-phosphate isomerase fully suppresses the "Warburg effect" and activates OXPHOS with minimal impact on tumor growth except in hypoxia. *Oncotarget.* 8, 87623–87637.

55. de Padua, M. C., Delodi, G., Vučetić, M., Durivault, J., Vial, V., Bayer, P., Noleto, G. R., Mazure, N. M., Ždralević, M., and Pouysségur, J. (2017) Disrupting glucose-6-phosphate isomerase fully suppresses the "Warburg effect" and activates OXPHOS with minimal impact on tumor growth except in hypoxia. *Oncotarget.* 8, 87623–87637.

56. Pouysségur, J., Franchi, A., Salomon, J. C., and Silvestre, P. (1980) Isolation of a Chinese hamster fibroblast mutant defective in hexose transport and aerobic glycolysis: its use to dissect the malignant phenotype. *Proc. Natl. Acad. Sci. USA.* 77, 2698–2701. Pusapati, R. V., Daemen, A., Wilson, C., Sandoval, W., Gao, M., Haley, B., Baudy, A. R., Hatzivassiliou, G., Evangelista, M., and Settleman, J. (2016) MTORC1-dependent metabolic reprogramming underlies escape from glycolysis addiction in cancer cells. *Cancer Cell.* 29, 548–562.

57. Pusapati, R. V., Daemen, A., Wilson, C., Sandoval, W., Gao, M., Haley, B., Baudy, A. R., Hatzivassiliou, G., Evangelista, M., and Settleman, J. (2016) MTORC1-dependent metabolic reprogramming underlies escape from glycolysis addiction in cancer cells. *Cancer Cell.* 29, 548–562.

58. Goldberg, E. (1971) Immunochemical specificity of lactate dehydrogenase-X. *Proc. Natl. Acad. Sci. USA.* 68, 349–352.

59. Krieg, A., Rosenblum, L., and Henry, J. (1967) Lactate dehydrogenase. *Clin. Chem.* 13, 196–203. Gallo, M., Sapio, L., Spina, A., Naviglio, D., Calogero, A., and Naviglio, S. (2015) Lactic dehydrogenase and cancer: an overview. *Front. Biosci.* 20, 1234–1249.

60. Gallo, M., Sapio, L., Spina, A., Naviglio, D., Calogero, A., and Naviglio, S. (2015) Lactic dehydrogenase and cancer: an overview. *Front. Biosci.* 20, 1234–1249.

61. Koen, A. L., and Goodman, M. (1969) Lactate dehydrogenase isozymes: qualitative and quantitative changes during primate evolution. *Biochem. Genet.* 3, 457–474.

62. Gallo, M., Sapio, L., Spina, A., Naviglio, D., Calogero, A., and Naviglio, S. (2015) Lactic dehydrogenase and cancer: an overview. *Front. Biosci.* 20, 1234–1249.

63. Brooks, G. A. (2009) Cell-cell and intracellular lactate shuttles. *J. Physiol.* 587, 5591–5600.

64. Brooks, G. A., Dubouchaud, H., Brown, M., Sicurello, J. P., and Butz, C. E. (1999) Role of mitochondrial lactate dehydrogenase and lactate oxidation in the intracellular lactate shuttle. *Proc. Natl. Acad. Sci. USA.* 96, 1129–1134.

65. Grosse, F., Nasheuer, H. P., Scholtissek, S., and Schomburg, U. (1986) Lactate dehydrogenase and glyceraldehyde-phosphate dehydrogenase are single-stranded DNA-binding proteins that affect the DNA-polymerase-alpha-primase complex. *Eur. J. Biochem.* 160, 459–467.

66. Uhlén, M., Fagerberg, L., Hallström, B. M., Lindskog, C., Oksvold, P., Mardinoglu, A., Sivertsson, Å., Kampf, C., Sjöstedt, E., Asplund, A., Olsson, I. M., Edlund, K., Lundberg, E., Navani, S., Szigyarto, C. A. K., Odeberg, J., Djureinovic, D., Takanen, J. O., Hober, S., Alm, T., Edqvist, P. H., Berling, H., Tegel, H., Mulder, J., Rockberg, J., Nilsson, P., Schwenk, J. M., Hamsten, M., Von Feilitzen, K., Forsberg, M., Persson, L., Johansson, F., Zwahlen, M., Von Heijne, G., Nielsen, J., and Pontén, F. (2015) Tissue-based map of the human proteome. *Science* 347, 1260419.

67. Liu, Y., Guo, J. Z., Liu, Y., Wang, K., Ding, W., Wang, H., Liu, X., Zhou, S., Lu, X. C., Yang, H. Bin, Xu, C., Gao, W., Zhou, L., Wang, Y. P., Hu, W., Wei, Y., Huang, C., and Lei, Q. Y. (2018) Nuclear lactate dehydrogenase A senses ROS to produce α-hydroxybutyrate for HPV-induced cervical tumor growth. *Nat. Commun.* 9, 4429.

68. Matrisian, L. M., Rautmann, G., Magun, B. E., and Breathnach, R. (1985) Epidermal growth factor or serum stimulation of rat fibroblasts induces an elevation in mRNA levels for lactate dehydrogenase and other glycolytic

enzymes. *Nucleic Acids Res.* 13, 711–726. Boussouar, F., Grataroli, R., Ji, N., and Benahmed, M. (1999) Tumor necrosis factor-α stimulates lactate dehydrogenase a expression in porcine cultured sertoli cells: mechanisms of action. *Endocrinology.* 140, 3054–3062. Tian, D., Huang, D., Brown, R. C., and Jungmann, R. A. (1998) Protein kinase A stimulates binding of multiple proteins to a U-rich domain in the 3'-untranslated region of lactate dehydrogenase a mRNA that is required for the regulation of mRNA stability. *J. Biol. Chem.* 273, 28454–28460. Boussouar, F., and Benahmed, M. (1999) Epidermal growth factor regulates glucose metabolism through lactate dehydrogenase A messenger ribonucleic acid expression in cultured porcine Sertoli cells. *Biol. Reprod.* 61, 1139–1145. Huang, D., and Jungmann, R. A. (1995) Transcriptional regulation of the lactate dehydrogenase A subunit gene by the phorbol ester 12-O-tetradecanoylphorbol-13-acetate. *Mol. Cell. Endocrinol.* 108, 87–94.

69. Shim, H., Dolde, C., Lewis, B. C., Wu, C.-S., Dang, G., Jungmann, R. A., Dalla-Favera, R., and Dang, C. V. (1997) c-Myc transactivation of LDH-A: implications for tumor metabolism and growth. *Biochemistry.* 94, 6658–6663.

70. Short, M. L., Huang, D., Milkowski, D. M., Short, S., Kunstman, K., Soong, C. J., Chung, K. C., and Jungmann, R. A. (1994) Analysis of the rat lactate dehydrogenase A subunit gene promoter/regulatory region. *Biochem J.* 304(Pt 2), 391–398.

71. Koukourakis, M. I., Giatromanolaki, A., Sivridis, E., Bougioukas, G., Didilis, V., Gatter, K. C., and Harris, A. L. (2003) Lactate dehydrogenase-5 (LDH-5) overexpression in non-small-cell lung cancer tissues is linked to tumour hypoxia, angiogenic factor production and poor prognosis. *Br J Cancer.* 89, 877–885. Semenza, G. L., Jiang, B. H., Leung, S. W., Passantino, R., Concordet, J. P., Maire, P., and Giallongo, A. (1996) Hypoxia response elements in the aldolase A, enolase 1, and lactate dehydrogenase A gene promoters contain essential binding sites for hypoxia-inducible factor 1. *J. Biol. Chem.* 271, 32529–32537.

72. Fan, J., Hitosugi, T., Chung, T.-W., Xie, J., Ge, Q., Gu, T.-L., Polakiewicz, R. D., Chen, G. Z., Boggon, T. J., Lonial, S., Khuri, F. R., Kang, S., and Chen, J. (2011) Tyrosine phosphorylation of lactate dehydrogenase A is important for NADH/NAD+ redox homeostasis in cancer cells. *Mol. Cell. Biol.* 31, 4938–4950.

73. Zhao, D., Xiong, Y., Lei, Q.-Y., Guan, K.-L., and Mito, N. (2013) LDH-A acetylation: implication in cancer. *Oncotarget.* 4, 802–803.

74. Fan, J., Hitosugi, T., Chung, T.-W., Xie, J., Ge, Q., Gu, T.-L., Polakiewicz, R. D., Chen, G. Z., Boggon, T. J., Lonial, S., Khuri, F. R., Kang, S., and Chen, J. (2011) Tyrosine phosphorylation of lactate dehydrogenase A is important for NADH/NAD+ redox homeostasis in cancer cells. *Mol. Cell. Biol.* 31, 4938–4950.

75. Zhao, D., Xiong, Y., Lei, Q.-Y., Guan, K.-L., and Mito, N. (2013) LDH-A acetylation: implication in cancer. *Oncotarget.* 4, 802–803.

76. Augoff, K., Hryniewicz-Jankowska, A., and Tabola, R. (2015) Lactate dehydrogenase 5: an old friend and a new hope in the war on cancer. *Cancer Lett.* 358, 1–7.

77. Ferraris, A. M., Giuntini, P., and Gaetani, G. F. (1979) Serum lactic dehydrogenase as a prognostic tool for non-Hodgkin lymphomas. *Blood.* 54, 928–932.

78. García, R., Hernández, J. M., Caballero, M. D., González, M., Galende, J., del Cañizo, M. C., Vázquez, L., and San Miguel, J. F. (1993) Serum lactate dehydrogenase level as a prognostic factor in Hodgkin's disease. *Br. J. Cancer.* 68, 1227–1231.

79. Gkotzamanidou, M., Kastritis, E., Roussou, M., Migkou, M., Gavriatopoulou, M., Nikitas, N., Gika, D., Mparmparousi, D., Matsouka, C., Terpos, E., and Dimopoulos, M. A. (2011) Increased serum lactate dehydrogenase should be included among the variables that define very-high-risk multiple myeloma. *Clin. Lymphoma, Myeloma Leuk.* 11, 409–413.

80. Bouafia, F., Drai, J., Bienvenu, J., Thieblemont, C., Espinouse, D., Salles, G., and Coiffier, B. (1999) Profiles and prognostic values of serum LDH isoenzymes in patients with haematopoietic malignancies. *Bull. Cancer.* 91, E229–E240.

81. Lossos, I. S., Intrator, O., Berkman, N., and Breuer, R. (1999) Lactate dehydrogenase isoenzyme analysis for the diagnosis of pleural effusion in haemato-oncological patients. *Respir. Med.* 93, 338–341. Sevinc, A., Sari, R., and Fadillioglu, E. (2005) The utility of lactate dehydrogenase isoenzyme pattern in the diagnostic evaluation of malignant and nonmalignant ascites. *J. Natl. Med. Assoc.* 97, 79–84. Nishikawa, A., Tanaka, T., Takeuchi, T., Fujihiro, S., and Mori, H. (1991) The diagnostic significance of lactate dehydrogenase isoenzymes in urinary cytology. *Br. J. Cancer.* 63, 819–821. Carda-Abella, P., Perez-Cuadrado, S., Lara-Baruque, S., Gil-Grande, L., and Nuñez-Puertas, A. (1982) LDH isoenzyme patterns in tumors, polyps, and uninvolved mucosa of human cancerous colon. *Cancer.* 49, 80–83. Langvad, E., and Jemec, B. (1975) Prediction of local recurrence in colorectal

carcinoma: an LDH isoenzymatic assay. *Br. J. Cancer*. 31, 661–664. Nishitani, K., Namba, M., and Kimoto, T. (1981) Lactate dehydrogenase isozyme patterns of normal human fibroblasts and their in vitro-transformed counterparts obtained by treatment with CO-60 gamma-rays, SV40 or 4-nitroquinoline 1-oxide. *Gan.* 72, 300–304.

82. Kolev, Y., Uetake, H., Takagi, Y., and Sugihara, K. (2008) Lactate dehydrogenase-5 (LDH-5) expression in human gastric cancer: association with hypoxia-inducible factor (HIF-1alpha) pathway, angiogenic factors production and poor prognosis. *Ann. Surg. Oncol.* 15, 2336–2344. Timperley, W. R. (1971) Lactate dehydrogenase isoenzymes in tumours of the nervous system. *Acta Neuropathol.* 19, 20–24. Caltrider, N. D., and Lehman, J. M. (1975) Changes in lactate dehydrogenase enzyme pattern in chinese hamster cells infected and transformed with simian virus 40. *Cancer Res.* 35, 1944–1949.

83. Kolev, Y., Uetake, H., Takagi, Y., and Sugihara, K. (2008) Lactate dehydrogenase-5 (LDH-5) expression in human gastric cancer: association with hypoxia-inducible factor (HIF-1alpha) pathway, angiogenic factors production and poor prognosis. *Ann. Surg. Oncol.* 15, 2336–2344.

84. Giatromanolaki, A., Sivridis, E., Gatter, K. C., Turley, H., Harris, A. L., and Koukourakis, M. I. (2006) Lactate dehydrogenase 5 (LDH-5) expression in endometrial cancer relates to the activated VEGF/VEGFR2(KDR) pathway and prognosis. *Gynecol. Oncol.* 103, 912–918.

85. Koukourakis, M. I., Giatromanolaki, A., Sivridis, E., Bougioukas, G., Didilis, V., Gatter, K. C., and Harris, A. L. (2003) Lactate dehydrogenase-5 (LDH-5) overexpression in non-small-cell lung cancer tissues is linked to tumour hypoxia, angiogenic factor production and poor prognosis. *Br J Cancer*. 89, 877–885.

86. Koukourakis, M. I., Giatromanolaki, A., Simopoulos, C., Polychronidis, A., and Sivridis, E. (2005) Lactate dehydrogenase 5 (LDH5) relates to up-regulated hypoxia inducible factor pathway and metastasis in colorectal cancer. *Clin. Exp. Metastasis*. 22, 25–30.

87. Lu, R., Jiang, M., Chen, Z., Xu, X., Hu, H., Zhao, X., Gao, X., and Guo, L. (2013) Lactate dehydrogenase 5 expression in non-Hodgkin lymphoma is associated with the induced hypoxia regulated protein and poor prognosis. *PLoS One*. 8, 1–8.

88. Hirschhaeuser, F., Sattler, U. G. A., and Mueller-Klieser, W. (2011) Lactate: a metabolic key player in cancer. *Cancer Res.* 71, 6921–6925. Parks, S. K., Mueller-Klieser, W., and Pouyssegur, J. (2020) Lactate in the cancer microenvironment. *Annu. Rev. Cancer Biol.* 4, 141–158.

89. Kim, H. S., Lee, H. E., Yang, H. K., and Kim, W. H. (2014) High lactate dehydrogenase 5 expression correlates with high tumoral and stromal vascular endothelial growth factor expression in gastric cancer. *Pathobiology*. 81, 78–85.

90. Yao, F., Zhao, T., Zhong, C., Zhu, J., and Zhao, H. (2013) LDHA is necessary for the tumorigenicity of esophageal squamous cell carcinoma. *Tumor Biol.* 34, 25–31.

91. Grimm, M., Alexander, D., Munz, A., Hoffmann, J., and Reinert, S. (2013) Increased LDH5 expression is associated with lymph node metastasis and outcome in oral squamous cell carcinoma. *Clin. Exp. Metastasis*. 30, 529–540.

92. Boudreau, A., Purkey, H. E., Hitz, A., Robarge, K., Peterson, D., Labadie, S., Kwong, M., Hong, R., Gao, M., Del Nagro, C., Pusapati, R., Ma, S., Salphati, L., Pang, J., Zhou, A., Lai, T., Li, Y., Chen, Z., Wei, B., Yen, I., Sideris, S., McCleland, M., Firestein, R., Corson, L., Vanderbilt, A., Williams, S., Daemen, A., Belvin, M., Eigenbrot, C., Jackson, P. K., Malek, S., Hatzivassiliou, G., Sampath, D., Evangelista, M., and O'Brien, T. (2016) Metabolic plasticity underpins innate and acquired resistance to LDHA inhibition. *Nat. Chem. Biol.* 12, 779–786. Fantin, V. R., St-Pierre, J., and Leder, P. (2006) Attenuation of LDH-A expression uncovers a link between glycolysis, mitochondrial physiology, and tumor maintenance. *Cancer Cell*. 9, 425–434. Le, A., Cooper, C. R., Gouw, A. M., Dinavahi, R., Maitra, A., Deck, L. M., Royer, R. E., Vander Jagt, D. L., Semenza, G. L., and Dang, C. V. (2010) Inhibition of lactate dehydrogenase A induces oxidative stress and inhibits tumor progression. *Proc. Natl. Acad. Sci.* 107, 2037–2042. Sheng, S. L., Liu, J. J., Dai, Y. H., Sun, X. G., Xiong, X. P., and Huang, G. (2012) Knockdown of lactate dehydrogenase A suppresses tumor growth and metastasis of human hepatocellular carcinoma. *FEBS J.* 279, 3898–3910. Wang, Z. Y., Loo, T. Y., Shen, J. G., Wang, N., Wang, D. M., Yang, D. P., Mo, S. L., Guan, X. Y., and Chen, J. P. (2012) LDH-A silencing suppresses breast cancer tumorigenicity through induction of oxidative stress mediated mitochondrial pathway apoptosis. *Breast Cancer Res. Treat.* 131, 791–800. Xie, H., Hanai, J. I., Ren, J. G., Kats, L., Burgess, K., Bhargava, P., Signoretti, S., Billiard, J., Duffy, K. J., Grant, A., Wang, X., Lorkiewicz, P. K., Schatzman, S., Bousamra, M., Lane, A. N., Higashi, R. M., Fan, T. W. M., Pandolfi, P. P., Sukhatme, V. P., and Seth, P. (2014) Targeting lactate dehydrogenase-A inhibits tumorigenesis and tumor progression in mouse models of lung cancer and impacts tumor-initiating cells. *Cell Metab.* 19, 795–809.

93. Sheng, S. L., Liu, J. J., Dai, Y. H., Sun, X. G., Xiong, X. P., and Huang, G. (2012) Knockdown of lactate dehydrogenase A suppresses tumor growth and metastasis of human hepatocellular carcinoma. *FEBS J.* 279, 3898–3910. Le, A., Cooper, C. R., Gouw, A. M., Dinavahi, R., Maitra, A., Deck, L. M., Royer, R. E., Vander Jagt, D. L., Semenza, G. L., and Dang, C. V. (2010) Inhibition of lactate dehydrogenase A induces oxidative stress and inhibits tumor progression. *Proc. Natl. Acad. Sci.* 107, 2037–2042. Wang, Z. Y., Loo, T. Y., Shen, J. G., Wang, N., Wang, D. M., Yang, D. P., Mo, S. L., Guan, X. Y., and Chen, J. P. (2012) LDH-A silencing suppresses breast cancer tumorigenicity through induction of oxidative stress mediated mitochondrial pathway apoptosis. *Breast Cancer Res. Treat.* 131, 791–800.

94. Fantin, V. R., St-Pierre, J., and Leder, P. (2006) Attenuation of LDH-A expression uncovers a link between glycolysis, mitochondrial physiology, and tumor maintenance. *Cancer Cell.* 9, 425–434.

95. Le, A., Cooper, C. R., Gouw, A. M., Dinavahi, R., Maitra, A., Deck, L. M., Royer, R. E., Vander Jagt, D. L., Semenza, G. L., and Dang, C. V. (2010) Inhibition of lactate dehydrogenase A induces oxidative stress and inhibits tumor progression. *Proc. Natl. Acad. Sci. USA.* 107, 2037–2042.

96. Rani, R., and Kumar, V. (2016) Recent update on human lactate dehydrogenase enzyme 5 (h LDH5) inhibitors: a promising approach for cancer chemotherapy. *J. Med. Chem.* 59, 487–496.

97. Rani, R., and Kumar, V. (2016) Recent update on human lactate dehydrogenase enzyme 5 (h LDH5) inhibitors: a promising approach for cancer chemotherapy. *J. Med. Chem.* 59, 487–496.

98. Boudreau, A., Purkey, H. E., Hitz, A., Robarge, K., Peterson, D., Labadie, S., Kwong, M., Hong, R., Gao, M., Del Nagro, C., Pusapati, R., Ma, S., Salphati, L., Pang, J., Zhou, A., Lai, T., Li, Y., Chen, Z., Wei, B., Yen, I., Sideris, S., McCleland, M., Firestein, R., Corson, L., Vanderbilt, A., Williams, S., Daemen, A., Belvin, M., Eigenbrot, C., Jackson, P. K., Malek, S., Hatzivassiliou, G., Sampath, D., Evangelista, M., and O'Brien, T. (2016) Metabolic plasticity underpins innate and acquired resistance to LDHA inhibition. *Nat. Chem. Biol.* 12, 779–786.

99. Boudreau, A., Purkey, H. E., Hitz, A., Robarge, K., Peterson, D., Labadie, S., Kwong, M., Hong, R., Gao, M., Del Nagro, C., Pusapati, R., Ma, S., Salphati, L., Pang, J., Zhou, A., Lai, T., Li, Y., Chen, Z., Wei, B., Yen, I., Sideris, S., McCleland, M., Firestein, R., Corson, L., Vanderbilt, A., Williams, S., Daemen, A., Belvin, M., Eigenbrot, C., Jackson, P. K., Malek, S., Hatzivassiliou, G., Sampath, D., Evangelista, M., and O'Brien, T. (2016) Metabolic plasticity underpins innate and acquired resistance to LDHA inhibition. *Nat. Chem. Biol.* 12, 779–786.

100. Ždralević, M., Brand, A., Di Ianni, L., Dettmer, K., Reinders, J., Singer, K., Peter, K., Schnell, A., Bruss, C., Decking, S.-M., Koehl, G., Felipe-Abrio, B., Durivault, J., Bayer, P., Evangelista, M., O'Brien, T., Oefner, P. J., Renner, K., Pouysségur, J., and Kreutz, M. (2018) Double genetic disruption of lactate dehydrogenases A and B is required to ablate the "Warburg effect" restricting tumor growth to oxidative metabolism. *J. Biol. Chem.* 293, 15947–15961.

101. Ždralević, M., Brand, A., Di Ianni, L., Dettmer, K., Reinders, J., Singer, K., Peter, K., Schnell, A., Bruss, C., Decking, S.-M., Koehl, G., Felipe-Abrio, B., Durivault, J., Bayer, P., Evangelista, M., O'Brien, T., Oefner, P. J., Renner, K., Pouysségur, J., and Kreutz, M. (2018) Double genetic disruption of lactate dehydrogenases A and B is required to ablate the "Warburg effect" restricting tumor growth to oxidative metabolism. *J. Biol. Chem.* 293, 15947–15961.

102. Ždralević, M., Brand, A., Di Ianni, L., Dettmer, K., Reinders, J., Singer, K., Peter, K., Schnell, A., Bruss, C., Decking, S.-M., Koehl, G., Felipe-Abrio, B., Durivault, J., Bayer, P., Evangelista, M., O'Brien, T., Oefner, P. J., Renner, K., Pouysségur, J., and Kreutz, M. (2018) Double genetic disruption of lactate dehydrogenases A and B is required to ablate the "Warburg effect" restricting tumor growth to oxidative metabolism. *J. Biol. Chem.* 293, 15947–15961.

103. Boudreau, A., Purkey, H. E., Hitz, A., Robarge, K., Peterson, D., Labadie, S., Kwong, M., Hong, R., Gao, M., Del Nagro, C., Pusapati, R., Ma, S., Salphati, L., Pang, J., Zhou, A., Lai, T., Li, Y., Chen, Z., Wei, B., Yen, I., Sideris, S., McCleland, M., Firestein, R., Corson, L., Vanderbilt, A., Williams, S., Daemen, A., Belvin, M., Eigenbrot, C., Jackson, P. K., Malek, S., Hatzivassiliou, G., Sampath, D., Evangelista, M., and O'Brien, T. (2016) Metabolic plasticity underpins innate and acquired resistance to LDHA inhibition. *Nat. Chem. Biol.* 12, 779–786.

104. Zha, X., Wang, F., Wang, Y., He, S., Jing, Y., Wu, X., and Zhang, H. (2011) Lactate dehydrogenase B is critical for hyperactive mTOR-mediated tumorigenesis. *Cancer Res.* 71, 13–18.

105. Zha, X., Wang, F., Wang, Y., He, S., Jing, Y., Wu, X., and Zhang, H. (2011) Lactate dehydrogenase B is critical for hyperactive mTOR-mediated tumorigenesis. *Cancer Res.* 71, 13–18.

106. Zha, X., Wang, F., Wang, Y., He, S., Jing, Y., Wu, X., and Zhang, H. (2011) Lactate dehydrogenase B is critical for hyperactive mTOR-mediated tumorigenesis. *Cancer Res.* 71, 13–18. McCleland, M. L., Adler, A. S.,

Shang, Y., Hunsaker, T., Truong, T., Peterson, D., Torres, E., Li, L., Haley, B., Stephan, J.-P., Belvin, M., Hatzivassiliou, G., Blackwood, E. M., Corson, L., Evangelista, M., Zha, J., and Firestein, R. (2012) An integrated genomic screen identifies LDHB as an essential gene for triple-negative breast cancer. *Cancer Res.* 72, 5812–23. McCleland, M. L., Adler, A. S., Deming, L., Cosino, E., Lee, L., Blackwood, E. M., Solon, M., Tao, J., Li, L., Shames, D., Jackson, E., Forrest, W. F., and Firestein, R. (2013) Lactate dehydrogenase B is required for the growth of KRAS-dependent lung adenocarcinomas. *Clin. Cancer Res.* 19, 773–784. Kim, J. H., Kim, E. L., Lee, Y. K., Park, C. B., Kim, B. W., Wang, H. J., Yoon, C. H., Lee, S. J., and Yoon, G. (2011) Decreased lactate dehydrogenase B expression enhances claudin 1-mediated hepatoma cell invasiveness via mitochondrial defects. *Exp. Cell Res.* 317, 1108–1118.

107. Leiblich, A., Cross, S. S., Catto, J. W. F., Phillips, J. T., Leung, H. Y., Hamdy, F. C., and Rehman, I. (2006) Lactate dehydrogenase-B is silenced by promoter hypermethylation in human prostate cancer. *Oncogene.* 25, 2953–2960.

108. Dennison, J. B., Molina, J. R., Mitra, S., Gonzalez-Angulo, A. M., Balko, J. M., Kuba, M. G., Sanders, M. E., Pinto, J. A., Gomez, H. L., Arteaga, C. L., Brown, R. E., and Mills, G. B. (2013) Lactate dehydrogenase B: a metabolic marker of response to neoadjuvant chemotherapy in breast cancer. *Clin. Cancer Res.* 19, 3703–3713.

109. Cui, J., Quan, M., Jiang, W., Hu, H., Jiao, F., Li, N., Jin, Z., Wang, L., Wang, Y., and Wang, L. (2015) Suppressed expression of LDHB promotes pancreatic cancer progression via inducing glycolytic phenotype. *Med. Oncol.* 32, 143.

110. McCleland, M. L., Adler, A. S., Shang, Y., Hunsaker, T., Truong, T., Peterson, D., Torres, E., Li, L., Haley, B., Stephan, J.-P., Belvin, M., Hatzivassiliou, G., Blackwood, E. M., Corson, L., Evangelista, M., Zha, J., and Firestein, R. (2012) An integrated genomic screen identifies LDHB as an essential gene for triple-negative breast cancer. *Cancer Res.* 72, 5812–5823.

111. McCleland, M. L., Adler, A. S., Deming, L., Cosino, E., Lee, L., Blackwood, E. M., Solon, M., Tao, J., Li, L., Shames, D., Jackson, E., Forrest, W. F., and Firestein, R. (2013) Lactate dehydrogenase B is required for the growth of KRAS-dependent lung adenocarcinomas. *Clin. Cancer Res.* 19, 773–784.

112. Kinoshita, T., Nohata, N., Yoshino, H., Hanazawa, T., Kikawa, N., Fujimura, L., Chiyomaru, T., Kawakami, K., Enokida, H., Nakagawa, M., Okamoto, Y., and Seki, N. (2012) Tumor suppressive microRNA-375 regulates lactate dehydrogenase B in maxillary sinus squamous cell carcinoma. *Int. J. Oncol.* 40, 185–193.

113. Li, C., Chen, Y., Bai, P., Wang, J., Liu, Z., Wang, T., and Cai, Q. (2016) LDHB may be a significant predictor of poor prognosis in osteosarcoma. *Am. J. Transl. Res.* 8, 4831–4843.

114. Luo, Y., Yang, Z., Li, D., Liu, Z., Yang, L., Zou, Q., and Yuan, Y. (2017) LDHB and FABP4 are associated with progression and poor prognosis of pancreatic ductal adenocarcinomas. *Appl. Immunohistochem. Mol. Morphol.* 25, 351–357.

115. Koh, Y. W., Lee, S. J., and Park, S. Y. (2017) Prognostic significance of lactate dehydrogenase B according to histologic type of non-small-cell lung cancer and its association with serum lactate dehydrogenase. *Pathol. Res. Pract.* 213, 1134–1138.

116. Chen, R., Zhou, X., Yu, Z., Liu, J., and Huang, G. (2015) Low expression of LDHB correlates with unfavorable survival in hepatocellular carcinoma: strobe-compliant article. *Med. (United States).* 94, e1583.

117. Leiblich, A., Cross, S. S., Catto, J. W. F., Phillips, J. T., Leung, H. Y., Hamdy, F. C., and Rehman, I. (2006) Lactate dehydrogenase-B is silenced by promoter hypermethylation in human prostate cancer. *Oncogene.* 25, 2953–2960.

118. Dennison, J. B., Molina, J. R., Mitra, S., Gonzalez-Angulo, A. M., Balko, J. M., Kuba, M. G., Sanders, M. E., Pinto, J. A., Gomez, H. L., Arteaga, C. L., Brown, R. E., and Mills, G. B. (2013) Lactate dehydrogenase B: a metabolic marker of response to neoadjuvant chemotherapy in breast cancer. *Clin. Cancer Res.* 19, 3703–3713.

119. Sun, W., Zhang, X., Ding, X., Li, H., Geng, M., Xie, Z., Wu, H., and Huang, M. (2015) Lactate dehydrogenase B is associated with the response to neoadjuvant chemotherapy in oral squamous cell carcinoma. *PLoS One.* 10, 1–19.

120. Pavlides, S., Whitaker-Menezes, D., Castello-Cros, R., Flomenberg, N., Witkiewicz, A. K., Frank, P. G., Casimiro, M. C., Wang, C., Fortina, P., Addya, S., Pestell, R. G., Martinez-Outschoorn, U. E., Sotgia, F., and Lisanti, M. P. (2009) The reverse Warburg effect: aerobic glycolysis in cancer associated fibroblasts and the tumor stroma. *Cell Cycle.* 8, 3984–4001. Sonveaux, P., Végran, F., Schroeder, T., Wergin, M. C., Verrax, J.,

Rabbani, Z. N., De Saedeleer, C. J., Kennedy, K. M., Diepart, C., Jordan, B. F., Kelley, M. J., Gallez, B., Wahl, M. L., Feron, O., and Dewhirst, M. W. (2008) Targeting lactate-fueled respiration selectively kills hypoxic tumor cells in mice. *J. Clin. Invest.* 118, 3930–3942.

121. Ždralević, M., Brand, A., Di Ianni, L., Dettmer, K., Reinders, J., Singer, K., Peter, K., Schnell, A., Bruss, C., Decking, S.-M., Koehl, G., Felipe-Abrio, B., Durivault, J., Bayer, P., Evangelista, M., O'Brien, T., Oefner, P. J., Renner, K., Pouysségur, J., and Kreutz, M. (2018) Double genetic disruption of lactate dehydrogenases A and B is required to ablate the "Warburg effect" restricting tumor growth to oxidative metabolism. *J. Biol. Chem.* 293, 15947–15961.

122. Ždralević, M., Brand, A., Di Ianni, L., Dettmer, K., Reinders, J., Singer, K., Peter, K., Schnell, A., Bruss, C., Decking, S.-M., Koehl, G., Felipe-Abrio, B., Durivault, J., Bayer, P., Evangelista, M., O'Brien, T., Oefner, P. J., Renner, K., Pouysségur, J., and Kreutz, M. (2018) Double genetic disruption of lactate dehydrogenases A and B is required to ablate the "Warburg effect" restricting tumor growth to oxidative metabolism. *J. Biol. Chem.* 293, 15947–15961.

123. Ždralević, M., Brand, A., Di Ianni, L., Dettmer, K., Reinders, J., Singer, K., Peter, K., Schnell, A., Bruss, C., Decking, S.-M., Koehl, G., Felipe-Abrio, B., Durivault, J., Bayer, P., Evangelista, M., O'Brien, T., Oefner, P. J., Renner, K., Pouysségur, J., and Kreutz, M. (2018) Double genetic disruption of lactate dehydrogenases A and B is required to ablate the "Warburg effect" restricting tumor growth to oxidative metabolism. *J. Biol. Chem.* 293, 15947–15961.

124. Ždralević, M., Brand, A., Di Ianni, L., Dettmer, K., Reinders, J., Singer, K., Peter, K., Schnell, A., Bruss, C., Decking, S.-M., Koehl, G., Felipe-Abrio, B., Durivault, J., Bayer, P., Evangelista, M., O'Brien, T., Oefner, P. J., Renner, K., Pouysségur, J., and Kreutz, M. (2018) Double genetic disruption of lactate dehydrogenases A and B is required to ablate the "Warburg effect" restricting tumor growth to oxidative metabolism. *J. Biol. Chem.* 293, 15947–15961.

125. Ždralević, M., Brand, A., Di Ianni, L., Dettmer, K., Reinders, J., Singer, K., Peter, K., Schnell, A., Bruss, C., Decking, S.-M., Koehl, G., Felipe-Abrio, B., Durivault, J., Bayer, P., Evangelista, M., O'Brien, T., Oefner, P. J., Renner, K., Pouysségur, J., and Kreutz, M. (2018) Double genetic disruption of lactate dehydrogenases A and B is required to ablate the "Warburg effect" restricting tumor growth to oxidative metabolism. *J. Biol. Chem.* 293, 15947–15961.

126. Parks, S. K., Chiche, J., and Pouysségur, J. (2013) Disrupting proton dynamics and energy metabolism for cancer therapy. *Nat. Rev. Cancer.* 13, 611–623.

127. Parks, S. K., Mueller-Klieser, W., and Pouysségur, J. (2020) Lactate in the cancer microenvironment. *Annu. Rev. Cancer Biol.* 4, 141–158. San-Millán, I., and Brooks, G. A. (2016) Reexamining cancer metabolism: lactate production for carcinogenesis could be the purpose and explanation of the Warburg effect. *Carcinogenesis.* 38, bgw127. Brooks, G. A. (2002) Lactate shuttles in nature. *Biochem. Soc. Trans.* 30, 258–264. Groussard, C., Morel, I., Chevanne, M., Monnier, M., Cillard, J., and Delamarche, A. (2000) Free radical scavenging and antioxidant effects of lactate ion: an in vitro study. *J. Appl. Physiol.* 89, 169–175.

128. Halestrap, A. P. (2013) monocarboxylic acid transport. *Compr. Physiol.* 3, 1611–1643.

129. Halestrap, A. P. (2013) Monocarboxylic acid transport. *Compr. Physiol.* 3, 1611–1643.

130. Dhup, S., Dadhich, R. K., Porporato, P. E., and Sonveaux, P. (2012) Multiple biological activities of lactic acid in cancer: influences on tumor growth, angiogenesis and metastasis. *Curr. Pharm. Des.* 18, 1319–1330. Chiche, J., Fur, Y. Le, Vilmen, C., Frassineti, F., Daniel, L., Halestrap, A. P., Cozzone, P. J., Pouysségur, J., and Lutz, N. W. (2012) In vivo pH in metabolic-defective Ras-transformed fibroblast tumors: key role of the mono-carboxylate transporter, MCT4, for inducing an alkaline intracellular pH. *Int. J. Cancer.* 130, 1511–1520. Parks, S. K., Chiche, J., and Pouysségur, J. (2011) pH control mechanisms of tumor survival and growth. *J. Cell. Physiol.* 226, 299–308. Ullah, M. S., Davies, A. J., and Halestrap, A. P. (2006) The plasma membrane lactate transporter MCT4, but not MCT1, is up-regulated by hypoxia through a HIF-1α-dependent mechanism. *J. Biol. Chem.* 281, 9030–9037.

131. Gallagher, S. M., Castorino, J. J., Wang, D., and Philp, N. J. (2007) Monocarboxylate transporter 4 regulates maturation and trafficking of CD147 to the plasma membrane in the metastatic breast cancer cell line MDA-MB-231. *Cancer Res.* 67, 4182–4189. Kirk, P., Wilson, M. C., Heddle, C., Brown, M. H., Barclay, A. N., and Halestrap, A. P. (2000) CD147 is tightly associated with lactate transporters MCT1 and MCT4 and facilitates their cell surface expression. *EMBO J.* 19, 3896–904. Le Floch, R., Chiche, J., Marchiq, I., Naiken, T., Naïken, T., Ilc, K., Ilk, K., Murray, C. M., Critchlow, S. E., Roux, D., Simon, M.-P., and Pouysségur, J. (2011) CD147

subunit of lactate/H+ symporters MCT1 and hypoxia-inducible MCT4 is critical for energetics and growth of glycolytic tumors. *Proc. Natl. Acad. Sci. USA.* 108, 16663–16668.

132. Dhup, S., Dadhich, R. K., Porporato, P. E., and Sonveaux, P. (2012) Multiple biological activities of lactic acid in cancer: influences on tumor growth, angiogenesis and metastasis. *Curr. Pharm. Des.* 18, 1319–30. Sonveaux, P., Végran, F., Schroeder, T., Wergin, M. C., Verrax, J., Rabbani, Z. N., De Saedeleer, C. J., Kennedy, K. M., Diepart, C., Jordan, B. F., Kelley, M. J., Gallez, B., Wahl, M. L., Feron, O., and Dewhirst, M. W. (2008) Targeting lactate-fueled respiration selectively kills hypoxic tumor cells in mice. *J. Clin. Invest.* 118, 3930–3942.

133. Brahimi-Horn, M. C., Bellot, G., and Pouysségur, J. (2011) Hypoxia and energetic tumour metabolism. *Curr. Opin. Genet. Dev.* 21, 67–72. Chiche, J., Fur, Y. Le, Vilmen, C., Frassineti, F., Daniel, L., Halestrap, A. P., Cozzone, P. J., Pouysségur, J., and Lutz, N. W. (2012) In vivo pH in metabolic-defective Ras-transformed fibroblast tumors: key role of the monocarboxylate transporter, MCT4, for inducing an alkaline intracellular pH. *Int. J. Cancer.* 130, 1511–1520. Parks, S. K., Chiche, J., and Pouyssegur, J. (2011) pH control mechanisms of tumor survival and growth. *J. Cell. Physiol.* 226, 299–308.

134. Baenke, F., Dubuis, S., Brault, C., Weigelt, B., Dankworth, B., Griffiths, B., Jiang, M., Mackay, A., Saunders, B., Spencer-Dene, B., Ros, S., Stamp, G., Reis-Filho, J. S., Howell, M., Zamboni, N., and Schulze, A. (2015) Functional screening identifies MCT4 as a key regulator of breast cancer cell metabolism and survival. *J. Pathol.* 237, 152–165. Baltazar, F., Pinheiro, C., Morais-Santos, F., Azevedo-Silva, J., Queirós, O., Preto, A., and Casal, M. (2014) Monocarboxylate transporters as targets and mediators in cancer therapy response. *Histol. Histopathol.* 29, 1511–1524. Doyen, J., Trastour, C., Ettore, F., Peyrottes, I., Toussant, N., Gal, J., Ilc, K., Roux, D., Parks, S. K., Ferrero, J. M., and Pouysségur, J. (2014) Expression of the hypoxia-inducible monocarboxylate transporter MCT4 is increased in triple negative breast cancer and correlates independently with clinical outcome. *Biochem. Biophys. Res. Commun.* 451, 54–61. Kong, S. C., Nøhr-Nielsen, A., Zeeberg, K., Reshkin, S. J., Hoffmann, E. K., Novak, I., and Pedersen, S. F. (2016) Monocarboxylate transporters MCT1 and MCT4 regulate migration and invasion of pancreatic ductal adenocarcinoma cells. *Pancreas.* 45, 1036–1047. Pinheiro, C., Longatto-Filho, A., Scapulatempo, C., Ferreira, L., Martins, S., Pellerin, L., Rodrigues, M., Alves, V. A. F., Schmitt, F., and Baltazar, F. (2008) Increased expression of monocarboxylate transporters 1, 2, and 4 in colorectal carcinomas. *Virchows Arch.* 452, 139–146.

135. Le Floch, R., Chiche, J., Marchiq, I., Naiken, T., Naïken, T., Ilc, K., Ilk, K., Murray, C. M., Critchlow, S. E., Roux, D., Simon, M.-P., and Pouysségur, J. (2011) CD147 subunit of lactate/H+ symporters MCT1 and hypoxia-inducible MCT4 is critical for energetics and growth of glycolytic tumors. *Proc. Natl. Acad. Sci. USA.* 108, 16663–16668. Kaelin, W. G., and Thompson, C. B. (2010) Clues from cell metabolism. *Nature.* 465, 562–564. Schulze, A., and Harris, A. L. (2012) How cancer metabolism is tuned for proliferation and vulnerable to disruption. *Nature.* 491, 364–373. Vander Heiden, M. G. (2011) Targeting cancer metabolism: a therapeutic window opens. *Nat. Rev. Drug Discov.* 10, 671–684.

136. Doherty, J. R., Yang, C., Scott, K. E. N., Cameron, M. D., Fallahi, M., Li, W., Hall, M. A., Amelio, A. L., Mishra, J. K., Li, F., Tortosa, M., Genau, H. M., Rounbehler, R. J., Lu, Y., Dang, C. V, Kumar, K. G., Butler, A. A., Bannister, T. D., Hooper, A. T., Unsal-Kacmaz, K., Roush, W. R., and Cleveland, J. L. (2014) Blocking lactate export by inhibiting the Myc target MCT1 disables glycolysis and glutathione synthesis. *Cancer Res.* 74, 908–920. Granja, S., Marchiq, I., Le Floch, R., Moura, C. S., Baltazar, F., and Pouysségur, J. (2014) Disruption of BASIGIN decreases lactic acid export and sensitizes non-small cell lung cancer to biguanides independently of the LKB1 status. *Oncotarget.* 6, 1–14. Le Floch, R., Chiche, J., Marchiq, I., Naiken, T., Naïken, T., Ilc, K., Ilk, K., Murray, C. M., Critchlow, S. E., Roux, D., Simon, M.-P., and Pouysségur, J. (2011) CD147 subunit of lactate/H+ symporters MCT1 and hypoxia-inducible MCT4 is critical for energetics and growth of glycolytic tumors. *Proc. Natl. Acad. Sci. USA.* 108, 16663–16668. Marchiq, I., Le Floch, R., Roux, D., Simon, M. P., and Pouyssegur, J. (2015) Genetic disruption of lactate/H+ symporters (MCTs) and their subunit CD147/BASIGIN sensitizes glycolytic tumor cells to phenformin. *Cancer Res.* 75, 171–180.

137. Parks, S. K., Chiche, J., and Pouysségur, J. (2013) Disrupting proton dynamics and energy metabolism for cancer therapy. *Nat. Rev. Cancer.* 13, 611–623. Pouysségur, J., Dayan, F., and Mazure, N. M. (2006) Hypoxia signalling in cancer and approaches to enforce tumour regression. *Nature.* 441, 437–443.

138. Le Floch, R., Chiche, J., Marchiq, I., Naiken, T., Naïken, T., Ilc, K., Ilk, K., Murray, C. M., Critchlow, S. E., Roux, D., Simon, M.-P., and Pouysségur, J. (2011) CD147 subunit of lactate/H+ symporters MCT1 and hypoxia-inducible MCT4 is critical for energetics and growth of glycolytic tumors. *Proc. Natl. Acad. Sci. USA.* 108, 16663–16668.

139. Le Floch, R., Chiche, J., Marchiq, I., Naiken, T., Naïken, T., Ilc, K., Ilk, K., Murray, C. M., Critchlow, S. E., Roux, D., Simon, M.-P., and Pouysségur, J. (2011) CD147 subunit of lactate/H+ symporters MCT1 and hypoxia-

inducible MCT4 is critical for energetics and growth of glycolytic tumors. *Proc. Natl. Acad. Sci. USA.* 108, 16663–16668.

140. Chiche, J., Fur, Y. Le, Vilmen, C., Frassineti, F., Daniel, L., Halestrap, A. P., Cozzone, P. J., Pouysségur, J., and Lutz, N. W. (2012) In vivo pH in metabolic-defective Ras-transformed fibroblast tumors: key role of the monocarboxylate transporter, MCT4, for inducing an alkaline intracellular pH. *Int. J. Cancer.* 130, 1511–1520.

141. Granja, S., Marchiq, I., Le Floch, R., Moura, C. S., Baltazar, F., and Pouysségur, J. (2014) Disruption of BASIGIN decreases lactic acid export and sensitizes non-small cell lung cancer to biguanides independently of the LKB1 status. *Oncotarget.* 6, 1–14. Marchiq, I., Le Floch, R., Roux, D., Simon, M. P., and Pouyssegur, J. (2015) Genetic disruption of lactate/H+ symporters (MCTs) and their subunit CD147/BASIGIN sensitizes glycolytic tumor cells to phenformin. *Cancer Res.* 75, 171–180.

142. Marchiq, I., Le Floch, R., Roux, D., Simon, M. P., and Pouyssegur, J. (2015) Genetic disruption of lactate/H+ symporters (MCTs) and their subunit CD147/BASIGIN sensitizes glycolytic tumor cells to phenformin. *Cancer Res.* 75, 171–180.

143. Marchiq, I., Le Floch, R., Roux, D., Simon, M. P., and Pouyssegur, J. (2015) Genetic disruption of lactate/H+ symporters (MCTs) and their subunit CD147/BASIGIN sensitizes glycolytic tumor cells to phenformin. *Cancer Res.* 75, 171–180.

144. Le Floch, R., Chiche, J., Marchiq, I., Naiken, T., Naïken, T., Ilc, K., Ilk, K., Murray, C. M., Critchlow, S. E., Roux, D., Simon, M.-P., and Pouysségur, J. (2011) CD147 subunit of lactate/H+ symporters MCT1 and hypoxia-inducible MCT4 is critical for energetics and growth of glycolytic tumors. *Proc. Natl. Acad. Sci. USA.* 108, 16663–16668. Baltazar, F., Pinheiro, C., Morais-Santos, F., Azevedo-Silva, J., Queirós, O., Preto, A., and Casal, M. (2014) Monocarboxylate transporters as targets and mediators in cancer therapy response. *Histol. Histopathol.* 29, 1511–1524. Granja, S., Marchiq, I., Le Floch, R., Moura, C. S., Baltazar, F., and Pouysségur, J. (2014) Disruption of BASIGIN decreases lactic acid export and sensitizes non-small cell lung cancer to biguanides independently of the LKB1 status. *Oncotarget.* 6, 1–14. Marchiq, I., Le Floch, R., Roux, D., Simon, M. P., and Pouysségur, J. (2015) Genetic disruption of lactate/H+ symporters (MCTs) and their subunit CD147/BASIGIN sensitizes glycolytic tumor cells to phenformin. *Cancer Res.* 75, 171–180.

145. Marchiq, I., Le Floch, R., Roux, D., Simon, M. P., and Pouysségur, J. (2015) Genetic disruption of lactate/H+ symporters (MCTs) and their subunit CD147/BASIGIN sensitizes glycolytic tumor cells to phenformin. *Cancer Res.* 75, 171–180.

146. Le Floch, R., Chiche, J., Marchiq, I., Naiken, T., Naïken, T., Ilc, K., Ilk, K., Murray, C. M., Critchlow, S. E., Roux, D., Simon, M.-P., and Pouysségur, J. (2011) CD147 subunit of lactate/H+ symporters MCT1 and hypoxia-inducible MCT4 is critical for energetics and growth of glycolytic tumors. *Proc. Natl. Acad. Sci. USA.* 108, 16663–16668.

147. Ždralević, M., Vučetić, M., Daher, B., Marchiq, I., Parks, S. K., and Pouysségur, J. (2018) Disrupting the 'Warburg effect' re-routes cancer cells to OXPHOS offering a vulnerability point via 'ferroptosis'-induced cell death. *Adv. Biol. Regul.* 68, 55–63.

148. Chambard, J. C., and Pouyssegur, J. (1986) Intracellular pH controls growth factor-induced ribosomal protein S6 phosphorylation and protein synthesis in the G0→G1 transition of fibroblasts. *Exp. Cell Res.* 164, 282–294. Marchiq, I., Le Floch, R., Roux, D., Simon, M. P., and Pouysségur, J. (2015) Genetic disruption of lactate/H+ symporters (MCTs) and their subunit CD147/BASIGIN sensitizes glycolytic tumor cells to phenformin. *Cancer Res.* 75, 171–180. Balgi, A. D., Diering, G. H., Donohue, E., Lam, K. K. Y., Fonseca, B. D., Zimmerman, C., Numata, M., and Roberge, M. (2011) Regulation of mTORC1 signaling by pH. *PLoS One.* 6, e21549.

149. Boudreau, A., Purkey, H. E., Hitz, A., Robarge, K., Peterson, D., Labadie, S., Kwong, M., Hong, R., Gao, M., Del Nagro, C., Pusapati, R., Ma, S., Salphati, L., Pang, J., Zhou, A., Lai, T., Li, Y., Chen, Z., Wei, B., Yen, I., Sideris, S., McCleland, M., Firestein, R., Corson, L., Vanderbilt, A., Williams, S., Daemen, A., Belvin, M., Eigenbrot, C., Jackson, P. K., Malek, S., Hatzivassiliou, G., Sampath, D., Evangelista, M., and O'Brien, T. (2016) Metabolic plasticity underpins innate and acquired resistance to LDHA inhibition. *Nat. Chem. Biol.* 12, 779–786. Pusapati, R. V., Daemen, A., Wilson, C., Sandoval, W., Gao, M., Haley, B., Baudy, A. R., Hatzivassiliou, G., Evangelista, M., and Settleman, J. (2016) MTORC1-dependent metabolic reprogramming underlies escape from glycolysis addiction in cancer cells. *Cancer Cell.* 29, 548–562.

150. Parks, S. K., Chiche, J., and Pouysségur, J. (2013) Disrupting proton dynamics and energy metabolism for cancer therapy. *Nat. Rev. Cancer.* 13, 611–623.

151. Ždralević, M., Brand, A., Di Ianni, L., Dettmer, K., Reinders, J., Singer, K., Peter, K., Schnell, A., Bruss, C., Decking, S.-M., Koehl, G., Felipe-Abrio, B., Durivault, J., Bayer, P., Evangelista, M., O'Brien, T., Oefner,

P. J., Renner, K., Pouysségur, J., and Kreutz, M. (2018) Double genetic disruption of lactate dehydrogenases A and B is required to ablate the "Warburg effect" restricting tumor growth to oxidative metabolism. *J. Biol. Chem.* 293, 15947–15961.

152. Benjamin, D., Robay, D., Hindupur, S. K., Pohlmann, J., Colombi, M., El-Shemerly, M. Y., Maira, S. M., Moroni, C., Lane, H. A., and Hall, M. N. (2018) Dual inhibition of the lactate transporters MCT1 and MCT4 is synthetic lethal with metformin due to NAD+ depletion in cancer cells. *Cell Rep.* 25, 3047–3058.e4. Parks, S. K., Chiche, J., and Pouysségur, J. (2013) Disrupting proton dynamics and energy metabolism for cancer therapy. *Nat. Rev. Cancer.* 13, 611–623. Marchiq, I., Le Floch, R., Roux, D., Simon, M. P., and Pouysségur, J. (2015) Genetic disruption of lactate/H+ symporters (MCTs) and their subunit CD147/BASIGIN sensitizes glycolytic tumor cells to phenformin. *Cancer Res.* 75, 171–180.

153. Colegio, O. R., Chu, N.-Q., Szabo, A. L., Chu, T., Rhebergen, A. M., Jairam, V., Cyrus, N., Brokowski, C. E., Eisenbarth, S. C., Phillips, G. M., Cline, G. W., Phillips, A. J., and Medzhitov, R. (2014) Functional polarization of tumour-associated macrophages by tumour-derived lactic acid. *Nature.* 513, 559–563. Brand, A., Singer, K., Koehl, G. E., Kolitzus, M., Schoenhammer, G., Thiel, A., Matos, C., Bruss, C., Klobuch, S., Peter, K., Kastenberger, M., Bogdan, C., Schleicher, U., Mackensen, A., Ullrich, E., Fichtner-Feigl, S., Kesselring, R., Mack, M., Ritter, U., Schmid, M., Blank, C., Dettmer, K., Oefner, P. J., Hoffmann, P., Walenta, S., Geissler, E. K., Pouyssegur, J., Villunger, A., Steven, A., Seliger, B., Schreml, S., Haferkamp, S., Kohl, E., Karrer, S., Berneburg, M., Herr, W., Mueller-Klieser, W., Renner, K., and Kreutz, M. (2016) LDHA-associated lactic acid production blunts tumor immunosurveillance by T and NK cells. *Cell Metab.* 24, 657–671. Bohn, T., Rapp, S., Luther, N., Klein, M., Bruehl, T. J., Kojima, N., Aranda Lopez, P., Hahlbrock, J., Muth, S., Endo, S., Pektor, S., Brand, A., Renner, K., Popp, V., Gerlach, K., Vogel, D., Lueckel, C., Arnold-Schild, D., Pouyssegur, J., Kreutz, M., Huber, M., Koenig, J., Weigmann, B., Probst, H. C., von Stebut, E., Becker, C., Schild, H., Schmitt, E., and Bopp, T. (2018) Tumor immunoevasion via acidosis-dependent induction of regulatory tumor-associated macrophages. *Nat. Immunol.* 19, 1319–1329. Damgaci, S., Ibrahim-Hashim, A., Enriquez-Navas, P. M., Pilon-Thomas, S., Guvenis, A., and Gillies, R. J. (2018) Hypoxia and acidosis: immune suppressors and therapeutic targets. *Immunology.* 154, 354–362.

154. Pillai, S. R., Damaghi, M., Marunaka, Y., Spugnini, E. P., Fais, S., and Gillies, R. J. (2019) Causes, consequences, and therapy of tumors acidosis. *Cancer Metastasis Rev.* 38, 205–222.

155. Vučetić, M., Cormerais, Y., Parks, S. K., and Pouysségur, J. (2017) The central role of amino acids in cancer redox homeostasis: vulnerability points of the cancer redox code. *Front. Oncol.* 7, 319. Dixon, S. J., Lemberg, K. M., Lamprecht, M. R., Skouta, R., Zaitsev, E. M., Gleason, C. E., Patel, D. N., Bauer, A. J., Cantley, A. M., Yang, W. S., Morrison, B., and Stockwell, B. R. (2012) Ferroptosis: an iron-dependent form of nonapoptotic cell death. *Cell.* 149, 1060–1072.

156. Wang, W., Green, M., Choi, J. E., Gijón, M., Kennedy, P. D., Johnson, J. K., Liao, P., Lang, X., Kryczek, I., Sell, A., Xia, H., Zhou, J., Li, G., Li, J., Li, W., Wei, S., Vatan, L., Zhang, H., Szeliga, W., Gu, W., Liu, R., Lawrence, T. S., Lamb, C., Tanno, Y., Cieslik, M., Stone, E., Georgiou, G., Chan, T. A., Chinnaiyan, A., and Zou, W. (2019) CD8 + T cells regulate tumour ferroptosis during cancer immunotherapy. *Nature.* 569, 270–274.

157. Friedmann Angeli, J. P., Krysko, D. V., and Conrad, M. (2019) Ferroptosis at the crossroads of cancer-acquired drug resistance and immune evasion. *Nat. Rev. Cancer.* 19, 405–414.

13 Corrupted Vascular Tumor Niches Confer Aggressiveness and Chemoresistance to Neoplastic Cells

Luca Vincenzo Cappelli, Liron Yoffe, and Giorgio Inghirami

Overview

In the past two decades, we have witnessed an increasing appreciation of the biological role of host elements in controlling tumor survival and progression. It is now evident that, besides intrinsic genetic alterations in tumor cells, a variety of signals from the microenvironment facilitate tumor proliferation and promote resistance to chemotherapeutic agents. In hematologic malignancies (cancers of the blood), studies have proven that a pool of stem-like leukemic cells serves as a reservoir for disease progression. These leukemia-initiating cells (LICs) reside preferentially within specific microenvironmental niches and share common features with normal hematopoietic stem cells (HSCs), such as multipotency, dormancy, and self-renewal capability. During carcinogenesis, a relationship of mutual dependency becomes established between tumor cells and their surrounding tissue niche trough, in which both compartments can modulate each other. Tumors send out aberrant signals to the surrounding microenvironment, misinstructing it to set up a pathologic niche that promotes the neoplastic phenotype and cancer progression. Conversely, this abnormal niche microenvironment provides protumorigenic signals to cancer cells, which help them to overcome stress conditions and allow them to survive and, eventually, escape and metastasize to different tissues. Numerous studies have asserted that vascular endothelial cells (ECs), a specialized component of the tumor microenvironment, play a critical role in the survival of tumor cells. Here, we review recent findings concerning the interplay between tumor cells and their microenvironment in leukemia/lymphoma, focusing on the specific interactions between ECs and neoplastic cells. Understanding the molecular mechanisms that govern these interactions will provide the rationale for designing novel kinds of therapies that target not only tumor cells but also tumor niche components. We anticipate that such host-based treatment strategies will help improve clinical outcomes for cancer patients in the near future.

13.1 Endothelial Cells: Structure and Functions

ECs are the fundamental units of the microvascular system, which includes terminal arterioles, metarterioles, capillaries, and venules that connect arteries to veins. The adult human body contains 10 to 60 trillion ECs, collectively covering an overall surface area of approximately 100 m² (corresponding to 95 percent of the whole circulatory surface area), arborizing into almost every cellular compartment of all organs.[1] ECs not only form the structural basis of a tubular vessel but also actively control coagulation and thrombosis, trafficking of circulating cells, and body compartmentalization.[2] These vascular beds perform a variety of unique activities: delivery of oxygen and nutrients, modulation of coagulation, controlling the entry of inflammatory cells, and gatekeeping cellular metabolism.[3] In recent years, it has become evident that the capillary ECs play a vital role in maintaining the homeostasis of resident stem cells in various tissues, guiding the regeneration and repair of adult organs, and preventing fibrosis. Microanatomical observations, as well as genetic and biochemical studies, have demonstrated that stem cells of epithelial, mesenchymal, hematopoietic, and neuronal origin reside near capillary ECs. At this location, they create "instructive" niches that control the homeostasis and the metabolism of surrounding elements and, in case of injury, coordinate organ regeneration in a perfusion-independent manner. This critical function is orchestrated by an array of paracrine factors, known as angiocrine factors. The coordinated secretion of angiocrine factors controls the temporal and spatial order of organ regeneration in normal tissues.[4] Conversely, in cancer, misregulated angiocrine factors orchestrate the balance between dormancy and cell growth, as well as promote tumor cell survival and metastasis.

13.2   Endothelial Cell Heterogeneity and Tissue Homeostasis

13.2.1   Epigenetic and Tissue-Mediated Heterogeneity of Endothelial Cells

Normal quiescent ECs are highly heterogeneous and display unique morphological phenotypes, properties, and functions in different tissues (e.g., blood-brain barrier, lung, and liver ECs).[5] This heterogeneity is critical for establishing exquisite anatomic structures that fulfill distinct functions within different tissues (e.g., fenestrations within glomerular endothelium facilitate renal filtration, tight junctions and the enzymatic machinery of the blood-brain barrier are necessary to protect the brain, and fenestrations and lack of basement membrane of hepatic sinusoids enable the passage of nutrients in the liver). Moreover, EC heterogeneity enables coping with very specific microenvironments characterized by very different molecular concentrations (e.g., high levels of oxygen in the lung, hypoxia in the kidney and bone marrow [BM], hyperosmolar and hyperkalemic environment in the kidney). Additionally, ECs have unique functions modulating host innate immune responses and, under certain conditions, adaptive immune responses as well.[6] The mechanisms dictating EC heterogeneity in adult organisms have been shown to depend on (1) a "fixed" epigenome, impervious to changes in the extracellular environment,[7] and (2) signals provided by the respective tissue microenvironments[8] (figure 13.1A).

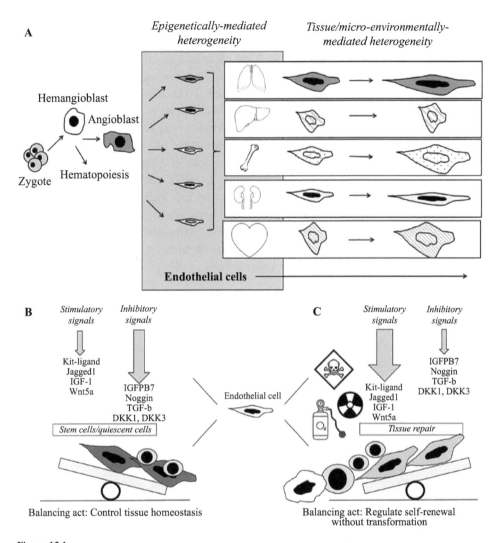

Figure 13.1
Endothelial heterogeneity. (A) During embryogenesis and early endothelial differentiation, the chromatin organization of endothelial cells becomes specified. ECs in different organs are then exposed to their respective tissue microenvironments and undergo phenotypic changes, gaining unique phenotypes and functions, as well as adapting to the unique requirements of each tissue. (B, C) ECs play a vital role in controlling resident stem cells in various tissues, creating instructive niches that regulate the homeostasis and the metabolism of surrounding elements and in coordinating organ regeneration in a perfusion-independent manner, a balance that can be undermined by tissue stress events counteracted by the instructive signals of the ECs that control tissue regeneration.

13.2.1.1   The epigenome of ECs

Initial studies have shown that the expression of endothelial-specific genes is regulated at the chromatin level. During embryogenesis, the chromatin organization of the endothelial nitric oxide synthase gene (endothelial isoform of NO synthase, eNOS [alias: NOS3]), a central player in angiogenesis signaling in ECs, is in a relaxed state and transcriptionally permissive. Conversely, the same locus in nonendothelial cells is in a repressive configuration.[9] Another example is the histone 3 lysine 4 demethylase, JARID1B, which occupies and reduces the histone 3 lysine 4 methylation levels at the *HOXA5* promoter and blocks its antiangiogenic activity.[10] Related findings have extended this notion, demonstrating that endothelium-restricted genes, including VWF, NOTCH4, and EPHB4, are controlled via chromatin-based mechanisms, supporting a model that highlights the critical role of epigenetic pathways in determining cell-restricted gene expression in the vascular endothelium.[11] Notably, high-throughput chromatin immunoprecipitation sequencing (ChIP-seq) approaches have demonstrated the preferential engagement of selected transcription factors to different genomic regions, with exclusive cell-type specificity and a strong correlation to crucial EC functions.[12]

13.2.1.2   Microenvironmental signals

It is well established that endothelial heterogeneity is also achieved via microenvironmental signals. Studies in human embryos have shown that at three months of gestation, ECs of the liver, lung, heart, and kidneys acquire distinct barrier properties, angiogenic potential, and metabolic rate, thus obtaining a unique phenotype to provide specific organ functions.[13] Moreover, the specific tissue microenvironment plays a critical role in providing a plethora of signals to ECs driving vascular maturation. ECs receive positive and negative signals via the basement membrane of the extracellular matrix (ECM)[14] and constituent molecules such as collagen type IV and laminins, the two major basement membrane components that confer, respectively, structural stability and biological activity. Vascular ECs also interact with the basement membrane via laminin α4 and α5 chains, in combination with laminin β1 and γ1 chains, to modulate cell migration, survival, proliferation, and differentiation.[15] Nevertheless, stabilized and mature vessels require the support of pericytes and vascular smooth muscle cells, which are attracted to ECs via PDGF-B- and TGF-β-mediated signals.[16] Repulsive and attractive guidance cues modulate the elaboration of vessels during angiogenesis. Mechanistically, this is achieved via multiple receptor-ligand pairs, including Eph/ephrin, semaphorin/neuropilin/plexin, netrin/UNC, and Slit/Robo.[17] Further tissue specific identity and specializations are dictated by the activation of the Notch signaling pathway: upregulation of Notch ligands via vascular endothelial growth factor (VEGF) produced by surrounding somatic tissue leads to the selective expression of ephrinB2 and EphB4 on arterial or venous ECs, respectively.[18] Earlier models predicted the actively "instructed" specialization of arterial ECs but anticipated that the differentiation into venule ECs was simply a default process. This hypothesis was later challenged by the discovery of COUP-TFII, an orphan nuclear receptor, proven to be an active vein specification factor.[19]

Over the years, various studies trying to elucidate how endothelial heterogeneity is achieved and maintained in distinct tissues have faced technical challenges when examining ways in which particular features of ECs are modulated, as loss-of-function approaches cannot easily be conducted in various tissues. Specifically, loss of function of essential EC genes leads to general loss of vascularization. Gene expression profiling studies on intravitally labeled ECs have shown that ECs from the brain, heart, lung, and muscle tissues tend to cluster tightly together, while BM, liver, and spleen ECs cluster apart from the aforementioned group.[20] A number of studies have shown that the TGF-β and VEGF produced by myocardium[21] or by astrocytes[22] are critical for the formation of ECs in the heart, vascular fenestrations in the pancreatic islets, and the kidney,[23] possibly through internalization of vascular endothelial (VE)–cadherin[24]; the same is true for establishing/maintaining the blood-brain barrier phenotype. Additional studies have finally shown that in the placenta, trophoblasts can induce an epithelial-to-endothelial transition of vascular epithelial cells to ECs via αvβ3 integrin, VE-cadherin.[25] Conversely, sFlt-1 appears to be a critical factor to inhibit the vascularization of the corneal epithelium.[26] Importantly, with a view to carcinogenesis, perivascular stromal cells such as pericytes also contribute to maintaining a stable EC phenotype by secreting angiopoietin 1.[27]

13.2.2   Angiocrine Factors: Tissue Homeostasis and Stress Mitigation

ECs can regulate organ homeostasis and repair through the production of a variety of angiocrine factors in an angiogenesis-independent manner. There is a plethora of angiocrine factors, which comprise both inhibitory and stimulatory growth factors, trophogens, chemokines, adhesion molecules such as intercellular adhesion molecule 1 (ICAM1), vascular cell adhesion molecule 1 (VCAM1), E-selectin, P-selectin, and hyaluronan, as well as chemokines, such as interleukin (IL) 8, monocyte chemotactic protein 1 (MCP1; also known as CCL2), and stromal cell–derived factor 1 (SDF1; also known as CXCL12).

Upon stress conditions (e.g., ionizing radiation, chemotherapy, tissue injury, and hypoxic status) or tissue damage/loss, angiocrine factors initiate a series of instructive programs to counteract these insults (figure 13.1B). These factors activate quiescent tissue-specific stem cells and start the execution of regenerative programs with the aim to reestablish homeostatic conditions. These properties are common to many ECs in different tissues, although the role of angiocrine factors in the organization and maintenance of tissue niches has been so far extensively studied only in the context of neural, spermatogonial, and hematopoietic stem and progenitor cell (HSPC) compartments.[28] Initial studies demonstrated that under physiological conditions, endothelial growth factors including VEGFA and FGF2 were capable of transiently maintaining human CD34+ hematopoietic stem and progenitor cells for several days.[29] Furthermore, it was discovered that insulin-like growth factor 1 (IGF1) engagement of the IGF1R in ECs promotes endothelium regeneration, migration, and tubule formation,[30] and IGF1/IGF1R-mediated signaling was subsequently proven to facilitate in vitro and in vivo angiogenesis.[31] Mechanistically, IGF1R signaling is controlled by opposing molecules controlling positive or negative inputs.[32] Negative regulators include members of the IGF-binding protein (IGFBP) family, which comprises different proteins[33] that display a higher affinity for IGF than for IGFR, acting therefore as modulators of IGF availability and activity. Gene expression profiling then defined distinct EC transcriptome signatures, specific for different organs, also linked to selective expression of angiocrine factors. Interestingly, angiocrine factors are either uniquely expressed or absent in distinct tissues, suggesting that their signatures, within a vascular niche in each organ, attain their specificity through a combinatorial expression of numerous angiocrine factors rather than any specific one.[34] In regenerating/injured tissues (e.g., BM after myeloablation and liver after partial hepatectomy), the transcriptional signature of regenerating ECs displayed profound tissue-specific alterations in their angiocrine profiles.[35] Specifically, NOTCH members were altered in BM but not in liver ECs; conversely, HGF and WNT2 were specifically upregulated in liver ECs, and angiopoietin 2 (ANG2), R-spondin 3 (RSPO3), and WNT9B produced by the liver ECs sustained hepatic homeostasis.[36] Collectively, these findings support the notion that selective angiocrine factors are expressed at steady state in different tissues, while in case of injury, tissue-unique factors are appropriately modulated to promote tissue repair.

13.3   Normal versus Tumor Endothelial Cells

13.3.1   Structural, Genomic, and Gene Expression Differences

During carcinogenesis, ECs undergo structural and functional changes. In addition to angiogenetic changes, in which tumor endothelial cells (TECs) undergo rapid cell division in concert with a variety of tumor-associated stromal cell changes, they acquire unique properties shaped by the unique requirements of the tumor cells and the tumor-specific tissue environments.[37] Interestingly, TECs display genomic and recurrent structural abnormalities. Both human and murine TECs have been shown to bear abnormal centrosomes and chromosomal abnormalities,[38] features that are not, however, shared by the entire TEC population within any given tumor, suggesting a heterogeneous rather than a clonal origin. The mechanisms leading to these phenotypes are still under investigation, but in brain tumors and lymphoma, supporting evidence has been reported in favor of a transdifferentiation of tumor cells into TECs.[39] In patients with myeloproliferative neoplasms negative for Philadelphia chromosome (Ph, originating from chromosomal translocation t(9;22), which involves the tyrosine kinase ABL gene and the breakpoint cluster region–BCR gene), TECs have been proven to share JAK mutations with tumor cells, suggesting a common origin with the myeloid neoplastic elements.[40] Controversial reports have proposed the presence of the BCR/ABL fusion in TECs from individuals with chronic myelogenous leukemia (CML).[41] Collectively, these data rather support a model that predicts that abnormal stem cell precursors may contribute to both TECs and neoplastic elements.

Over the years, many studies have investigated the transcriptional profile of TECs in solid and liquid neoplasms. In one of the first studies, St Croix et al. compared normal ECs and TECs in colon cancer using serial analysis of gene expression (SAGE). They discovered forty-six transcripts specifically elevated in tumor-associated endothelium, many of which encoded ECM proteins, while the function of many other genes was unknown.[42] Some of these tumor endothelial markers were originally thought to be selectively restricted to TEC, a finding that was not confirmed in subsequent investigations.[43] More recent studies have confirmed that many of the genes differentially expressed between TECs and normal counterpart ECs are members of the ECM signal pathways.[44] Some exceptions are MMP9, which is preferentially expressed in breast and ovarian cancers; HEYL, which is expressed in breast and colon carcinomas; and SPARC, which is seen in breast, colon, and brain tumors. Data that suggest that TECs of different tumors reflect tumor stage and type-specific profiles.[45] Moreover, there is compelling evidence that the phenotype of TECs is driven in part by an epigenetic program. For example, by comparing normal organ-matched ECs versus TECs, the low expression of ICAM1, ICAM2, and CD34 genes has been linked to the hypermethylation of these loci,[46] suggesting a role for demethylating agents in targeting TECs.[47] Downstream of the majority of growth factor–induced signals, the role of the PI3K/Akt/mTOR pathway is critical in ECs. This pathway regulates EC survival and cell migration[48] and modulates key cellular processes like cell metabolism, cell cycle and apoptosis, and overall gene transcription.[49] The constitutive activation of AKT has also been proven to be sufficient to recapitulate the abnormal structural and functional features of tumor blood vessels in nontumor tissues[50] and to promote tumor growth in an ovarian cancer model[51] where its activation has been linked to profound protumorigenic transcriptional changes. Bussolati et al. have shown that the constitutive activation of AKT led to the transcriptional expression of TSP-1, a potent inhibitor of angiogenesis, a phenotype that could be reverted by selective PI3K or mTOR inhibitors or by dominant negative Akt.[52] Interestingly, the PI3K/Akt/mTOR axis also plays a critical role in the reconstitution of hematopoietic stem and progenitor cells mediated by the vascular niche. In this setting, it upregulates specific angiocrine factors that support the expansion of cKit+Lineage−Sca1+ (KLS) HSPCs (not expressing CD34 and FLT3) with long-term hematopoietic stem cell (LT-HSC) repopulation capacity through the recruitment of mTOR (but not the FoxO pathway). Conversely, coactivation of Akt-stimulated endothelial cells, together with p42/44 MAPK, shifts the balance toward maintenance and differentiation of the HSPCs.[53]

13.3.2   Single-Cell RNA Sequencing Allows Distinguishing Normal versus Tumor EC Heterogeneity

Single-cell RNA-sequencing (scRNA-seq) technology has provided new opportunities to explore cell heterogeneity at unprecedented resolution. Specifically, scRNA-seq enabled investigating distinct populations and subpopulations of ECs and TECs as well as their relationships with other cells in the normal tissue or the tumor microenvironment. Tikhonova et al. first utilized scRNA-seq to investigate the normal BM microenvironment,[54] providing compelling data on BM ECs. In their study, the authors confirmed the presence of two distinct EC subpopulations corresponding to sinusoidal capillaries (which constitute the large majority of vessels in the BM) and to a smaller proportion of arteries, represented by a low number of side branches often longitudinally aligned along the diaphysis. Both vessel subpopulations had unique gene expression profiles: VE-Cad+ cells expressed high levels of the Notch ligands DLL1 and DLL4, which were rapidly downregulated under stress conditions, while vascular and perivascular populations characteristically expressed low levels of Jag1. Interestingly, the loss of DLL4 by ECs was linked to a decreased frequency of common lymphoid progenitors in association with an expansion of myeloid progenitor populations. Collectively, these data show that Notch engagement via its ligand expressed by ECs plays a critical role in modulating cell differentiation and maturation of multiple elements of the hematopoietic system. In a similar study, Baryawno et al. characterized BM ECs using scRNA-seq in control and leukemic patients.[55] They identified three EC subpopulations characterized by distinct gene expression signatures: sinusoidal ECs, arteriolar ECs, and arterial ECs. Arterial ECs displayed similar expression of arteriolar gene markers (e.g., *Vwf*, *CD34*, and lack of expression of *Il6st*) and hence were defined as a subset of arteriolar ECs. The sinusoidal ECs were characterized by high expression of Flt4 (Vegfr-3) and low expression of Ly6a (Sca-1), while arteriolar ECs displayed the reverse pattern. Interestingly, hematopoietic stem cell (HSC) niche factors *Kitl* and *Cxcl12* were mostly expressed by arteriolar ECs with significantly higher expression of *Kitl* in the arterial EC subset. Comparing ECs between normal and leukemic BM, they found a reduced number of sinusoidal ECs and an increased number of arteriolar ECs. Recently, ECs from eleven healthy murine tissues were characterized at single-cell resolution, demonstrating the vast heterogeneity of the EC transcriptome.[56] A comparison of the different ECs showed that tissue type is more dominant than the vessel type in determining ECs expression signatures (for a similar phenomenon, see chapter 15, this volume, regarding tissue-type dominance in tumor phenotype determination). Focusing on ECs of solid tumors, Zhao et al. utilized scRNA-seq to identify three TEC subpopulations in colon carcinoma xenograft mouse models.[57] These subpopulations corresponded to endothelial tip-like, transition, and stalk-like cells and were identified by known and novel markers. Additionally, the authors were able to explore the role of VEGF and DLL4-Notch signaling in specifying TEC expression profiles. These studies were recently expanded to investigate breast cancer TECs versus control ECs.[58] The results showed overexpression of ECM-associated genes in TECs, suggesting a critical role of the ECM in EC biology during the development of breast cancer. TEC-EC differences were also explored in gastric cancer[59] and colorectal cancer liver metastases,[60] revealing widespread transcriptional changes that might introduce novel molecular targets for cancer therapy.

13.4 Interactions between Endothelial Cells and Cancer Cells

In the past two decades, many groups have attempted to specify the pathogenetic contribution of ECs to the development and progression of solid and liquid neoplasms. As mentioned in the previous sections, in normal as well as in cancer microenvironments,[61] there is an equilibrium between stimulatory and inhibitory angiocrine factors. In tumors, TEC can release angiocrine factors that have been proven to sustain tumor cells.[62] More remarkably, the microenvironment can modulate drug responses and contribute to drug resistance,[63] trigger metabolic changes,[64] and induce nonredundant immunosuppressive mechanisms.[65]

13.4.1 Angiocrine Factors and Cancer

The intrinsic and extrinsic signals that control the secretion of angiocrine factors can be hijacked by cancer cells to promote their own survival, expansion, and tissue invasion. In the context of acute lymphoblastic leukemia (ALL), stromal cells counteract the cytotoxic effect of many therapeutic agents (steroids, cytarabine, CXCR4 inhibitors, etc.), frequently via chemokine/lymphokine disregulated signals.[66] Elevated expression of positive angiocrine factors has been documented after chemotherapy or radiation in both mice and humans.[67] Mechanistically, the neoplastic lymphoid cells produce VEGFA,[68] and circulating blood levels of VEGFA have been proven to be an informative biomarker for predicting clinical outcome in children with leukemia and lymphoma.[69] Through the production of VEGFA, leukemic cells are able to activate TECs, enhancing the release of leukemic trophogens, which support leukemia expansion via growth factors (IL6, IL3, granulocyte-colony-stimulating factor [G-CSF; also known as CSF3]), granulocyte-macrophage-CSF (GM-CSF), IL1, and nitric oxide (NO).[70] Moreover, high expression levels of VEGF correlate with the degree of meninges infiltration in nonobese diabetic/severe combined immunodeficiency mice, closely mimicking the phenotype observed in patients with ALL.[71] VEGF/VEFGR1 signaling has also been shown to play a critical role in tumor angiogenesis,[72] and expression of VEGF is required for stimulating the maintenance of ECs via an autocrine loop.[73] Notably, Notch signaling, likely through autocrine activation via Notch ligands on ECs, regulates the expression of VEGFR1 in human umbilical vein endothelial cells (HUVECs).[74] In addition to VEGF, basic fibroblast growth factor (bFGF), angiopoietins, hepatocyte growth factor, epidermal growth factor (EGF), platelet-derived growth factor (PDGF), and placental-derived growth factor can also act as positive regulators of tumor survival and growth. In both acute myeloid leukemia (AML) and ALL, the neoplastic cells secrete bFGF, which modulates the ECs,[75] and urine levels of bFGF correlate with the density of BM vessels. This is consistent with the observation that the FGF4 produced by B-cell lymphoma cells upregulates the Notch ligand Jag1 on neighboring ECs through the activation of FGFR1. Increased production of FGF4 has also been proposed as an alternative mechanism of radiation resistance in advanced-stage solid tumors: indeed, radiation therapy dramatically increases the quiescent cancer cell fraction through the upregulation of multiple quiescence-promoting angiocrine factors produced by the BM microenvironment. To counteract this effect, irradiated cancer cells produce FGF4, which engages the FGFR1 receptor expressed on TECs, leading to a nuclear factor (NF)–κB-mediated release of lymphokines, sustaining the survival of the neoplastic elements.[76]

Supernatants from liver, lung, and brain microvascular ECs have also proven to enhance the growth and migration of lymphoma cells,[77] and the upregulation of CXCL8 in VEGF-treated ECs can promote invasion of oral squamous cell carcinoma.[78] Soluble circulating Jagged1, secreted by tumor cells or myeloid cells, has been found to drive the pre-TEC to secrete protumorigenic angiocrine factors, which then facilitate tumor extravasation and survival.[79] In breast cancers, TECs, expressing high levels of EphA2, have low expression of Slit2, a potent suppressive angiocrine factor. Activation of the EphA2 receptor, possibly via ephrin ligands expressed on adjacent TECs or tumor cells, represses the transcription of Slit2, alleviating its antigrowth and chemo-repulsive effects in breast tumor cells.[80] In aggressive colon cancers, the loss of expression of another negative EC-derived factor, SPARCL1, has been linked to tumor aggressiveness. In contrast, physiological levels of SPARCL1 in normal ECs promote an antitumorigenic microenvironment by inducing cell quiescence and limiting angiogenesis and by stabilizing mural cell coverage of mature vessels.[81] Additional findings have been provided by Singh et al., who have demonstrated the role of ECs and pericytes in cancer maintenance. In these studies, the authors have shown that in young animals, pericytes are highly represented and control the quiescent state of stem cells and micro-metastases via PDFG-B signaling. Once they decrease over time, a permissive environment, rich in interleukins (IL1b, IL6, IL27, and IL1f9; CCL4 and Ccl5; TNF superfamily member 14 [Tnfsf14]; and lymphotoxin β [Ltb]) is established, and micro-disseminated tumor cells are unleashed.[82] Additional, negative regulators of tumor growth include thrombospondin 1 (TSP-1), Notch ligand delta-like 4 (DLL4), vasohibin 1 (VASH1), and Down syndrome critical region 1 (DSCR1) (figure 13.2A).

This brief review of recent relevant literature concerning the crucial interactions between ECs and TECs and their respective microenvironments through angiocrine signaling clearly demonstrates the importance of continuous feedback between ECs and their tissue microenvironment in carcinogenesis. These results also made clear that a close collaboration between the fields of vascular medicine and oncology should be very productive in delivering clinical results. Understanding the mechanisms that control the delivery of angiocrine factors that regulate the relationship between ECs and the surrounding host elements is the foundation for innovative clinical trials in regenerative medicine, organ transplantation, and the design and implementation of new chemotherapeutic approaches for disrupting the interactions of ECs with cancer cells.

13.4.2   Pathways Modulating the Interactions between Tumor Cells and Endothelial Cells

Vascular ECs modulate tumor aggressiveness through multiple and synergistic mechanisms, via well-characterized bona fide signaling axes (SDF1α/CXCR4, DLL4-Jag1–2/

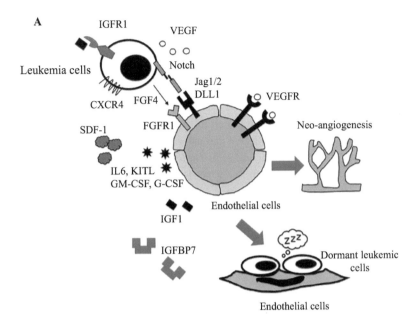

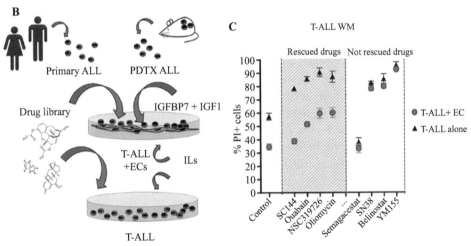

Figure 13.2

Leukemia-endothelial cell interactions. (A) Tumor and ECs display complex and bilateral interactions, which shape their phenotype and functions. Via cell-cell interactions and angiocrines, ECs modulate the growth, survival, and responses to chemotherapies against leukemia cells. Positive and negative regulators are controlling the fate of both tumor and host cells. (B) In vitro dual models to test properties of drugs and angiocrines. Leukemia cells from patients or patient-derived tumor xenograft (PDTX) lines are isolated and cultured without or with ECs. Cells are challenged with drugs, and the viability/apoptosis rates are determined over time. Angiocrines can be investigated to define their properties and contribution to leukemia growth/survival and chemosensitivity. (C) T-cell lymphoblastic acute leukemias (T-ALL) PDTX cells were cultured with and without ECs in the presence of the indicated drugs, and rates of apoptosis was assessed (72 hours). Leukemia cells were partially rescued when cultured in the presence of ECs, which also protected tumor cells when challenged by selected drugs.

NOTCH, IGFBP7/IGF1R) and additional, unknown signals, leading to the activation of various prosurvival pathways and homing pathways[83] (figure 13.2).

CXCR4 axis. CXCR4 is a member of the chemokine receptor family.[84] Chemokine receptors are a family of seven transmembrane domain G protein–coupled cell surface receptors that are able to bind chemokines. Chemokines are small, secreted proteins that can be divided into two main subfamilies, based on conserved cysteine residues, CXC or CC. These molecules play a critical function in cell trafficking between blood and tissues. Chemokines on the luminal endothelial surface can activate and engage their correspondent chemokine receptors on both normal and pathological cells. These interactions ultimately trigger multiple pathways, including PI3K and Ras, NF-κB, and MYC,[85] and lead to the activation of membrane surface integrins.[86] Basal levels of CXCR4 are increased by lymphokines (i.e., IL2, IL4, IL7, IL10, IL15), growth factors (e.g., bFGF, VEGF), and TCR activation and decreased by inflammatory cytokines like TNFα, IFN-gamma, and IL1β. The majority of T-cell lymphoblastic acute leukemias (T-ALLs) and B-cell acute lymphoblastic leukemias (B-ALLs) express increased levels of CXCR4 receptors, which influence tissue distribution of tumor cells (e.g., BM, liver, central nervous system), stroma-mediated drug resistance, and clinical outcome in both mouse models and humans.[87] Passaro et al. have also demonstrated that CXCR4 is regulated in a cortactin-dependent manner, and it is essential to the leukemia-initiating cell (LIC) activity in NOTCH-induced T-ALL mouse models as well as human T-ALL xenograft models.[88] These findings were further expanded by Pitt et al. using genetically modified mouse models and antagonist antibodies, showing that the loss/inhibition of CXCR4 prolongs clinical responses and sustains remission in their in vivo models.[89]

NOTCH axis. Notch and Notch ligand interactions can occur in *cis* or in *trans* and show opposite effects, leading to either inhibition or activation signals, respectively.[90] The inhibition in *cis* establishes and maintains the signaling polarity that is necessary for Notch-dependent cell fate determination.[91] Notch transactivation plays a critical role in cancer biology promoting growth, survival, and, in some tumor types, dormancy and metastasis.[92] Moreover, the delicate balance between negative and positive signals controls sprouting angiogenesis, a mechanism that depends on the opposing effects of DLL4 and Jag1 engagements.[93] Stromal cells and ECs effectively engage the Notch receptors on the target cells via DLLs, Jag1 and Jag2 Notch ligands, and fire prosurvival signals.[94] Remarkably, these pleiotropic Notch-mediated signals are in addition modulated by signal strength and dynamics.[95] In T-ALL (i.e., "common T-ALL"), the constitutive activation of Notch has been proven to play a critical role in preventing blast cell apoptosis and dramatically upregulates the metabolisms of the leukemic cells. The deregulated activation of Notch in T-ALL is frequently achieved via activating somatic mutations, which occur in more than 50 percent of all T-ALL cases.[96] However, canonical Notch mutations are absent in many other types of tumors in which, however, there is a robust activation of the Notch pathway. Mechanistically, it is believed that in these neoplasms, Notch is activated through Notch ligand interactions, a scenario that shapes phenotypic changes as well as cell fate.[97] Our group has also recently shown that the crosstalk between TECs and lymphoma cells can be modulated by Notch signaling. As mentioned above, the B-lymphoma cell can engage the FGFR1 on TECs by producing FGF4, a signal that leads to the upregulation of Jag1 and the further activation of the Notch2-Hey1 pathway. Increased levels of Jag1 on the ECs are sufficient to induce Notch2-Hey signaling in the lymphoma cells, leading to a phenotypic change (CD44+IGF1R+CSF1R+) associated with refractory phenotype and chemoresistance.[98] In aggressive glioblastoma, the refractory phenotype has also been linked to Notch activation, which modulates the stem cell niche.[99] Additionally, Notch signaling activation was correlated to tumor aggressiveness, tumor cell growth, survival, and invasive properties in various other types of neoplasms.[100] These findings are also in line with several studies that found a positive correlation between the activation of Notch via TECs and melanoma stem-like cells (MSLCs) and epithelial-mesenchymal transition (EMT).[101] Moreover, the soluble Notch ligand, secreted by TECs, promotes cancer stem cells in colorectal cancer,[102] and blocking antibodies to Notch ligand DLL4 were proven to inhibit tumor growth.[103] By contrast, in TECs, the deregulated activation of Notch signaling can be associated with growth arrest in ECs and tumor cell intravasation. In this latter context, TECs acquire a senescent-like phenotype and actively express proinflammatory genes, such as VCAM1. This leads to increased interactions between tumor cells and TECs, impaired EC junctions by decreasing VE-cadherin, and enhanced tumor transendothelial migration.[104]

IGFR axis. As mentioned above, IGFs play an important role in EC activation: within an array of physiological and pathological conditions, they modulate the signaling of their cognate receptor tyrosine kinase (IGF1R), and this interaction regulates cell migration, tube formation, production of nitric oxide, and by enhancing the number and function of endothelial progenitor cells. In cancer, IGF1/IGF1R signaling leads to the expression of HIF1α and its targets GPER and VEGF by engaging the ERK1/2 and AKT transduction pathways.[105] Pharmacological inhibition of IGF1/IGF1R signaling has been envisioned as a potential strategy to treat cancers,[106] although clinical trials have failed so far to prove any significant clinical advantage. We and others have demonstrated that angiomodulin/ IGFBP7 signaling is upregulated in the angiogenic vasculature during both physiological and pathological angiogenesis[107] and suppresses VEGF-induced tube formation, proliferation, and the phosphorylation of mitogen-activated protein kinase (MEK) and extracellular signal-regulated kinase (ERK).[108] Along the same line, increased levels of IGFBP7 signaling have been documented in TECs and linked either to capillary-like tube formation and lymphangiogenesis.[109] We have recently demonstrated that during tumorigenesis, IGFBP7 blocks IGF1 and inhibits expansion and aggressiveness of tumor stem-like cells (TSCs) expressing IGF1 receptor (IGF1R). However, under stress conditions and during chemotherapy, IGFBP7 levels are decreased and associated with the higher levels of FGF4 produced by tumor cells. In this context, FGF4 interacting with its cognate receptor

(FGFR1) can effectively upregulate ETS2, which in turn controls the expression of IGF1 (upregulation) and IGFBP7 (downregulation), overall promoting cancer invasiveness and progression. Low-expression IGFBP7 has been considered a feature of LSCs in AML and associated with reduced chemotherapy sensitivity and outcome.[110] Notably, rhIGFBP7 reduced LSCs and/or progenitor cell survival and reversed a stem-like gene signature, but it did not influence normal hematopoietic stem cell survival.[111]

13.5   In Vitro Models of Endothelial Cells as Testing Platforms for Novel Cancer Treatment Approaches Exploiting Angiocrine Signaling

A variety of mature ECs from different species (cattle, pigs, dogs, and humans) have been used to develop in vitro vascular models. As the importance of the tissue origin of ECs and EC heterogeneity has been recognized earlier, human umbilical cord ECs are now frequently used as a source for many of the studies in vascular biology. Intravitally stained mouse ECs can be used as a source of normal and TEC cells after appropriate digestion and flow sorting.[112] These elements can then be cultured with angiogenic factors in the absence of pituitary extract and serum to avoid the loss of tissue-specific signatures and to block the transition from endothelial to mesenchymal cells. However, even under favorable conditions, the expansion of angiocrine-competent ECs for extended periods of time remains highly problematic. Nevertheless, almost 9,000 studies have been published using HUVECs, contributing to our collective knowledge. Unfortunately, these cells can be propagated only for a few passages (five to ten times), as cells eventually acquire a senescent phenotype characterized by giant cells that ultimately undergo cell death. Multiple strategies have been employed to maintain the ECs for a longer period in culture. These include an array of media with different constituents, gelatin-coated vessels, and immortalized cells after SV40 viral infection or after transfection with human telomerase reverse transcriptase (hTERT) constructs.[113] Alternatively, mesenchymal stem cells cultured in the presence of VEGF and/or endothelial growth media have been used as a source of ECs. More recently, induced pluripotent stem (iPS) cells were used to generate both two-dimensional and three-dimensional vascular cultures.[114] Despite these strategies, HUVECs or other normal ECs cannot be maintained for a long period of time, and the artificial generation of complex three-dimensional functional vascular beds remains hardly achievable to date. To overcome these challenges, engineered ECs were constructed after transduction with lentiviral vectors that express myristoylated AKT or the nononcogenic adenoviral gene E4ORF1.[115] The activation of the AKT-mTOR pathway via myristoylated AKT or E4ORF1 sustains the survival of ECs, which can be propagated serially for multiple passages in serum-free and xenobiotic-free conditions, maintaining the features of primary ECs. ECs that express E4ORF1 are not transformed and, when implanted into immunocompromised animals, do not generate tumors, even after a long period of observation (more than twelve months). In vitro drug screening platforms are today common tools used to test the response of tumor cells to different compounds, with both agnostic and/or targeted approaches. However, several challenges currently hamper the predictive capabilities of these platforms, including the fact that they cannot fully recapitulate the complexity of tumor systems in conjunction with the microenvironment. In vivo high-throughput drug screening (HTS) using patient-derived tumor xenografts (PDTXs) has recently improved preclinical evaluation of treatment modalities and prediction of responses.[116] Nevertheless, in vivo HTS approaches are demanding, costly, and time-consuming. To mitigate some of these factors, coculture studies using E4ORF1$^+$ ECs have been shown to provide an effective platform for drug discovery and for identifying angio-crine pathways that foster the expansion and proliferation of both parenchymal and stem cells. Indeed, E4ORF1+ ECs have been proven to establish an ex vivo platform that recapitulates organotypic niches[117] capable of sustaining their organotypic pro-stem-cell functions, contrary to SV40 ECs.[118]

As further evidence for the feasibility of such in vitro approaches, we have recently implemented E4ORF1+ ECs in a coculture platform and interrogated this system for its ability to sustain T-ALL cells in vitro. We first discovered that in two-dimensional and three-dimensional cultures, T-ALL cells could establish close contact with the ECs, displaying unique relationships when individual T-ALL PDTXs were cocultured with the ECs. Next, we used these platforms to interrogate by single-cell RNA-seq the profiles of E4ORF+ ECs alone or cocultured with leukemic elements ("educated" ECs). This approach led to the recognition of unique transcription profiles demonstrating that both T-ALL and EC cells displayed well-defined transcriptional changes, driven by this reciprocal interaction. Remarkably, the phenotype of ECs cultured with T-ALL in vitro closely resembled those seen in mouse or human TEC. Conversely, when we interrogated the transcriptomics of T-ALL cocultured with ECs, we found that a plethora of genes targeting MYC, or downstream of Notch1 signaling, and E2F, G2M-checkpoint genes, were significantly enriched but displayed significantly lower levels in T-ALL culture alone. Remarkably, these changes closely resembled those seen in T-ALL in PDTX models. Collectively, these findings demonstrate that there are bilateral signals between ECs and T-ALL, which reciprocally modulate their respective phenotypes.

Taking advantage of these findings, we subsequently designed a flexible in vitro coculture platform to test primary T-ALL PDTX-derived cells with a library of compounds (figure 13.2B). Using this two-dimensional platform, we could prove that the toxicity of selected compounds was mitigated by E4ORF+ ECs, and we discovered that individual T-ALL cells displayed unique responses, suggesting that each leukemia has unique vulnerabilities that can be selectively modulated by the ECs (figure 13.2C). The combination of PDTX-derived tumor cells and engineered ECs is an invaluable experimental resource to investigate the crosstalk between tumor cells and the microenvironment, as well as to develop new therapeutic compounds aimed at disrupting this interaction.

13.6   Final Remarks

Over the past twenty years, we have been witnessing the discoveries of fundamental findings defining the role and function of endothelial cells. From the simplistic view that ECs form just a passive lining of the capillaries or were just passive conduits for delivering blood nutrients, we moved on to the notion that ECs are active players in complex microenvironments, in which they control and regulate tissue integrity, regeneration, and tissue homeostasis. The pioneering contributions of Judah Folkman led to the recognition that in cancer, the angiogenic switch controls neovascularization and thus tumor growth. This transformative discovery was soon followed by the realization that tumor endothelial cells have unique properties, primarily epitomized by their specific gene expression signatures and unique functions. Indeed, within the tumor microenvironment, the ECs, stressed by a plethora of aberrant signals, get activated and release angiocrines. These factors, modulated both intrinsically and extrinsically, control the complex bilateral interactions between ECs and cancer cells through specific activation pathways. Remarkably, this fine balance of angiocrines is known to control cell differentiation, stem cell quiescence, host cell responses, and tumor dormancy/growth and ultimately foster refractory and metastatic phenotypes. Collectively, these studies have revolutionized our understanding of the physiological role of ECs and opened up new avenues to dissect the mechanisms that drive tumorigenesis. We anticipate that new "ad hoc two-dimensional and three-dimensional" models, fully recapitulating tumorigenic niches, will develop into powerful tools for performing innovative drug discovery and will allow for the implementation of personalized treatments. We predict that these models will further our understanding of the mechanisms regulating host and tumor interactions and will drive the design of more effective and potentially truly cancer-eradicating therapies.

Notes

1. Jaffe, E. A. Cell biology of endothelial cells. *Hum Pathol* 18, 234–239, doi:10.1016/s0046–8177(87)80005–9 (1987).

2. Pearson, J. D. Endothelial cell biology. *Radiology* 179, 9–14, doi:10.1148/radiology.179.1.2006310 (1991).

3. Carmeliet, P. & Jain, R. K. Molecular mechanisms and clinical applications of angiogenesis. *Nature* 473, 298–307, doi:10.1038/nature10144 (2011). Ghesquiere, B., Wong, B. W., Kuchnio, A. & Carmeliet, P. Metabolism of stromal and immune cells in health and disease. *Nature* 511, 167–176, doi:10.1038/nature13312 (2014).

4. Rafii, S., Butler, J. M. & Ding, B. S. Angiocrine functions of organ-specific endothelial cells. *Nature* 529, 316–325, doi:10.1038/nature17040 (2016).

5. Aird, W. C. Endothelial cell heterogeneity. *Cold Spring Harbor Perspect Med* 2, a006429, doi:10.1101/csh perspect.a006429 (2012).

6. Mai, J., Virtue, A., Shen, J., Wang, H. & Yang, X. F. An evolving new paradigm: endothelial cells—conditional innate immune cells. *J Hematol Oncol* 6, 61, doi:10.1186/1756-8722-6-61 (2013).

7. Chi, J. T. et al. Endothelial cell diversity revealed by global expression profiling. *Proc Natl Acad Sci USA* 100, 10623–10628, doi:10.1073/pnas.1434429100 (2003). Lacorre, D. A. et al. Plasticity of endothelial cells:

rapid dedifferentiation of freshly isolated high endothelial venule endothelial cells outside the lymphoid tissue microenvironment. *Blood* 103, 4164–4172, doi:10.1182/blood-2003-10-3537 (2004).

8. Marcu, R. et al. Human organ-specific endothelial cell heterogeneity. *iScience* 4, 20–35, doi:10.1016/j.isci .2018.05.003 (2018). Aird, W. C. et al. Vascular bed-specific expression of an endothelial cell gene is programmed by the tissue microenvironment. *J Cell Biol* 138, 1117–1124, doi:10.1083/jcb.138.5.1117 (1997).

9. Yan, M. S., Matouk, C. C. & Marsden, P. A. Epigenetics of the vascular endothelium. *J Appl Physiol* 109, 916–926, doi:10.1152/japplphysiol.00131.2010 (2010). Fish, J. E. & Marsden, P. A. Endothelial nitric oxide synthase: insight into cell-specific gene regulation in the vascular endothelium. *Cell Mol Life Sci* 63, 144–162, doi:10.1007 /s00018-005-5421-8 (2006).

10. Fork, C. et al. Epigenetic regulation of angiogenesis by JARID1B-induced repression of HOXA5. *Arterioscler Thromb Vasc Biol* 35, 1645–1652, doi:10.1161/ATVBAHA.115.305561 (2015).

11. Shirodkar, A. V. et al. A mechanistic role for DNA methylation in endothelial cell (EC)-enriched gene expression: relationship with DNA replication timing. *Blood* 121, 3531–3540, doi:10.1182/blood-2013-01-479170 (2013).

12. Nakato, R. et al. Comprehensive epigenome characterization reveals diverse transcriptional regulation across human vascular endothelial cells. *Epigenetics Chromatin* 12, 77, doi:10.1186/s13072-019-0319-0 (2019).

13. Marcu, R. et al. Human organ-specific endothelial cell heterogeneity. *iScience* 4, 20–35, doi:10.1016/j.isci .2018.05.003 (2018).

14. Sun, Z. et al. Single-cell RNA sequencing reveals gene expression signatures of breast cancer-associated endothelial cells. *Oncotarget* 9, 10945–10961, doi:10.18632/oncotarget.23760 (2018). Sun, L., Vitolo, M. & Passaniti, A. Runt-related gene 2 in endothelial cells: inducible expression and specific regulation of cell migration and invasion. *Cancer Res* 61, 4994–5001 (2001). Ria, R. et al. Gene expression profiling of bone marrow endothelial cells in patients with multiple myeloma. *Clin Cancer Res* 15, 5369–5378, doi:10.1158/1078–0432. CCR-09–0040 (2009).

15. Witjas, F. M. R., van den Berg, B. M., van den Berg, C. W., Engelse, M. A. & Rabelink, T. J. Concise review: the endothelial cell extracellular matrix regulates tissue homeostasis and repair. *Stem Cells Transl Med* 8, 375–382, doi:10.1002/sctm.18–0155 (2019).

16. Ding, R., Darland, D. C., Parmacek, M. S. & D'Amore, P. A. Endothelial-mesenchymal interactions in vitro reveal molecular mechanisms of smooth muscle/pericyte differentiation. *Stem Cells Dev* 13, 509–520, doi:10.1089 /scd.2004.13.509 (2004). Lindahl, P., Johansson, B. R., Leveen, P. & Betsholtz, C. Pericyte loss and microaneurysm formation in PDGF-B-deficient mice. *Science* 277, 242–245, doi:10.1126/science.277.5323.242 (1997). Gerhardt, H. & Betsholtz, C. Endothelial-pericyte interactions in angiogenesis. *Cell Tissue Res* 314, 15–23, doi:10.1007/s00441-003-0745-x (2003). Hirschi, K. K., Burt, J. M., Hirschi, K. D. & Dai, C. Gap junction communication mediates transforming growth factor-beta activation and endothelial-induced mural cell differentiation. *Circ Res* 93, 429–437, doi:10.1161/01.RES.0000091259.84556.D5 (2003).

17. Sweeney, M. & Foldes, G. It takes two: endothelial-perivascular cell cross-talk in vascular development and disease. *Front Cardiovasc Med* 5, 154, doi:10.3389/fcvm.2018.00154 (2018).

18. Adams, R. H. et al. Roles of ephrinB ligands and EphB receptors in cardiovascular development: demarcation of arterial/venous domains, vascular morphogenesis, and sprouting angiogenesis. *Genes Dev* 13, 295–306, doi:10.1101/gad.13.3.295 (1999). Wang, H. U., Chen, Z. F. & Anderson, D. J. Molecular distinction and angiogenic interaction between embryonic arteries and veins revealed by ephrin-B2 and its receptor Eph-B4. *Cell* 93, 741–753, doi:10.1016/s0092–8674(00)81436–1 (1998).

19. You, L. R. et al. Suppression of Notch signalling by the COUP-TFII transcription factor regulates vein identity. *Nature* 435, 98–104, doi:10.1038/nature03511 (2005).

20. Nolan, D. J. et al. Molecular signatures of tissue-specific microvascular endothelial cell heterogeneity in organ maintenance and regeneration. *Dev Cell* 26, 204–219, doi:10.1016/j.devcel.2013.06.017 (2013).

21. Brown, C. B., Boyer, A. S., Runyan, R. B. & Barnett, J. V. Requirement of type III TGF-beta receptor for endocardial cell transformation in the heart. *Science* 283, 2080–2082, doi:10.1126/science.283.5410.2080 (1999). Dor, Y. et al. A novel role for VEGF in endocardial cushion formation and its potential contribution to congenital heart defects. *Development* 128, 1531–1538 (2001).

22. Tran, N. D., Correale, J., Schreiber, S. S. & Fisher, M. Transforming growth factor-beta mediates astrocyte-specific regulation of brain endothelial anticoagulant factors. *Stroke* 30, 1671–1678, doi:10.1161/01.str.30.8.1671 (1999).

23. Gerber, H. P. et al. Vascular endothelial growth factor regulates endothelial cell survival through the phosphatidylinositol 3'-kinase/Akt signal transduction pathway. Requirement for Flk-1/KDR activation. *J Biol Chem* 273, 30336–30343, doi:10.1074/jbc.273.46.30336 (1998). Kamba, T. et al. VEGF-dependent plasticity of fenestrated capillaries in the normal adult microvasculature. *Am J Physiol Heart Circ Physiol* 290, H560–576, doi:10.1152/ajpheart.00133.2005 (2006).

24. Gavard, J. & Gutkind, J. S. VEGF controls endothelial-cell permeability by promoting the beta-arrestin-dependent endocytosis of VE-cadherin. *Nat Cell Biol* 8, 1223–1234, doi:10.1038/ncb1486 (2006).

25. Zhou, Y., Damsky, C. H. & Fisher, S. J. Preeclampsia is associated with failure of human cytotrophoblasts to mimic a vascular adhesion phenotype. One cause of defective endovascular invasion in this syndrome? *J Clin Invest* 99, 2152–2164, doi:10.1172/JCI119388 (1997). Zhou, Y. et al. Human cytotrophoblasts adopt a vascular phenotype as they differentiate. A strategy for successful endovascular invasion? *J Clin Invest* 99, 2139–2151, doi:10.1172/JCI119387 (1997).

26. Chang, J. H., Gabison, E. E., Kato, T. & Azar, D. T. Corneal neovascularization. *Curr Opin Ophthalmol* 12, 242–249, doi:10.1097/00055735-200108000-00002 (2001).

27. Jeansson, M. et al. Angiopoietin-1 is essential in mouse vasculature during development and in response to injury. *J Clin Invest* 121, 2278–2289, doi:10.1172/JCI46322 (2011).

28. Rafii, S., Butler, J. M. & Ding, B. S. Angiocrine functions of organ-specific endothelial cells. *Nature* 529, 316–325, doi:10.1038/nature17040 (2016).

29. Rafii, S. et al. Isolation and characterization of human bone marrow microvascular endothelial cells: hematopoietic progenitor cell adhesion. *Blood* 84, 10–19 (1994). Rafii, S. et al. Human bone marrow microvascular endothelial cells support long-term proliferation and differentiation of myeloid and megakaryocytic progenitors. *Blood* 86, 3353–3363 (1995).

30. Imrie, H. et al. Novel role of the IGF-1 receptor in endothelial function and repair: studies in endothelium-targeted IGF-1 receptor transgenic mice. *Diabetes* 61, 2359–2368, doi:10.2337/db11–1494 (2012). Holzenberger, M. et al. IGF-1 receptor regulates lifespan and resistance to oxidative stress in mice. *Nature* 421, 182–187, doi:10.1038/nature01298 (2003). Kahn, M. B. et al. Insulin resistance impairs circulating angiogenic progenitor cell function and delays endothelial regeneration. *Diabetes* 60, 1295–1303, doi:10.2337/db10–1080 (2011). Yuldasheva, N. Y. et al. Haploinsufficiency of the insulin-like growth factor-1 receptor enhances endothelial repair and favorably modifies angiogenic progenitor cell phenotype. *Arterioscler Thromb Vasc Biol* 34, 2051–2058, doi:10.1161/ATVBAHA.114.304121 (2014).

31. Shigematsu, S. et al. IGF-1 regulates migration and angiogenesis of human endothelial cells. *Endocrine J* 46 Suppl, S59–S62, doi:10.1507/endocrj.46.suppl_s59 (1999).

32. Ekyalongo, R. C. & Yee, D. Revisiting the IGF-1R as a breast cancer target. *NPJ Precision Oncol* 1, doi:10.1038/s41698-017-0017-y (2017).

33. Hwa, V., Oh, Y. & Rosenfeld, R. G. The insulin-like growth factor-binding protein (IGFBP) superfamily. *Endocrine Rev* 20, 761–787, doi:10.1210/edrv.20.6.0382 (1999).

34. Nolan, D. J. et al. Molecular signatures of tissue-specific microvascular endothelial cell heterogeneity in organ maintenance and regeneration. *Dev Cell* 26, 204–219, doi:10.1016/j.devcel.2013.06.017 (2013).

35. Marcu, R. et al. Human organ-specific endothelial cell heterogeneity. *iScience* 4, 20–35, doi:10.1016/j.isci.2018.05.003 (2018). Nolan, D. J. et al. Molecular signatures of tissue-specific microvascular endothelial cell heterogeneity in organ maintenance and regeneration. *Dev Cell* 26, 204–219, doi:10.1016/j.devcel.2013.06.017 (2013). Jambusaria, A. et al. Endothelial heterogeneity across distinct vascular beds during homeostasis and inflammation. *Elife* 9, e51413, doi:10.7554/eLife.51413 (2020).

36. Nolan, D. J. et al. Molecular signatures of tissue-specific microvascular endothelial cell heterogeneity in organ maintenance and regeneration. *Dev Cell* 26, 204–219, doi:10.1016/j.devcel.2013.06.017 (2013). Geraud, C. et al. GATA4-dependent organ-specific endothelial differentiation controls liver development and embryonic hematopoiesis. *J Clin Invest* 127, 1099–1114, doi:10.1172/JCI90086 (2017).

37. De Palma, M., Biziato, D. & Petrova, T. V. Microenvironmental regulation of tumour angiogenesis. *Nat Rev Cancer* 17, 457–474, doi:10.1038/nrc.2017.51 (2017).

38. Hida, K. et al. Tumor-associated endothelial cells with cytogenetic abnormalities. *Cancer Res* 64, 8249–8255, doi:10.1158/0008–5472.CAN-04–1567 (2004). Akino, T. et al. Cytogenetic abnormalities of tumor-associated endothelial cells in human malignant tumors. *Am J Pathol* 175, 2657–2667, doi:10.2353

/ajpath.2009.090202 (2009). Lin, P. P., Gires, O., Wang, D. D., Li, L. & Wang, H. Comprehensive in situ co-detection of aneuploid circulating endothelial and tumor cells. *Sci Rep* 7, 9789, doi:10.1038/s41598-017-10763-7 (2017).

39. Streubel, B. et al. Lymphoma-specific genetic aberrations in microvascular endothelial cells in B-cell lymphomas. *N Engl J Med* 351, 250–259, doi:10.1056/NEJMoa033153 (2004). Wang, R. et al. Glioblastoma stem-like cells give rise to tumour endothelium. *Nature* 468, 829–833, doi:10.1038/nature09624 (2010).

40. Teofili, L. et al. Endothelial progenitor cells are clonal and exhibit the JAK2(V617F) mutation in a subset of thrombotic patients with Ph-negative myeloproliferative neoplasms. *Blood* 117, 2700–2707, doi:10.1182/blood-2010-07-297598 (2011). Rosti, V. et al. Spleen endothelial cells from patients with myelofibrosis harbor the JAK2V617F mutation. *Blood* 121, 360–368, doi:10.1182/blood-2012-01-404889 (2013).

41. Fang, B. et al. Identification of human chronic myelogenous leukemia progenitor cells with hemangioblastic characteristics. *Blood* 105, 2733–2740, doi:10.1182/blood-2004-07-2514 (2005). Otten, J. et al. Blood outgrowth endothelial cells from chronic myeloid leukaemia patients are BCR/ABL1 negative. *Br J Haematol* 142, 115–118, doi:10.1111/j.1365–2141.2008.07195.x (2008).

42. St Croix, B. et al. Genes expressed in human tumor endothelium. *Science* 289, 1197–1202 (2000).

43. Seaman, S. et al. Genes that distinguish physiological and pathological angiogenesis. *Cancer Cell* 11, 539–554, doi:10.1016/j.ccr.2007.04.017 (2007). MacFadyen, J., Savage, K., Wienke, D. & Isacke, C. M. Endosialin is expressed on stromal fibroblasts and CNS pericytes in mouse embryos and is downregulated during development. *Gene Expression Patterns* 7, 363–369, doi:10.1016/j.modgep.2006.07.006 (2007). Lee, H. K. et al. Cloning, characterization and neuronal expression profiles of tumor endothelial marker 7 in the rat brain. *Brain Res Mol Brain Res* 136, 189–198, doi:10.1016/j.molbrainres.2005.02.010 (2005). Cullen, M. et al. Host-derived tumor endothelial marker 8 promotes the growth of melanoma. *Cancer Res* 69, 6021–6026, doi:10.1158/0008–5472.CAN-09–1086 (2009).

44. Sun, Z. et al. Single-cell RNA sequencing reveals gene expression signatures of breast cancer-associated endothelial cells. *Oncotarget* 9, 10945–10961, doi:10.18632/oncotarget.23760 (2018). Ria, R. et al. Gene expression profiling of bone marrow endothelial cells in patients with multiple myeloma. *Clin Cancer Res* 15, 5369–5378, doi:10.1158/1078–0432.CCR-09–0040 (2009).

45. Aird, W. C. Molecular heterogeneity of tumor endothelium. *Cell Tissue Res* 335, 271–281, doi:10.1007/s00441-008-0672-y (2009).

46. Bussolati, B., Deambrosis, I., Russo, S., Deregibus, M. C. & Camussi, G. Altered angiogenesis and survival in human tumor-derived endothelial cells. *FASEB J* 17, 1159–1161, doi:10.1096/fj.02–0557fje (2003). Hellebrekers, D. M. et al. Epigenetic regulation of tumor endothelial cell anergy: silencing of intercellular adhesion molecule-1 by histone modifications. *Cancer Res* 66, 10770–10777, doi:10.1158/0008–5472.CAN-06–1609 (2006).

47. Hellebrekers, D. M. et al. Identification of epigenetically silenced genes in tumor endothelial cells. *Cancer Res* 67, 4138–4148, doi:10.1158/0008–5472.CAN-06–3032 (2007).

48. Chin, Y. R. & Toker, A. Function of Akt/PKB signaling to cell motility, invasion and the tumor stroma in cancer. *Cell Signal* 21, 470–476, doi:10.1016/j.cellsig.2008.11.015 (2009).

49. Gerber, H. P. et al. Vascular endothelial growth factor regulates endothelial cell survival through the phosphatidylinositol 3'-kinase/Akt signal transduction pathway. Requirement for Flk-1/KDR activation. *J Biol Chem* 273, 30336–30343, doi:10.1074/jbc.273.46.30336 (1998). Chen, J. et al. Akt1 regulates pathological angiogenesis, vascular maturation and permeability in vivo. *Nat Med* 11, 1188–1196, doi:10.1038/nm1307 (2005). Alon, T. et al. Vascular endothelial growth factor acts as a survival factor for newly formed retinal vessels and has implications for retinopathy of prematurity. *Nat Med* 1, 1024–1028, doi:10.1038/nm1095–1024 (1995). Morales-Ruiz, M. et al. Vascular endothelial growth factor-stimulated actin reorganization and migration of endothelial cells is regulated via the serine/threonine kinase Akt. *Circ Res* 86, 892–896, doi:10.1161/01.res.86.8.892 (2000).

50. Phung, T. L. et al. Pathological angiogenesis is induced by sustained Akt signaling and inhibited by rapamycin. *Cancer Cell* 10, 159–170, doi:10.1016/j.ccr.2006.07.003 (2006).

51. Hoarau-Vechot, J. et al. Akt-activated endothelium promotes ovarian cancer proliferation through notch activation. *J Transl Med* 17, 194, doi:10.1186/s12967-019-1942-z (2019).

52. Bussolati, B., Assenzio, B., Deregibus, M. C. & Camussi, G. The proangiogenic phenotype of human tumor-derived endothelial cells depends on thrombospondin-1 downregulation via phosphatidylinositol 3-kinase/Akt pathway. *J Mol Med* 84, 852–863, doi:10.1007/s00109-006-0075-z (2006).

53. Kobayashi, H. et al. Angiocrine factors from Akt-activated endothelial cells balance self-renewal and differentiation of haematopoietic stem cells. *Nat Cell Biol* 12, 1046–1056, doi:10.1038/ncb2108 (2010).

54. Tikhonova, A. N. et al. The bone marrow microenvironment at single-cell resolution. *Nature* 569, 222–228, doi:10.1038/s41586-019-1104-8 (2019).

55. Baryawno, N. et al. A cellular taxonomy of the bone marrow stroma in homeostasis and leukemia. *Cell* 177, 1915–1932 e1916, doi:10.1016/j.cell.2019.04.040 (2019).

56. Kalucka, J. et al. Single-cell transcriptome atlas of murine endothelial cells. *Cell* 180, 764–779 e720, doi:10.1016/j.cell.2020.01.015 (2020).

57. Zhao, Q. et al. Single-cell transcriptome analyses reveal endothelial cell heterogeneity in tumors and changes following antiangiogenic treatment. *Cancer Res* 78, 2370, doi:10.1158/0008-5472.CAN-17-2728 (2018).

58. Sun, Z. et al. Single-cell RNA sequencing reveals gene expression signatures of breast cancer-associated endothelial cells. *Oncotarget* 9, 10945–10961, doi:10.18632/oncotarget.23760 (2018).

59. Sathe, A. et al. Single cell genomic characterization reveals the cellular reprogramming of the gastric tumor microenvironment. *bioRxiv*, 783027, doi:10.1101/783027 (2019).

60. Zhang, Y. et al. Single-cell transcriptome analysis reveals tumor immune microenvironment heterogenicity and granulocytes enrichment in colorectal cancer liver metastases. *Cancer Lett* 470, 84–94, doi:10.1016/j.canlet .2019.10.016 (2020).

61. Butler, J. M., Kobayashi, H. & Rafii, S. Instructive role of the vascular niche in promoting tumour growth and tissue repair by angiocrine factors. *Nat Rev Cancer* 10, 138–146, doi:10.1038/nrc2791 (2010). Hida, K., Maishi, N., Annan, D. A. & Hida, Y. Contribution of tumor endothelial cells in cancer progression. *Int J Mol Sci* 19, 1272, doi:10.3390/ijms19051272 (2018).

62. Folkman, J. Angiogenesis in cancer, vascular, rheumatoid and other disease. *Nat Med* 1, 27–31, doi:10.1038 /nm0195-27 (1995).

63. Ayala, F., Dewar, R., Kieran, M. & Kalluri, R. Contribution of bone microenvironment to leukemogenesis and leukemia progression. *Leukemia* 23, 2233–2241, doi:10.1038/leu.2009.175 (2009). McMillin, D. W., Negri, J. M. & Mitsiades, C. S. The role of tumour-stromal interactions in modifying drug response: challenges and opportunities. *Nat Rev Drug Discov* 12, 217–228, doi:10.1038/nrd3870 (2013). Shafat, M. S., Gnaneswaran, B., Bowles, K. M. & Rushworth, S. A. The bone marrow microenvironment—home of the leukemic blasts. *Blood Rev* 31, 277–286, doi:10.1016/j.blre.2017.03.004 (2017).

64. Vijayan, D., Young, A., Teng, M. W. L. & Smyth, M. J. Targeting immunosuppressive adenosine in cancer. *Nat Rev Cancer* 17, 709–724, doi:10.1038/nrc.2017.86 (2017).

65. Mantovani, A. & Sica, A. Macrophages, innate immunity and cancer: balance, tolerance, and diversity. *Curr Opin Immunol* 22, 231–237, doi:10.1016/j.coi.2010.01.009 (2010).

66. Bakker, E., Qattan, M., Mutti, L., Demonacos, C. & Krstic-Demonacos, M. The role of microenvironment and immunity in drug response in leukemia. *Biochim Biophys Acta* 1863, 414–426, doi:10.1016/j.bbamcr.2015 .08.003 (2016).

67. Butler, J. M., Kobayashi, H. & Rafii, S. Instructive role of the vascular niche in promoting tumour growth and tissue repair by angiocrine factors. *Nat Rev Cancer* 10, 138–146, doi:10.1038/nrc2791 (2010). Cao, Z. et al. Angiocrine factors deployed by tumor vascular niche induce B cell lymphoma invasiveness and chemoresistance. *Cancer Cell* 25, 350–365, doi:10.1016/j.ccr.2014.02.005 (2014). Drusbosky, L. et al. Endothelial cell derived angiocrine support of acute myeloid leukemia targeted by receptor tyrosine kinase inhibition. *Leukemia Res* 39, 984–989, doi:10.1016/j.leukres.2015.05.015 (2015). Choi, S. H. et al. Tumour-vasculature development via endothelial-to-mesenchymal transition after radiotherapy controls CD44v6(+) cancer cell and macrophage polarization. *Nat Commun* 9, 5108, doi:10.1038/s41467-018-07470-w (2018). Singh, A. et al. Angiocrine signals regulate quiescence and therapy resistance in bone metastasis. *JCI Insight* 4, e125679, doi:10.1172/jci.insight.125679 (2019).

68. Krejsgaard, T. et al. A novel xenograft model of cutaneous T-cell lymphoma. *Exp Dermatol* 19, 1096–1102, doi:10.1111/j.1600-0625.2010.01138.x (2010). Chen, H. et al. In vitro and in vivo production of vascular endothelial growth factor by chronic lymphocytic leukemia cells. *Blood* 96, 3181–3187 (2000). Ribatti, D., Nico, B., Ranieri, G., Specchia, G. & Vacca, A. The role of angiogenesis in human non-Hodgkin lymphomas. *Neoplasia* 15, 231–238, doi:10.1593/neo.121962 (2013).

69. Faderl, S. et al. Angiogenic factors may have a different prognostic role in adult acute lymphoblastic leukemia. *Blood* 106, 4303–4307, doi:10.1182/blood-2005-03-1010 (2005). Avramis, I. A. et al. Correlation between

high vascular endothelial growth factor-A serum levels and treatment outcome in patients with standard-risk acute lymphoblastic leukemia: a report from Children's Oncology Group Study CCG-1962. *Clin Cancer Res* 12, 6978–6984, doi:10.1158/1078–0432.CCR-06–1140 (2006).

70. Dias, S. et al. Autocrine stimulation of VEGFR-2 activates human leukemic cell growth and migration. *J Clin Invest* 106, 511–521, doi:10.1172/JCI8978 (2000). Dias, S. et al. Inhibition of both paracrine and autocrine VEGF/ VEGFR-2 signaling pathways is essential to induce long-term remission of xenotransplanted human leukemias. *Proc Natl Acad Sci USA* 98, 10857–10862, doi:10.1073/pnas.191117498 (2001). Dias, S., Shmelkov, S. V., Lam, G. & Rafii, S. VEGF(165) promotes survival of leukemic cells by Hsp90-mediated induction of Bcl-2 expression and apoptosis inhibition. *Blood* 99, 2532–2540, doi:10.1182/blood.v99.7.2532 (2002).

71. Munch, V. et al. Central nervous system involvement in acute lymphoblastic leukemia is mediated by vascular endothelial growth factor. *Blood* 130, 643–654, doi:10.1182/blood-2017-03-769315 (2017).

72. Simons, M., Gordon, E. & Claesson-Welsh, L. Mechanisms and regulation of endothelial VEGF receptor signalling. *Nat Rev Mol Cell Biol* 17, 611–625, doi:10.1038/nrm.2016.87 (2016).

73. Domigan, C. K. et al. Autocrine VEGF maintains endothelial survival through regulation of metabolism and autophagy. *J Cell Sci* 128, 2236–2248, doi:10.1242/jcs.163774 (2015).

74. Funahashi, Y. et al. Notch regulates the angiogenic response via induction of VEGFR-1. *J Angiogenesis Res* 2, 3, doi:10.1186/2040-2384-2-3 (2010).

75. Perez-Atayde, A. R. et al. Spectrum of tumor angiogenesis in the bone marrow of children with acute lymphoblastic leukemia. *Am J Pathol* 150, 815–821 (1997). Fiedler, W. et al. Vascular endothelial growth factor, a possible paracrine growth factor in human acute myeloid leukemia. *Blood* 89, 1870–1875 (1997). Nguyen, M. et al. Elevated levels of an angiogenic peptide, basic fibroblast growth factor, in the urine of patients with a wide spectrum of cancers. *J Natl Cancer Inst* 86, 356–361, doi:10.1093/jnci/86.5.356 (1994).

76. Roy-Luzarraga, M. & Hodivala-Dilke, K. Molecular pathways: endothelial cell FAK-A target for cancer treatment. *Clin Cancer Res* 22, 3718–3724, doi:10.1158/1078–0432.CCR-14–2021 (2016).

77. Hamada, J., Cavanaugh, P. G., Lotan, O. & Nicolson, G. L. Separable growth and migration factors for large-cell lymphoma cells secreted by microvascular endothelial cells derived from target organs for metastasis. *Br J Cancer* 66, 349–354, doi:10.1038/bjc.1992.269 (1992).

78. Warner, K. A. et al. Endothelial cells enhance tumor cell invasion through a crosstalk mediated by CXC chemokine signaling. *Neoplasia* 10, 131–139, doi:10.1593/neo.07815 (2008).

79. Cao, Z. et al. Angiocrine factors deployed by tumor vascular niche induce B cell lymphoma invasiveness and chemoresistance. *Cancer Cell* 25, 350–365, doi:10.1016/j.ccr.2014.02.005 (2014).

80. Brantley-Sieders, D. M. et al. Angiocrine factors modulate tumor proliferation and motility through EphA2 repression of Slit2 tumor suppressor function in endothelium. *Cancer Res* 71, 976–987, doi:10.1158/0008–5472 .CAN-10–3396 (2011).

81. Naschberger, E. et al. Matricellular protein SPARCL1 regulates tumor microenvironment-dependent endothelial cell heterogeneity in colorectal carcinoma. *J Clin Invest* 126, 4187–4204, doi:10.1172/JCI78260 (2016).

82. Singh, A. et al. Angiocrine signals regulate quiescence and therapy resistance in bone metastasis. *JCI Insight* 4, e125679, doi:10.1172/jci.insight.125679 (2019).

83. Pitt, L. A. et al. CXCL12-producing vascular endothelial niches control acute t cell leukemia maintenance. *Cancer Cell* 27, 755–768, doi:10.1016/j.ccell.2015.05.002 (2015). Passaro, D. et al. CXCR4 is required for leukemia-initiating cell activity in T cell acute lymphoblastic leukemia. *Cancer Cell* 27, 769–779, doi:10.1016 /j.ccell.2015.05.003 (2015). Sipkins, D. A. et al. In vivo imaging of specialized bone marrow endothelial micro-domains for tumour engraftment. *Nature* 435, 969–973, doi:10.1038/nature03703 (2005). Cao, Z. et al. Molecular checkpoint decisions made by subverted vascular niche transform indolent tumor cells into chemoresistant cancer stem cells. *Cancer Cell* 31, 110–126, doi:10.1016/j.ccell.2016.11.010 (2017).

84. Burger, J. A. & Kipps, T. J. CXCR4: a key receptor in the crosstalk between tumor cells and their microenvironment. *Blood* 107, 1761–1767, doi:10.1182/blood-2005-08-3182 (2006).

85. Spinosa, P. C. et al. Short-term cellular memory tunes the signaling responses of the chemokine receptor CXCR4. *Sci Signal* 12, eaaw4204, doi:10.1126/scisignal.aaw4204 (2019).

86. Springer, T. A. Traffic signals for lymphocyte recirculation and leukocyte emigration: the multistep paradigm. *Cell* 76, 301–314, doi:10.1016/0092–8674(94)90337–9 (1994).

87. Pitt, L. A. et al. CXCL12-producing vascular endothelial niches control acute T cell leukemia maintenance. *Cancer Cell* 27, 755–768, doi:10.1016/j.ccell.2015.05.002 (2015). Tsaouli, G., Ferretti, E., Bellavia, D., Vacca, A. & Felli, M. P. Notch/CXCR4 partnership in acute lymphoblastic leukemia progression. *J Immunol Res* 2019, 5601396, doi:10.1155/2019/5601396 (2019).

88. Passaro, D. et al. CXCR4 is required for leukemia-initiating cell activity in T cell acute lymphoblastic leukemia. *Cancer Cell* 27, 769–779, doi:10.1016/j.ccell.2015.05.003 (2015).

89. Pitt, L. A. et al. CXCL12-producing vascular endothelial niches control acute T cell leukemia maintenance. *Cancer Cell* 27, 755–768, doi:10.1016/j.ccell.2015.05.002 (2015).

90. D'Souza, B., Meloty-Kapella, L. & Weinmaster, G. Canonical and non-canonical Notch ligands. *Curr Topics Dev Biol* 92, 73–129, doi:10.1016/S0070–2153(10)92003–6 (2010).

91. Sprinzak, D. et al. Cis-interactions between Notch and Delta generate mutually exclusive signalling states. *Nature* 465, 86–90, doi:10.1038/nature08959 (2010).

92. Kuhnert, F. et al. Dll4 blockade in stromal cells mediates antitumor effects in preclinical models of ovarian cancer. *Cancer Res* 75, 4086–4096, doi:10.1158/0008–5472.CAN-14–3773 (2015). Lu, J. et al. Endothelial cells promote the colorectal cancer stem cell phenotype through a soluble form of Jagged-1. *Cancer Cell* 23, 171–185, doi:10.1016/j.ccr.2012.12.021 (2013). Zhu, T. S. et al. Endothelial cells create a stem cell niche in glioblastoma by providing NOTCH ligands that nurture self-renewal of cancer stem-like cells. *Cancer Res* 71, 6061–6072, doi:10.1158/0008–5472.CAN-10–4269 (2011). Indraccolo, S. et al. Cross-talk between tumor and endothelial cells involving the Notch3-Dll4 interaction marks escape from tumor dormancy. *Cancer Res* 69, 1314–1323, doi:10.1158/0008–5472.CAN-08–2791 (2009). Sonoshita, M. et al. Suppression of colon cancer metastasis by Aes through inhibition of Notch signaling. *Cancer Cell* 19, 125–137, doi:10.1016/j.ccr.2010.11.008 (2011). Wieland, E. et al. Endothelial Notch1 activity facilitates metastasis. *Cancer Cell* 31, 355–367, doi:10.1016/j.ccell.2017.01.007 (2017).

93. Benedito, R. et al. The notch ligands Dll4 and Jagged1 have opposing effects on angiogenesis. *Cell* 137, 1124–1135, doi:10.1016/j.cell.2009.03.025 (2009).

94. Nwabo Kamdje, A. H. & Krampera, M. Notch signaling in acute lymphoblastic leukemia: any role for stromal microenvironment? *Blood* 118, 6506–6514, doi:10.1182/blood-2011-08-376061 (2011). Meurette, O. & Mehlen, P. Notch signaling in the tumor microenvironment. *Cancer Cell* 34, 536–548, doi:10.1016/j.ccell.2018.07.009 (2018).

95. Gama-Norton, L. et al. Notch signal strength controls cell fate in the haemogenic endothelium. *Nat Commun* 6, 8510, doi:10.1038/ncomms9510 (2015). Nandagopal, N. et al. Dynamic ligand discrimination in the Notch signaling pathway. *Cell* 172, 869–880 e819, doi:10.1016/j.cell.2018.01.002 (2018). Nandagopal, N. et al. Dynamic ligand discrimination in the Notch signaling pathway. *Cell* 172, 869–880 e819, doi:10.1016/j.cell.2018.01.002 (2018).

96. Ferrando, A. A. The role of NOTCH1 signaling in T-ALL. *Hematology Am Soc Hematol Educ Program* 2009, 353–361, doi:10.1182/asheducation-2009.1.353 (2009). Aster, J. C., Pear, W. S. & Blacklow, S. C. The varied roles of Notch in cancer. *Annu Rev Pathol* 12, 245–275, doi:10.1146/annurev-pathol-052016–100127 (2017).

97. Lim, J. S. et al. Intratumoural heterogeneity generated by Notch signalling promotes small-cell lung cancer. *Nature* 545, 360–364, doi:10.1038/nature22323 (2017).

98. Cao, Z. et al. Angiocrine factors deployed by tumor vascular niche induce B cell lymphoma invasiveness and chemoresistance. *Cancer Cell* 25, 350–365, doi:10.1016/j.ccr.2014.02.005 (2014).

99. Zhu, T. S. et al. Endothelial cells create a stem cell niche in glioblastoma by providing NOTCH ligands that nurture self-renewal of cancer stem-like cells. *Cancer Res* 71, 6061–6072, doi:10.1158/0008–5472.CAN-10–4269 (2011).

100. Wang, Z., Li, Y., Kong, D. & Sarkar, F. H. The role of Notch signaling pathway in epithelial-mesenchymal transition (EMT) during development and tumor aggressiveness. *Curr Drug Targets* 11, 745–751, doi:10.2174/138945010791170860 (2010).

101. Hsu, M. Y. et al. Notch3 signaling-mediated melanoma-endothelial crosstalk regulates melanoma stem-like cell homeostasis and niche morphogenesis. *Lab Invest* 97, 725–736, doi:10.1038/labinvest.2017.1 (2017).

102. Lu, J. et al. Endothelial cells promote the colorectal cancer stem cell phenotype through a soluble form of Jagged-1. *Cancer Cell* 23, 171–185, doi:10.1016/j.ccr.2012.12.021 (2013).

103. Kuhnert, F. et al. Dll4 blockade in stromal cells mediates antitumor effects in preclinical models of ovarian cancer. *Cancer Res* 75, 4086–4096, doi:10.1158/0008–5472.CAN-14–3773 (2015). Meurette, O. & Mehlen, P. Notch signaling in the tumor microenvironment. *Cancer Cell* 34, 536–548, doi:10.1016/j.ccell.2018.07.009 (2018).

104. Wieland, E. et al. Endothelial Notch1 activity facilitates metastasis. *Cancer Cell* 31, 355–367, doi:10.1016/j.ccell.2017.01.007 (2017).

105. De Francesco, E. M. et al. GPER mediates the angiocrine actions induced by IGF1 through the HIF-1alpha/VEGF pathway in the breast tumor microenvironment. *Breast Cancer Res* 19, 129, doi:10.1186/s13058-017-0923-5 (2017). Wang, X. et al. Crosstalk between TEMs and endothelial cells modulates angiogenesis and metastasis via IGF1-IGF1R signalling in epithelial ovarian cancer. *Br J Cancer* 117, 1371–1382, doi:10.1038/bjc.2017.297 (2017).

106. Bid, H. K., Zhan, J., Phelps, D. A., Kurmasheva, R. T. & Houghton, P. J. Potent inhibition of angiogenesis by the IGF-1 receptor-targeting antibody SCH717454 is reversed by IGF-2. *Mol Cancer Ther* 11, 649–659, doi:10.1158/1535–7163.MCT-11–0575 (2012). Weroha, S. J. & Haluska, P. IGF-1 receptor inhibitors in clinical trials—early lessons. *J Mammary Gland Biol Neoplasia* 13, 471–483, doi:10.1007/s10911-008-9104-6 (2008).

107. Hooper, A. T. et al. Angiomodulin is a specific marker of vasculature and regulates vascular endothelial growth factor-A-dependent neoangiogenesis. *Circ Res* 105, 201–208, doi:10.1161/CIRCRESAHA.109.196790 (2009).

108. Tamura, K. et al. Insulin-like growth factor binding protein-7 (IGFBP7) blocks vascular endothelial cell growth factor (VEGF)-induced angiogenesis in human vascular endothelial cells. *Eur J Pharmacol* 610, 61–67, doi:10.1016/j.ejphar.2009.01.045 (2009).

109. Zhao, W. et al. IGFBP7 functions as a potential lymphangiogenesis inducer in non-small cell lung carcinoma. *Oncol Rep* 35, 1483–1492, doi:10.3892/or.2015.4516 (2016).

110. Heesch, S. et al. BAALC-associated gene expression profiles define IGFBP7 as a novel molecular marker in acute leukemia. *Leukemia* 24, 1429–1436, doi:10.1038/leu.2010.130 (2010). Laranjeira, A. B. et al. IGFBP7 participates in the reciprocal interaction between acute lymphoblastic leukemia and BM stromal cells and in leukemia resistance to asparaginase. *Leukemia* 26, 1001–1011, doi:10.1038/leu.2011.289 (2012).

111. Cao, Z. et al. Molecular checkpoint decisions made by subverted vascular niche transform indolent tumor cells into chemoresistant cancer stem cells. *Cancer Cell*, 31, 110–126, doi:10.1016/j.ccell.2016.11.010 (2016). Ferrarelli, L. K. Tumors direct vessels to feed growth. *Sci Signal* 10, eaam9091, doi:10.1126/scisignal.aam9091 (2017). Verhagen, H. et al. IGFBP7 induces differentiation and loss of survival of human acute myeloid leukemia stem cells without affecting normal hematopoiesis. *Cell Rep* 25, 3021–3035 e3025, doi:10.1016/j.celrep.2018.11.062 (2018).

112. Nolan, D. J. et al. Molecular signatures of tissue-specific microvascular endothelial cell heterogeneity in organ maintenance and regeneration. *Dev Cell* 26, 204–219, doi:10.1016/j.devcel.2013.06.017 (2013).

113. Bouis, D., Hospers, G. A., Meijer, C., Molema, G. & Mulder, N. H. Endothelium in vitro: a review of human vascular endothelial cell lines for blood vessel-related research. *Angiogenesis* 4, 91–102, doi:10.1023/a:1012259529167 (2001).

114. Masuda, S., Matsuura, K. & Shimizu, T. Preparation of iPS cell-derived CD31(+) endothelial cells using three-dimensional suspension culture. *Regenerative Ther* 9, 1–9, doi:10.1016/j.reth.2018.06.004 (2018).

115. Seandel, M. et al. Generation of a functional and durable vascular niche by the adenoviral E4ORF1 gene. *Proc Natl Acad Sci USA* 105, 19288–19293, doi:10.1073/pnas.0805980105 (2008).

116. Gao, H. et al. High-throughput screening using patient-derived tumor xenografts to predict clinical trial drug response. *Nat Med* 21, 1318–1325, doi:10.1038/nm.3954 (2015).

117. Nolan, D. J. et al. Molecular signatures of tissue-specific microvascular endothelial cell heterogeneity in organ maintenance and regeneration. *Dev Cell* 26, 204–219, doi:10.1016/j.devcel.2013.06.017 (2013).

118. Kobayashi, H. et al. Angiocrine factors from Akt-activated endothelial cells balance self-renewal and differentiation of haematopoietic stem cells. *Nat Cell Biol* 12, 1046–1056, doi:10.1038/ncb2108 (2010).

14 Metastasis as a Tug of War between Cell Autonomy and Microenvironmental Control

Readdressing Unresolved Questions in Cancer Metastasis

Courtney König and Christoph A. Klein

Overview

How locally growing cancers generate systemic disease is apparently more complex than previously thought. Traditionally, cancer has been viewed as a progressive, cell-autonomous disease caused by the accumulation of oncogenic mutations, leading to loss of cellular control, unrestricted proliferation, and distant organ invasion that, when reaching a total tumor mass of 1 to 2 kg, will eventually kill the patient. This concept of malignancy has underpinned the search for improved cancer therapies over the past five decades with a view that the primary tumor is also the perfect surrogate model for metastatic growth and that the most abnormal cellular phenotypes should be the prime targets of therapeutic intervention. However, the course of disease varies significantly between patients, and consequently, our focus on the most aggressively growing cancers, cell lines, or mouse models is certainly too narrow and blurs treatment opportunities for less aggressive but nevertheless lethal disease variants. Importantly, as cancer mortality is mainly caused by metastatic rather than local cancer, tumor aggressiveness needs to be defined not only by its ability to proliferate locally but also by its ability to disseminate into, adapt at, and proliferate successfully in various tissue environments. Mechanisms of selection and adaptation at distant metastatic sites as well as organ defense against invading cells are therefore likely to affect disease courses. Here we discuss the origination of malignant phenotypes and focus on unresolved questions that need to be readdressed in order to find novel treatment options for metastatic cancer or to prevent lethal metastasis spread in the first place.

14.1 What Makes a Cancer Malignant?

The U.S. National Cancer Institute's dictionary of cancer terms defines malignancy as "a term for diseases in which abnormal cells divide without control and can invade nearby tissues. Malignant cells can also spread to other parts of the body through the blood and lymph systems" (https://www.cancer.gov/publications/dictionaries/cancer-terms/def

/malignancy). This three-part definition (comprising [1] uncontrolled proliferation, [2] cellular abnormality, and [3] invasion, dissemination, and ectopic growth as functional consequence), while appearing clear and straightforward, is however open to interpretation, given a large body of empirical evidence that might suggest a rather different and possibly more complex view of malignancy. We summarize here such relevant findings and conclude that the term "malignancy" cannot be assigned to an individual cell but is context dependent and therefore describes a dynamic phenotypic potentiality and not an intrinsic, binary property of cancer cells. In the following, we revisit the classical characteristics of malignancy with this in mind when presenting relevant evidence from the literature.

14.1.1   Uncontrolled Proliferation

If uncontrolled proliferation were cancer defining, many "malignant" diseases shouldn't be called so. In breast cancer, the 5 percent fastest growing tumors reach a size of 1 to 2 cm within twelve months, whereas the 5 percent slowest growing need more than fifty years.[1] Obviously, control of cellular proliferation is different from case to case even within a specific cancer type. Yet, fully autonomous proliferation has traditionally been regarded as the main defining hallmark of cancer,[2] a view that has become entrenched since the 1960s as the first cancer cell lines grown in culture were highly proliferative.

Uncontrolled proliferation (i.e., the ability to generate unlimited progeny within a short period of time) is considered the result of an evolutionary process within diseased tissue, based on mutation and selection. Benign tumors are distinguished from malignant tumors by their lower rate of proliferation and by their inability to invade and spread beyond their tissue of origin. It was therefore expected that benign tumors would differ from malignant tumors in their mutational load, specifically in oncogenes. However, somatic, "oncogenic" mutations have been implicated in many benign lesions.[3] Activating oncogene mutations in *FGFR3* or *PIK3CA*, for instance, have been found in seborrheic keratosis that poses no risk of malignancy.[4] Interestingly, constitutive high FGFR3 kinase activity was seen more frequently in benign lesions than in urothelial carcinoma, suggesting that even harboring mutations with strong oncogenic potential is not sufficient for causing malignancy.[5] In addition, even normal skin cells harbor high numbers of mutations, including oncogenic mutations,[6] which were apparently caused by sun exposure. Similarly, chronic inflammation led to mutations in the TP53 gene in synovial tissues from patients with rheumatoid arthritis identical to the ones found in certain malignancies.[7] This phenomenon was also seen in benign nevi and melanoma, where frequencies of BRAF and NRAS mutations were similar, but only the presence of concurrent genetic hits such as PTEN activity enabled transformation into a malignant disease.[8] While occasionally a single so-called driver mutation in a normal epithelial cell can provide a selective growth advantage, often additional mutations are needed to unleash uncontrolled proliferation in order to generate a malignant cancer.[9] Consequently, one concept holds that advantageous *combinations* of oncogenic driver mutations differentiate normal cells, benign tumor, and malignant tumor cells, and mathematical modeling has identified a full set of multihit combinations that differentiate tumor from normal tissue samples.[10] At least in some cancers, the acquisition of genomic instability is a better differentiating feature to distinguish between malignant versus benign tumors. The little data available for benign proliferative breast disease suggest that although hyperplasia may display limited allelic losses, ductal carcinoma in situ (DCIS) is principally distinguished by the emergence of genome instability.[11] Aneuploidy, a result of chromosome missegregation and errors in DNA replication and repair, was proposed to be an important initiating event in the conversion of normal cells into cancer cells.[12] However, mouse model experiments failed to prove chromosomal instability to be the only driver for tumorigenesis as additional genetic alterations in genes advantageous for tumor development were needed.[13] Thus, genomic instability appears to drive tumor initiation and progression only in the context of tissue-specific changes in cellular differentiation and other genetic and epigenetic alterations. Conversely, while mutations occur also in benign lesions and in healthy tissue, an additional increase in genomic instability is required to promote cancer formation.

In summary, acquisition of malignant traits, and in particular of uncontrolled proliferation, is currently considered a process of progressive accumulation of genomic and (epi) genetic alterations, including chromosomal translocations, point mutations, deletions, aneuploidy, and gene amplifications that enable the transition from a benign to a malignant phenotype. Once genomic instability has been acquired in proliferating cells, it becomes the basis for the rapid generation of variant cell phenotypes upon which tumor-specific selection can act—leading to the selection of increasingly malignant and autonomous cells. But how, when, and where are fully autonomous, malignant cells selected, and why do cancers differ so dramatically in their aggressiveness? In the traditional concept, acquisition of proliferative autonomy within the primary tumor precedes the event of dissemination—an aspect of malignancy that we will address below again in more detail.

14.1.2 Phenotypic Abnormality

Since oncogenic mutations per se do not allow one to distinguish clearly between benign and malignant tumors, other cellular abnormalities associated with cancers have been used for characterizing malignancy. It has been noted since the second half of the nineteenth and the early twentieth centuries that very particular changes in cellular morphology and metabolism characterize malignant cancers.

Tissue morphology is often diagnostic and hence used in pathological routine. However, cytological assessment (i.e., morphological changes of individual cells) is much less informative. Malignant cells display, for example, an increase in nuclear size, which results from ploidy changes or in part from missegregation of chromosomes during mitosis. If this occurs, aneuploidy may (further) increase chromosomal instability and drive further malignant progression.[14] Morphological changes also result from altered interactions with the microenvironment, supporting proliferation and motility and thereby promoting malignant traits.[15] In addition, metabolic alterations accompany morphological changes. Cancer cells display altered metabolic pathways enabling metabolic autonomy adapted to the specific conditions of the tumor microenvironment (usually low pH, low oxygen, altered paracrine signaling, and a changed mechanical context) in order to fulfill the high bioenergetic demands required for proliferation and movement.[16] However, changes in a cell's metabolism are not solely a consequence of oncogenic mutational transformation but may be achieved by epigenetic and other nongenetic modes of metabolic switching (e.g., activation of epithelial-mesenchymal transition transcription factors) that cells activate under certain stress conditions. Such metabolic adaptations can then be selected upon by the specific selective pressures within an abnormal, stressful tumor microenvironment.[17] Altered metabolite concentration, for instance, can activate enzymes that modify DNA and/or histones, resulting in large-scale changes of transcriptional programs leading to tumor initiation. Induction of this epigenetic remodeling often occurs through mutations in metabolic genes such as the tricarboxylic acid cycle enzymes isocitrate dehydrogenase, succinate dehydrogenase, and fumarate hydratase.[18] (See also chapters 11 and 12, this volume.) As a consequence, different metabolites accumulate that inhibit the ten-eleven translocation methylcytosine dioxygenase (TET) enzyme's activity by preventing α-ketoglutarate to succinate conversion, thereby often causing genome-wide DNA demethylation.[19] At least two additional abnormal metabolic phenotypes are observed throughout tumor progression. First, disseminating cancer cells that have lost matrix attachment while in circulation upregulate antioxidants through activating the pentose phosphate pathway, which improves their survival.[20] Second, disseminated cancer cells require increased energy metabolism and upregulate mitochondrial and proline catabolism during metastatic colonization.[21] These examples demonstrate the plasticity of cancer cells, enabling them to adapt to varying conditions, acquire additional changes, and increase their autonomy. However, to our knowledge, a minimal definition of which phenotype of an individual cell is to be considered still normal and which defines its malignant state has not yet been determined. As noted for uncontrolled proliferation at the primary and distant sites and its driving mechanisms, cellular abnormality cannot be defined in a binary fashion by either the presence or the absence of traditional phenotypic parameters such as cancerous morphology and tumor metabolism. Hence, their acquisition needs to be readdressed in the context of systemic cancer progression.

14.1.3   Dissemination to and Growth at Distant Sites

Local invasion and subsequent dissemination are two defining characteristics of malignant cells. Cancer cells disseminate through blood and lymphatic vessels. Malignant tumors can stimulate (neo) angiogenesis, ensuring nutrition supply to the tumor, enabling growth beyond a certain size. As an indirect consequence, tumor angiogenesis promotes vascular invasion and metastatic spread. However, some malignant diseases, such as pancreatic ductal adenocarcinoma (PDAC), display only poor vascularization and overall perfusion but are considered highly aggressive, which indicates that establishment of neoangiogenesis cannot be used as a surrogate marker for the degree to which cancers disseminate.[22] Injection of cancer cells into experimental metastasis models, on the other hand, reveals that not all injected cells are successful. It was concluded from such experiments that only a handful of injected cancer cells are capable of invading foreign tissue, and even fewer manage to proliferate and form metastatic colonies at distant sites.[23] However, experimental metastasis models probably fail to identify all the characteristics of malignant cells since active dissemination and migration through different tissue contexts may equip cancer cells along their way with additional features needed for successful metastasis.

Therefore, the question pertains to whether active dissemination suffices to qualify an epithelial cell as malignant. There is generally agreement that the presence of circulating tumor cells (CTCs; i.e., cancer cells that circulate in the bloodstream) per se is not sufficient to generate metastasis; neither is the presence of disseminated cancer cells (DCCs; i.e., cancer cells that homed in a target organ). DCCs can be detected in patients who have not yet developed distant metastasis using epithelial markers in organs that normally do not comprise epithelial cells, such as bone marrow or lymph nodes. A large study of 4,700 breast cancer patients revealed that detection of DCCs at the time of diagnosis, but also DCC persistence after complete surgical resection and therapeutic intervention, was associated with poor clinical outcome with respect to disease-free and overall survival.[24] Despite being prognostic, not all DCCs are able to grow into metastases, since not all patients with DCCs at the time point of surgery will develop manifest metastases, indicating that DCCs cannot be unequivocally defined as "fully" malignant cells.

The malignant character of cells has therefore been deduced from successful engraftment of tumors in mice (i.e., via proof of substantial cell autonomy that can tolerate a xeno-environment). However, engraftment rates are strongly dependent on the choice of implantation site, mouse strain, and the tumor cells or cell line used, since copy-number alterations (CNAs) and other genomic alterations can differ significantly in cell lines compared to their counterpart tumors.[25] While xenograft models of human cancer cell lines often display a take rate of more than 90 percent, engraftment rates of primary patient-derived tumor xenografts are only about 25 percent.[26] However, nobody would call the remaining 75 percent of cancers, which have killed or will inevitably kill their hosts, "benign." Furthermore, we have investigated engraftment of DCCs, which, per definition, should fulfill the characteristics of malignant cells as they were derived from a malignant melanoma that has already invaded the lymph node.[27] Surprisingly, tumor formation depended on whether or not the cells were isolated from a lymph node colony consisting of a larger group of cells or had lodged there as single cells. DCCs from lymph node colonies engrafted and formed tumors in mice, whereas individual DCCs did not.[28] This finding was particularly informative as the engraftment ability of DCCs was directly linked to the presence of specific alterations, such as mutations in *BRAF* or *NRAS*. In other cancers, such as breast cancers, engraftment seems to be related to the presence of certain stem cell traits.[29] We recently found that this phenotype can be ectopically acquired, similar to mutations that are acquired outside the tumor at colony formation.[30] In this study, the bone marrow microenvironment, rich in IL-6 and soluble IL6 receptor, likely induced stemness in breast cancer DCCs, facilitating the formation of metastatic colonies. Therefore, malignant cell deregulation results from intrinsic or extrinsic mechanisms that need to occur during metastasis formation.

In summary, this short overview of the three classical malignancy-defining hallmarks demonstrates that the dichotomy of malignant versus benign/normal has conceptual weaknesses that have become quite apparent with recent advances in single-cell and genomic technologies. It might be useful for the clinical routine, in particular for outcome and therapy response prediction, to include within the concept of malignancy the interdependencies of intrinsic cellular deregulation and extrinsic factors that promote or inhibit progression. To this end, it seems essential to identify the defining cellular features of a "malignant potential" of abnormal cells with respect to a specific environment (i.e., the patient) that drives progression and deterioration in a context-dependent manner. If such an inclusive definition could be applied early on, treatment decisions could be better adapted to the individual needs of the patient.

14.2   When Do Cancer Cells Disseminate?

Traditionally, dissemination has been considered to occur late, possibly the day before surgery of the primary tumor. This view grounds in the TNM staging system, which links anatomical disease extension to outcome (T is a descriptor of primary tumor size and local invasiveness, N describes to what degree local lymph nodes are affected, and M describes the degree of metastatic spread to distant sites). For most cancers, anatomical staging predicts that the larger tumors have grown or spread, the lower the chances to cure the patient by surgery or other treatments. Since early surgery (i.e., at low TNM stages) may cure a patient, it was concluded that (1) dissemination starts late and (2) disseminating cancer cells represent the genomically most altered, most aggressive clones and therefore are equipped to colonize distant organs and kill the host. Moreover, it was thought that evolution toward cell autonomy and uncontrolled growth occurs within the primary tumor *before* metastatic spread.[31]

However, with the development of highly reliable whole-genome amplification protocols for single cells and the subsequent analysis of disseminated cancer cells isolated from cancer patients years before metastatic manifestation, there is now strong evidence for early dissemination, early in regard to lesion size and to "genomic time" (i.e., the accumulation of genomic alterations).[32] Since then, overwhelming evidence that cancerous lesions seed cells early on has been obtained from mouse models and patients.[33] Two recent publications provided first mechanistic insight into how DCCs disseminate from the mammary gland to metastatic sites during very early stages of tumor development. Using Her-2-driven transgenic mouse models, it was found that early DCCs are more metastasis competent than cells that leave the tumor at later stages, and a mechanism was identified that explained significantly reduced migration and dissemination from large tumors.[34] In patients, two lines of evidence support early dissemination of cancer cells and their relevance as metastases founders, direct assessment of DCCs by immunocytology and genomic analyses. Direct testing for DCCs demonstrated that in breast cancer, in situ carcinomas (i.e., lesions that are by definition not invasive) already seed cancer cells.[35] It was further found in mice that the number of early DCCs was independent of the tumor mass present, contradicting the traditional view that metastatic seeding would occur only once tumors have grown to a certain size.[36] Likewise, in melanoma, only a marginal increase in DCC numbers per million sentinel lymph node cells was noted over the time span from T1 to T4 stages, and the same was true for DCCs in bone marrow of breast cancer patients.[37] For both cancers, one can calculate that initial dissemination of cancer cells starts already from lesions of very small thickness (about 500 μm in melanoma; see below) or diameter (<4 mm in breast cancer) and that the risk of de novo dissemination decreases the larger or thicker the primary tumor becomes.[38]

The observed starting point of dissemination from very small lesions is also supported by genomics studies of DCCs. Genetic analyses of DCCs suggest that these cells have disseminated early also in cancer evolutionary time as estimated by the number of genomic alterations acquired throughout the time of progression (table 14.1). In breast, prostate, and esophageal cancer, DCCs recovered from bone marrow generally display fewer genetic abnormalities than their matched primary tumors, implying that DCCs seed the bone marrow early in disease progression.[39] It was therefore concluded in these studies that DCCs disseminate early, evolve independently from the primary tumor at metastatic sites, and thus display high genetic divergence from their primary tumor.[40] For melanoma, primary tumor thickness and genomic state of DCCs were directly compared. Melanoma DCCs were found to often leave the primary tumor at a size of less than 0.5 mm, home to the lymph node, while still lacking important genetic drivers and to acquire them during formation of early metastatic colonies.[41] Similarly, DCCs of M1 stage breast cancer patients displayed a higher genetic concordance with the primary tumor[42] than early M0 stage DCCs. This indicates that also in breast cancer, DCCs lacking typical "malignant" genetic changes disseminate and acquire them at the distant sites during formation of manifest metastases and thereby become more similar to the primary tumor. One possible interpretation of these observations from melanoma and breast cancer is that successful generation of metastatic colonies requires acquisition of the typical driver changes (i.e., loss of proliferation control) and that this process exploits genetic or epigenetic vulnerabilities of individual genotypes. A study in esophageal cancer demonstrated greater genetic similarity of lymph node (LN) DCCs with manifest LN metastasis than with bone marrow

Table 14.1

Genome-wide comparison of solid primary tumors and their DCCs.

aCGH, array comparative genomic hybridization; BM, bone marrow; CGH, comparative genomic hybridization; CNA, copy-number alteration; DCC, disseminated cancer cell; LN, lymph node; mCGH, metaphase comparative genomic hybridization; PT, primary tumor; WES, whole-exome sequencing; WGS, whole-genome sequencing.

Study (year)	Primary tumor type	No. of patients (No. of matched samples)	Method	Genetic relationship between primary tumor and DCCs	Proposed time point of dissemination
Hosseini et al. (2016)[a]	Breast	1,637* (PT from database), 27* (94 M0 DCCs), 21* (91 M1 DCCs)	aCGH	M0 DCCs less aberrant than PT; M1 DCCs more similar to PT	Early dissemination
Demeulemeester et al. (2016)[b]	Breast	3 (3 PTs, 1 LN, 8 DCCs)	WES (PT and LN bulk), WGS (DCCs)	53 percent of DCCs show similarity with PT	Late dissemination
Weckermann et al. (2009)[c]	Prostate	20 (32 PTs, 38 DCCs)	CGH	1/8 patients (M0) showed 8 percent shared CNA, 6/9 M1 patients shared 8.3–25.0 percent	Early dissemination
Holocomb et al. (2008)[d]	Prostate	9 (9 PTs, 9 pools of 10–20 DCCs)	aCGH	1/3 of DCC deviations shared with PT	Early dissemination
Werner-Klein et al. (2018)[e]	Melanoma	19 (23 PTs, 24 DCCs)	CGH	Disparity between PTs and matched DCCs	Early dissemination
Schumacher et al. (2017)[f]	Esophageal	17 (17 PTs, 34 BMs/LN-DCCs)	mCGH	38 percent of BM-DCCs, 75 percent of LN-DCCs show PT similarity	Not specified
Stoecklein et al. (2008)[g]	Esophageal	8 (8 PTs, 12 DCCs [5 LNs, 7 BMs])	mCGH	PT and DTCs grouped remotely from each other	Early dissemination

* Nonmatched samples.

a. Hosseini, H. et al. Early dissemination seeds metastasis in breast cancer. *Nature* **540**, 552–558, doi:10.1038/nature20785 (2016).

b. Demeulemeester, J. et al. Tracing the origin of disseminated tumor cells in breast cancer using single-cell sequencing. *Genome Biol* **17**, 250, doi:10.1186/s13059-016-1109-7 (2016).

c. Weckermann, D. et al. Perioperative activation of disseminated tumor cells in bone marrow of patients with prostate cancer. *J Clin Oncol* **27**, 1549–1556, doi:10.1200/JCO.2008.17.0563 (2009).

d. Holcomb, I. N. et al. Genomic alterations indicate tumor origin and varied metastatic potential of disseminated cells from prostate cancer patients. *Cancer Res* **68**, 5599–5608, doi:10.1158/0008–5472.CAN-08–0812 (2008).

e. Werner-Klein, M. et al. Genetic alterations driving metastatic colony formation are acquired outside of the primary tumour in melanoma. *Nat Commun* **9**, 595, doi:10.1038/s41467-017-02674-y (2018).

f. Schumacher, S. et al. Disseminated tumour cells with highly aberrant genomes are linked to poor prognosis in operable oesophageal adenocarcinoma. *Br J Cancer* **117**, 725–733, doi:10.1038/bjc.2017.233 (2017).

g. Stoecklein, N. H. et al. Direct genetic analysis of single disseminated cancer cells for prediction of outcome and therapy selection in esophageal cancer. *Cancer Cell* **13**, 441–453, doi:10.1016/j.ccr.2008.04.005 (2008).

DCCs. These shared LN-DCC/metastasis alterations were not detected in the primary tumor, indicating site-specific alterations in lymphatic metastases,[43] a finding that is fully consistent with independent acquisition of alterations after dissemination and ongoing evolution outside the primary tumor.

14.3   Which Cells Are Founders of Metastases?

In transgenic mouse models, it can be shown by genomic analysis that about 80 percent of lung metastases grow out from early DCCs, whereas only 20 percent are derived from cells that left the primary tumor after more than 50 percent of its genomic alterations were acquired.[44] Bulk sequencing data of human cancers comparing matched primary tumor–metastases pairs, however, do not generate such a clear picture. Therefore, the debate about the cell of origin of a metastasis has gained considerable new momentum. Are early DCCs the founder cells of metastasis or are late DCCs the culprits? Besides the argument that high T stage (= larger tumor size) is associated with poor outcome (see above), indicating that late DCCs might be better equipped to form a metastasis, the results of sequencing studies comparing primary tumors and metastases (see table 14.2) are mostly interpreted in favor of late DCCs as founder cells of metastases.

While the association with high T stage may simply reflect the fact that a late-diagnosed cancer had more time to grow and therefore also a DCC had more time to form a metastasis, the genome-wide or targeted comparison of primary tumors with metastases should reveal their evolutionary relationship. However, almost all studies suffer from a significant sample selection bias. Most analyzed samples came from two kinds of sources, namely, either from autopsy studies in which late-stage patients agreed to donate their body for analysis or from patients who presented with synchronous metastatic disease and therefore biopsies from primary tumors and metastases were collected at the same time. The most common cases, however, in which metastases arose years after primary tumor resection are currently underrepresented. A further confounding factor—namely, the types and number of courses of systemic therapies—is not always reported, and therefore the data must be viewed with caution, as therapy itself has specific effects on the survival of specific clones and on the evolution of genomic alterations. Finally, the clonal structure of primary tumors and metastases was in most cases deduced from bulk sequencing data, whereas only few studies have performed single-cell analyses to reveal clonal architecture (see below). We therefore summarize in the following available studies in relation to these caveats.

14.3.1   Synchronous versus Metachronous Metastasis Formation

The frequency of patients presenting with metastasis at primary diagnosis varies between 7 percent (breast cancer) and 70 percent (small cell lung cancer), clearly depending on tumor type. For example, in breast (7 percent) and colorectal cancer (24 percent; data from the Munich cancer registry, https://www.tumorregister-muenchen.de/), primary metastatic disease is rather rare. Nevertheless, matched primary tumor–metastases pairs from such patients currently predominate in available data sets. There are at least two reasons why pairs of primary tumors and metastases from synchronous metastasis are more similar than pairs of primary tumors and metachronous metastasis. First, primary metastatic disease could be more aggressive and hence be diagnosed in cases where a clone emerged that was a priori highly autonomous, in contrast to cases where metastases arose years after primary tumor resection. Second, many patients with primary metastatic disease are treated with nonsurgical therapies involving systemic drug administration. Here, the effects of first- to third-line (or more) therapies may select for aggressive, highly resistant clones and hence homogenize the clonal architecture. Therefore, clinical information regarding past treatments has to be considered when conclusions are drawn from sequencing data.

Several studies on synchronous primary tumor–metastasis pairs in melanoma, breast, colon, and renal cancer revealed a high genomic similarity of primary tumors (PTs) and their matched metastasis (Met).[45] Currently, colon cancers display a higher percentage (median of four studies[46]: 67.5 percent) of similarity between primary tumors and their metastases, as determined by single-nucleotide variants (SNVs) and overall mutations or copy-number alterations than breast cancers (median of four studies[47]: 55 percent). However, also for colorectal cancer, several studies report a lower overlap of 20 to 54 percent of shared SNVs, where disparate mutations outnumber shared changes.[48] Similarly, the breast cancer studies by Ding et al.,[49] Navin et al.,[50] and Yates et al.[51] deduce linear progression of late disseminating cells from low primary tumor–metastases disparity, whereas others concluded that after early dissemination, clones within the primary tumor and the metastasis progress in parallel.[52] In prostate cancer, two recent studies claimed early divergence of PT and metastases.[53] Gundem et al. analyzed ten metastatic prostate cancer patients who died of the disease by whole-genome sequencing and found that multiple metastases were more closely related to each other than any of them were to the corresponding primary tumor. Interestingly, metastases located in the same tissue were more closely related than those in different tissues, suggesting local spread within an organ.[54]

The available data indicate that synchronous metastases display higher concordance with the paired primary tumor than metachronous metastases,[55] but comparison data are yet too scarce to draw strong conclusions. Therefore, the question awaits systematic analysis. For example, a recent study on renal cancer with 90 percent synchronous and 10 percent metachronous primary tumor–metastases pairs did not assess their clonal relation according to disease and treatment courses.[56]

14.3.2   Bulk Sequencing Data versus Single-Cell Analysis

Studies listed in table 14.1 (single-cell analysis) and table 14.2 (mostly bulk DNA) differ in that analysis of single cells captures significant differences between (individual) DCCs and bulk primary tumor DNA, whereas comparison of bulk primary tumor DNA and bulk metastasis DNA identifies a high percentage of shared aberrations. Therefore, the data sets are difficult to reconcile if DCCs that have been analyzed so far should be the precursor cells of metastases. However, phylogenetic reconstruction of the clonal architecture from bulk DNA is based on several fundamental premises. Of these, the infinite site assumption, used in almost all phylogenetic models, is certainly more than questionable. The infinite site hypothesis formulated by Kimura in 1969 posits that the probability for a mutation event to occur at the same site more than once would be infinitely small and therefore assumes that each mutation is generated at most once within a tumor, and therefore if it is found in the primary tumor *and* the metastasis, it must have been inherited from the primary tumor. This premise excludes a priori the possibility that two cancer cells are independently selected for the same change. However, for example, in the case of *BRAF V600E* mutations that are shared between 40 percent of melanoma patients, this assumption for an individual tumor makes no sense. Since we do not assume that 40 percent of melanoma patients are clonally related, why should we do so within a large cancer cell population of an individual patient? On the other hand, while this argument seems highly adequate for driver mutations that confer fitness advantages, the infinite site assumption may be plausibly applied to noncoding somatic mutations. Another weak point of bulk analyses is the adequate determination of the number of clones that evidently impacts on the retrieval of a presumed clonal architecture.[57] A clone is defined by full genetic identity. However, in our single-cell analysis, we have never observed two cells of a patient to be fully identical. In many phylogenetic models from bulk data, clone number is estimated as a practical compromise. Therefore, to better understand the impact of current bioinformatics approaches and premises, single-cell analysis of primary tumors, DCCs, and metastases will be helpful to test, validate, or refute currently published phylogenetic models from bulk data.

A first example of single-cell exome sequencing was generated on two colorectal cancer patients and revealed high PT-Met concordance, with 89 percent similarity for patient 1 and 64 percent similarity for patient 2.[58] Although the primary and metastatic cells in this study shared many alterations, the primary tumor cells as well as the metastasis cells also contained additional private mutations. Notably, both patients presented with metastatic disease. In contrast, no study has been published so far that had addressed the clonal architecture of primary tumor and metachronous metastasis pairs by single-cell sequencing.

In summary, currently we cannot unequivocally determine which scenario is more frequent in human patients: either early DCCs evolving over years to form a metastasis or late disseminating DCCs growing rapidly and killing the patient *after* they have acquired all necessary oncogenic mutations and a substantial autonomy within the primary tumor. However, recent insights into the importance of tissue niches, in which metastatic founder cells may lodge, could blur this distinction since DCCs arriving late in the niche might need the prior presence of early DCCs to prime the niche for successful metastatic colonization,

Table 14.2

Genome-wide comparison of solid primary tumors and their metastases.

aLN, axillary lymph node; asyn, asynchronous; CNAs, copy-number alterations; d, days; EAC, esophageal adenocarcinoma; LN, lymph node; LNMET, lymph node metastasis; met, metastasis; meta, metachronous; MSK-IMPACT, Memorial Sloan Kettering–Integrated Mutation Profiling of Actionable Cancer Targets; m, months; mut, mutation; NA, not assessed; PT, primary tumor; sc, single cell; scSeq, single-cell sequencing; SNVs, single-nucleotide variants; syn, synchronous; tSeq, targeted sequencing; WES, whole-exome sequencing; WGS, whole-genome sequencing; y, years.

Study (year)	Primary tumor type	Analyzed metastatic site	No. of patients (no. of samples)	Method	Sample type	Time between primary tumor resection and first metastases	Genetic relationship between primary tumor and metastases			Proposed time point of dissemination
							Met specific	Shared between PT-Met	PT specific	
Turajlic et al. (2012)[a]	Melanoma	LN	1 (1 PT, 1 LN)	WGS	Bulk DNA	Syn with PT	2 mut	Majority of the coding SNVs and CNAs shared	1 mut	NA
Sanborn et al. (2015)[b]	Melanoma	LN, locoregional, and distant skin	8 (NA)	WES	Bulk DNA	1 syn LN met (2–4 months later distant met), others: between 6 and 46 months	NA	Majority of SNVs shared in 7/8 patients	NA	Late dissemination
Leung et al. (2017)[c]	Colon	Liver	2 (2 PTs, 2 met, 372 sc)	scSeq, WES, and ultra-deep tSeq	Bulk and sc DNA	Syn	5- to 14-point mut (in sc)	88.8 to 63.5 percent SNVs (bulk DNA)	1- to 2-point mut (in sc)	Late dissemination
Kim et al. (2015)[d]	Colon	Liver	5 (35 mixed PT and met biopsies)	WES	Bulk DNA	3/5 syn; 2/5 meta	2.4 to 40.7 percent SNVs	19.8 to 53.9 percent SNVs	13.8 to 56.0 percent SNVs	NA
Ishaque et al. (2018)[e]	Colon	Liver, lung	12 (PT: 12; met: 11 liver, 1 lung)	WGS	Bulk DNA	NA	19 percent	65 percent SNVs	15 percent	11 patients late, 1 patient early dissemination

Study	Cancer type	Metastatic sites	Samples	Method	DNA	Timing		Shared SNVs		Evolutionary pattern
Hu et al., 2019[f]	Colon	Liver, brain, lung, LN	23 (10 brain, including 1 liver, 1 lung, and 4 LNs; 13 liver met; total: 118 samples)	MuTect (v.1.1.7)	Bulk DNA	17 syn, 6 asyn	NA	70 percent SNVs	NA	Early by tumor size; late by evolutionary stage
Hong et al. (2015)[g]	Prostate	Bone	4 (26 samples)	WGS, ultra-deep tSeq	Bulk DNA	meta	NA	Great diversity in underlying mutational processes	NA	Parallel progression, early divergence
Gundem et al. (2015)[h]	Prostate	LNs, liver, bladder, adrenal gland, rib, bone marrow, others	10 (5 PTs, 10 met; 51 samples)	WGS	Bulk DNA	syn (at autopsy)	NA	Multiple mets were more closely related to each other than any of them to PT	NA	Parallel progression and linear met-to-met seeding
Yates et al. (2017)[i]	Breast	Lung, liver, distant skin region, contralateral breast, distant LN	17 (40 samples; met: 1 lung, 1 liver, 2 distant skin regions, 1 contralateral breast, 2 distant LNs)	WGS	Bulk DNA	meta (8–158 m), syn aLN	NA	12 to 98 percent SNVs	NA	Late dissemination
Kroigård et al. (2017)[j]	Breast	aLN, bone, LN, liver	3 (9 samples: 4 PTs, 2 aLNs, 2 livers, 1 bone)	WES and tSeq	Bulk DNA	syn aLN; meta distant mets (median 3.08 y)	4.5 to 35.5 percent SNVs	69 to 99 percent SNVs	6.2 to 29.5 percent SNVs	NA
Ullah et al. (2018)[k]	Breast	aLN, brain, bone, skin, liver, lung, colon, uterus, ovary	20 (93 samples: 33 PTs, 13 aLNs, 47 mets)	WES	Bulk DNA	syn LN, meta (median: 33.5 m)	Not quantified	9 to 88 percent SNVs	Not quantified	NA
Schrijver et al. (2018)[l]	Breast	Brain, skin	17 (PT: 17, skin: 8, brain: 9)	MSK-IMPACT (341 cancer genes)	Bulk DNA	3 syn and 14 meta (brain 468 d; skin 760 d)	43 percent SNVs	23 to 100 percent SNVs in putative driver genes	12 percent SNVs	NA

(continued)

Table 14.2 (continued)

Study (year)	Primary tumor type	Analyzed metastatic site	No. of patients (samples)	Method	Sample type	Time between primary tumor resection and first metastases	Genetic relationship between primary tumor and metastases			Proposed time of dissemination
							Met specific	Shared between PT-Met	PT specific	
Schumacher et al. (2017)[m]	Esophageal	LN	17 (PT: 30 sc, LN: 37 sc)	mCGH	sc DNA	NA	13 CNAs	Typical EAC alterations present in all PT and LNMET samples	10 CNAs	Early dissemination
Turajlic et al. (2018)[n]	Kidney	LN, others	38 (59 PT samples with 462 regions; 110 met)	tSeq (driver-gene panel)	Bulk DNA	89.5 percent syn, 10.5 percent meta	5 percent driver events	63 percent of driver events	32 percent driver events	NA

a. Turajlic, S. et al. Whole genome sequencing of matched primary and metastatic acral melanomas. *Genome Res* **22**, 196–207, doi:10.1101/gr.125591.111 (2012).

b. Sanborn, J. Z. et al. Phylogenetic analyses of melanoma reveal complex patterns of metastatic dissemination. *Proc Natl Acad Sci USA* **112**, 10995–11000, doi:10.1073/pnas.1508074112 (2015).

c. Leung, M. L. et al. Single-cell DNA sequencing reveals a late-dissemination model in metastatic colorectal cancer. *Genome Res* **27**, 1287–1299, doi:10.1101/gr.209973.116 (2017).

d. Kim, T. M. et al. Subclonal genomic architectures of primary and metastatic colorectal cancer based on intratumoral genetic heterogeneity. *Clin Cancer Res* **21**, 4461–4472, doi:10.1158/1078–0432.CCR-14–2413 (2015).

e. Ishaque, N. et al. Whole genome sequencing puts forward hypotheses on metastasis evolution and therapy in colorectal cancer. *Nat Commun* **9**, 4782, doi:10.1038/s41467-018-07041-z (2018).

f. Hu, Z. et al. Quantitative evidence for early metastatic seeding in colorectal cancer. *Nat Genet* **51**, 1113–1122, doi:10.1038/s41588-019-0423-x (2019).

g. Hong, M. K. et al. Tracking the origins and drivers of subclonal metastatic expansion in prostate cancer. *Nat Commun* **6**, 6605, doi:10.1038/ncomms7605 (2015).

h. Gundem, G. et al. The evolutionary history of lethal metastatic prostate cancer. *Nature* **520**, 353–357, doi:10.1038/nature14347 (2015).

i. Yates, L. R. et al. Genomic evolution of breast cancer metastasis and relapse. *Cancer Cell* **32**, 169–184 e167, doi:10.1016/j.ccell.2017.07.005 (2017).

j. Kroigård, A. B. et al. Genomic analyses of breast cancer progression reveal distinct routes of metastasis emergence. *Sci Rep* **7**, 43813, doi:10.1038/srep43813 (2017).

k. Ullah, I. et al. Evolutionary history of metastatic breast cancer reveals minimal seeding from axillary lymph nodes. *J Clin Invest* **128**, 1355–1370, doi:10.1172/JCI96149 (2018).

l. Schrijver, W. et al. Mutation profiling of key cancer genes in primary breast cancers and their distant metastases. *Cancer Res* **78**, 3112–3121, doi:10.1158/0008–5472.CAN-17–2310 (2018).

m. Schumacher, S. et al. Disseminated tumour cells with highly aberrant genomes are linked to poor prognosis in operable oesophageal adenocarcinoma. *Br J Cancer* **117**, 725–733, doi:10.1038/bjc.2017.233 (2017).

n. Turajlic, S. et al. Tracking cancer evolution reveals constrained routes to metastases: TRACERx renal. *Cell* **173**, 581–594 e512, doi:10.1016/j.cell.2018.03.057 (2018).

a process that could start years before primary tumor diagnosis. It is foreseeable that more careful cell lineage analysis will shed light onto this process. As technologies are emerging that will enable better resolution than bulk sequencing, we expect to gain substantial insights within the coming five to ten years. These insights may become fundamentally important for attempts to prevent metastasis formation.

14.4   What Happens during the Invisible Phase of Metastasis?

Whether metastases arise from early DCCs continuously evolving during colonization or from explosively growing late DCCs cannot be investigated directly in patients (figure 14.1). Clinical imaging technologies can resolve metastases earliest at a diameter of about 4 to 6 mm, and the diagnosis often needs to be confirmed by longitudinal size measurements (i.e., by the fact that the lesion is growing). Once a metastasis can be visualized, growth rates can be measured. For most carcinomas, it has been found that the tumor volume doubling time (TVDT) is rather slow (i.e., between 100 and 200 days for most primary tumors and their metastases). This means that a single cell grows to a 1-cm lesion within a decade. From this it was concluded that growth of metastases from late DCCs is very unlikely since disease courses with early relapse (less than five years after diagnosis and surgery) can hardly be reconciled with such kinetics. However, this reasoning is only true if the growth rate is constant and therefore similar between the visible and the invisible phase (for a discussion of specific growth kinetics of tumors, see chapter 10, this volume).

Early DCCs, undetectable by clinical imaging, might have the same growth rate as visible primary tumors or metastases. In this case, the time point of metastasis detection after primary tumor surgery would reflect the lag time between cancer initiation and dissemination. For example, in a cancer that grows to a 1-cm lesion within ten years and whose metastasis is diagnosed three years after primary treatment, DCCs could have left the tiny primary lesion after three years. Assuming constant, exponential growth, such a primary lesion would have the size of about 500 cells (i.e., after nine of the thirty volume doublings) when the metastasis founder is disseminating. Given such doubling times, late DCCs that disseminated shortly before surgery would not be able to form a metastasis within five years from a single cell. Consequently, if late DCCs are metastasis founders, growth rates must be significantly higher during the invisible phase of metastatic colonization than during the visible phase.

On the other hand, early DCCs lack many typical genomic aberrations and need to acquire additional changes to become autonomous from microenvironmental control. Since DNA changes are mostly acquired during cell division, one would need to postulate that early DCCs are actively dividing rather than remaining dormant.

In both scenarios, proliferation is a sine qua non, but its regulation might be fundamentally different. Late DCCs could therefore have a higher intrinsic propensity to proliferate, whereas early DCCs would need microenvironmental stimulation or a breakdown of

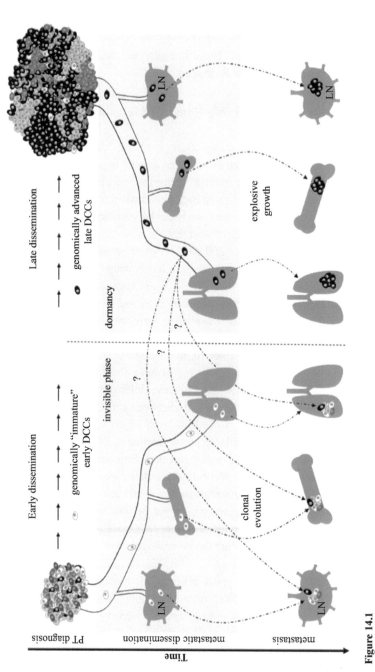

Figure 14.1

Alternative pathways to metastasis formation. Currently available data are compatible with at least two models and their combination. Disseminated cancer cells (DCCs) leaving the primary tumor (PT) early in a genomically "immature" state evolve at metastatic sites and acquire metastasis-forming capacity. Alternatively, "fully malignant" DCCs disseminate from advanced tumor stages, possibly get growth-arrested, and resume rapid proliferation after reactivation. The first concept could explain cases of high genotypic and phenotypic disparity between primary tumors and matched metastases and would be consistent with the frequently observed early dissemination. In contrast, the latter concept would agree with data showing substantial similarity between primary tumors and their metastases. Early DCCs could facilitate the colonization of late DCCs by "preparing" a metastatic niche and clonal cooperation.

tumor-suppressive regulatory signals from the tissue niche that the early DCC inhabits. Since we usually observe a rather low number of manifest metastasis, most intrinsically active late DCCs might be arrested by the microenvironment, whereas many early DCCs that lodge in stimulatory niches fail to acquire the right genetic alterations that render them more autonomous. For these fundamental questions to be resolved, the biology of DCCs during the invisible phase of metastasis formation needs to be studied at greater resolution and in larger numbers of patients than is currently achievable. In the following, we will discuss selected aspects of these questions based on currently available data.

14.4.1   Survival at Distant Sites

Although circulating tumor cell (CTC) counts and DCC numbers are prognostically relevant and correlate with disease-free and overall survival, not every DCC is capable of surviving and forming a metastasis at distant sites.[59] One mechanism of tumor cells to survive and adapt to conditions that are initially hostile for them at secondary sites is entering a state of cellular dormancy, which is defined as cell cycle arrest. As these non-proliferating dormant DCCs are protected by their cell cycle–arrested state from targeted and cytotoxic therapies,[60] cellular dormancy represents a survival mechanism for cells under stress. The temporary cell cycle arrest in the G0 to G1 phase also hides a cell from immune surveillance by activating components of the unfolded protein response (UPR) pathway that in turn downregulates major histocompatibility complex (MHC) class I molecules.[61] Several cell-intrinsic signaling pathways are involved in the complex process of dormancy regulation, with p38 stress signaling and the UPR pathway being the most prominent. p38 phosphorylation activates the UPR pathway that, in concert with the RAS-MEK-ERK pathway, induces a dormant phenotype in cells when the ratio of p38 over extracellular regulated kinase (ERK1/2) is increased.[62] The UPR also kickstarts ATF6/Rheb/mTOR signaling, leading to the induction of various dormancy-associated transcription factors DEC2/Sharp1, p27Kip1, p21, and NR2F1.[63] These intrinsic cell-signaling pathways are often linked to microenvironmental traits that in concert play a role in inducing cellular dormancy. For example, the balance of ERK1/2 and p38 signaling is regulated by fibronectin and urokinase-type plasminogen activator (uPA) signaling via the uPA receptor and specific integrins.[64] However, further microenvironment-derived factors such as TGFβ2, Bmp-4, and Bmp-7, as well as retinoic acid, can activate p38 and NR2F1 signaling via inhibiting ERK1/2 and inducing a dormant state in DCCs.[65]

Although most data on cell cycle–arrested DCCs are derived from animal models, clinical evidence for cellular dormancy and awakening of dormant DCCs comes from inadvertent transmission of cancer during organ transplantations.[66] Tissues from donors who were clinically evaluated as cancer free for a latency period of more than ten years but had a history of cutaneous malignant melanoma, or renal cell carcinoma, gave rise to metastasis in recipients only shortly after organ transplantation. These metastases were proven to be of donor origin and were found in various organs, including bone marrow, lungs, livers, and lymph nodes.[67] Evidence that DCCs may persist over time but do not progress was also found in cancer patients directly. For example, a cohort of nonprogressing prostate cancer patients who agreed to annual bone marrow sampling after radical prostatectomy was found to harbor bone marrow DCCs at a constant level of 20 percent positivity for almost a decade.[68] The genomic aberrations of the DCCs that were isolated from repeatedly positive patients (note that in patients without manifest metastasis, the median number of DCCs per million bone marrow cells is two) were highly heterogeneous, indicating that in nonprogressing patients, no clonal expansion of DCCs occurs—in contrast to patients who progress.[69] Which genetic programs enable hibernation of cancer cells for such long periods or which organ defense mechanisms control these cells is under active investigation.

14.4.2 Supportive and Repressive Niches

Besides escaping from immune control and activating survival programs, DCCs need to adapt to the new tissue microenvironments they have invaded as those may provide growth factors, ECM-related factors, and metabolic stress factors that are quite different from the ones at the site of their origin.[70] As a consequence, DCCs respond to these extrinsic stress signals with dormancy, aiding cells to cope with the new tissue context.[71] For example, hypoxic areas in the PT were found to give rise to DCCs that in turn expressed dormancy genes and were more resistant to chemotherapy.[72] Furthermore, dormant DCCs often lodge in close proximity to vascular basement membranes in the so-called perivascular niche.[73] Here DCCs associate with the endothelial cell–derived thrombospondin 1–enriched, stable microvasculature and exit the cell cycle, whereas actively growing tumor cell clusters reside on neovascular sprouting tips, which lack thrombospondin 1.[74] Further examples on how the microenvironment regulates DCC proliferation and dormancy emerge from in vitro studies that include extracellular matrix (ECM) components in three-dimensional culture systems.[75] The interaction of the ECM component fibronectin with integrin β1 on DCCs leads to focal adhesion kinase (FAK) activation through Src and subsequent downstream activation of ERK. While these signals switched cells from a dormant to proliferative phenotype and conferred therapy resistance, inhibition of integrin β1 maintained cells in a dormant state and also enhanced chemotherapeutic response.[76] Engineering of well-defined DCC niches provides further possibilities to study tumor cell dormancy, including the dynamics of reawakening.[77] This will be important as increasing evidence suggests that many factors influence the activity of DCCs. For example, primary tumors may contribute to create a tissue microenvironment for DCCs to settle, proliferate, and grow. Mechanisms include the secretion of growth factors or exosomes (small vesicles containing proteins, RNA, and DNA from the primary tumor) that can prime the microenvironment of distant sites for DCC survival and thereby create (pre)metastatic niches.[78] In addition, inflammation can cause awakening of dormant DCCs through a change of microenvironmental cues due to infiltration of immune cells.[79] In response to inflammation, bone marrow–derived myeloid cells are recruited to the target organs, causing the microenvironment to change and downregulate potent dormancy inducers such as thrombospondin 1.[80] In addition, inflammation induces the formation of neutrophil extracellular traps (NETs), which together with their released proteases (neutrophil elastase and matrix metalloproteinase 9) in turn enhance laminin-111 processing and activate FAK signaling in dormant DCCs. As a result, DCCs may awake, start proliferating, and form a metastasis.[81] Importantly, immune-mediated activation of early DCCs might also be caused by surgery-induced cytokine release.[82]

The importance of understanding the invisible phase of metastasis formation is closely related to the question of when the founder cells of metastasis begin to disseminate from the primary tumor. Early DCCs may require different conditions than late DCCs, and the microenvironmental signals that suppress or activate early DCCs may differ from those controlling late DCCs. Immune escape mechanisms may differ as well, as the neoantigen repertoire of early versus late DCCs is expected to be rather different. Moreover, all this might be tissue specific for the different metastatic sites, such as the brain, bone marrow, lungs, and others, as well as for the cancer types lodging there. Systematic comparisons using advanced in vitro systems or in vivo imaging guided by findings from the direct analysis of patient-derived DCCs will therefore become crucial in the coming years. It is quite possible that despite this diversity of factors, important underlying master mechanisms for early and late DCCs will be identified that can either drive or delay progression.

14.5   What Are the Treatment Options for a Highly Individualized Disease?

Cancer may represent one of the most extreme forms of individualized progression of any disease. Genetically and epigenetically unstable and plastic cells that arise within individual genetic backgrounds (determined by the germline of each patient), in different organs, home to different ectopic sites, and are exposed to changing systemic therapies, follow the rules of evolution (i.e., mutation and selection). As a consequence, it seems impossible to define the term "malignant" as a specific, concrete molecular or phenotypic realization or a unique cellular program. Rather, malignancy results from the potential to evolve and to escape tissue-specific control mechanisms. Therefore, therapeutic intervention needs to take into account and exploit different evolutionary (and phenotypic) stages of cancer cells and possibly different tissue microenvironmental conditions as well. Below we summarize the main aspects of disease progression that are currently not well addressed in cancer treatments.

14.5.1   Targeting Early versus Late Systemic Cancer

Routinely, novel drugs are first tested in patients with manifest metastasis and mostly end-stage disease. In a subsequent step, successful anticancer drugs are then administered to patients in adjuvant or neoadjuvant settings. However, conditions for early and late disease are fundamentally different. Early systemic cancer may present with single disseminated cancer cells, small micrometastases that comprise few to thousands of cells. The micro-environment of early DCCs is clearly different from the microenvironment that epithelial cancers form when they establish manifest metastasis.[83] Once formed, the metastatic microenvironment is microscopically indistinguishable from the primary tumor, although the cellular and molecular composition may differ. However, resistance mechanisms of late-stage metastasis to treatment have been extensively investigated, with major determinants being the ability of a drug to penetrate the tumor tissue and the cellular, genetic, and epigenetic resistance mechanisms.[84] Therefore, tissue penetration, metabolism, and elimination of the drug are prominent determinants in assessing effectiveness of treatment of manifest metastasis, whereas cell-intrinsic and acquired mechanisms, such as drug efflux by cancer stem cells, drug inactivation, mutation of the drug target, activation of alternative signaling pathways, DNA-damage repair, evasion from cell death, and immune evasion, affect early as well as late metastatic colonies. In early metastasis, the organ microenvironment defines the direct cellular and noncellular neighborhood of DCCs.[85] This means that for adjuvant therapies, tumor tissue penetration is less important, whereas the cellular, genetic, and epigenetic resistance mechanisms are certainly relevant, as are the organ characteristics of the metastatic niche.[86]

As pointed out above, also cell-intrinsic characteristics might differ substantially between early and late DCCs. This may be one reason why EGFR-targeting therapies have so far failed in the adjuvant therapy setting.[87] Moreover, oncogene addiction might differ between early colonies and late manifest metastases.

14.5.2   Targeting Proliferating versus Quiescent Disseminated Cancer Cells

Neoadjuvant and adjuvant chemotherapies in early breast or androgen deprivation therapy in localized prostate cancer significantly decreased bone marrow DCCs but did not fully eradicate them.[88] Small clinical trials targeting DCCs with docetaxel (an antiproliferative drug) as a second-line treatment also showed promising results with increased survival for breast cancer patients.[89] However, dormant DCCs might be inherently difficult to treat, as they are chemotherapy resistant and might not express target molecules of targeted therapies.[90]

14.5.3   Targeting Either Disseminated Cancer Cells or Their Microenvironment

Targeting the microenvironment of dormant DCCs may be a reasonable alternative.[91] Preclinical in vivo studies inhibiting integrin-mediated interactions between DCCs and the perivascular niche sensitized dormant and proliferative DCCs to chemotherapy and resulted in reduced DCC numbers and increased survival in mice.[92] Similarly, antiangiogenic drugs that target vascular endothelial cells and drugs targeting the immune system (such as checkpoint blockade inhibitors) do not attack the cancer cells but alter the microenvironment. Most of these studies have been performed in metastatic patients.[93] It will need additional efforts to exploit this concept for the interactions of early DCCs to prevent the establishment of metastasis in the first place, rather than starting treatment when they are already manifest. Therapy with bisphosphonates may serve as a first example that can effectively reduce DCC numbers and prevent bone metastasis in breast cancer patients.[94]

Finally, it will be interesting to see whether prolonging dormancy of DCCs through targeting the metastatic niche is possible in patients and whether this will result in a delay or even prevention of metastasis.[95]

14.5.4 Addressing Challenges in Diagnosing and Treating Individual Disease Stages

The numbers of detectable DCCs and CTCs have predictive value for patient outcome.[96] In addition, their genomic and genetic alterations apparently correlate with progression and possibly therapy response.[97] Thus, novel methods are needed to monitor the evolution of DCCs over time to assess the genomic landscape of these early disseminated tumor cells.[98] Unfortunately, currently applied imaging techniques for tumor detection, such as positron emission tomography, computed tomography, and magnetic resonance imaging, have limited resolution and are unable to characterize early DCCs at the molecular level. Novel strategies of DCC or CTC enrichment through, for example, liquid biopsy sampling or lymph node disaggregation combined with other novel technologies may become important in the near future the more molecular single-cell technologies such as single-cell sequencing become commonplace.[99] In addition, circulating cell-free (cf) DNA is detectable at very early stages of metastatic disease, carrying CNAs and signs of loss of heterozygosity representative of the primary tumor and its CTCs.[100] It remains to be seen whether detection of cfDNA is sensitive and specific enough to aid in characterizing early stage (or relapsing) cancer patients and can be used as a surrogate end point for metastatic spread in clinical trials. The challenge here is certainly to detect cfDNA originating from few and rare cancer cells within the plethora of cfDNA originating from nonmalignant cells (i.e., to find the needle in the haystack).

Finally, improved in vivo and in vitro models mimicking early DCCs are needed in order to investigate potential novel therapeutic strategies. Such models could become decisive for the development of rational systemic therapies to prevent manifest metastasis. They would cover all the shades of "malignancy" that are currently not represented by our "conventional" cell lines that have been selected for highest aggressiveness. Such models would then reflect—and possibly target—the dependencies early DCCs have in their metastatic niche and inform effective and possibly less toxic adjuvant therapies. Modeling early versus late DCCs in their metastatic niches, addressing the various systemic effects that cancer cells induce, is part of a conceptual rethinking of cancer as a truly systemic disease that begins long before it becomes detectable and may progress long after initial treatment. In our search for therapies, such a view also shifts our attention away from the tumor to the integrated system that the whole of the body is, in order to make use of its local and systemic defense mechanisms to prevent and cure cancer. Evidence is already accumulating that such an understanding of cancer will save more lives in the future.

Notes

1. Klein, C. A. Framework models of tumor dormancy from patient-derived observations. *Curr Opin Genet Dev* 21, 42–49, doi:10.1016/j.gde.2010.10.011 (2011).

2. Hanahan, D. & Weinberg, R. A. Hallmarks of cancer: the next generation. *Cell* 144, 646–674, https://doi.org /10.1016/j.cell.2011.02.013 (2011).

3. See table 1&S1 in Marino-Enriquez, A. & Fletcher, C. D. Shouldn't we care about the biology of benign tumours? *Nat Rev Cancer* 14, 701–702, doi:10.1038/nrc3845 (2014).

4. Hafner, C. et al. Oncogenic PIK3CA mutations occur in epidermal nevi and seborrheic keratoses with a characteristic mutation pattern. *Proc Natl Acad Sci USA* 104, 13450–13454, doi:10.1073/pnas.0705218104 (2007). Hafner, C. et al. Multiple oncogenic mutations and clonal relationship in spatially distinct benign human epidermal tumors. *Proc Natl Acad Sci USA* 107, 20780–20785, doi:10.1073/pnas.1008365107 (2010).

5. Naski, M. C., Wang, Q., Xu, J. & Ornitz, D. M. Graded activation of fibroblast growth factor receptor 3 by mutations causing achondroplasia and thanatophoric dysplasia. *Nat Genet* 13, 233–237, doi:10.1038/ng0696-233 (1996).

6. Martincorena, I. et al. Tumor evolution: high burden and pervasive positive selection of somatic mutations in normal human skin. *Science* 348, 880–886, doi:10.1126/science.aaa6806 (2015).

7. Firestein, G. S., Echeverri, F., Yeo, M., Zvaifler, N. J. & Green, D. R. Somatic mutations in the p53 tumor suppressor gene in rheumatoid arthritis synovium. *Proc Natl Acad Sci USA* 94, 10895–10900, doi:10.1073 /pnas.94.20.10895 (1997). Reme, T. et al. Mutations of the p53 tumour suppressor gene in erosive rheumatoid synovial tissue. *Clin Exp Immunol* 111, 353–358, doi:10.1046/j.1365-2249.1998.00508.x (1998).

8. Kumar, R., Angelini, S., Snellman, E. & Hemminki, K. BRAF mutations are common somatic events in melanocytic nevi. *J Invest Dermatol* 122, 342–348, doi:10.1046/j.0022-202X.2004.22225.x (2004). Dankort, D. et al. Braf(V600E) cooperates with Pten loss to induce metastatic melanoma. *Nat Genet* 41, 544–552, doi:10.1038 /ng.356 (2009).

9. Fearon, E. R. & Vogelstein, B. A genetic model for colorectal tumorigenesis. *Cell* 61, 759–767, doi:10.1016/0092 –8674(90)90186-i (1990). Vogelstein, B. et al. Cancer genome landscapes. *Science* 339, 1546–1558, doi:10.1126 /science.1235122 (2013).

10. Dash, S. et al. Differentiating between cancer and normal tissue samples using multi-hit combinations of genetic mutations. *Sci Rep* 9, 1005, doi:10.1038/s41598-018-37835-6 (2019).

11. Boecker, W. et al. Ductal epithelial proliferations of the breast: a biological continuum? Comparative genomic hybridization and high-molecular-weight cytokeratin expression patterns. *J Pathol* 195, 415–421, doi:10.1002/path.982 (2001). Chin, K. et al. In situ analyses of genome instability in breast cancer. *Nat Genet* 36, 984–988, doi:10.1038/ng1409 (2004).

12. Bakhoum, S. F. & Swanton, C. Chromosomal instability, aneuploidy, and cancer. *Front Oncol* 4, 161, doi:10.3389/fonc.2014.00161 (2014). Weaver, B. A. & Cleveland, D. W. Does aneuploidy cause cancer? *Curr Opin Cell Biol* 18, 658–667, doi:10.1016/j.ceb.2006.10.002 (2006).

13. Fujiwara, T. et al. Cytokinesis failure generating tetraploids promotes tumorigenesis in p53-null cells. *Nature* 437, 1043–1047, doi:10.1038/nature04217 (2005). Weaver, B. A., Silk, A. D., Montagna, C., Verdier-Pinard, P. & Cleveland, D. W. Aneuploidy acts both oncogenically and as a tumor suppressor. *Cancer Cell* 11, 25–36, doi:10.1016/j.ccr.2006.12.003 (2007).

14. Weaver, B. A. & Cleveland, D. W. Does aneuploidy cause cancer? *Curr Opin Cell Biol* 18, 658–667, doi:10.1016/j.ceb.2006.10.002 (2006). Cimini, D. Merotelic kinetochore orientation, aneuploidy, and cancer. *Biochim Biophys Acta* 1786, 32–40, doi:10.1016/j.bbcan.2008.05.003 (2008). Schvartzman, J. M., Sotillo, R. & Benezra, R. Mitotic chromosomal instability and cancer: mouse modelling of the human disease. *Nat Rev Cancer* 10, 102–115, doi:10.1038/nrc2781 (2010).

15. Liotta, L. A., Rao, C. N. & Barsky, S. H. Tumor invasion and the extracellular matrix. *Lab Invest* 49, 636–649 (1983). Baba, A. I. & Catoi, C. Tumor cell morphology. In *Comparative Oncology* chap. 3 (Publishing House of the Romanian Academy, 2007).

16. Baba, A. I. & Catoi, C. Tumor cell morphology. In *Comparative Oncology* chap. 3 (Publishing House of the Romanian Academy, 2007).

17. Thomson, T. M., Balcells, C. & Cascante, M. Metabolic plasticity and epithelial-mesenchymal transition. *J Clin Med* 8, 967, doi:10.3390/jcm8070967 (2019).

18. Cohen, A. L., Holmen, S. L. & Colman, H. IDH1 and IDH2 mutations in gliomas. *Curr Neurol Neurosci Rep* 13, 345, doi:10.1007/s11910-013-0345-4 (2013). Frezza, C. & Gottlieb, E. Mitochondria in cancer: not just innocent bystanders. *Semin Cancer Biol* 19, 4–11, doi:10.1016/j.semcancer.2008.11.008 (2009). Linehan, W. M., Srinivasan, R. & Schmidt, L. S. The genetic basis of kidney cancer: a metabolic disease. *Nat Rev Urol* 7, 277–285, doi:10.1038/nrurol.2010.47 (2010).

19. Thomson, T. M., Balcells, C. & Cascante, M. Metabolic plasticity and epithelial-mesenchymal transition. *J Clin Med* 8, 967, doi:10.3390/jcm8070967 (2019). Laukka, T. et al. Fumarate and succinate regulate expression of hypoxia-inducible genes via TET enzymes. *J Biol Chem* 291, 4256–4265, doi:10.1074/jbc.M115.688762 (2016). Lu, C. et al. IDH mutation impairs histone demethylation and results in a block to cell differentiation. *Nature* 483, 474–478, doi:10.1038/nature10860 (2012). Xiao, M. et al. Inhibition of alpha-KG-dependent histone and DNA demethylases by fumarate and succinate that are accumulated in mutations of FH and SDH tumor suppressors. *Genes Dev* 26, 1326–1338, doi:10.1101/gad.191056.112 (2012).

20. Schafer, Z. T. et al. Antioxidant and oncogene rescue of metabolic defects caused by loss of matrix attachment. *Nature* 461, 109–113, doi:10.1038/nature08268 (2009).

21. Andrzejewski, S. et al. PGC-1alpha promotes breast cancer metastasis and confers bioenergetic flexibility against metabolic drugs. *Cell Metab* 26, 778–787 e775, doi:10.1016/j.cmet.2017.09.006 (2017). Elia, I. et al. Proline metabolism supports metastasis formation and could be inhibited to selectively target metastasizing cancer cells. *Nature Commun* 8, 15267, doi:10.1038/ncomms15267 (2017).

22. Koong, A. C. et al. Pancreatic tumors show high levels of hypoxia. *Int J Radiat Oncol Biol Phys* 48, 919–922, doi:10.1016/s0360–3016(00)00803–8 (2000). Neesse, A. et al. Stromal biology and therapy in pancreatic cancer. *Gut* 60, 861–868, doi:10.1136/gut.2010.226092 (2011).

23. Massague, J. & Obenauf, A. C. Metastatic colonization by circulating tumour cells. *Nature* 529, 298–306, doi:10.1038/nature17038 (2016).

24. Braun, S. et al. A pooled analysis of bone marrow micrometastasis in breast cancer. *N Engl J Med* 353, 793–802, doi:10.1056/NEJMoa050434 (2005). Janni, W. et al. Persistence of disseminated tumor cells in the bone marrow of breast cancer patients predicts increased risk for relapse—a European pooled analysis. *Clin Cancer Res* 17, 2967–2976, doi:10.1158/1078–0432.CCR-10–2515 (2011).

25. Snyder, J., Duchamp, O., Paz, K. & Sathyan, P. Role of companies and corporations in the cevelopment and utilization of PDX models. In *Patient Derived Tumor Xenograft Models* (eds Rajesh Uthamanthil & Peggy Tinkey) 409–426 (Academic Press, 2017). Domcke, S., Sinha, R., Levine, D. A., Sander, C. & Schultz, N. Evaluating cell lines as tumour models by comparison of genomic profiles. *Nat Commun* 4, 2126, doi:10.1038/ncomms3126 (2013). Neve, R. M. et al. A collection of breast cancer cell lines for the study of functionally distinct cancer subtypes. *Cancer Cell* 10, 515–527, doi:10.1016/j.ccr.2006.10.008 (2006). Ross, D. T. & Perou, C. M. A comparison of gene expression signatures from breast tumors and breast tissue derived cell lines. *Dis Markers* 17, 99–109, doi:10.1155/2001/850531 (2001). Quintana, E. et al. Efficient tumour formation by single human melanoma cells. *Nature* 456, 593–598, doi:10.1038/nature07567 (2008).

26. Jung, J. Human tumor xenograft models for preclinical assessment of anticancer drug development. *Toxicol Res* 30, 1–5, doi:10.5487/TR.2014.30.1.001 (2014).

27. Werner-Klein, M. et al. Genetic alterations driving metastatic colony formation are acquired outside of the primary tumour in melanoma. *Nat Commun* 9, 595, doi:10.1038/s41467-017-02674-y (2018).

28. Werner-Klein, M. et al. Genetic alterations driving metastatic colony formation are acquired outside of the primary tumour in melanoma. *Nat Commun* 9, 595, doi:10.1038/s41467-017-02674-y (2018).

29. Al-Hajj, M., Wicha, M. S., Benito-Hernandez, A., Morrison, S. J. & Clarke, M. F. Prospective identification of tumorigenic breast cancer cells. *Proc Natl Acad Sci USA* 100, 3983–3988, doi:10.1073/pnas.0530291100 (2003).

30. Werner-Klein, M. et al. Interleukin-6 trans-signaling is a candidate mechanism to drive progression of human DCCs during clinical latency. *Nat Commun* 11, 4977, doi:10.1038/s41467-020-18701-4 (2020).

31. Fearon, E. R. & Vogelstein, B. A genetic model for colorectal tumorigenesis. *Cell* 61, 759–767, doi:10.1016/0092–8674(90)90186-i (1990).

32. Klein, C. A. et al. Comparative genomic hybridization, loss of heterozygosity, and DNA sequence analysis of single cells. *Proc Natl Acad Sci USA* 96, 4494–4499, doi:10.1073/pnas.96.8.4494 (1999). Schardt, J. A. et al. Genomic analysis of single cytokeratin-positive cells from bone marrow reveals early mutational events in breast cancer. *Cancer Cell* 8, 227–239, doi:10.1016/j.ccr.2005.08.003 (2005). Schmidt-Kittler, O. et al. From latent disseminated cells to overt metastasis: genetic analysis of systemic breast cancer progression. *Proc Natl Acad Sci USA* 100, 7737–7742, doi:10.1073/pnas.1331931100 (2003).

33. Klein, C. A. Selection and adaptation during metastatic cancer progression. *Nature* 501, 365–372, doi:10.1038/nature12628 (2013). Stoecklein, N. H. & Klein, C. A. Genetic disparity between primary tumours, disseminated tumour cells, and manifest metastasis. *Int J Cancer* 126, 589–598, doi:10.1002/ijc.24916 (2010). Eyles, J. et al. Tumor cells disseminate early, but immunosurveillance limits metastatic outgrowth, in a mouse model of melanoma. *J Clin Invest* 120, 2030–2039, doi:10.1172/JCI42002 (2010). Husemann, Y. et al. Systemic spread is an early step in breast cancer. *Cancer Cell* 13, 58–68, doi:10.1016/j.ccr.2007.12.003 (2008). Rhim, A. D. et al. EMT and dissemination precede pancreatic tumor formation. *Cell* 148, 349–361, doi:10.1016/j.cell.2011.11.025 (2012).

34. Harper, K. L. et al. Mechanism of early dissemination and metastasis in Her2(+) mammary cancer. *Nature* 540, 588–592, doi:10.1038/nature20609 (2016). Hosseini, H. et al. Early dissemination seeds metastasis in breast cancer. *Nature* 540, 552–558, doi:10.1038/nature20785 (2016).

35. Hartkopf, A. D. et al. Disseminated tumor cells from the bone marrow of patients with nonmetastatic primary breast cancer are predictive of locoregional relapse. *Ann Oncol* 26, 1155–1160, doi:10.1093/annonc/mdv148 (2015). Sanger, N. et al. Disseminated tumor cells in the bone marrow of patients with ductal carcinoma in situ. *Int J Cancer* 129, 2522–2526, doi:10.1002/ijc.25895 (2011).

36. Husemann, Y. et al. Systemic spread is an early step in breast cancer. *Cancer Cell* 13, 58–68, doi:10.1016/j.ccr.2007.12.003 (2008).

37. Werner-Klein, M. et al. Genetic alterations driving metastatic colony formation are acquired outside of the primary tumour in melanoma. *Nat Commun* 9, 595, doi:10.1038/s41467-017-02674-y (2018). Hosseini, H. et al. Early dissemination seeds metastasis in breast cancer. *Nature* 540, 552–558, doi:10.1038/nature20785 (2016).

38. Werner-Klein, M. et al. Genetic alterations driving metastatic colony formation are acquired outside of the primary tumour in melanoma. *Nat Commun* 9, 595, doi:10.1038/s41467-017-02674-y (2018).

39. Reviewed in Klein, C. A. Parallel progression of primary tumours and metastases. *Nat Rev Cancer* 9, 302–312, doi:10.1038/nrc2627 (2009).

40. Werner-Klein, M. et al. Genetic alterations driving metastatic colony formation are acquired outside of the primary tumour in melanoma. *Nat Commun* 9, 595, doi:10.1038/s41467-017-02674-y (2018). Hosseini, H. et al. Early dissemination seeds metastasis in breast cancer. *Nature* 540, 552–558, doi:10.1038/nature20785 (2016). Stoecklein, N. H. et al. Direct genetic analysis of single disseminated cancer cells for prediction of outcome and therapy selection in esophageal cancer. *Cancer Cell* 13, 441–453, doi:10.1016/j.ccr.2008.04.005 (2008). Weckermann, D. et al. Perioperative activation of disseminated tumor cells in bone marrow of patients with prostate cancer. *J Clin Oncol* 27, 1549–1556, doi:10.1200/JCO.2008.17.0563 (2009).

41. Werner-Klein, M. et al. Genetic alterations driving metastatic colony formation are acquired outside of the primary tumour in melanoma. *Nat Commun* 9, 595, doi:10.1038/s41467-017-02674-y (2018).

42. Reviewed in Klein, C. A. Selection and adaptation during metastatic cancer progression. *Nature* 501, 365–372, doi:10.1038/nature12628 (2013).

43. Schumacher, S. et al. Disseminated tumour cells with highly aberrant genomes are linked to poor prognosis in operable oesophageal adenocarcinoma. *Br J Cancer* 117, 725–733, doi:10.1038/bjc.2017.233 (2017).

44. Hosseini, H. et al. Early dissemination seeds metastasis in breast cancer. *Nature* 540, 552–558, doi:10.1038/nature20785 (2016).

45. Sanborn, J. Z. et al. Phylogenetic analyses of melanoma reveal complex patterns of metastatic dissemination. *Proc Natl Acad Sci USA* 112, 10995–11000, doi:10.1073/pnas.1508074112 (2015). Turajlic, S. et al. Whole genome sequencing of matched primary and metastatic acral melanomas. *Genome Res* 22, 196–207, doi:10.1101/gr.125591.111 (2012). Schrijver, W. et al. Mutation profiling of key cancer genes in primary breast cancers and their distant metastases. *Cancer Res* 78, 3112–3121, doi:10.1158/0008-5472.CAN-17-2310 (2018). Ullah, I. et al. Evolutionary history of metastatic breast cancer reveals minimal seeding from axillary lymph nodes. *J Clin Invest* 128, 1355–1370, doi:10.1172/JCI96149 (2018). Leung, M. L. et al. Single-cell DNA sequencing reveals a late-dissemination model in metastatic colorectal cancer. *Genome Res* 27, 1287–1299, doi:10.1101/gr.209973.116 (2017). Kim, T. M. et al. Subclonal genomic architectures of primary and metastatic colorectal cancer based on intratumoral genetic heterogeneity. *Clin Cancer Res* 21, 4461–4472, doi:10.1158/1078-0432.CCR-14-2413 (2015). Turajlic, S. et al. Tracking cancer evolution reveals constrained routes to metastases: TRACERx renal. *Cell* 173, 581–594 e512, doi:10.1016/j.cell.2018.03.057 (2018).

46. Leung, M. L. et al. Single-cell DNA sequencing reveals a late-dissemination model in metastatic colorectal cancer. *Genome Res* 27, 1287–1299, doi:10.1101/gr.209973.116 (2017). Kim, T. M. et al. Subclonal genomic architectures of primary and metastatic colorectal cancer based on intratumoral genetic heterogeneity. *Clin Cancer Res* 21, 4461–4472, doi:10.1158/1078-0432.CCR-14-2413 (2015). Hu, Z. et al. Quantitative evidence

for early metastatic seeding in colorectal cancer. *Nat Genet* 51, 1113–1122, doi:10.1038/s41588-019-0423-x (2019). Ishaque, N. et al. Whole genome sequencing puts forward hypotheses on metastasis evolution and therapy in colorectal cancer. *Nat Commun* 9, 4782, doi:10.1038/s41467-018-07041-z (2018).

47. Schrijver, W. et al. Mutation profiling of key cancer genes in primary breast cancers and their distant metastases. *Cancer Res* 78, 3112–3121, doi:10.1158/0008–5472.CAN-17–2310 (2018). Ullah, I. et al. Evolutionary history of metastatic breast cancer reveals minimal seeding from axillary lymph nodes. *J Clin Invest* 128, 1355–1370, doi:10.1172/JCI96149 (2018). Krøigård, A. B. et al. Genomic analyses of breast cancer progression reveal distinct routes of metastasis emergence. *Sci Rep* 7, 43813, doi:10.1038/srep43813 (2017). Yates, L. R. et al. Genomic evolution of breast cancer metastasis and relapse. *Cancer Cell* 32, 169–184 e167, doi:10.1016/j.ccell .2017.07.005 (2017).

48. Kim, T. M. et al. Subclonal genomic architectures of primary and metastatic colorectal cancer based on intratumoral genetic heterogeneity. *Clin Cancer Res* 21, 4461–4472, doi:10.1158/1078–0432.CCR-14–2413 (2015). Ishaque, N. et al. Whole genome sequencing puts forward hypotheses on metastasis evolution and therapy in colorectal cancer. *Nat Commun* 9, 4782, doi:10.1038/s41467-018-07041-z (2018).

49. Ding, L. et al. Genome remodelling in a basal-like breast cancer metastasis and xenograft. *Nature* 464, 999–1005, doi:10.1038/nature08989 (2010).

50. Navin, N. et al. Tumour evolution inferred by single-cell sequencing. *Nature* 472, 90–94, doi:10.1038/ nature09807 (2011).

51. Yates, L. R. et al. Genomic evolution of breast cancer metastasis and relapse. *Cancer Cell* 32, 169–184 e167, doi:10.1016/j.ccell.2017.07.005 (2017).

52. Ullah, I. et al. Evolutionary history of metastatic breast cancer reveals minimal seeding from axillary lymph nodes. *J Clin Invest* 128, 1355–1370, doi:10.1172/JCI96149 (2018). Krøigård, A. B. et al. Genomic analyses of breast cancer progression reveal distinct routes of metastasis emergence. *Sci Rep* 7, 43813, doi:10.1038/srep43813 (2017).

53. Gundem, G. et al. The evolutionary history of lethal metastatic prostate cancer. *Nature* 520, 353–357, doi:10.1038/nature14347 (2015). Hong, M. K. et al. Tracking the origins and drivers of subclonal metastatic expansion in prostate cancer. *Nat Commun* 6, 6605, doi:10.1038/ncomms7605 (2015).

54. Gundem, G. et al. The evolutionary history of lethal metastatic prostate cancer. *Nature* 520, 353–357, doi:10.1038/nature14347 (2015).

55. Schrijver, W. et al. Mutation profiling of key cancer genes in primary breast cancers and their distant metastases. *Cancer Res* 78, 3112–3121, doi:10.1158/0008–5472.CAN-17–2310 (2018). Ullah, I. et al. Evolutionary history of metastatic breast cancer reveals minimal seeding from axillary lymph nodes. *J Clin Invest* 128, 1355–1370, doi:10.1172/JCI96149 (2018).

56. Turajlic, S. et al. Tracking cancer evolution reveals constrained routes to metastases: TRACERx renal. *Cell* 173, 581–594 e512, doi:10.1016/j.cell.2018.03.057 (2018).

57. Kuipers, J., Jahn, K. & Beerenwinkel, N. Advances in understanding tumour evolution through single-cell sequencing. *Biochim Biophys Acta Rev Cancer* 1867, 127–138, doi:10.1016/j.bbcan.2017.02.001 (2017). Kuipers, J., Jahn, K., Raphael, B. J. & Beerenwinkel, N. Single-cell sequencing data reveal widespread recurrence and loss of mutational hits in the life histories of tumors. *Genome Res* 27, 1885–1894, doi:10.1101/gr.220707.117 (2017).

58. Leung, M. L. et al. Single-cell DNA sequencing reveals a late-dissemination model in metastatic colorectal cancer. *Genome Res* 27, 1287–1299, doi:10.1101/gr.209973.116 (2017).

59. Janni, W. et al. Persistence of disseminated tumor cells in the bone marrow of breast cancer patients predicts increased risk for relapse—a European pooled analysis. *Clin Cancer Res* 17, 2967–2976, doi:10.1158/1078–0432. CCR-10–2515 (2011). Werner-Klein, M. et al. Genetic alterations driving metastatic colony formation are acquired outside of the primary tumour in melanoma. *Nat Commun* 9, 595, doi:10.1038/s41467-017-02674-y (2018). Schumacher, S. et al. Disseminated tumour cells with highly aberrant genomes are linked to poor prognosis in operable oesophageal adenocarcinoma. *Br J Cancer* 117, 725–733, doi:10.1038/bjc.2017.233 (2017). Chemi, F. et al. Pulmonary venous circulating tumor cell dissemination before tumor resection and disease relapse. *Nat Med* 25, 1534–1539, doi:10.1038/s41591-019-0593-1 (2019). Ignatiadis, M. et al. Molecular detection and prognostic value of circulating cytokeratin-19 messenger RNA-positive and HER2 messenger RNA-positive cells in the peripheral blood of women with early-stage breast cancer. *Clin Breast Cancer* 7, 883–889, doi:10.3816/CBC.2007.n.054 (2007). Lucci, A. et al. Circulating tumour cells in non-metastatic breast cancer: a prospective study. *Lancet Oncol* 13, 688–695, doi:10.1016/S1470–2045(12)70209–7 (2012). Paterlini-Brechot, P. & Benali, N. L. Circulating tumor cells (CTC) detection: clinical impact and future directions. *Cancer Lett*

253, 180–204, doi:10.1016/j.canlet.2006.12.014 (2007). Xenidis, N. et al. Cytokeratin-19 mRNA-positive circulating tumor cells after adjuvant chemotherapy in patients with early breast cancer. *J Clin Oncol* 27, 2177–2184, doi:10.1200/JCO.2008.18.0497 (2009).

60. Reviewed in Dasgupta, A., Lim, A. R. & Ghajar, C. M. Circulating and disseminated tumor cells: harbingers or initiators of metastasis? *Mol Oncol* 11, 40–61, doi:10.1002/1878–0261.12022 (2017).

61. Pommier, A. et al. Unresolved endoplasmic reticulum stress engenders immune-resistant, latent pancreatic cancer metastases. *Science* 360, eaao4908, doi:10.1126/science.aao4908 (2018).

62. Aguirre-Ghiso, J. A., Ossowski, L. & Rosenbaum, S. K. Green fluorescent protein tagging of extracellular signal-regulated kinase and p38 pathways reveals novel dynamics of pathway activation during primary and metastatic growth. *Cancer Res* 64, 7336–7345, doi:10.1158/0008–5472.CAN-04–0113 (2004). Aguirre-Ghiso, J. A., Estrada, Y., Liu, D. & Ossowski, L. ERK(MAPK) activity as a determinant of tumor growth and dormancy; regulation by p38(SAPK). *Cancer Res* 63, 1684–1695 (2003).

63. Bragado, P. et al. TGF-beta2 dictates disseminated tumour cell fate in target organs through TGF-beta-RIII and p38alpha/beta signalling. *Nat Cell Biol* 15, 1351–1361, doi:10.1038/ncb2861 (2013). Schewe, D. M. & Aguirre-Ghiso, J. A. ATF6alpha-Rheb-mTOR signaling promotes survival of dormant tumor cells in vivo. *Proc Natl Acad Sci USA* 105, 10519–10524, doi:10.1073/pnas.0800939105 (2008). Sosa, M. S. et al. NR2F1 controls tumour cell dormancy via SOX9- and RARbeta-driven quiescence programmes. *Nat Commun* 6, 6170, doi:10.1038/ncomms7170 (2015).

64. Aguirre Ghiso, J. A., Kovalski, K. & Ossowski, L. Tumor dormancy induced by downregulation of urokinase receptor in human carcinoma involves integrin and MAPK signaling. *J Cell Biol* 147, 89–104, doi:10.1083/jcb.147.1.89 (1999). Aguirre-Ghiso, J. A., Liu, D., Mignatti, A., Kovalski, K. & Ossowski, L. Urokinase receptor and fibronectin regulate the ERK(MAPK) to p38(MAPK) activity ratios that determine carcinoma cell proliferation or dormancy in vivo. *Mol Biol Cell* 12, 863–879, doi:10.1091/mbc.12.4.863 (2001).

65. Bragado, P. et al. TGF-beta2 dictates disseminated tumour cell fate in target organs through TGF-beta-RIII and p38alpha/beta signalling. *Nat Cell Biol* 15, 1351–1361, doi:10.1038/ncb2861 (2013). Sosa, M. S. et al. NR2F1 controls tumour cell dormancy via SOX9- and RARbeta-driven quiescence programmes. *Nat Commun* 6, 6170, doi:10.1038/ncomms7170 (2015). Kobayashi, A. et al. Bone morphogenetic protein 7 in dormancy and metastasis of prostate cancer stem-like cells in bone. *J Exp Med* 208, 2641–2655, doi:10.1084/jem.20110840 (2011).

66. Reviewed in Klein, C. A. Framework models of tumor dormancy from patient-derived observations. *Curr Opin Genet Dev* 21, 42–49, doi:10.1016/j.gde.2010.10.011 (2011).

67. MacKie, R. M., Reid, R. & Junor, B. Fatal melanoma transferred in a donated kidney 16 years after melanoma surgery. *N Engl J Med* 348, 567–568, doi:10.1056/NEJM200302063480620 (2003). Martin, D. C., Rubini, M. & Rosen, V. J. Cadaveric renal homotransplantation with inadvertent transplantation of carcinoma. *JAMA* 192, 752–754, doi:10.1001/jama.1965.03080220016003 (1965). Penn, I. Transmission of cancer from organ donors. *Ann Transplant* 2, 7–12 (1997).

68. Weckermann, D. et al. Perioperative activation of disseminated tumor cells in bone marrow of patients with prostate cancer. *J Clin Oncol* 27, 1549–1556, doi:10.1200/JCO.2008.17.0563 (2009).

69. Weckermann, D. et al. Perioperative activation of disseminated tumor cells in bone marrow of patients with prostate cancer. *J Clin Oncol* 27, 1549–1556, doi:10.1200/JCO.2008.17.0563 (2009). Klein, C. A. et al. Genetic heterogeneity of single disseminated tumour cells in minimal residual cancer. *Lancet* 360, 683–689, doi:10.1016/S0140–6736(02)09838–0 (2002).

70. Sosa, M. S., Bragado, P. & Aguirre-Ghiso, J. A. Mechanisms of disseminated cancer cell dormancy: an awakening field. *Nat Rev Cancer* 14, 611–622, doi:10.1038/nrc3793 (2014).

71. Sosa, M. S., Bragado, P. & Aguirre-Ghiso, J. A. Mechanisms of disseminated cancer cell dormancy: an awakening field. *Nat Rev Cancer* 14, 611–622, doi:10.1038/nrc3793 (2014). Giancotti, F. G. Mechanisms governing metastatic dormancy and reactivation. *Cell* 155, 750–764, doi:10.1016/j.cell.2013.10.029 (2013). Linde, N., Fluegen, G. & Aguirre-Ghiso, J. A. The relationship between dormant cancer cells and their microenvironment. *Adv Cancer Res* 132, 45–71, doi:10.1016/bs.acr.2016.07.002 (2016).

72. Fluegen, G. et al. Phenotypic heterogeneity of disseminated tumour cells is preset by primary tumour hypoxic microenvironments. *Nat Cell Biol* 19, 120–132, doi:10.1038/ncb3465 (2017).

73. Ghajar, C. M. et al. The perivascular niche regulates breast tumour dormancy. *Nat Cell Biol* 15, 807–817, doi:10.1038/ncb2767 (2013). Kienast, Y. et al. Real-time imaging reveals the single steps of brain metastasis formation. *Nat Med* 16, 116–122, doi:10.1038/nm.2072 (2010). Price, T. T. et al. Dormant breast cancer micro-

metastases reside in specific bone marrow niches that regulate their transit to and from bone. *Sci Transl Med* 8, 340ra373, doi:10.1126/scitranslmed.aad4059 (2016).

74. Ghajar, C. M. et al. The perivascular niche regulates breast tumour dormancy. *Nat Cell Biol* 15, 807–817, doi:10.1038/ncb2767 (2013).

75. Barkan, D. et al. Inhibition of metastatic outgrowth from single dormant tumor cells by targeting the cytoskeleton. *Cancer Res* 68, 6241–6250, doi:10.1158/0008–5472.CAN-07–6849 (2008).

76. Barkan, D. et al. Inhibition of metastatic outgrowth from single dormant tumor cells by targeting the cytoskeleton. *Cancer Res* 68, 6241–6250, doi:10.1158/0008–5472.CAN-07–6849 (2008). Carlson, P. et al. Targeting the perivascular niche sensitizes disseminated tumour cells to chemotherapy. *Nat Cell Biol* 21, 238–250, doi:10.1038/s41556-018-0267-0 (2019).

77. Ghajar, C. M. et al. The perivascular niche regulates breast tumour dormancy. *Nat Cell Biol* 15, 807–817, doi:10.1038/ncb2767 (2013). Marlow, R. et al. A novel model of dormancy for bone metastatic breast cancer cells. *Cancer Res* 73, 6886–6899, doi:10.1158/0008–5472.CAN-13–0991 (2013).

78. Hoshino, A. et al. Tumour exosome integrins determine organotropic metastasis. *Nature* 527, 329–335, doi:10.1038/nature15756 (2015). Kaplan, R. N. et al. VEGFR1-positive haematopoietic bone marrow progenitors initiate the pre-metastatic niche. *Nature* 438, 820–827, doi:10.1038/nature04186 (2005). Peinado, H. et al. Melanoma exosomes educate bone marrow progenitor cells toward a pro-metastatic phenotype through MET. *Nat Med* 18, 883–891, doi:10.1038/nm.2753 (2012). Psaila, B. & Lyden, D. The metastatic niche: adapting the foreign soil. *Nat Rev Cancer* 9, 285–293, doi:10.1038/nrc2621 (2009).

79. Dasgupta, A., Lim, A. R. & Ghajar, C. M. Circulating and disseminated tumor cells: harbingers or initiators of metastasis? *Mol Oncol* 11, 40–61, doi:10.1002/1878–0261.12022 (2017).

80. El Rayes, T. et al. Lung inflammation promotes metastasis through neutrophil protease-mediated degradation of Tsp-1. *Proc Natl Acad Sci USA* 112, 16000–16005, doi:10.1073/pnas.1507294112 (2015).

81. Albrengues, J. et al. Neutrophil extracellular traps produced during inflammation awaken dormant cancer cells in mice. *Science* 361, eaao4227, doi:10.1126/science.aao4227 (2018).

82. Krall, J. A. et al. The systemic response to surgery triggers the outgrowth of distant immune-controlled tumors in mouse models of dormancy. *Sci Transl Med* 10, eaan3464, doi:10.1126/scitranslmed.aan3464 (2018).

83. Joyce, J. A. & Pollard, J. W. Microenvironmental regulation of metastasis. *Nat Rev Cancer* 9, 239–252, doi:10.1038/nrc2618 (2009).

84. Holohan, C., Van Schaeybroeck, S., Longley, D. B. & Johnston, P. G. Cancer drug resistance: an evolving paradigm. *Nat Rev Cancer* 13, 714–726, doi:10.1038/nrc3599 (2013). Minchinton, A. I. & Tannock, I. F. Drug penetration in solid tumours. *Nat Rev Cancer* 6, 583–592, doi:10.1038/nrc1893 (2006).

85. Croucher, P. I., McDonald, M. M. & Martin, T. J. Bone metastasis: the importance of the neighbourhood. *Nat Rev Cancer* 16, 373–386, doi:10.1038/nrc.2016.44 (2016).

86. Werner-Klein, M. & Klein, C. A. Therapy resistance beyond cellular dormancy. *Nat Cell Biol* 21, 117–119, doi:10.1038/s41556-019-0276-7 (2019).

87. Alberts, S. R. et al. Effect of oxaliplatin, fluorouracil, and leucovorin with or without cetuximab on survival among patients with resected stage III colon cancer: a randomized trial. *JAMA* 307, 1383–1393, doi:10.1001/jama.2012.385 (2012). Goss, G. D. et al. Randomized, double-blind trial of carboplatin and paclitaxel with either daily oral cediranib or placebo in advanced non–small-cell lung cancer: NCIC Clinical Trials Group BR24 Study. *J Clin Oncol* 28, 49–55, doi:10.1200/jco.2009.22.9427 (2010).

88. Becker, S., Becker-Pergola, G., Wallwiener, D., Solomayer, E. F. & Fehm, T. Detection of cytokeratin-positive cells in the bone marrow of breast cancer patients undergoing adjuvant therapy. *Breast Cancer Res Treat* 97, 91–96, doi:10.1007/s10549-005-9095-6 (2006). Becker, S., Solomayer, E., Becker-Pergola, G., Wallwiener, D. & Fehm, T. Primary systemic therapy does not eradicate disseminated tumor cells in breast cancer patients. *Breast Cancer Res Treat* 106, 239–243, doi:10.1007/s10549-006-9484-5 (2007). Kollermann, M. W. et al. Supersensitive PSA-monitored neoadjuvant hormone treatment of clinically localized prostate cancer: effects on positive margins, tumor detection and epithelial cells in bone marrow. *Eur Urol* 34, 318–324, doi:10.1159/000019748 (1998).

89. Naume, B. et al. Clinical outcome with correlation to disseminated tumor cell (DTC) status after DTC-guided secondary adjuvant treatment with docetaxel in early breast cancer. *J Clin Oncol* 32, 3848–3857, doi:10.1200/JCO.2014.56.9327 (2014).

90. Braun, S. et al. Lack of effect of adjuvant chemotherapy on the elimination of single dormant tumor cells in bone marrow of high-risk breast cancer patients. *J Clin Oncol* 18, 80–86, doi:10.1200/JCO.2000.18.1.80 (2000). Polzer, B. & Klein, C. A. Metastasis awakening: the challenges of targeting minimal residual cancer. *Nat Med* 19, 274–275, doi:10.1038/nm.3121 (2013).

91. Carlson, P. et al. Targeting the perivascular niche sensitizes disseminated tumour cells to chemotherapy. *Nat Cell Biol* 21, 238–250, doi:10.1038/s41556-018-0267-0 (2019). Werner-Klein, M. & Klein, C. A. Therapy resistance beyond cellular dormancy. *Nat Cell Biol* 21, 117–119, doi:10.1038/s41556-019-0276-7 (2019).

92. Carlson, P. et al. Targeting the perivascular niche sensitizes disseminated tumour cells to chemotherapy. *Nat Cell Biol* 21, 238–250, doi:10.1038/s41556-018-0267-0 (2019).

93. Sounni, N. E. & Noel, A. Targeting the tumor microenvironment for cancer therapy. *Clin Chem* 59, 85–93, doi:10.1373/clinchem.2012.185363 (2013). Ebos, J. M. & Kerbel, R. S. Antiangiogenic therapy: impact on invasion, disease progression, and metastasis. *Nat Rev Clin Oncol* 8, 210–221, doi:10.1038/nrclinonc.2011.21 (2011). Farhood, B., Najafi, M. & Mortezaee, K. CD8(+) cytotoxic T lymphocytes in cancer immunotherapy: a review. *J Cell Physiol* 234, 8509–8521, doi:10.1002/jcp.27782 (2019).

94. Banys, M. et al. Influence of zoledronic acid on disseminated tumor cells in bone marrow and survival: results of a prospective clinical trial. *BMC Cancer* 13, 480, doi:10.1186/1471-2407-13-480 (2013). Solomayer, E. F. et al. Influence of zoledronic acid on disseminated tumor cells in primary breast cancer patients. *Ann Oncol* 23, 2271–2277, doi:10.1093/annonc/mdr612 (2012).

95. Reviewed in Ghajar, C. M. Metastasis prevention by targeting the dormant niche. *Nat Rev Cancer* 15, 238–247, doi:10.1038/nrc3910 (2015).

96. Naume, B. et al. Clinical outcome with correlation to disseminated tumor cell (DTC) status after DTC-guided secondary adjuvant treatment with docetaxel in early breast cancer. *J Clin Oncol* 32, 3848–3857, doi:10.1200/JCO.2014.56.9327 (2014). Gruber, I. et al. Disseminated tumor cells as a monitoring tool for adjuvant therapy in patients with primary breast cancer. *Breast Cancer Res Treat* 144, 353–360, doi:10.1007/s10549-014-2853-6 (2014). Hartkopf, A. D. et al. The presence and prognostic impact of apoptotic and nonapoptotic disseminated tumor cells in the bone marrow of primary breast cancer patients after neoadjuvant chemotherapy. *Breast Cancer Res* 15, R94, doi:10.1186/bcr3496 (2013). Ilie, M. et al. "Sentinel" circulating tumor cells allow early diagnosis of lung cancer in patients with chronic obstructive pulmonary disease. *PLoS One* 9, e111597, doi:10.1371/journal.pone.0111597 (2014). Reid, A. L. et al. Markers of circulating tumour cells in the peripheral blood of patients with melanoma correlate with disease recurrence and progression. *Br J Dermatol* 168, 85–92, doi:10.1111/bjd.12057 (2013). Rhim, A. D. et al. Detection of circulating pancreas epithelial cells in patients with pancreatic cystic lesions. *Gastroenterology* 146, 647–651, doi:10.1053/j.gastro.2013.12.007 (2014).

97. Stoecklein, N. H. et al. Direct genetic analysis of single disseminated cancer cells for prediction of outcome and therapy selection in esophageal cancer. *Cancer Cell* 13, 441–453, doi:10.1016/j.ccr.2008.04.005 (2008). Schumacher, S. et al. Disseminated tumour cells with highly aberrant genomes are linked to poor prognosis in operable oesophageal adenocarcinoma. *Br J Cancer* 117, 725–733, doi:10.1038/bjc.2017.233 (2017).

98. Werner-Klein, M. et al. Genetic alterations driving metastatic colony formation are acquired outside of the primary tumour in melanoma. *Nat Commun* 9, 595, doi:10.1038/s41467-017-02674-y (2018). Hosseini, H. et al. Early dissemination seeds metastasis in breast cancer. *Nature* 540, 552–558, doi:10.1038/nature20785 (2016). Stoecklein, N. H. et al. Direct genetic analysis of single disseminated cancer cells for prediction of outcome and therapy selection in esophageal cancer. *Cancer Cell* 13, 441–453, doi:10.1016/j.ccr.2008.04.005 (2008). Schumacher, S. et al. Disseminated tumour cells with highly aberrant genomes are linked to poor prognosis in operable oesophageal adenocarcinoma. *Br J Cancer* 117, 725–733, doi:10.1038/bjc.2017.233 (2017). Holcomb, I. N. et al. Genomic alterations indicate tumor origin and varied metastatic potential of disseminated cells from prostate cancer patients. *Cancer Res* 68, 5599–5608, doi:10.1158/0008-5472.CAN-08-0812 (2008). Guzvic, M. et al. Combined genome and transcriptome analysis of single disseminated cancer cells from bone marrow of prostate cancer patients reveals unexpected transcriptomes. *Cancer Res* 74, 7383–7394, doi:10.1158/0008-5472.CAN-14-0934 (2014).

99. Chudziak, J. et al. Clinical evaluation of a novel microfluidic device for epitope-independent enrichment of circulating tumour cells in patients with small cell lung cancer. *Analyst* 141, 669–678, doi:10.1039/c5an02156a (2016). Gorges, T. M. et al. Accession of tumor heterogeneity by multiplex transcriptome profiling of single circulating tumor cells. *Clin Chem* 62, 1504–1515, doi:10.1373/clinchem.2016.260299 (2016). Hvichia, G. E. et al. A novel microfluidic platform for size and deformability based separation and the subsequent molecular characterization of viable circulating tumor cells. *Int J Cancer* 138, 2894–2904, doi:10.1002/ijc.30007 (2016).

Xu, L. et al. Optimization and evaluation of a novel size based circulating tumor cell isolation system. *PLoS One* 10, e0138032, doi:10.1371/journal.pone.0138032 (2015). Weidele, K. et al. Microfluidic enrichment, isolation and characterization of disseminated melanoma cells from lymph node samples. *Int J Cancer* 145, 232–241, doi:10.1002/ijc.32092 (2019).

100. Chemi, F. et al. Pulmonary venous circulating tumor cell dissemination before tumor resection and disease relapse. *Nature Medicine* 25, 1534–1539, doi:10.1038/s41591-019-0593-1 (2019). Shaw, J. A. et al. Genomic analysis of circulating cell-free DNA infers breast cancer dormancy. *Genome Res* 22, 220–231, doi:10.1101/gr.123497.111 (2012).

15 Niche Reconstruction to Revert or Transcend the Cancer State

Emmy W. Verschuren

To suppose that the evolution of the wonderfully adapted biological mechanisms has depended only on a selection out of a haphazard set of variations, each produced by blind chance, is like suggesting that if we went on throwing bricks together into heaps, we should eventually be able to choose ourselves the most desirable house.
—Conrad H. Waddington in *The Listener* (1952)

Overview

The cancer state can be described as a disease not only of the material aspects of cells and tissues but also of the evolutionary mechanisms that govern their development. While this confounds direct clinical approaches to treatment, study of the evolutionary process of niche construction in the cancer context has recently gained momentum and may provide new ways to both conceptualize fundamental mechanisms of carcinogenesis and clinically manage cancer progression. The cancer state can be regarded as an open dynamic system within a spatially and temporally variable but defined, contextual niche, where cancers behave like complex ecosystems. While precursor lesions are fairly common in healthy tissues, the detectable malignant cancer state is rare, due to cellular repair and resilience mechanisms that help enhance the overall fitness of the organism. Within an evolutionary theory framework, convergent cancer phenotypes may represent currently ill-defined "units of selection" upon which evolutionary forces have acted to overcome the natural resilience mechanisms of tissues at both material and immaterial levels, thus explaining both the rareness and the durability of cancer states. Applying this conceptual framework, I propose here a figure-ground shift, from targeting the proliferation of cancer cells to targeting the establishment of niches in which cancer cells can thrive. The breaking down of the healthy tissue niche and its transformation into a cancer niche is compared with processes acting at other scales of organization, such as the global cancer pandemic. Ramifications of this perceptual shift for basic research and clinical applications are discussed, also in light of work my team has undertaken in lung cancer. Finally, niche reconstruction is discussed as an approach to reverting, or transcending, the cancer state.

15.1   Introduction

In this chapter, the concept of niche construction is applied to describe cancer as a dysfunction of evolutionary processes. Recent survey data suggest that global cancer incidence will nearly double by 2040.[1] Diverse technical advances in recent decades have provided new tools to study, explain, and manipulate the biological and physiological, or material, properties of cancers. This has precipitated a boom in pharmaceutical research and driven the development of personalized treatments. However, due to either lack of efficacy or adaptive changes in the tumor tissue during treatment, the health benefits of most cancer drugs have remained rather modest.[2] This situation justifies addressing, with increasing urgency, whether we are missing a fundamental piece of the cancer puzzle. Currently, machine learning over large data sets is being applied in search of what this missing piece could be. Yet, from an evolutionary perspective of biological complexity, it seems unlikely that such computational extensions of the existing paradigms in cancer research can develop the tools and methods to reliably cure or prevent cancer at physiological levels in the near future. As Robert Weinberg lamented in 2014,[3] we seem again to have misjudged the complexity of the problem; with cancer incidences on the rise and very limited progress in finding cancer cures over the past four decades, perhaps we have also misjudged *how* to address this complexity. A revised theory of cancer may be needed to shift the current paradigm, a theory that incorporates the immaterial processes of evolutionary dysfunction that create the niches in which the cancer state arises.

The extended evolutionary synthesis (EES) is a set of recent extensions to the modern synthesis of evolutionary theory. Where the modern synthesis is based on the classical concepts of genetic variability and natural selection, the EES emphasizes the role of developmental processes that occur in an organism's lifetime and how dysfunctions in evolutionary programs cannot be reduced to dysfunctions in genes.[4] To paraphrase Fritjof Capra, these dysfunctions are "systemic, interdependent, and not to be understood in isolation," and the problems of our time "are all different facets of one and the same crisis, which is essentially a crisis of perception."[5] This chapter explores the consequences of a perceptual shift to viewing the cancer state as arising in an ecosystem of interdependent niches. In this approach, cancer is prevented, detected, and treated through reverting or transcending the niches in which the cancer state arises. As the process of neoplastic growth seems to constitute a systemic perturbation in the formation of organized tissue, we start by summarizing current theories of tissue organization in healthy contexts.

15.2   Conceptualizing Normal Tissue Organization

The development of a fertilized egg into an adult organism is a prototypical marvel of biology. This process, known as morphogenesis, establishes the complex anatomical shapes and cellular organization unique to each species. In Alan Turing's seminal 1952 study, he suggested that a simple reaction-diffusion system of two chemicals, which he named morphogens, could account for the main phenomena of morphogenesis.[6] The simplicity of his mathematical model is striking. In this model, an activator catalyzes synthesis of itself and an inhibitor while diffusing slowly, followed by the inhibitor catalyzing destruction of both morphogens while diffusing faster. In other words, Turing predicted that a simple interaction between pairs of positive and negative signals, acting at different distances and speeds, might underlie pattern formation in biology.

Decades later, Turing's simple predictions have to some extent been validated. Secreted signaling molecules of different protein families, such as Hedgehog (Hh), Wnt, Hox, EGF, FGF, and TGFβ, are now known to act as morphogens by forming concentration gradients that spatially instruct cell identities within tissues.[7] In many instances, a negative regulator is transcriptionally regulated by the action of its paired morphogens, as in the classical example of the regulation of the Hh receptor patched 1, which is both a Hh target gene and a negative pathway regulator.[8] Development of a variety of periodic tissue patterns, such as branching morphogenesis,[9] digit formation,[10] hair follicle spacing,[11] and the orientation of tissue stripes,[12] has been proposed, and sometimes experimentally validated, to arise via Turing-like mechanisms.

Not surprisingly, this straightforward model of morphogen activity based on free diffusion of molecules has also been challenged, notably in attempts to incorporate aspects of the material and three-dimensional properties of real tissues. Inhibitory morphogens have not always been identified; alternative mechanisms, such as physical depletion of the activator, can substitute for the inhibitory morphogen,[13] and it is not easy to incorporate time or tissue growth aspects into spatiotemporal morphogenesis models based on reaction-diffusion mechanisms.[14] A number of studies have shown that tissue mechanics can influence the distribution and persistence of morphogen gradients, suggesting an extension of the model is needed.[15] A central role for tissue mechanics in regulating morphogenesis is similarly evidenced by the self-organization of colon cells into crypt-like structures in microfluidic systems that mimic the gut's physical flow cues[16] and in the transformation of stem cells into gut-shaped organoids in growth factor–supplemented matrix.[17] These findings emphasize the central role of the extracellular matrix and tissue microenvironment, uncovered by Mina Bissell and colleagues, who recognized that "cell morphology and the expression of tissue-specific genes are both intimately related to the nature of the substrata."[18]

What ensues in normal tissue organization is an interplay between opposing chemicals, amplified by intracellular signaling responses, coordinated by extracellular, mechanical, and other physical and chemical cues in the microenvironment. This interplay is further perpetuated and stabilized by various feedback loops through the deposition of extracellular matrix by newly formed cells. Interestingly, this dichotomous nature of morphogenesis was already described by Turing, as well as others before him, such as D'Arcy Thompson and Erich Blechschmidt. In Turing's 1952 article, the embryo transitions from one "state of the system" to the next, where "the state consists of two parts, the mechanical and the chemical. The mechanical part of the system describes the positions, masses, velocities and elastic properties of the cells, and the forces between them."[19]

In retrospect, it is easy to see how theoretical models of such a system, with one set of factors building on the next, eventually diverged from our factual knowledge of the mechanical and biochemical properties of real cells and tissues. For the physical components, measurements and derivative calculations or predictive models are needed, while the dissection of biochemical components requires methods to extract, purify, and analyze various biochemical cues. This methodological dichotomy mirrors the conceptual divergence between fields of science that apply, on the one hand, a systems view and, on the other, a parts view of biology. We now turn to how these two fields approach the problem of cancer origination, which appears from the outset to be linked to tissue organization.

15.3   Carcinogenesis from an Experimental Biology Perspective

At the tissue level, cancer can be conceptualized as a wound that does not heal, characterized by a leaky vasculature network, and the persistent secretion of fibronectin and fibrin-rich matrix typical of injured tissue,[20] or as a newly formed cancerous "organ" that has co-opted the local tissue bed.[21] Cancer growth elicits inflammation, either as a direct response to the injury or in response to damaged or cytosolic DNA,[22] resulting in the recruitment of innate immune cells to the tumor microenvironment. Cancer-promoting inflammation can also originate in response to viral infections or other chronic conditions, eliciting a systemic inflammatory cascade that can be further modified by commensal microbiota.[23] While the tissue injury model best fits carcinomas that grow outside the confines of the basement membrane, inflammatory molecules also play an important role in soft tissue sarcomas and blood cancers.

In addition to inflammation, cancer lesions exhibit altered pleiotropic interactions between stromal and cancer cells, co-opting both the adaptive and innate arms of the immune system in amplifying and sustaining the cancer state.[24] Interestingly, while tumors show considerable differences in their precise immune contextures, immune landscape profiling has revealed trends toward an association of increased tumor-resident immune-suppressive cells, particularly myeloid macrophages or neutrophils, with poor cancer prognosis, while increased T- or B-lymphocyte infiltration tends to associate with improved survival.[25] Thus, at a very general level, cancerous growth initiation constitutes an intricate interplay between niche-localized tumor cells *and* stromal cells, physiological and mechanical aspects of the extracellular matrix (see chapter 11, this volume), and systemic roles of both the innate arm of the immune system in tumor promotion and the adaptive arm in tumor surveillance. Cancer acts as a wound that does not heal, then, not so much because of the perturbed tissue but because of a *systemic perturbation* that continuously interferes with the process of tissue regeneration.

Cancerous growth goes hand-in-hand with the acquisition of molecular and genetic alterations in cancer cells, leading to the activation of oncogenes or the loss of tumor suppressor genes that can affect oncogenic signaling activities, tumor cell survival, or cancer metabolism.[26] The signatures of these genetic alterations relate to the mechanisms of the mutagenic process[27] and thus can mark a tumor's environmental exposure.[28] However, rather than acting as causative cancer agents in their own right, it is increasingly evident that genetic drivers exert their effects only in a context- or niche-dependent manner. Lineage-tracing studies in murine models indicate that disease etiology is influenced by the cell and tissue type in which the tumor originates.[29] This is corroborated by research from my own group: in non–small cell lung cancer (NSCLC), the precise cell of origin defines what type of histopathology lesions arise, and histopathology subtypes can have different immune microenvironments[30] or utilize different signaling networks,[31] which can drive sensitivity to targeted pathway inhibition.[32] Similarly, combined expression of oncogenic *Kras* and *Myc* elicits tumor growth by suppression of adaptive immunity and stromal engagement via different, seemingly tissue-selective, mechanisms in lung and pancreas tissue.[33] Thus, phenotypic diversity in signaling and immune contexture aligns more closely with histotype and tissue type than driver genotype. Tumor phenotypic diversity therefore appears to involve both oncogenic driver functions, as a form of "fuel" for cancerous growth, and progenitor- or tissue-selective functional programs, which may comprise a form of "epigenetic memory" of the cell lineage in which the cancer arises.

The paradigm of lineage-selective tumor heterogeneity, as deciphered in mouse model systems, has recently gained momentum in the study of human cancers, where pan-cancer genome atlas data analyses revealed that the anatomic origin of tumors constitutes a major factor in tumor classification.[34] In human tumors, phenotypic diversity is further amplified by long lifetime exposure to environmental mutagens and pathogens, contributing to extensive molecular heterogeneity. A key question for future research, then, is in how far organ-selective programs, in particular the co-option of both regenerative inflammatory processes and developmental lineage programs, generate convergent phenotypes. A recent working model by Gerard Evan and colleagues proposes that oncogenicity involves the co-option of latent tissue-specific regenerative repair mechanisms, suggesting that transcriptional networks that define ontogeny may be targetable in therapeutic approaches.[35] One might then ask whether cancer formation involves the reactivation of morphogen-coordinated programs, albeit in a dysfunctional manner. In support of this hypothesis, morphogen pathways are widely affected in tumors, as shown by, for example, a hierarchy of wnt-secreting and wnt-responding niches driving lung cancer,[36] as well as germ cell–derived teratomas, which sometimes produce discernable organ pieces. This highlights the relevance of exploring cancer growth as a process of morphogenesis gone awry, as well as the importance of reuniting the molecular and systems approaches to cancer biology.

15.4   Carcinogenesis from a Systems Perspective

While experimentation-based cancer studies focus on the functions of molecules and cellular phenotypes, systems theoretical approaches aim instead to understand the dynamic relationships between cellular, organismic, and environmental factors, as well as the evolutionary processes that govern cancer development. In any systems approach, molecules and phenotypes are viewed as integral, emergent entities of an interactive network. The idea that the collective behavior of complex systems has "emergent features,"[37] typically overlooked when studying causative driver genes, is central to systems theory. Another systems concept, "stigmergy," originally developed in the study of social insects. Stigmergy underlies the self-organization of collective communities of agents into the emergent structures of complex systems. While not formally the same, both reaction-diffusion mechanisms in morphogenesis and stigmergy are underpinned by very simple rules of interaction, operating adaptively in complex environments.

Biological systems theory began in the first half of the twentieth century, through studies on biological topology[38] and relational biology,[39] among other subjects. The field remained underappreciated by mainstream biology until the early twenty-first century, when, for example, Denis Noble recodified and popularized the principles of systems theory as *The Music of Life*.[40] Noble argued that genes do not act in isolation from each other, or from their environment, and called for the replacement of "the selfish gene" metaphor with metaphors that shift emphasis to the processes and relations involved, rather than the parts. Noble contended that, in integrating the information that explains living systems, there is "no privileged level of causality," which is the foundation of "scale relativity."[41] In other words, living systems constitute a continuum of interactions and interdependencies. Noble's work is related to other systems theories, such as the "tissue organization field theory," which views cancer as a tissue-based disease of altered reciprocal interactions between cells and matrix, with no privileged level of causation,[42] and to the process-ontological view that explains cancer as a consequence of dysfunctional process coupling across scales (see also chapter 2, this volume).[43] These and other systems approaches have begun to be adopted in cancer research, to some extent countering the gene- and mutation-centric "somatic mutation theory." Thus, in a systems view, the cancer state is an inherent part of a dynamically evolving living system of pleiotropic interactions between functional components across scales.

The application of systems thinking to cancer biology revives the legacy of Conrad Waddington. Relating embryology and developmental trajectories to gene regulatory networks, Waddington synthesized a conceptual framework in which cell lineages develop via canalization, referring to the constancy of a phenotype moving through a landscape of evolving relationships. This generates the so-called epigenetic landscape: a topographical map in which hilltops and valleys represent systems-level stable states, and the trajectories of paths to and from these states are guided by the topography of the hill slopes defined through dynamic feedback between the environment, phenotype, and genotype.[44] In Waddington's theory, environmental pressures can result in the inheritance of genetic characteristics, providing plasticity and adaptability to any particular trait. Other network components provide buffering of genetic variation.[45] Stated differently, genotypes are favored that confer the phenotypic *adaptability* required for survival in dynamically changing environments. This departs from the neo-Darwinistic "selfish gene" paradigm in that the evolution of acquired traits is not strictly mutation driven, but rather inheritance is attributed to the functional interplay between genes, phenotypes, and the environment.

In an implementation of Waddington's epigenetic landscape analysis, Stuart Kauffman, Sui Huang, Donald Ingber, and colleagues applied mathematical modeling of gene regulatory networks to describe cell state transitions using so-called attractor landscape analysis.[46] In this kind of analysis, attractor states are modeled as valleys representing energetically stable states, with cancer cells described as being trapped in an abnormal valley, while hilltops represent undifferentiated or pluripotent cell fates[47] (for details, see chapters 4 and 5, this volume). Adopting the attractor paradigm, the fundamental experimental questions become if and how the state space can be modified to navigate cancer cells out of the valley or, better, prevent them from entering it in the first place.[48] Attractor modeling has been used to address a number of fundamental questions relevant to our understanding of carcinogenesis; for example, Boolean logic was applied to describe functional nodes and interactions related to p53-dependent cell fate decisions[49] or colorectal tumorigenesis,[50] or to predict the outcomes of genetic and drug-induced perturbations. However, not all components of a gene regulatory network architecture are usually known, and tumors harbor heterogeneous cell states that may respond differently to perturbations, rendering modeling approximate at best.[51] Thus, while networks provide powerful conceptualizations of living systems, they remain reductionist in practice through their reliance on often linear and biased input data, which limits inclusion of often unknown dynamic interactions with factors—material as well as immaterial—in the local tissue and larger macroenvironment surrounding the tumor niche.

15.5 Conceptualizing Resilience Mechanisms against Cancer in Healthy Tissues

Conceptualizing the cancer state as the result of homeostatic imbalances, eventually culminating in malignant disease, aligns with a systems view of biological phenomena, in which pleiotropic tissue and environmental factors elicit its manifestation. Indeed, if only a few mutations were required to drive clonal expansion and malignancy, then cancers would not be so rare.[52] From the systems standpoint, the relative rareness of cancers is best explained by the progressive loss of tissue homeostasis[53] (with the possible exception of childhood cancers), rather than by an increasingly destructive potential of cancer cells. Such a view of cancer progression affirms Theodor Boveri's original ideas on tumor suppression,[54] reiterated in Conrad Waddington's 1935 statement, "The fundamental fact of cancerous tissue is that it has escaped from the normal growth-controlling agents of the body."[55]

Various mechanisms contributing to fitness-enhancing cancer resilience have been described. In the case of primary tumor progression, these include (1) cell-intrinsic protection against mutagenesis, such as DNA repair and epigenetic mechanisms; (2) cell-intrinsic growth suppression mechanisms following oncogenic triggers, such as apoptosis and senescence; (3) suppressive signals provided by the local tissue microenvironment, such as the extracellular matrix, macrophages, and antioxidants; and (4) immune surveillance by the adaptive immune system.[56] The high oxidative stress environment of the blood provides another physiological barrier against carcinogenesis, possibly explaining why metabolic adaptations, such as mutations in the antioxidant NRF2/KEAP1 system, are common upon metastatic progression.[57] Finally, when a metastatic tumor cell successfully reaches its secondary site, dormancy is common, indicating further resilience elicited by the microenvironment in the metastatic niche[58] (see also chapter 14, this volume). Spatio-temporal and tissue-selective mechanisms operate across these resilience layers, most likely related to the developmental and tissue-regenerative programs mentioned in section 15.3, adding complexity. This understanding attests to the progress made in recent decades in investigating multistage tumor progression and underscores the numerous ways in which extensive tumor heterogeneity can arise.

Most cancer resilience mechanisms have been studied using model systems, most often in rodents carrying high tumor burdens, and in at least partially immune-suppressed inbred animals. Encouragingly, accumulating clinical data imply that similar mechanisms keep cancer progression under control in humans. For example, ultrasensitive sequencing applications reveal significant driver gene mutation rates in *healthy* tissues and liquid biopsies from aging individuals, including leukocytes, skin, esophagus, colon, and uterine lavage samples.[59] In the esophagus, TP53 and NOTCH1 mutations can respectively cover a remarkable 37 percent to 80 percent of healthy epithelia, and cancer-associated TP53 mutations expand during aging, implying their growth advantage.[60] Corroborating these findings, computational estimates suggested that at least half of all somatic mutations arise prior to tumor initiation in self-renewing tissues.[61] Finally, tissue analyses from individuals lacking symptomatic cancer, often at autopsy, have revealed that both premalignant and malignant lesions are also surprisingly common, with, for example, breast cancers detected in almost half of all middle-aged women.[62] Consensus is therefore building that, while the accumulation of cancer-associated mutations and the onset of cancerous growth are common aging-related phenomena, they do not necessarily lead to malignant cancers due to tissue-inherent resilience mechanisms, which actively suppress them.

Significant clinical questions follow from these insights, such as whether cancerous lesions can undergo spontaneous regression due to tissue resilience mechanisms and what causes the transition to aggressive disease. In the lung, an elegant bronchoscopy surveillance study compared the genomic profiles of preinvasive carcinoma biopsies, sampled just prior to progression to squamous carcinoma or regression to low-grade or normal epithelia. This revealed a strong association of chromosome instability (CIN)–related gene expression and copy-number alterations with malignant conversion.[63] Indeed, CIN, defined as persistent genomic alteration caused by aberrant chromosome integrity or dysfunctional cell cycle checkpoints, is a common feature of malignancy.[64] Evidence from multicenter lung cancer evolution studies supports a model in which adaptive immune surveillance constrains preinvasive lesion growth and in which metastatic conversion is associated with various types of immune evasion, such as expression of T-cell inhibitory checkpoint molecules or suppressive interleukins, or immunoediting escape through HLA loss of heterozygosity or reduced major histocompatibility complex class I (MHC-I) presentation due to oncogenic mutations.[65] Similar immune surveillance mechanisms operating across the metastatic cascade have been identified in other tumor types, including colorectal and breast.[66] More recently, metastatic niches have been shown to undergo immune surveillance, with most aggressive lesions exhibiting immune privilege, again emphasizing remarkable tumor heterogeneity, with evidence that different immune evasion mechanisms can operate in different metastases in the same patient.[67]

Taken together, we may derive from this that a phase of resilience to oncogenic mutations and cancerous growth, imposed by the local tissue microenvironment and adaptive immune activity, precedes a phase of gradual loss of these mechanisms, causing the onset of malignant transformation. Indeed, cancer progression overall seems better explained by a cumulative loss in resilience factors than by a gradual increase in the fitness of mutation-carrying cells. This in turn raises the question of whether we can develop the technical capabilities to transcend progression to the cancer state. In the below, I will explore a figure-ground shift from targeting cancer cells to targeting the establishment of the niches in which cancer cells can thrive, using the evolutionary process of niche construction to conceptualize the approach.

15.6   An Extended Evolutionary Concept for Understanding the Cancer Niche

Recent decades have seen the rise of the field of cancer evolution, adopting metaphors and concepts from evolutionary theory to describe the developmental trajectories of cancer mutations and cells.[68] At the same time, evolutionary theory itself has been in flux, with further ramifications for cancer theory, which we will address in the next section. The modern synthesis, the evolutionary theory developed in the early twentieth century based on Darwin's work and Mendel's genetics, integrated with population-level modeling, considers genetic variation, inheritance, and natural selection as the basis of evolutionary biology. In recent decades, some evolutionary biologists have argued for an extension of the modern synthesis, named the extended evolutionary synthesis (EES).[69] Proponents of the EES maintain that the processes that take place during an organism's lifetime, as well as their reciprocal influence on the inheritance of phenotypic traits, have been underappreciated. Their views echo Karl Popper, who argued against the emphasis on adaptation following random mutations. Popper claimed that developmental processes have goal-directedness, with genes being followers rather than leaders in driving the evolution of acquired traits.[70] The EES theory reaffirms Waddington's epigenetic paradigm, Denis Noble's Music of Life, and related systems theories, in that there is "no privileged level of causality,"[71] and in calling for discarding the "selfish gene" paradigm.

Advocates of the EES emphasize the reciprocal causality of developmental processes in evolution, in particular, how the rate and direction of evolution are influenced by developmental bias and plasticity, nongenetic inclusive inheritance, and niche construction.[72] This view has triggered a debate in which skeptics argue that these processes have long been studied and are known to be both outcomes and causes of evolution.[73] The debate reflects a perceptual difference: the modern synthesis is anchored in the notion that alterations in gene frequencies enact evolution, while the EES stresses the relative roles that developmental context and organismic interactions with the environment play in feeding back and regulating these alterations. This perceptual difference mirrors the original divergence between systems and molecular biology, holding out the possibility of the reintegration of these knowledge domains into a revised theory of evolution that can be applied to cancer.

Of the processes that influence evolution, niche construction is the activity that most intuitively conveys the instructive role that organisms play in natural selection. Niche construction is defined as the processes by which organisms modify their environments, via their metabolism, activities, and choices.[74] This provides a feedback mechanism through which organisms affect both their own and other species' evolution, with important transgenerational ramifications. Animals implement niche construction in a variety of ways, ranging from the building of habitats and the manufacturing of artifacts to waste generation and organismal death. Through shaping local-to-global niches by their behavior, agency is thus added to the ways in which organisms influence selection pressures. The prime planetary example is human culture and behavior, as these directly impact organismal and ecosystem evolution in various ways.[75] Importantly, niche construction can be maladaptive, not only for the entity that builds a niche but also for others, as seen in both the pollution of shared ecological environments and disease niches in tissues that affect the whole organism.

Subsequent questions are in what ways organismal biology pertains to natural selection and at what levels of ecosystem hierarchy it acts. Interestingly, akin to the process-ontological view of cancer theory mentioned earlier,[76] a number of evolutionary theorists propose that, rather than through selection on genes, individuals, or communities, *the actual functions implemented by these entities represent units of selection*. Their recently formulated theory, called "It's the song not the singer" by Ford Doolittle (ITSNTS; pronounced "it's nuts"), is inspired by "the observation that the collective *functions* of microbial communities (the songs) are more stably conserved and ecologically relevant than are the taxa that implement them (the singers)."[77] According to ITSNTS theory, evolution is explained by the *differential persistence of a process or pattern of interaction* (song) executed by multispecies collectives (singers); this only indirectly fosters the differential reproduction of the contributing species. Examples of such communal processes are the global biogeochemical cycles that distribute elements and molecules, interconnecting food and nutrient cycles that operate in ecosystems.[78] Interestingly, while Doolittle originally critiqued the idea that global homeostatic mechanisms evolve by natural selection,[79] ITSNTS is formulated to align with James Lovelock's and Lynn Margulis's Gaia hypothesis,[80] as well as Margulis's endosymbiosis theory,[81] which have both recently received support from genomics studies showing widespread horizontal gene transfer across species and phyla.[82] These theories jointly support the view that the exchange of heritable information between organisms is common and dynamic and thus has the potential to facilitate the differential persistence of processes relevant to life. In this cooperative paradigm, processual phenotypes represent units of selection, within the context of a global homeostatic ecosystem, called, in the case of Lovelock's theory, Gaia. The renowned developmental biologist Scott Gilbert has also recently embraced the notion that living entities constitute "holobionts," defined as assemblages of eukaryotes and symbiotic microorganisms,[83] making what counts as "self" both dynamic and highly context dependent.[84]

In summary, we see an increased appreciation for the important role of environmental cues, which are often generated by the organisms themselves, in altering developmental and evolutionary trajectories. An important application of empirical niche construction methodologies may be to integrate ecosystem biology and evolutionary theory to understand how niche construction has ramifications for transgenerational ecological inheritance and the corresponding processes in carcinogenesis. This kind of theoretical framework, initially used to describe ecological and evolutionary phenomena at the species level, is increasingly being applied to the cancer problem, as a number of publications have recently shown.[85] In the following, I will transpose some of these concepts to the cancer niche and explore how they might be applied in treatment.

15.7   Reconceptualizing the Stages of Cancer Progression from an Evolutionary Perspective

The EES extended evolutionary concept, the function-based evolution defined by ITSNTS, and the global self-regulation defined by the Gaia hypothesis all agree that organisms respond in dynamic, context-dependent ways to their natural environment, which itself constitutes a global web of homeostatic functional relations. Or, as Scott Gilbert states, "We have never been individuals."[86] The similar idea that cancers behave as open complex systems is central to an ecological framework of cancer developed by Robert Gatenby and colleagues. In their perspective, temporal fluctuations in environmental conditions promote phenotypes that can adjust to changing circumstances, termed "adaptive phenotypic plasticity."[87] It follows that genotypes that endow their cells with the capacity to survive in changing conditions, such as altered metabolic or oxygen conditions, are favored, making genes the followers in cancer evolution, aligning with Waddington's views. Thus, similarly to how regenerative programs can be co-opted by cancer cells, genetic mutations can also be co-opted in response to environmental demands, rendering mutations the passengers, rather than the drivers, in cancer evolution.

Many characteristics of the cancer state, such as heterogeneity and plasticity of phenotypes, spatial diversity in genetic mutations, temporal fluctuations in blood flow and oxygenation, and spatial variations in intratumor pH, show that cancer cells are exposed to variable, stochastically changing microenvironments. Cancer cells also deposit their own matrix and secrete ligands and metabolites, constructing local niches that can further evolution, as seen, for example, in the promotion of tumor invasion by acidic environments generated by aerobic glycolysis.[88] Similarly, we reported pronounced spatial diversity in the activities of oncogenic signaling pathways in lung tumor tissues, suggesting that oncogenic activities are influenced by intratumor heterogeneity in ligand gradients and other microenvironmental factors influencing receptor tyrosine kinase activities.[89] Thus, once resilience mechanisms are defeated, the neoplastic process is uncoupled from the fitness of the homeostatic organism, and fitness starts to evolve as an adaptive response to variable local microenvironments instead. This gradual shift, from vital cooperation between cells integrated in a healthy organismal "self" to the opportunistic "selfish" behavior of individual tumor cells, is indicative of a loss of buffering capacity. Is this shift a consequence of persistent environmental challenges imposed on tissues, such as wounding and inflammatory responses and high interstitial pressures, effectively forcing cells into a new topological attractor in the cellular phenotype attractor landscape? This conceptualization aligns with Paul Davies's atavism theory, which postulates that tumor cells have ancestral traits, part of a more rudimentary evolutionary program, evolved and optimized for unicellular premetazoan life.[90] What follows is that neoplasms appear to constitute an inherent vulnerability of multicellularity when single cells become trapped in a state, or niche, that promotes the persistence of processes that benefit *single-cell* behavior and survival, rather than cell cooperation and integration within a metazoan tissue context (see also chapter 9, this volume).

From the previous, a coherent picture starts to emerge in which the cancer state is rare, and convergent cancer phenotypes, such as metabolic adaptations, oncogenic signaling, or immune evasion, represent *units of selection against natural resilience mechanisms*. Adopting the metaphor that processual functions, part of a scaled continuum of interdependencies, are like songs, Denis Noble's "Music of Life"[91] takes on new meaning. Cancer, as a consequence of lifetime exposure to environmental stresses, alters the "musical dialogue" between intracellular and extracellular cues, disrupting the harmony regulated by simple rules of interaction within complex environments, as seen in morphogenesis and self-organization. Cells in the cancer state lose this harmony, but instead of falling silent, they switch to playing independently. The neoplastic style is an older form of the music of life, but to complex organisms, it is cacophony. Similarly, the process of carcinogenesis could be described as differential functional persistence in accordance with the ITSNTS paradigm (i.e., a song) with the following musical stages (figure 15.1): (0) symphony—healthy tissue conducted by the mechanisms of tissue homeostasis; (1) out-of-tune instruments—cells accumulate carrying cancer-associated mutations, but cell-intrinsic and tissue environment–based growth suppression mechanisms maintain tissue harmony; (2) dissonance—cancerous songs emerge as pre- or early stage cancer lesions, with the loss of harmony bringing additional surveillance by the innate and adaptive arms of the immune system; and (3) cacophony—the players disregard the conductor and each follow their own tune, which drowns out adaptive immune surveillance and eventually elicits metastatic progression, accompanied by adaptations to overcome (e.g., blood oxidative stress levels or metastatic soil-selective growth suppression). The symphony has transcribed itself into a neoplastic cacophony of individual adaptations, and now the sequence repeats, as new instruments go out of tune, and new strains of the disharmonious ancestral cancer song take root and grow within the metastatic niche.

In process-based evolution, cancer progression results from altered functional interactions between a set of biological entities (including cells, matrix, and microbiota) residing in a specific tissue niche, within the context of variable environmental demands. Interestingly, in addition to the well-established roles of viral and mutagenic agents such as tobacco in generating tumor heterogeneity, it was recently shown that genes can form de novo from noncoding DNA,[92] raising the intriguing possibility that cell-intrinsic functions are also in continuous flux. This view therefore shifts the emphasis to understanding how environmental and intracellular cues within the precancerous niche connect to immaterial

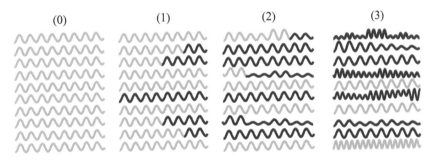

Figure 15.1
The musical stages of carcinogenesis. (0) Symphony: homeostasis; (1) out-of-tune instruments: accumulation of mutations and suppression of cell growth via tissue maintenance mechanisms; (2) dissonance: loss of harmony due to growth of cancer lesions, controlled by immune surveillance mechanisms; (3) cacophony: overt neoplastic growth and adaptive phenotypic plasticity, culminating in metastatic seeding and repeat of the musical sequence in select metastatic niches. Each curved line represents a cell clone (or instrument), with darker lines indicating mutant clones (or out-of-tune instruments) and irregular curves indicating ancient types of neoplastic tunes or songs.

process-based cycles: if functional units in cancer evolution are *not* single cells but parts of higher-order, systemic factors, what is it that cells perceive? These theories raise new questions regarding the nature of the neoplastic tune, such as, what communally regulated processes regulate the cancer state? Which biological entities influence them and how? How do selective forces cause the emergence and persistence of cancer-associated phenotypes? What are such phenotypic units of selection? In section 15.9, we will speculate on how these questions can be experimentally pursued.

Important additional questions follow from such a framework, such as, what constitutes the tipping point in cancer progression, and at which point is bodily resilience against the cancer state defeated and single cancer cell-based evolution ensues? Does cancer emerge once the number of tissue maintenance inputs, constantly monitored by cells, drops below a "malignancy threshold"? Do the perceived units of selection change once cells enter the malignant state? Multiple lines of evidence suggest that clinical cancer progression is far from a linear process. In many cancers, there is evidence that metastatic seeding can arise very early in tumor progression, in parallel with primary tumor progression[93] (and see chapter 14, this volume). The process of "field cancerization," in which recurrent tumor lesions arise in multiple affected organs, suggests that systemic factors can regulate cancer progression in parallel niches.[94] Furthermore, parallel tumor evolution can foster phenotypic diversity in spatially diverse niches within the same tumor, as seen in lung cancer in the association of spatial mutant *KRAS* allelic amplification, with malignancy mediated by a glycolytic phenotype switch.[95] Hence, while tumor progression overall follows relatively well-defined stages, the process of malignant conversion in a human body is spatially heterogeneous and stochastic, complicating temporal monitoring. This indicates the need for a shift of emphasis from the treatment of end-stage cancers to effective interventions in early stage cancer evolution, via an approach that considers the cancer state as part of a dynamic system of interdependent, coevolving relationships.

In the following section, I posit that by embracing niche reconstruction as a central, cooperative approach to influencing neoplastic conversion, we have the best shot at reverting the cancer state to nonmalignancy and perhaps transcending progression to the malignant cancer state altogether.

15.8   Reversion or Transcendence of the Cancer State through Niche Reconstruction

Despite major progress in our biological understanding of cancer and continued support from funding agencies, taxpayers, and the private sector, the recent AACR Cancer Progress Report 2019[96] is sobering. Cancer in the United States is projected to pose increasing health challenges over the next two decades, reflected in a 1.35-fold increase in cancer incidence related to population aging; increases in harmful youth e-cigarette use, which is likely to lead to a subsequent increase in addictive tobacco use; and an increase in the prevalence of obesity, which is causally linked to carcinogenesis.[97] Globally, the situation looks even worse, with cancer incidence projected to double by 2040.[98] At the same time, the past decades have taught us that long-term cancer treatment efficacy is often compromised by adaptive plasticity, a feature inherent to developing complex systems. Together, these projections and learnings point not only to a need for more research to gain further insights in the contextual behaviors of cancer states but also to the need to better *target or manage the environmental challenges that cause the cancer state to persist.* Environments are directly modified by organisms via the process of niche construction, and humans have developed a broad range of often far-reaching capabilities by which they shape their own and other organisms' niches. This indicates that, in order to control what will otherwise become an escalating health problem, we must identify ways to adjust our own behavior to revert maladaptive cancer niche construction. In the following, we'll look at how interdependent material and immaterial processes influence the cancer state, presenting both obstacles and opportunities to reverting cancer-promoting niches.

(1) Personal cancer niches: Following tissue organization during embryogenesis, each person is born with a unique genetic makeup, which constitutes the heritable basis of an individual's cancer risk. Heritability extends also to the distinct behavior of a person through the actions of signaling molecules such as hormones and neurotransmitters; for example, in his book *Behave*, Robert Sapolsky described how the lifestyles of our prehistoric ancestors still shape our behaviors.[99] From childhood until adulthood, a person's cancer risk and behavioral traits are both increasingly regulated by the environment. These environmental cues range from parental care, now considered to contribute durable epigenetic effects that can affect health later in life,[100] to a person's dietary habits and exposures to microbes and environmental carcinogens. The cancer state thus evolves as part of a holobiotic milieu that is both dynamic and unique to a particular individual. How can such complexity in the causative factors of cancer be managed? Personalized approaches to cancer diagnostics and treatment constitute one important step toward addressing this, on which I will touch upon in the next section. However, these remain *reactive* approaches in which a patient is typically treated as a case study and largely ignore the fact that disease develops as part of a multilevel complex system. Systematic enhancement of tissue resilience, for example, is not part of the clinical repertoire, although the partial success of checkpoint inhibitors in eliciting antitumor T-cell adaptive responses[101] could be considered an approach to boosting natural resilience. Understanding cancer evolution as a dynamically adaptive process, in principle adjustable by resilience-promoting environments, underscores a need to *shift emphasis to interventions in early stage cancer evolution.*

The current lack of approaches to strengthening tissue maintenance mechanisms is partly explained by the fact that these are functions of healthy homeostasis and not diagnostically analyzed. Furthermore, tissue microenvironments are regulated in pleiotropic ways by a variety of physiological systems, making it difficult to discern precisely how health-related lifestyle interventions, such as exercise and dietary adjustments, affect the cancer niche. Accumulating epidemiological evidence shows that behavioral factors such as chronic stress and social isolation can associate with worse cancer outcomes.[102] However, the molecular roles of the peripheral autonomic nervous system (ANS) and hypothalamic-pituitary-adrenal neuroendocrine signaling in tumor progression have emerged only recently. The ANS is now known to regulate hematopoietic homeostasis, inflammation, and progression or metastasis of various solid tumor types through pleiotropic effects on, among others, tumor, adipose, and immune cells. These insights have led to the piloting of novel therapies, either via direct nerve stimulation or pharmacological means. Intake of β-adrenergic blockers may, for example, benefit breast or prostate cancer patients, although not melanoma or colorectal cancer patients, indicating a need to personalize therapy.[103] Interestingly, akin to how fitness-enhancing mechanisms conspire to prevent cancer progression in healthy tissues, stress-associated physiologies also act adaptively to achieve homeostasis, a process termed "allostasis."[104] Hence, one way to conceptualize the reversion of cancer-promoting niches is via the relief of chronic stress mediators and reduction of the "allostatic load" of cumulative external stressors through long-term positive behavioral lifestyle changes.

(2) Health system niches: From historical roots in assisting with childbirth, illness, and trauma, medicine has grown into what is now recognized as a complex system in its own right. As such, it is difficult or impossible to predict the effectiveness, safety, and even the behavior of the system as a whole.[105] Health system niches comprise "networks of components (hospitals, clinics, nursing homes, rehabilitation units, patient homes, families, and patients) that interact nonlinearly on different scales (the patient, family, medical center, and government), and often produce unintended consequences (adverse drug reactions, nosocomial infections, rehospitalizations, and functional decline)."[106] Recognition of this in the 1970s led to the term "medicalization," defined in Ivan Illich's 1975 book *Limits of Medicine* as the process whereby the health care system unintentionally develops adverse, iatrogenic effects, leading to the goals of medicine becoming lost and even reversed in the complexity of the system.[107] As a consequence of the monopoly that health professionals have over medical technologies, industrial rather than personal growth receives emphasis.

Some of the iatrogenic effects that persons diagnosed with cancer are confronted with include durable health deterioration following treatment, fears of disease recurrence, difficult medical jargon, conflicting web advice, financial liabilities, fatigue-related workforce reentry issues that risk job loss,[108] and psychological burdens amplified by the distress of loved ones.[109] On top of this, expensive new cancer treatments often result only in marginal extensions of patient lives,[110] and quality-of-life improvements are not central to medical follow-up.[111] A 2015 National Health Service report concluded that pharmaceutical sales, when priced by commercial interests, harm around five times more lives than are helped, because resources are consequently directed away from services such as early cancer diagnosis and mental health support.[112] The already problematic negotiation of control between the patient and the medical systems that, as Illich and others have described, need to act authoritatively and decisively is exacerbated by the complexity of tumor biology and medical capabilities. Finally, since cancer evolution occurs prior to hospitalization and medical care, it is not surprising that cancer treatment is largely palliative. In the face of the synergistic complexities of cancer evolution and the medical system itself, medical niches thus need to be rebuilt via a *fundamental change in the perception and promotion of health*, both prior to symptomatic disease and following cancer diagnosis.

Health designates an adaptive process in which the neoplastic process is coupled to the fitness of the organism, sustaining homeostasis until anticancer resilience is defeated. Cancer state reversion is therefore best facilitated by *the allocation of resources that equip individuals with the knowledge, tools, and behaviors to help shape their environments and promote their own health*. This view aligns with the 1948 World Health Organization (WHO) definition of health, which states that "health is a state of complete physical, mental and social well-being and not merely the absence of disease or infirmity."[113] Sir Harry Burns, previously chief medical officer of Scotland, argues that this requires improvements not only in health care but also in socioeconomic conditions, as this can ameliorate the diets and reduce the chronic stress levels of the poor.[114] This calls for public policy actions that enhance the ability for people to self-manage their health, for example, via "saluto-genesis." This process, first termed by Aron Antonovksy,[115] recognizes that a person's "sense of coherence" can reduce the impact of chronic stress by providing a person with the assets needed to adapt their behavior to challenging environments. Salutogenic solutions may thus reverse the lack of control people can experience as a medicalized patient. A framework that incorporates such a participatory approach is the "P4 Health Continuum" model, which proposes how the current reactive care paradigm can become "Predictive, Preventive, Personalized and Participatory."[116]

(3) Cultural and social niches: The recentralization of health systems on prospective and actual cancer patients implies changes in the roles of health professionals, scientists, and the public. This can be facilitated by expanding the public reach of research activities, a process in principle supported by biobanking efforts that share biomedical data and samples with researchers, enabling population studies and evidence-based prevention.[117] However, the biomedical data infrastructure is also a heavily financialized element of the rising knowledge-based bioeconomy,[118] raising ethical questions related to value and accountability.[119] Given the powerful private interests invested in biomedical data and intellectual property, it is critical that researchers and the public continually reassess whether health data practices are centered on patient and public benefit. Currently, cancer research primarily endeavors to utilize patient samples and data, and it assumes that the initiation of clinical trials, or sharing of information with health professionals and the public via research publication and other dissemination routes, suffices in benefiting people. This inherently puts researchers and the public in a passive role with respect to the big data movement, whereas one implication of the abovementioned niches is that

simply the reversal of loss of agency on behalf of the patient would be beneficial to cancer prognosis. The highest medical value of personal biomedical data, then, is gained by each person, in search not of personalized palliation but personalized prevention. A niche-oriented approach to prevention would identify systemic salutogenic solutions, rooted in medical science, that will help individuals to truly self-manage their own wellness.

Paths to such salutogenic solutions include an increased involvement of the public in biomedical research, for example, by connecting students with cancer patients[120] or co-creative teamwork that centralizes the human dimensions of an illness, involving social and behavioral scientists, nurses, and bioethicists. As social animals, humans are particularly sensitive to sociocultural cues, which, via the ANS and neuroendocrine signals, affect tissue homeostatic physiologies and immune competency. While it is not well understood how such immaterial factors affect tissue resilience in cancerous niches, an increasing number of studies demonstrate how social solutions can foster positive behavioral and health outcomes. For example, smoking cessation, itself known to ameliorate cancer outcomes,[121] is particularly effective when nicotine replacement therapy is combined with group-based psychosocial therapy[122] and even benefits schizophrenic patients.[123] Similarly, a behavioral program designed to help breast cancer patients deal with stress showed improvements in mood and physical activity, as well as suggested improved immunity and reduced recurrence risk.[124] By taking a qualitative approach, Ulrika Sandén, herself a cancer patient,[125] developed an old Nordic coping strategy based on momentary contentment to understand how to live with a cancer diagnosis. Her study describes how contentment can be found in nature, art, and humor and how communities can provide a sense of safety.[126] The recent WHO Europe first report on arts and health similarly concluded that arts and social programs are to be prioritized as health-promoting activities, supplementing biomedical treatments.[127] These indicators point at how placing patients and their needs at the center of health care approaches can add a missing integrative perspective to health care innovation.[128]

A new focus on *health-promoting sociocultural environments that complement biomedical niches* thus appears to represent a missing piece of the cancer puzzle. Humans are distinguished from other animals by the prefrontal cortex, which regulates complex cognitive behavior, including problem solving, long-term planning, and social behavior, via orchestrating executive functions that allow a person to differentiate conflicting opposites. Interestingly, while it was long assumed that neurons do not grow following birth, recent imaging studies have in fact revealed that brain development takes place into adulthood and that adolescent risk-seeking and exploratory creative behaviors go hand-in-hand with active pruning of synaptic connections and differential maturation of major brain regions.[129] Reciprocal adaptive interactions between genes and social environments take place until adulthood and actively sculpt neuronal connections via neuroplasticity. This highlights the dependence of adolescent youth on their peer networks, as well as social and educational environments, as developmental niches. Importantly, there is ample evidence that adverse adolescent environments define later risk for mental and other medical illnesses,[130] including inflammatory processes and addictions such as tobacco smoking and alcohol use, *the prime risk factors of most cancers.* Here we see the sociocultural conditions surrounding adolescent education and working environments *acting as immaterial processes* that may affect cancer state niches later in life.

(4) National niches: Widespread prevention through reconstructing the niches that increase cancer risks can in principle be achieved by governance actions and policy changes. However, the state of prevention efforts against lung cancer reveals a glaring shortfall: around 90 percent of lung cancers, the leading cause of cancer mortality with 1.7 million deaths in 2018,[131] are caused by tobacco exposure. Yet lung cancer persists as a global health issue, even though its causal link to tobacco was established in the 1950s.[132] The epidemiology caused by this link goes back to the historical entanglement of plantation farming and the tobacco industry with state profits, particularly in the United States, and protobacco lobbying continues to this day.[133] To balance revenues, reductions in tobacco use following regulatory programs and public smoking bans in Western countries were offset by increased exports to vulnerable regions of the world.[134] With new risk factors such as obesity and air pollution emerging (air pollution now accounts for 10 percent of lung cancers in the United Kingdom), the increased cancer trends in youths and in resource-limited countries are worrying.[135] At the same time, it is estimated that increased primary prevention, such as by curbing carcinogen exposures, promoting lifestyle changes, or expanding vaccination programs, could reduce cancer deaths by 30 to 50 percent.[136] Yet funding for prevention amounts to a fraction (<10 percent) of overall cancer funding, in part due to the lower financial returns of prevention support compared with new therapeutics.[137] Such conflicts of interest between private-sector gains and public health outcomes are further exacerbated by pervasive corruption[138] and obstruct effective tackling of the cancer problem at its root. Thankfully, awareness of these systemic issues is mounting, with solutions being proposed, such as the uncoupling of research interests from "medical marketing"[139] or the design of incentives that create financial gains when insured people are healthy.[140] While working toward such fundamental changes, researchers can continue to build public perception of these new routes to prevention through communication, education, and collaborative practice in reconstructing sociocultural and health system niches.

(5) Global niches: According to the 2019 Intergovernmental Panel on Climate Change special report, humans have changed more than 70 percent of the Earth's surface through agriculture, deforestation, and resource extraction, among other means.[141] Nature's capacity to buffer these damages and sustain the homeostasis of global ecosystems is deteriorating, increasing the rate of biodiversity loss and risk of extinction.[142] While the epidemiology of ecological breakdown and cancer incidence remains to be analyzed, the evolutionary parallels between cancer and the biodiversity crisis, both "hostile takeovers" of self-regulating complex systems, imply interdependent causes.[143] Systemic toxicities in environments are causes of both ecological and tissue resilience breakdowns, and both cancer

and biodiversity loss involve the collapse of homeostatic niches. This fits with an ITSNTS view, where impacts on biogeochemical food and nutrient cycles may eventually impact the persistence of cancer phenotypes. Niche reconstruction is hence a necessary focus, both to reverse biodiversity loss and climate change in ecosystems, as well as to canalize cancer progression in organisms.

The latest Lancet Commission on obesity described the three pandemics of obesity, malnutrition, and climate change as combining into a synergistic epidemic, or "Global Syndemic," with emergent negative effects that may constitute new diseases in themselves.[144] Given the causal links between obesity and cancer,[145] as well as our current understanding of processual interdependencies across scale, the cancer pandemic is likely integral to the Syndemic dynamic. It is remarkable how human inflicted these toxicities are: obesity and malnutrition are rooted in agricultural practices and food industries that value corporate gains over public health,[146] and the tobacco and artificially sweetened drink industries have overlap in ownership and use similar marketing tactics.[147] This culminates in mounting obesity and cancer incidences, adding to imminent environmental threats in resource-limited regions, where "85% of people in extreme poverty live in the 20 countries most vulnerable to climate change."[148] These regions require resources and governance strategies adaptive to the imminent public health and ecological crises.[149] This indicates a need for preventative, resilience-fostering, and salutogenic approaches, which reconstruct and revert the health-deteriorating environments that ominously link cancer with other global pandemics.

Summary and brief perspectives on niche reconstruction: A period of rapid scientific, medical, and technological progress has yielded many insights in the molecular underpinnings of the neoplastic process, including how the evolutionary trajectories of cancer states are dictated by processes that operate in interrelated environments. While there is yet much to learn from seeking to treat tumors, we have to simultaneously face that these goals are not sufficiently effective in lessening the burdens of increasing cancer incidence. Addressing the maladaptive environments that permit cancer states to persist means we are entering a critical time in which the parts and systems views of cancer are to be united and applied to the reconstruction of local-to-global niches (figure 15.2). The era of a technological race for a cure was necessary to reveal errors in some of the approaches taken to solving the cancer puzzle, and now is the crucial moment to correct these errors.

Besides the escalating global cancer pandemic itself,[150] other processes that have recently been designated "crises" or "cancers" demanding "transformative" or "revolutionary" solutions include the causal-mechanistic medical model of disease,[151] health systems particularly in resource-limited regions,[152] corruption in governing bodies,[153] the Global Syndemic,[154] existential climate change threats,[155] and the scientific enterprise itself.[156] Waddington might remind us in this context that "it is the animal's behavior which to a considerable extent determines the nature of the environment to which it will submit itself and the character of the selective forces with which it will consent to wrestle. The various types of 'feedback' or

Figure 15.2
Niche reconstruction to revert or transcend the cancer state. Through a combination of reductionist pursuit and systems thinking, humans have eaten the fruit of knowledge and come to understand that we ourselves construct the maladaptive cancer niches in which the cancer state persists. Newly gained insights are now to be applied to the regeneration of environments that may emergently sustain homeostasis in cancer and global health. Niche reconstruction efforts are required in niches ranging from natural and artificial habitats, to tissue environments that can influence cancer progression at microscales. This requires a cooperative approach in which we focus on cancer state transcendence in both individuals and populations.
Center piece: *Paradise*—landscape showing various episodes from the story of Adam and Eve in paradise, symbolizing the creation and the fall of man; c. 1541–1550, by Herri met de Bles. Rijksmuseum Amsterdam collection, Amsterdam, The Netherlands.

circularity in the relation between an animal and its environment are rather generally neglected in present-day evolutionary theorizing."[157] It is easy to see how a neglect of behavioral feedback mechanisms in the EES niche construction paradigm underpins coexisting present-day crises. Tackling these requires a fundamental shift in the perception of how our choices regarding our social, economic, and ecological environments channel disease evolution during a person's lifetime. The systems view has long implied a "Declaration of Interdependence"— that "we can all work together in *a symphonic world*."[158] Human behavior itself appears to constitute another missing piece of the cancer puzzle.

Ethologists have long emphasized how, even though both cooperation and competition are innate animal behaviors, human aggression is particularly threatening. Nobel laureate Nikolaas Tinbergen wrote, "The study of the social behavior of animals may well help us serve—and possibly save—ourselves, and thus it may ultimately emerge as the most important science of all."[159] Today, the study of the social behavior of scientists, facing systemic crises, may help "deflect and sublimate our aggression" and provide goal-directedness in a communal cancer mission. To properly question whether we can prevent and treat cancer by reverting or transcending cancerous niches is also to ask if we need to improve the environments in which these questions are being asked. Repairing the culture and behavioral feedback mechanisms of the scientific ecosystem itself may nucleate role models for other domains. This can start by incentivizing cooperation over competition, rewarding communal over individual success, flattening organizational hierarchies, and uncoupling scientific output from perverse financial incentives. In this way, education might be better able to nurture the problem-solving behaviors of a budding generation of adolescent scientists and canalize efforts toward emergent discovery and global recovery of public and planetary health. Rather than a cancer moonshot or war, perhaps we need a second Renaissance[160] in which the organic health of the biomedical research community is a prerequisite for being able to help societies return to organic health.

15.9 Redesigning Experimental Cancer Research and Therapy

Research is needed to test the evolutionary framework proposed here in which convergent cancer phenotypes enact "units of selection" that progressively uncouple the neoplastic process from organismic fitness, addressing the nature of these units, the processes by which they regulate the cancer state, and their potential negation by rational reconstruction of tissue niches. This requires the development of assays and tools that can perceive the neoplastic "song," ideally long before the cacophony stage. While genomic and phenotypic profiles can serve as signatures, their study in isolation will unlikely suffice; rather, processual relationships and dependencies on dynamic niche-selective environments deserve focus. A central question then becomes, what collective processes cause cancer states to persist? As elaborated here and in chapter 9 (this volume), the answer is likely to be found in the reversion from multicellular cooperation to unicellular adaptive behaviors, in which tumor cells adopt premetazoan features. This connects to emerging evidence that not only bacteria but also fungi[161] can alter cancer progression via immune regulation. Lines of inquiry can address how collective processes relate to nutrient and metabolite cycles and the resetting of "stigmergic" fluxes by microbes and viruses in particular or to immunogenic antigen presentation in select tissue environments. Research might further explore the way in which such cycles are altered by ecological changes to ultimately link cancer evolution to biogeochemical processes affected by climate change.[162] An intriguing question is whether a common "detector" of the single-cell attractor exists, a state-change sensor that could be developed as a vaccine. A clue that this may be a viable direction is the recent discovery of a T-cell receptor that recognizes a metabolite ligand presented on a pan-cancer MHC-I-related protein.[163] These considerations thus shift emphasis to interrogating how precancerous niches connect to immaterial processes and the idea that living entities constitute "holobionts."

Suitable model systems are needed for these research directions. While representative culture models, such as "organ-on-chip" microfluidics,[164] can reveal fundamental principles of how, for example, alterations in fluxes of carbon, oxygen, nitrogen, and phosphorus can influence cancer evolution, it is the totality of the body's physiology that will crucially affect and define such fundamentals. On the other hand, in vivo models, such as rodents, permit conceptual elucidation of the relationships between niche alteration and neoplastic progression or systemic immune regulation, yet have limitations in that the tumor burden, artificial environment, history of cancer evolution, and precise immune functions do not mimic human cancers. Hence, it is likely that we will increasingly research human phenomes and people.[165] This underscores the importance of both guidelines for ethical oversight, currently not well defined for human embryo research,[166] for example, and a new understanding of how research is interdependently embedded in society. Such an understanding, mobilized into action as real-world experiments via randomized controlled trials in the field of development economics, was recently awarded the Nobel Prize in economics[167]: through deploying simple experiments among populations, researchers gained tangible results on what constitute effective measures to alleviate poverty, implement preventive health care, and improve health globally. Similarly, "developmental cancer research" could design salutogenic lifestyle solutions for people with high cancer risks and improve quality of life for cancer patients in ways embedded in divergent populations. This would finally mean heeding Tinbergen's warning from the discipline of ethology and leveraging the diversity of human behavior against cancer, instead of continuing to be confounded by it.

Analogous to how people can be locked into poverty traps, cancer attractors functionally trap cells following loss of a series of resilience mechanisms. A fundamental question for clinical therapy is whether cells can be navigated out of the attractor valley or, better, if timely interventions can prohibit cells from becoming trapped in the first place. Currently, precision medicine approaches increasingly use tumor-selective biomarkers and "-omics" profiles to match therapies to a patient's unique tumor. In addition, functional studies apply ex vivo drug profiling to match a patient's tumor phenotype to the most promising single or combination treatments.[168] Rather than aiming to restore tissue resilience, these personalized directions typically aim to kill tumor cells, with promising extensions exploiting the unique adaptive phenotypes of tumor cells within their ecological contexts, as so-called adaptive therapies.[169] Adaptive strategies shift focus to the evolutionary dynamics of treatments by suppressing the expansion of therapy-resistant cell clones and can indeed prolong time to progression, for example, in metastatic prostate cancer.[170] A subject for clinical exploration remains how natural health factors, such as improved diets and sleep patterns, alleviation of stress and addictive behaviors, exercise, micro- and mycobiome adjustments, and salutogenic "sense of coherence" solutions, can be combined with existing medical procedures, including promising immunotherapy directions, to navigate cancer states out of attractor valleys toward possible cures.

15.10   Conclusion

The field of biology is experiencing a revival of the epigenetic theory of Conrad Waddington, as well as research inspired by him, culminating in the reintegration of the systems and biological views of cancer. Following a century of reductionist inquiry, we have come to perceive cancers as part of an open, complex system of functional and processual relations. This perception is rooted in the view that organisms develop via simple rules of interaction, in harmony with the external environment, often using paired positive and negative feedback. We have learned that the process of cancer progression in some ways resembles regeneration following persistent tissue damage and wounding, is characterized by a progressive loss of bodily resilience to the cancer state, and behaves as an adaptive response to dynamically changing environmental stresses. Analogous to the view that functional processes or "songs" evolve in harmony with a functional web of interrelated niches, cancerous phenotypes may be the reemergence of ancestral songs, regressing cellular states into opportunistic single cell-like modes. To date, both clinical medicine and society at large attempt to treat cancer at the level of molecular pathogenesis. Our growing understanding of niche construction indicates that we must now also address the immaterial, cultural, and behavioral vectors directing cancer evolution, targeting the niches in which cancer arises. We only now have the empirical tools to do so. We also only now face the threat of synergistic epidemics of multiple globally arising diseases, which we may well expect to present new and worse progressions of cancer niches. Niche reconstruction merits greater focus in our approaches to the cancer pandemic and other micro- and macroenvironmental crises of overwhelming complexity. The need for research, peer mentoring, and education to develop approaches for preventing and reverting negative synergies, from cancers to syndemics, could not be more urgent.

Acknowledgments

Heartfelt gratitude to the organizers of the KLI workshop, Mina Bissell, Ingemar Ernberg, and Bernhard Strauss, for the opportunity to contribute to this body of science, and to Bernhard for essential edits. Thanks to all workshop participants, and Johannes Jäger and Gerd Müller for discussions; Thea Newman for continued inspiration; past and current FIMM directors, Olli Kallioniemi, Jaakko Kaprio, and Mark Daly, for demonstrating how an emergent scientific culture can thrive; John Hickman for challenging paradigms; Myles Byrne for idea canalization; and mentors and peers for guidance. Warm thanks to all past and current Verschuren team members, finding seeds of understanding amid complexity. Thanks also to funding agencies, as well as the scientists who supported them, for endorsing our research activities in recent years, particularly the Academy of Finland, IMI-JU, the Doctoral School of Health, and the Cancer Foundation Finland. The author apologizes for selectively referencing published works and gladly receives supplementing input.

Notes

1. Wilson, B. E., S. Jacob, M. L. Yap, J. Ferlay, F. Bray, and M. B. Barton. 2019. "Estimates of global chemotherapy demands and corresponding physician workforce requirements for 2018 and 2040: a population-based study." *Lancet Oncol* 20 (6):769–80. doi: 10.1016/S1470–2045(19)30163–9. Bray, F., J. Ferlay, I. Soerjomataram, R. L. Siegel, L. A. Torre, and A. Jemal. 2018. "Global cancer statistics 2018: GLOBOCAN estimates of incidence and mortality worldwide for 36 cancers in 185 countries." *CA Cancer J Clin* 68 (6):394–424. doi: 10.3322/caac.21492. The Lancet. 2018. "GLOBOCAN 2018: counting the toll of cancer." *Lancet* 392 (10152):985. doi: 10.1016/S0140–6736(18)32252–9.

2. Tannock, I. F., and J. A. Hickman. 2016. "Limits to personalized cancer medicine." *N Engl J Med* 375 (13):1289–94. doi: 10.1056/NEJMsb1607705. Prasad, V. 2016. "Perspective: The precision-oncology illusion." *Nature* 537 (7619):S63. doi: 10.1038/537S63a. Prasad, V. 2017. "Do cancer drugs improve survival or quality of life?" *BMJ* 359:j4528. doi: 10.1136/bmj.j4528.

3. Weinberg, R. A. 2014. "Coming full circle-from endless complexity to simplicity and back again." *Cell* 157 (1):267–71. doi: 10.1016/j.cell.2014.03.004.

4. Laland, K. N., T. Uller, M. W. Feldman, K. Sterelny, G. B. Muller, A. Moczek, E. Jablonka, and J. Odling-Smee. 2015. "The extended evolutionary synthesis: its structure, assumptions and predictions." *Proc Biol Sci* 282 (1813):20151019. doi: 10.1098/rspb.2015.1019. Pigliucci, M. and G. B. Müller. 2010. *Evolution: The Extended Synthesis* (MIT Press, ISBN 9780262513678).

5. Capra, F. 1982. *The Turning Point* (Bantam Books, ISBN 0553345729).

6. Turing, A. M. 1952. "The chemical basis of morphogenesis." *Philos Trans R Soc Lond* B 237, 37–72.

7. Kutejova, E., J. Briscoe, and A. Kicheva. 2009. "Temporal dynamics of patterning by morphogen gradients." *Curr Opin Genet Dev* 19 (4):315–22. doi: 10.1016/j.gde.2009.05.004. Meinhardt, H. 2015. "Models for patterning primary embryonic body axes: the role of space and time." *Semin Cell Dev Biol* 42:103–17. doi: 10.1016/j.semcdb.2015.06.005. Rogers, K. W., and A. F. Schier. 2011. "Morphogen gradients: from generation to interpretation." *Annu Rev Cell Dev Biol* 27:377–407. doi: 10.1146/annurev-cellbio-092910–154148. Gilbert, S. F. 1985. Developmental Biology (Sinauer Associates, ISBN 0878932461).

8. Briscoe, J., and P. P. Therond. 2013. "The mechanisms of Hedgehog signalling and its roles in development and disease." *Nat Rev Mol Cell Biol* 14 (7):416–29. doi: 10.1038/nrm3598.

9. Iber, D., and D. Menshykau. 2013. "The control of branching morphogenesis." *Open Biol* 3 (9):130088. doi: 10.1098/rsob.130088. Hannezo, E., and B. D. Simons. 2019. "Multiscale dynamics of branching morphogenesis." *Curr Opin Cell Biol* 60:99–105. doi: 10.1016/j.ceb.2019.04.008.

10. Sheth, R., L. Marcon, M. F. Bastida, M. Junco, L. Quintana, R. Dahn, M. Kmita, J. Sharpe, and M. A. Ros. 2012. "Hox genes regulate digit patterning by controlling the wavelength of a Turing-type mechanism." *Science* 338 (6113):1476–80. doi: 10.1126/science.1226804. Raspopovic, J., L. Marcon, L. Russo, and J. Sharpe. 2014. "Modeling digits. Digit patterning is controlled by a Bmp-Sox9-Wnt Turing network modulated by morphogen gradients." *Science* 345 (6196):566–70. doi: 10.1126/science.1252960.

11. Sick, S., S. Reinker, J. Timmer, and T. Schlake. 2006. "WNT and DKK determine hair follicle spacing through a reaction-diffusion mechanism." *Science* 314 (5804):1447–50. doi: 10.1126/science.1130088.

12. Hiscock, T. W., and S. G. Megason. 2015. "Orientation of Turing-like patterns by morphogen gradients and tissue anisotropies." *Cell Syst* 1 (6):408–16. doi: 10.1016/j.cels.2015.12.001.

13. Brinkmann, F., M. Mercker, T. Richter, and A. Marciniak-Czochra. 2018. "Post-Turing tissue pattern formation: advent of mechanochemistry." *PLoS Comput Biol* 14 (7):e1006259. doi: 10.1371/journal.pcbi.1006259.

14. Meinhardt, H. 2015. "Models for patterning primary embryonic body axes: the role of space and time." *Semin Cell Dev Biol* 42:103–17. doi: 10.1016/j.semcdb.2015.06.005. Dekanty, A., and M. Milan. 2011. "The interplay between morphogens and tissue growth." *EMBO Rep* 12 (10):1003–10. doi: 10.1038/embor.2011.172.

15. Brinkmann, F., M. Mercker, T. Richter, and A. Marciniak-Czochra. 2018. "Post-Turing tissue pattern formation: advent of mechanochemistry." *PLoS Comput Biol* 14 (7):e1006259. doi: 10.1371/journal.pcbi.1006259. Tabata, T., and Y. Takei. 2004. "Morphogens, their identification and regulation." *Development* 131 (4):703–12. doi: 10.1242/dev.01043. Recho, P., A. Hallou, and E. Hannezo. 2019. "Theory of mechanochemical patterning in biphasic biological tissues." *Proc Natl Acad Sci USA* 116 (12):5344–49. doi: 10.1073/pnas.1813255116.

16. Shin, W., C. D. Hinojosa, D. E. Ingber, and H. J. Kim. 2019. "Human intestinal morphogenesis controlled by transepithelial morphogen gradient and flow-dependent physical cues in a microengineered gut-on-a-chip." *iScience* 15:391–406. doi: 10.1016/j.isci.2019.04.037.

17. Sato, T., and H. Clevers. 2013. "Growing self-organizing mini-guts from a single intestinal stem cell: mechanism and applications." *Science* 340 (6137):1190–94. doi: 10.1126/science.1234852.

18. Barcellos-Hoff, M. H., J. Aggeler, T. G. Ram, and M. J. Bissell. 1989. "Functional differentiation and alveolar morphogenesis of primary mammary cultures on reconstituted basement membrane." *Development* 105 (2):223–35. Bissell, M. J., and H. G. Hall. 1987. "Form and function in the mammary gland: the role of extracellular matrix." In: *The Mammary Gland: Development, Regulation and Function*, edited by M. Neville and C. Daniel, pp. 97–146 (Springer US, 1489950451).

19. Turing, A. M. 1952. "The chemical basis of morphogenesis." *Philos Trans R Soc Lond B* 237:37–72.

20. Dvorak, H. F. 2015. "Tumors: wounds that do not heal-redux." *Cancer Immunol Res* 3 (1):1–11. doi: 10.1158/2326–6066.CIR-14–0209. Ribatti, D., and R. Tamma. 2018. "A revisited concept. Tumors: wounds that do not heal." *Crit Rev Oncol Hematol* 128:65–69. doi: 10.1016/j.critrevonc.2018.05.016. Balkwill, F., and A. Mantovani. 2001. "Inflammation and cancer: back to Virchow?" *Lancet* 357 (9255):539–45. doi: 10.1016/S0140–6736(00)04046–0.

21. Egeblad, M., E. S. Nakasone, and Z. Werb. 2010. "Tumors as organs: complex tissues that interface with the entire organism." *Dev Cell* 18 (6):884–901. doi: 10.1016/j.devcel.2010.05.012. Bissell, M. J., and D. Radisky. 2001. "Putting tumours in context." *Nat Rev Cancer* 1 (1):46–54. doi: 10.1038/35094059.

22. Motwani, M., S. Pesiridis, and K. A. Fitzgerald. 2019. "DNA sensing by the cGAS-STING pathway in health and disease." *Nat Rev Genet* 20 (11):657–74. doi: 10.1038/s41576-019-0151-1.

23. Balkwill, F. R., and A. Mantovani. 2012. "Cancer-related inflammation: common themes and therapeutic opportunities." *Semin Cancer Biol* 22 (1):33–40. doi: 10.1016/j.semcancer.2011.12.005. Elinav, E., R. Nowarski, C. A. Thaiss, B. Hu, C. Jin, and R. A. Flavell. 2013. "Inflammation-induced cancer: crosstalk between tumours, immune cells and microorganisms." *Nat Rev Cancer* 13 (11):759–71. doi: 10.1038/nrc3611.

24. Maman, S., and I. P. Witz. 2018. "A history of exploring cancer in context." *Nat Rev Cancer* 18 (6):359–76. doi: 10.1038/s41568-018-0006-7. Quail, D. F., and J. A. Joyce. 2013. "Microenvironmental regulation of tumor progression and metastasis." *Nat Med* 19 (11):1423–37. doi: 10.1038/nm.3394.

25. Galon, J., H. K. Angell, D. Bedognetti, and F. M. Marincola. 2013. "The continuum of cancer immunosurveillance: prognostic, predictive, and mechanistic signatures." *Immunity* 39 (1):11–26. doi: 10.1016/j.immuni.2013.07.008. Tsujikawa, T., S. Kumar, R. N. Borkar, V. Azimi, G. Thibault, Y. H. Chang, A. Balter, R. Kawashima, G. Choe, D. Sauer, E. El Rassi, D. R. Clayburgh, M. F. Kulesz-Martin, E. R. Lutz, L. Zheng, E. M. Jaffee, P. Leyshock, A. A. Margolin, M. Mori, J. W. Gray, P. W. Flint, and L. M. Coussens. 2017. "Quantita-

tive multiplex immunohistochemistry reveals myeloid-inflamed tumor-immune complexity associated with poor prognosis." *Cell Rep* 19 (1):203–17. doi: 10.1016/j.celrep.2017.03.037. Gentles, A. J., A. M. Newman, C. L. Liu, S. V. Bratman, W. Feng, D. Kim, V. S. Nair, Y. Xu, A. Khuong, C. D. Hoang, M. Diehn, R. B. West, S. K. Plevritis, and A. A. Alizadeh. 2015. "The prognostic landscape of genes and infiltrating immune cells across human cancers." *Nat Med* 21 (8):938–45. doi: 10.1038/nm.3909.

26. Vogelstein, B., and K. W. Kinzler. 2004. "Cancer genes and the pathways they control." *Nat Med* 10 (8):789–99. doi: 10.1038/nm1087. Weinberg, R. A. 2006. *The Biology of Cancer* (Garland Science, ISBN 0815342205).

27. Alexandrov, L. B., S. Nik-Zainal, D. C. Wedge, S. A. Aparicio, S. Behjati, A. V. Biankin, G. R. Bignell, N. Bolli, A. Borg, A. L. Borresen-Dale, S. Boyault, B. Burkhardt, A. P. Butler, C. Caldas, H. R. Davies, C. Desmedt, R. Eils, J. E. Eyfjord, J. A. Foekens, M. Greaves, F. Hosoda, B. Hutter, T. Ilicic, S. Imbeaud, M. Imielinski, N. Jager, D. T. Jones, D. Jones, S. Knappskog, M. Kool, S. R. Lakhani, C. Lopez-Otin, S. Martin, N. C. Munshi, H. Nakamura, P. A. Northcott, M. Pajic, E. Papaemmanuil, A. Paradiso, J. V. Pearson, X. S. Puente, K. Raine, M. Ramakrishna, A. L. Richardson, J. Richter, P. Rosenstiel, M. Schlesner, T. N. Schumacher, P. N. Span, J. W. Teague, Y. Totoki, A. N. Tutt, R. Valdes-Mas, M. M. van Buuren, L. van 't Veer, A. Vincent-Salomon, N. Waddell, L. R. Yates, Initiative Australian Pancreatic Cancer Genome, Icgc Breast Cancer Consortium, Icgc Mmml-Seq Consortium, Icgc PedBrain, J. Zucman-Rossi, P. A. Futreal, U. McDermott, P. Lichter, M. Meyerson, S. M. Grimmond, R. Siebert, E. Campo, T. Shibata, S. M. Pfister, P. J. Campbell, and M. R. Stratton. 2013. "Signatures of mutational processes in human cancer." *Nature* 500 (7463):415–21. doi: 10.1038/nature12477.

28. Kucab, J. E., X. Zou, S. Morganella, M. Joel, A. S. Nanda, E. Nagy, C. Gomez, A. Degasperi, R. Harris, S. P. Jackson, V. M. Arlt, D. H. Phillips, and S. Nik-Zainal. 2019. "A compendium of mutational signatures of environmental agents." *Cell* 177 (4):821–36 e16. doi: 10.1016/j.cell.2019.03.001.

29. Sutherland, K. D., and J. E. Visvader. 2015. "Cellular mechanisms underlying intertumoral heterogeneity." *Trends Cancer* 1 (1):15–23. doi: 10.1016/j.trecan.2015.07.003. Visvader, J. E. 2011. "Cells of origin in cancer." *Nature* 469 (7330):314–22. doi: 10.1038/nature09781. Blanpain, C. 2013. "Tracing the cellular origin of cancer." *Nat Cell Biol* 15 (2):126–34. doi: 10.1038/ncb2657.

30. Nagaraj, A. S., J. Lahtela, A. Hemmes, T. Pellinen, S. Blom, J. R. Devlin, K. Salmenkivi, O. Kallioniemi, M. I. Mayranpaa, K. Narhi, and E. W. Verschuren. 2017. "Cell of origin links histotype spectrum to immune microenvironment diversity in non-small-cell lung cancer driven by mutant Kras and loss of Lkb1." *Cell Rep* 18 (3):673–84. doi: 10.1016/j.celrep.2016.12.059.

31. Narhi, K., A. S. Nagaraj, E. Parri, R. Turkki, P. W. van Duijn, A. Hemmes, J. Lahtela, V. Uotinen, M. I. Mayranpaa, K. Salmenkivi, J. Rasanen, N. Linder, J. Trapman, A. Rannikko, O. Kallioniemi, T. M. Af Hallstrom, J. Lundin, W. Sommergruber, S. Anders, and E. W. Verschuren. 2018. "Spatial aspects of oncogenic signalling determine the response to combination therapy in slice explants from Kras-driven lung tumours." *J Pathol* 245 (1):101–13. doi: 10.1002/path.5059. Bao, J., M. Walliander, F. Kovacs, A. S. Nagaraj, A. Hemmes, V. K. Sarhadi, S. Knuutila, J. Lundin, P. Horvath, and E. W. Verschuren. 2019. "Spa-RQ: an image analysis tool to visualise and quantify spatial phenotypes applied to non-small cell lung cancer." *Sci Rep* 9 (1):17613. doi: 10.1038/s41598-019-54038-9.

32. Narhi, K., A. S. Nagaraj, E. Parri, R. Turkki, P. W. van Duijn, A. Hemmes, J. Lahtela, V. Uotinen, M. I. Mayranpaa, K. Salmenkivi, J. Rasanen, N. Linder, J. Trapman, A. Rannikko, O. Kallioniemi, T. M. Af Hallstrom, J. Lundin, W. Sommergruber, S. Anders, and E. W. Verschuren. 2018. "Spatial aspects of oncogenic signalling determine the response to combination therapy in slice explants from Kras-driven lung tumours." *J Pathol* 245 (1):101–13. doi: 10.1002/path.5059. Talwelkar, S. S., A. S. Nagaraj, J. R. Devlin, A. Hemmes, S. Potdar, E. A. Kiss, P. Saharinen, K. Salmenkivi, M. I. Mayranpaa, K. Wennerberg, and E. W. Verschuren. 2019. "Receptor tyrosine kinase signaling networks define sensitivity to ERBB inhibition and stratify Kras-mutant lung cancers." *Mol Cancer Ther* 18 (10):1863–74. doi: 10.1158/1535-7163.MCT-18-0573.

33. Kortlever, R. M., N. M. Sodir, C. H. Wilson, D. L. Burkhart, L. Pellegrinet, L. Brown Swigart, T. D. Littlewood, and G. I. Evan. 2017. "Myc cooperates with Ras by programming inflammation and immune suppression." *Cell* 171 (6):1301–15 e14. doi: 10.1016/j.cell.2017.11.013. Sodir, N. M., R. M. Kortlever, V. J. A. Barthet, T. Campos, L. Pellegrinet, S. Kupczak, P. Anastasiou, L. Brown Swigart, L. Soucek, M. J. Arends, T. D. Littlewood, and G. I. Evan. 2020. "Myc instructs and maintains pancreatic adenocarcinoma phenotype." *Cancer Discov* 10 (4):588–607. doi: 10.1158/2159-8290.CD-19-0435.

34. Hoadley, K. A., C. Yau, T. Hinoue, D. M. Wolf, A. J. Lazar, E. Drill, R. Shen, A. M. Taylor, A. D. Cherniack, V. Thorsson, R. Akbani, R. Bowlby, C. K. Wong, M. Wiznerowicz, F. Sanchez-Vega, A. G. Robertson, B. G. Schneider, M. S. Lawrence, H. Noushmehr, T. M. Malta, Network Cancer Genome Atlas, J. M. Stuart, C. C.

Benz, and P. W. Laird. 2018. "Cell-of-origin patterns dominate the molecular classification of 10,000 tumors from 33 types of cancer." *Cell* 173 (2):291–304 e6. doi: 10.1016/j.cell.2018.03.022.

35. Evan, G. I., N. Hah, T. D. Littlewood, N. M. Sodir, T. Campos, M. Downes, and R. M. Evans. 2017. "Re-engineering the pancreas tumor microenvironment: a 'regenerative program' hacked." *Clin Cancer Res* 23 (7):1647–55. doi: 10.1158/1078–0432.CCR-16–3275.

36. Tammela, T., F. J. Sanchez-Rivera, N. M. Cetinbas, K. Wu, N. S. Joshi, K. Helenius, Y. Park, R. Azimi, N. R. Kerper, R. A. Wesselhoeft, X. Gu, L. Schmidt, M. Cornwall-Brady, O. H. Yilmaz, W. Xue, P. Katajisto, A. Bhutkar, and T. Jacks. 2017. "A Wnt-producing niche drives proliferative potential and progression in lung adenocarcinoma." *Nature* 545 (7654):355–59. doi: 10.1038/nature22334.

37. Emergence refers to the development of collective features in a manner unpredictable from the individual parts. From the Latin "bringing to light," as if rising from a liquid by virtue of buoyancy.

38. Rashevsky, N. 1954. "Topology and life: in search of general mathematical principles in biology and sociology." *Bull Math Biophys* 16:317e384.

39. Rosen, R. 1958. "A relational theory of biological systems." *Bull Math Biophys* 20:245e260.

40. Noble, D. 2006. *The Music of Life: Biology beyond Genes* (Oxford University Press, ISBN 0199228362).

41. Noble, D. 2013. "A biological relativity view of the relationships between genomes and phenotypes." *Prog Biophys Mol Biol* 111 (2–3):59–65. doi: 10.1016/j.pbiomolbio.2012.09.004. Noble, D. 2012. "A theory of biological relativity: no privileged level of causation." *Interface Focus* 2 (1):55–64. doi: 10.1098/rsfs.2011.0067.

42. Soto, A. M., and C. Sonnenschein. 2011. "The tissue organization field theory of cancer: a testable replacement for the somatic mutation theory." *Bioessays* 33 (5):332–40. doi: 10.1002/bies.201100025.

43. Bertolaso, M., and J. Dupré. 2018. "A processual perspective on cancer." In: *Everything Flows: Towards a Processual Philosophy of Biology*, edited by D. J. Nicholson, and J. Dupré (Oxford Scholarship Online, ISBN 9780198779636). In addition, in work by Stanley Salthe, systems theory has been applied to describe the organization of complex systems as part of a biological hierarchy that explains how the Earth's biotic systems can operate away from a thermodynamic equilibrium: Salthe, S. N. 1985. *Evolving Hierarchical Systems: Their Structure and Representation* (Columbia University Press, ISBN 0231060173).

44. Waddington, C. H. 1957. *The Strategy of the Genes* (George Allen & Unwin Ltd, ISBN 1317657551).

45. Waddington, C. H. 1959. "[Evolutionary systems; animal and human]." *Nature* 183 (4676):1634–38. doi: 10.1038/1831634a0. Noble, D. 2015. "Conrad Waddington and the origin of epigenetics." *J Exp Biol* 218 (Pt 6):816–18. doi: 10.1242/jeb.120071. Hahlweg, K. 1981. "Progress through evolution? An inquiry into the thought of C. H. Waddington." *Acta Biother* 30:103–20. The buffering of a genotype here indicates that it can absorb a certain level of variation without affecting phenotypic development. In Waddington's words: "The appearance of the phenotype thus does not exhibit a genuine 'mapping' between genetic and phenotypic diversity and we find that 'identical phenotypes may have different genotypes, and identical genotypes may give rise to different phenotypes." From Waddington, C. H. 1975. *The Evolution of an Evolutionist* (Edinburgh Univ. Press, ISBN 0852242727).

46. Kauffman, S. 1971. "Differentiation of malignant to benign cells." *J Theor Biol* 31 (3):429–51. doi: 10.1016/0022–5193(71)90020–8. Huang, S., G. Eichler, Y. Bar-Yam, and D. E. Ingber. 2005. "Cell fates as high-dimensional attractor states of a complex gene regulatory network." *Phys Rev Lett* 94 (12):128701. doi: 10.1103/PhysRevLett.94.128701. Jaeger, J., and N. Monk. 2014. "Bioattractors: dynamical systems theory and the evolution of regulatory processes." *J Physiol* 592 (11):2267–81. doi: 10.1113/jphysiol.2014.272385.

47. Huang, S., and D. E. Ingber. 2006. "A non-genetic basis for cancer progression and metastasis: self-organizing attractors in cell regulatory networks." *Breast Dis* 26:27–54. doi: 10.3233/bd-2007–26104. Huang, S., I. Ernberg, and S. Kauffman. 2009. "Cancer attractors: a systems view of tumors from a gene network dynamics and developmental perspective." *Semin Cell Dev Biol* 20 (7):869–76. doi: 10.1016/j.semcdb.2009.07.003.

48. Huang, S., and S. Kauffman. 2013. "How to escape the cancer attractor: rationale and limitations of multi-target drugs." *Semin Cancer Biol* 23 (4):270–78. doi: 10.1016/j.semcancer.2013.06.003.

49. Choi, M., J. Shi, S. H. Jung, X. Chen, and K. H. Cho. 2012. "Attractor landscape analysis reveals feedback loops in the p53 network that control the cellular response to DNA damage." *Sci Signal* 5 (251):ra83. doi: 10.1126/scisignal.2003363.

50. Kim, Y., S. Choi, D. Shin, and K. H. Cho. 2017. "Quantitative evaluation and reversion analysis of the attractor landscapes of an intracellular regulatory network for colorectal cancer." *BMC Syst Biol* 11 (1):45. doi: 10.1186/s12918-017-0424-2.

51. Huang, S., and S. Kauffman. 2013. "How to escape the cancer attractor: rationale and limitations of multi-target drugs." *Semin Cancer Biol* 23 (4):270–78. doi: 10.1016/j.semcancer.2013.06.003.

52. Bissell, M. J., and W. C. Hines. 2011. "Why don't we get more cancer? A proposed role of the microenvironment in restraining cancer progression." *Nat Med* 17 (3):320–29. doi: 10.1038/nm.2328.

53. The loss of tissue homeostasis as the principle of disease was already referred to by Claude Bernard and his contemporary Rudolph Virchow during the middle of the nineteenth century; these physiologists posited that the maintenance of an internal environment ("le milieu intérieur" in Bernard's theory and cellular "economy of the body" in Virchow's theory), in which parts are in harmony with each other, is the condition of a healthy state.

54. Boveri, T. *The Origins of Malignant Tumors* (Transl. Boveri, M., 1929. Introduction by Metcalf, M. M. The Williams and Wilkins Company, 1914).

55. Waddington, C. H. 1935. "Cancer and the theory of organizers." *Nature* 135:606–8.

56. Quail, D. F., and J. A. Joyce. 2013. "Microenvironmental regulation of tumor progression and metastasis." *Nat Med* 19 (11):1423–37. doi: 10.1038/nm.3394. Bissell, M. J., and W. C. Hines. 2011. "Why don't we get more cancer? A proposed role of the microenvironment in restraining cancer progression." *Nat Med* 17 (3):320–29. doi: 10.1038/nm.2328. Klein, G., and E. Klein. 2005. "Surveillance against tumors—is it mainly immunological?" *Immunol Lett* 100 (1):29–33. doi: 10.1016/j.imlet.2005.06.024. Lowe, S. W., E. Cepero, and G. Evan. 2004. "Intrinsic tumour suppression." *Nature* 432 (7015):307–15. doi: 10.1038/nature03098.

57. Gill, J. G., E. Piskounova, and S. J. Morrison. 2016. "Cancer, oxidative stress, and metastasis." *Cold Spring Harb Symp Quant Biol* 81:163–75. doi: 10.1101/sqb.2016.81.030791. Rojo de la Vega, M., E. Chapman, and D. D. Zhang. 2018. "NRF2 and the hallmarks of cancer." *Cancer Cell* 34 (1):21–43. doi: 10.1016/j.ccell.2018.03.022.

58. Goddard, E. T., I. Bozic, S. R. Riddell, and C. M. Ghajar. 2018. "Dormant tumour cells, their niches and the influence of immunity." *Nat Cell Biol* 20 (11):1240–49. doi: 10.1038/s41556-018-0214-0. Klein-Goldberg, A., S. Maman, and I. P. Witz. 2014. "The role played by the microenvironment in site-specific metastasis." *Cancer Lett* 352 (1):54–58. doi: 10.1016/j.canlet.2013.08.029.

59. Salk, J. J., K. Loubet-Senear, E. Maritschnegg, C. C. Valentine, L. N. Williams, J. E. Higgins, R. Horvat, A. Vanderstichele, D. Nachmanson, K. T. Baker, M. J. Emond, E. Loter, M. Tretiakova, T. Soussi, L. A. Loeb, R. Zeillinger, P. Speiser, and R. A. Risques. 2019. "Ultra-sensitive TP53 sequencing for cancer detection reveals progressive clonal selection in normal tissue over a century of human lifespan." *Cell Rep* 28 (1):132–44 e3. doi: 10.1016/j.celrep.2019.05.109. Risques, R. A., and S. R. Kennedy. 2018. "Aging and the rise of somatic cancer-associated mutations in normal tissues." *PLoS Genet* 14 (1):e1007108. doi: 10.1371/journal.pgen.1007108. Martincorena, I., J. C. Fowler, A. Wabik, A. R. J. Lawson, F. Abascal, M. W. J. Hall, A. Cagan, K. Murai, K. Mahbubani, M. R. Stratton, R. C. Fitzgerald, P. A. Handford, P. J. Campbell, K. Saeb-Parsy, and P. H. Jones. 2018. "Somatic mutant clones colonize the human esophagus with age." *Science* 362 (6417):911–17. doi: 10.1126/science.aau3879. Martincorena, I., A. Roshan, M. Gerstung, P. Ellis, P. Van Loo, S. McLaren, D. C. Wedge, A. Fullam, L. B. Alexandrov, J. M. Tubio, L. Stebbings, A. Menzies, S. Widaa, M. R. Stratton, P. H. Jones, and P. J. Campbell. 2015. "Tumor evolution: high burden and pervasive positive selection of somatic mutations in normal human skin." *Science* 348 (6237):880–86. doi: 10.1126/science.aaa6806.

60. Salk, J. J., K. Loubet-Senear, E. Maritschnegg, C. C. Valentine, L. N. Williams, J. E. Higgins, R. Horvat, A. Vanderstichele, D. Nachmanson, K. T. Baker, M. J. Emond, E. Loter, M. Tretiakova, T. Soussi, L. A. Loeb, R. Zeillinger, P. Speiser, and R. A. Risques. 2019. "Ultra-sensitive TP53 sequencing for cancer detection reveals progressive clonal selection in normal tissue over a century of human lifespan." *Cell Rep* 28 (1):132–44 e3. doi: 10.1016/j.celrep.2019.05.109.

61. Tomasetti, C., B. Vogelstein, and G. Parmigiani. 2013. "Half or more of the somatic mutations in cancers of self-renewing tissues originate prior to tumor initiation." *Proc Natl Acad Sci USA* 110 (6):1999–2004. doi: 10.1073/pnas.1221068110.

62. Bissell, M. J., and W. C. Hines. 2011. "Why don't we get more cancer? A proposed role of the microenvironment in restraining cancer progression." *Nat Med* 17 (3):320–29. doi: 10.1038/nm.2328.

63. Teixeira, V. H., C. P. Pipinikas, A. Pennycuick, H. Lee-Six, D. Chandrasekharan, J. Beane, T. J. Morris, A. Karpathakis, A. Feber, C. E. Breeze, P. Ntolios, R. E. Hynds, M. Falzon, A. Capitanio, B. Carroll, P. F. Durrenberger, G. Hardavella, J. M. Brown, A. G. Lynch, H. Farmery, D. S. Paul, R. C. Chambers, N. McGranahan, N. Navani, R. M. Thakrar, C. Swanton, S. Beck, P. J. George, P. J. Campbell, C. Thirlwell, and S. M. Janes. 2019. "Deciphering the genomic, epigenomic, and transcriptomic landscapes of pre-invasive lung cancer lesions." *Nat Med* 25 (3):517–25. doi: 10.1038/s41591-018-0323-0.

64. Sansregret, L., B. Vanhaesebroeck, and C. Swanton. 2018. "Determinants and clinical implications of chromosomal instability in cancer." *Nat Rev Clin Oncol* 15 (3):139–50. doi: 10.1038/nrclinonc.2017.198.

65. Teixeira, V. H., C. P. Pipinikas, A. Pennycuick, H. Lee-Six, D. Chandrasekharan, J. Beane, T. J. Morris, A. Karpathakis, A. Feber, C. E. Breeze, P. Ntolios, R. E. Hynds, M. Falzon, A. Capitanio, B. Carroll, P. F. Durrenberger, G. Hardavella, J. M. Brown, A. G. Lynch, H. Farmery, D. S. Paul, R. C. Chambers, N. McGranahan, N. Navani, R. M. Thakrar, C. Swanton, S. Beck, P. J. George, A. Spira, P. J. Campbell, C. Thirlwell, and S. M. Janes. 2019. "Deciphering the genomic, epigenomic, and transcriptomic landscapes of pre-invasive lung cancer lesions." *Nat Med* 25 (3):517–25. doi: 10.1038/s41591-018-0323-0. McGranahan, N., R. Rosenthal, C. T. Hiley, A. J. Rowan, T. B. K. Watkins, G. A. Wilson, N. J. Birkbak, S. Veeriah, P. Van Loo, J. Herrero, C. Swanton, and T. RACERx Consortium. 2017. "Allele-specific HLA loss and immune escape in lung cancer evolution." *Cell* 171 (6):1259–71 e11. doi: 10.1016/j.cell.2017.10.001. Marty, R., S. Kaabinejadian, D. Rossell, M. J. Slifker, J. van de Haar, H. B. Engin, N. de Prisco, T. Ideker, W. H. Hildebrand, J. Font-Burgada, and H. Carter. 2017. "MHC-I genotype restricts the oncogenic mutational landscape." *Cell* 171 (6):1272–83 e15. doi: 10.1016/j.cell.2017.09.050.

66. Mlecnik, B., G. Bindea, A. Kirilovsky, H. K. Angell, A. C. Obenauf, M. Tosolini, S. E. Church, P. Maby, A. Vasaturo, M. Angelova, T. Fredriksen, S. Mauger, M. Waldner, A. Berger, M. R. 98. Speicher, F. Pages, V. Valge-Archer, and J. Galon. 2016. "The tumor microenvironment and Immunoscore are critical determinants of dissemination to distant metastasis." *Sci Transl Med* 8 (327):327ra26. doi: 10.1126/scitranslmed.aad6352. Teschendorff, A. E., A. Miremadi, S. E. Pinder, I. O. Ellis, and C. Caldas. 2007. "An immune response gene expression module identifies a good prognosis subtype in estrogen receptor negative breast cancer." *Genome Biol* 8 (8):R157. doi: 10.1186/gb-2007-8-8-r157. Thorsson, V., D. L. Gibbs, S. D. Brown, D. Wolf, D. S. Bortone, T. H. Ou Yang, E. Porta-Pardo, G. F. Gao, C. L. Plaisier, J. A. Eddy, E. Ziv, A. C. Culhane, E. O. Paull, I. K. A. Sivakumar, A. J. Gentles, R. Malhotra, F. Farshidfar, A. Colaprico, J. S. Parker, L. E. Mose, N. S. Vo, J. Liu, Y. Liu, J. Rader, V. Dhankani, S. M. Reynolds, R. Bowlby, A. Califano, A. D. Cherniack, D. Anastassiou, D. Bedognetti, Y. Mokrab, A. M. Newman, A. Rao, K. Chen, A. Krasnitz, H. Hu, T. M. Malta, H. Noushmehr, C. S. Pedamallu, S. Bullman, A. I. Ojesina, A. Lamb, W. Zhou, H. Shen, T. K. Choueiri, J. N. Weinstein, J. Guinney, J. Saltz, R. A. Holt, C. S. Rabkin, Network Cancer Genome Atlas Research, A. J. Lazar, J. S. Serody, E. G. Demicco, M. L. Disis, B. G. Vincent, and I. Shmulevich. 2018. "The immune landscape of cancer." *Immunity* 48 (4):812–30 e14. doi: 10.1016/j.immuni.2018.03.023.

67. Angelova, M., B. Mlecnik, A. Vasaturo, G. Bindea, T. Fredriksen, L. Lafontaine, B. Buttard, E. Morgand, D. Bruni, A. Jouret-Mourin, C. Hubert, A. Kartheuser, Y. Humblet, M. Ceccarelli, N. Syed, F. M. Marincola, D. Bedognetti, M. Van den Eynde, and J. Galon. 2018. "Evolution of metastases in space and time under immune selection." *Cell* 175 (3):751–65 e16. doi: 10.1016/j.cell.2018.09.018. De Mattos-Arruda, L., S. J. Sammut, E. M. Ross, R. Bashford-Rogers, E. Greenstein, H. Markus, S. Morganella, Y. Teng, Y. Maruvka, B. Pereira, O. M. Rueda, S. F. Chin, T. Contente-Cuomo, A. Mayor, A. Arias, H. R. Ali, W. Cope, D. Tiezzi, A. Dariush, T. Dias Amarante, D. Reshef, N. Ciriaco, E. Martinez-Saez, V. Peg, Y. Cajal S. Ramon, J. Cortes, G. Vassiliou, G. Getz, S. Nik-Zainal, M. Murtaza, N. Friedman, F. Markowetz, J. Seoane, and C. Caldas. 2019. "The genomic and immune landscapes of lethal metastatic breast cancer." *Cell Rep* 27 (9):2690–708 e10. doi: 10.1016/j.celrep.2019.04.098.

68. Greaves, M., and C. C. Maley. 2012. "Clonal evolution in cancer." *Nature* 481 (7381):306–13. doi: 10.1038/nature10762. McGranahan, N., and C. Swanton. 2017. "Clonal heterogeneity and tumor evolution: past, present, and the future." *Cell* 168 (4):613–28. doi: 10.1016/j.cell.2017.01.018. Graham, T. A., and A. Sottoriva. 2017. "Measuring cancer evolution from the genome." *J Pathol* 241 (2):183–91. doi: 10.1002/path.4821.

69. Laland, K. N., T. Uller, M. W. Feldman, K. Sterelny, G. B. Muller, A. Moczek, E. Jablonka, and J. Odling-Smee. 2015. "The extended evolutionary synthesis: its structure, assumptions and predictions." *Proc Biol Sci* 282 (1813):20151019. doi: 10.1098/rspb.2015.1019.

70. Jablonka, E. 2017. "The evolutionary implications of epigenetic inheritance." *Interface Focus* 7 (5):20160135. doi: 10.1098/rsfs.2016.0135.

71. Noble, D. 2013. "A biological relativity view of the relationships between genomes and phenotypes." *Prog Biophys Mol Biol* 111 (2–3):59–65. doi: 10.1016/j.pbiomolbio.2012.09.004.

72. Laland, K. N., T. Uller, M. W. Feldman, K. Sterelny, G. B. Muller, A. Moczek, E. Jablonka, and J. Odling-Smee. 2015. "The extended evolutionary synthesis: its structure, assumptions and predictions." *Proc Biol Sci* 282 (1813):20151019. doi: 10.1098/rspb.2015.1019.

73. Scott-Phillips, T. C., K. N. Laland, D. M. Shuker, T. E. Dickins, and S. A. West. 2014. "The niche construction perspective: a critical appraisal." *Evolution* 68 (5):1231–43. doi: 10.1111/evo.12332. Laland, K., T. Uller, M. Feldman, K. Sterelny, G. B. Muller, A. Moczek, E. Jablonka, J. Odling-Smee, G. A. Wray, H. E. Hoekstra, D. J. Futuyma, R. E. Lenski, T. F. Mackay, D. Schluter, and J. E. Strassmann. 2014. "Does evolutionary theory

need a rethink?" *Nature* 514 (7521):161–64. doi: 10.1038/514161a. Klug, H. 2014. "Evolution: students debate the debate." *Nature* 515 (7527):343. doi: 10.1038/515343a.

74. Laland, K. N., T. Uller, M. W. Feldman, K. Sterelny, G. B. Muller, A. Moczek, E. Jablonka, and J. Odling-Smee. 2015. "The extended evolutionary synthesis: its structure, assumptions and predictions." *Proc Biol Sci* 282 (1813):20151019. doi: 10.1098/rspb.2015.1019. Odling-Smee, F. J., K. N. Laland, and M. W. Feldman. 2003. *Niche Construction: The Neglected Process in Evolution* (Princeton University Press, ISBN 0691044384). Lewontin, R. C. 1983. "Gene, organism and environment." In: *Evolution from Molecules to Men*, edited by D. S. Bendall, pp. 273–85 (Cambridge University Press, ISBN 0521247535).

75. Laland, K. N., J. Odling-Smee, and M. W. Feldman. 2001. "Cultural niche construction and human evolution." *J Evol Biol* 14 (1):22–33. doi: 10.1046/j.1420–9101.2001.00262.x. Brewer, J., M. Gelfand, J. C. Jackson, I. F. MacDonald, P. N. Peregrine, P. J. Richerson, P. Turchin, H. Whitehouse, and D. S. Wilson. 2017. "Grand challenges for the study of cultural evolution." *Nat Ecol Evol* 1 (3):70. doi: 10.1038/s41559-017-0070.

76. Bertolaso, M., and J. Dupré. 2018. "A processual perspective on cancer." In: *Everything Flows: Towards a Processual Philosophy of Biology*, edited by D. J. Nicholson and J. Dupré (Oxford Scholarship Online, ISBN 9780198779636).

77. Doolittle, W. F., and S. A. Inkpen. 2018. "Processes and patterns of interaction as units of selection: an introduction to ITSNTS thinking." *Proc Natl Acad Sci USA* 115 (16):4006–14. doi: 10.1073/pnas.1722232115.

78. Hunter, P. 2017. "The role of biology in global climate change." *Embo Rep* 18 (5):673–76. doi: 10.15252/embr.201744260.

79. Doolittle, W. F. 2019. "Making evolutionary sense of Gaia." *Trends Ecol Evol* 34 (10):889–94. doi: 10.1016/j.tree.2019.05.001.

80. Lovelock, J. E. 1979. *Gaia: A New Look at Life on Earth* (Oxford University Press, ISBN 0192862189).

81. Sagan, L. 1967. "On the origin of mitosing cells." *J Theor Biol* 14 (3):255–74. doi: 10.1016/0022–5193(67)90079–3. The endosymbiosis theory articulated by Lynn Margulis (authored as Lynn Sagan) presented the origin of eukaryotic cells ("higher" cells that divide by classical mitosis). The abstract of the 1967 study reads, "By hypothesis, three fundamental organelles: the mitochondria, the photosynthetic plastids and the (9+2) basal bodies of flagella were themselves once free-living (prokaryotic) cells."

82. Doolittle, W. F. 2019. "Making evolutionary sense of Gaia." *Trends Ecol Evol* 34 (10):889–94. doi: 10.1016/j.tree.2019.05.001. Lovelock, J. 2003. "Gaia: the living Earth." *Nature* 426 (6968):769–70. doi: 10.1038/426769a. Gilbert, S. F., J. Sapp, and A. I. Tauber. 2012. "A symbiotic view of life: we have never been individuals." *Q Rev Biol* 87 (4):325–41. doi: 10.1086/668166.

83. Margulis, L., and L. Fester. 1991. *Symbiosis as a Source of Evolutionary Innovation: Speciation and Morphogenesis* (MIT Press, ISBN 0262519908).

84. Gilbert, S. F., J. Sapp, and A. I. Tauber. 2012. "A symbiotic view of life: we have never been individuals." *Q Rev Biol* 87 (4):325–41. doi: 10.1086/668166. Gilbert, S. F. 2016. "Developmental plasticity and developmental symbiosis: the return of eco-devo." *Curr Top Dev Biol* 116:415–33. doi: 10.1016/bs.ctdb.2015.12.006.

85. DeGregori, J. 2018. *Adaptive Oncogenesis—A New Understanding of How Cancer Evolves Inside Us* (Harvard University Press, ISBN 0674545397). Maley, C., and M. Graeves. 2018. *Frontiers in Cancer Research—Evolutionary Foundations, Revolutionary Directions* (Springer-Verlag New York, ISBN 1493964581). Ujvari, B., B. Roche, and F. Thomas. 2017. *Ecology and Evolution of Cancer* (Academic Press, ISBN 0128043105).

86. Gilbert, S. F., J. Sapp, and A. I. Tauber. 2012. "A symbiotic view of life: we have never been individuals." *Q Rev Biol* 87 (4):325–41. doi: 10.1086/668166.

87. Enriquez-Navas, P. M., J. W. Wojtkowiak, and R. A. Gatenby. 2015. "Application of evolutionary principles to cancer therapy." *Cancer Res* 75 (22):4675–80. doi: 10.1158/0008–5472.CAN-15-1337. Gatenby, R. A., and J. Brown. 2017. "Mutations, evolution and the central role of a self-defined fitness function in the initiation and progression of cancer." *Biochim Biophys Acta Rev Cancer* 1867 (2):162–66. doi: 10.1016/j.bbcan.2017.03.005. Gillies, R. J., D. Verduzco, and R. A. Gatenby. 2012. "Evolutionary dynamics of carcinogenesis and why targeted therapy does not work." *Nat Rev Cancer* 12 (7):487–93. doi: 10.1038/nrc3298. Maley, C. C., A. Aktipis, T. A. Graham, A. Sottoriva, A. M. Boddy, M. Janiszewska, A. S. Silva, M. Gerlinger, Y. Yuan, K. J. Pienta, K. S. Anderson, R. Gatenby, C. Swanton, D. Posada, C. I. Wu, J. D. Schiffman, E. S. Hwang, K. Polyak, A. R. A. Anderson, J. S. Brown, M. Greaves, and D. Shibata. 2017. "Classifying the evolutionary and ecological features of neoplasms." *Nat Rev Cancer* 17 (10):605–19. doi: 10.1038/nrc.2017.69.

88. Ibrahim-Hashim, A., M. Robertson-Tessi, P. M. Enriquez-Navas, M. Damaghi, Y. Balagurunathan, J. W. Wojtkowiak, S. Russell, K. Yoonseok, M. C. Lloyd, M. M. Bui, J. S. Brown, A. R. A. Anderson, R. J. Gillies, and R. A. Gatenby. 2017. "Defining cancer subpopulations by adaptive strategies rather than molecular properties provides novel insights into intratumoral evolution." *Cancer Res* 77 (9):2242–54. doi: 10.1158/0008–5472.CAN-16–2844. Estrella, V., T. Chen, M. Lloyd, J. Wojtkowiak, H. H. Cornnell, A. Ibrahim-Hashim, K. Bailey, Y. Balagurunathan, J. M. Rothberg, B. F. Sloane, J. Johnson, R. A. Gatenby, and R. J. Gillies. 2013. "Acidity generated by the tumor microenvironment drives local invasion." *Cancer Res* 73 (5):1524–35. doi: 10.1158/0008–5472. CAN-12–2796.

89. Narhi, K., A. S. Nagaraj, E. Parri, R. Turkki, P. W. van Duijn, A. Hemmes, J. Lahtela, V. Uotinen, M. I. Mayranpaa, K. Salmenkivi, J. Rasanen, N. Linder, J. Trapman, A. Rannikko, O. Kallioniemi, T. M. Af Hallstrom, J. Lundin, W. Sommergruber, S. Anders, and E. W. Verschuren. 2018. "Spatial aspects of oncogenic signalling determine the response to combination therapy in slice explants from Kras-driven lung tumours." *J Pathol* 245 (1):101–13. doi: 10.1002/path.5059. Bao, J., M. Walliander, F. Kovacs, A. S. Nagaraj, A. Hemmes, V. K. Sarhadi, S. Knuutila, J. Lundin, P. Horvath, and E. W. Verschuren. 2019. "Spa-RQ: an image analysis tool to visualise and quantify spatial phenotypes applied to non-small cell lung cancer." *Sci Rep* 9 (1):17613. doi: 10.1038 /s41598-019-54038-9.

90. Bussey, K. J., L. H. Cisneros, C. H. Lineweaver, and P. C. W. Davies. 2017. "Ancestral gene regulatory networks drive cancer." *Proc Natl Acad Sci USA* 114 (24):6160–62. doi: 10.1073/pnas.1706990114. Davies, P. C., and C. H. Lineweaver. 2011. "Cancer tumors as Metazoa 1.0: tapping genes of ancient ancestors." *Phys Biol* 8 (1):015001. doi: 10.1088/1478–3975/8/1/015001. Davies, P. C., L. Demetrius, and J. A. Tuszynski. 2011. "Cancer as a dynamical phase transition." *Theor Biol Med Model* 8:30. doi: 10.1186/1742-4682-8-30.

91. Noble, D. 2006. *The Music of Life: Biology beyond Genes* (Oxford University Press, ISBN 0199228362).

92. Levy, A. 2019. "How evolution builds genes from scratch." *Nature* 574 (7778):314–16. doi: 10.1038 /d41586-019-03061-x.

93. Klein, C. A. 2009. "Parallel progression of primary tumours and metastases." *Nat Rev Cancer* 9 (4):302–12. doi: 10.1038/nrc2627.

94. Dotto, G. P. 2014. "Multifocal epithelial tumors and field cancerization: stroma as a primary determinant." *J Clin Invest* 124 (4):1446–53. doi: 10.1172/JCI72589. Curtius, K., N. A. Wright, and T. A. Graham. 2018. "An evolutionary perspective on field cancerization." *Nat Rev Cancer* 18 (1):19–32. doi: 10.1038/nrc.2017.102.

95. Kerr, E. M., E. Gaude, F. K. Turrell, C. Frezza, and C. P. Martins. 2016. "Mutant Kras copy number defines metabolic reprogramming and therapeutic susceptibilities." *Nature* 531 (7592):110–13. doi: 10.1038/nature16967.\

96. https://cancerprogressreport.org/Pages/cpr19-contents.aspx.

97. Quail, D. F., and A. J. Dannenberg. 2019. "The obese adipose tissue microenvironment in cancer development and progression." *Nat Rev Endocrinol* 15 (3):139–54. doi: 10.1038/s41574-018-0126-x.

98. Wilson, B. E., S. Jacob, M. L. Yap, J. Ferlay, F. Bray, and M. B. Barton. 2019. "Estimates of global chemotherapy demands and corresponding physician workforce requirements for 2018 and 2040: a population-based study." *Lancet Oncol* 20 (6):769–80. doi: 10.1016/S1470–2045(19)30163–9. Bray, F., J. Ferlay, I. Soerjomataram, R. L. Siegel, L. A. Torre, and A. Jemal. 2018. "Global cancer statistics 2018: GLOBOCAN estimates of incidence and mortality worldwide for 36 cancers in 185 countries." *CA Cancer J Clin* 68 (6):394–424. doi: 10.3322 /caac.21492. The Lancet. 2018. "GLOBOCAN 2018: counting the toll of cancer." *Lancet* 392 (10152):985. doi: 10.1016/S0140–6736(18)32252–9.

99. Sapolsky, R. M. 2017. *Behave: The Biology of Humans at Our Best and Worst* (Penguin Press, ISBN 1594205078).

100. Eriksson, J. G. 2016. "Developmental origins of health and disease—from a small body size at birth to epigenetics." *Ann Med* 48 (6):456–67. doi: 10.1080/07853890.2016.1193786.

101. Sharma, P., and J. P. Allison. 2015. "Immune checkpoint targeting in cancer therapy: toward combination strategies with curative potential." *Cell* 161 (2):205–14. doi: 10.1016/j.cell.2015.03.030.

102. Hanoun, M., M. Maryanovich, A. Arnal-Estape, and P. S. Frenette. 2015. "Neural regulation of hematopoiesis, inflammation, and cancer." *Neuron* 86 (2):360–73. doi: 10.1016/j.neuron.2015.01.026. Umamaheswaran, S., S. K. Dasari, P. Yang, S. K. Lutgendorf, and A. K. Sood. 2018. "Stress, inflammation, and eicosanoids: an emerging perspective." *Cancer Metastasis Rev* 37 (2–3):203–11. doi: 10.1007/s10555-018-9741-1. Chida, Y., M. Hamer, J. Wardle, and A. Steptoe. 2008. "Do stress-related psychosocial factors contribute to cancer incidence and survival?" *Nat Clin Pract Oncol* 5 (8):466–75. doi: 10.1038/ncponc1134.

103. Hanoun, M., M. Maryanovich, A. Arnal-Estape, and P. S. Frenette. 2015. "Neural regulation of hemato-poiesis, inflammation, and cancer." *Neuron* 86 (2):360–73. doi: 10.1016/j.neuron.2015.01.026. Umamaheswaran, S., S. K. Dasari, P. Yang, S. K. Lutgendorf, and A. K. Sood. 2018. "Stress, inflammation, and eicosanoids: an emerging perspective." *Cancer Metastasis Rev* 37 (2–3):203–11. doi: 10.1007/s10555-018-9741-1. Cole, S. W., A. S. Nagaraja, S. K. Lutgendorf, P. A. Green, and A. K. Sood. 2015. "Sympathetic nervous system regulation of the tumour microenvironment." *Nat Rev Cancer* 15 (9):563–72. doi: 10.1038/nrc3978.

104. McEwen, B. S., and E. Stellar. 1993. "Stress and the individual: mechanisms leading to disease." *Arch Intern Med* 153 (18):2093–101. McEwen, B. S. 2002. "Sex, stress and the hippocampus: allostasis, allostatic load and the aging process." *Neurobiol Aging* 23 (5):921–39. doi: 10.1016/s0197-4580(02)00027-1.

105. Emanuel, L., D. Berwick, J. Conway, J. Combes, M. Hatlie, L. Leape, J. Reason, P. Schyve, C. Vincent, and M. Walton. 2008. "What exactly is patient safety?" In: *Advances in Patient Safety: New Directions and Alternative Approaches (Vol. 1: Assessment)*, edited by K. Henriksen, J. B. Battles, M. A. Keyes and M. L. Grady (Agency for Healthcare Research and Quality, Publication No. 08-0034-1).

106. Lipsitz, L. A. 2012. "Understanding health care as a complex system: the foundation for unintended con-sequences." *JAMA* 308 (3):243–44. doi: 10.1001/jama.2012.7551.

107. Illich, I. 1975. *Limits of Medicine. Medical Nemesis: The Expropriation of Health* (Marion Boyars, ISBN 0714525138).

108. van Muijen, P., S. F. A. Duijts, K. Bonefaas-Groenewoud, A. J. van der Beek, and J. R. Anema. 2017. "Predictors of fatigue and work ability in cancer survivors." *Occup Med (Lond)* 67 (9):703–11. doi: 10.1093/occmed/kqx165.

109. Sanden, U., F. Nilsson, H. Thulesius, M. Hagglund, and L. Harrysson. 2019. "Cancer, a relational disease exploring the needs of relatives to cancer patients." *Int J Qual Stud Health Well-being* 14 (1):1622354. doi: 10.1080/17482631.2019.1622354.

110. Fojo, T., and A. W. Lo. 2016. "Price, value, and the cost of cancer drugs." *Lancet Oncol* 17 (1):3–5. doi: 10.1016/S1470-2045(15)00564-1.

111. Schnipper, L. E., N. E. Davidson, D. S. Wollins, C. Tyne, D. W. Blayney, D. Blum, A. P. Dicker, P. A. Ganz, J. R. Hoverman, R. Langdon, G. H. Lyman, N. J. Meropol, T. Mulvey, L. Newcomer, J. Peppercorn, B. Polite, D. Raghavan, G. Rossi, L. Saltz, D. Schrag, T. J. Smith, P. P. Yu, C. A. Hudis, R. L. Schilsky, and American Society of Clinical Oncology. 2015. "American Society of Clinical Oncology statement: a conceptual framework to assess the value of cancer treatment options." *J Clin Oncol* 33 (23):2563–77. doi: 10.1200/JCO.2015.61.6706.

112. Claxton, K., S. Martin, M. Soares, N. Rice, E. Spackman, S. Hinde, N. Devlin, P. C. Smith, and M. Sculpher. 2015. "Methods for the estimation of the National Institute for Health and Care Excellence cost-effectiveness threshold." *Health Technol Assess* 19 (14):1–503, v–vi. doi: 10.3310/hta19140.

113. Preamble to the Constitution of WHO as adopted by the International Health Conference, New York, 19 June to 22 July 1946; signed on 22 July 1946 by the representatives of 61 States (Official Records of WHO, no. 2, p. 100).

114. Burns, H. 2014. "What causes health?" *J R Coll Physicians Edinb* 44 (2):103–5. doi: 10.4997/JRCPE.2014.202.

115. Antonovsky, A. 1979. *Health, Stress and Coping* (Jossey-Bass, ISBN 0875894127).

116. Sagner, M., A. McNeil, P. Puska, C. Auffray, N. D. Price, L. Hood, C. J. Lavie, Z. G. Han, Z. Chen, S. K. Brahmachari, B. S. McEwen, M. B. Soares, R. Balling, E. Epel, and R. Arena. 2017. "The P4 health spectrum—a predictive, preventive, personalized and participatory continuum for promoting healthspan." *Prog Cardiovasc Dis* 59 (5):506–21. doi: 10.1016/j.pcad.2016.08.002.

117. Brennan, P., M. Perola, G. J. van Ommen, E. Riboli, and Consortium European Cohort. 2017. "Chronic disease research in Europe and the need for integrated population cohorts." *Eur J Epidemiol* 32 (9):741–49. doi: 10.1007/s10654-017-0315-2.

118. Leonelli, S. 2016. *Data-Centric Biology: A Philosophical Study* (University of Chicago Press, ISBN 0226416472).

119. Leonelli, S. 2019. "Data—from objects to assets." *Nature* 574 (7778):317–20. doi: 10.1038/d41586-019-03062-w.

120. DelNero, P., and A. McGregor. 2017. "From patients to partners." *Science* 358 (6361):414. doi: 10.1126/science.358.6361.414.

121. Peisch, S. F., E. L. Van Blarigan, J. M. Chan, M. J. Stampfer, and S. A. Kenfield. 2017. "Prostate cancer progression and mortality: a review of diet and lifestyle factors." *World J Urol* 35 (6):867–74. doi: 10.1007 /s00345-016-1914-3. Johnston-Early, A., M. H. Cohen, J. D. Minna, L. M. Paxton, B. E. Fossieck, Jr., D. C. Ihde, P. A. Bunn Jr., M. J. Matthews, and R. Makuch. 1980. "Smoking abstinence and small cell lung cancer survival. An association." *JAMA* 244 (19):2175–79. Browman, G. P., G. Wong, I. Hodson, J. Sathya, R. Russell, L. McAlpine, P. Skingley, and M. N. Levine. 1993. "Influence of cigarette smoking on the efficacy of radiation therapy in head and neck cancer." *N Engl J Med* 328 (3):159–63. doi: 10.1056/NEJM199301213280302.

122. Tindle, H. A., and R. A. Greevy. 2018. "Smoking cessation pharmacotherapy, even without counseling, remains a cornerstone of treatment." *J Natl Cancer Inst* 110 (6):545–46. doi: 10.1093/jnci/djx246.

123. Evins, A. E., and C. Cather. 2015. "Effective cessation strategies for smokers with schizophrenia." *Int Rev Neurobiol* 124:133–47. doi: 10.1016/bs.irn.2015.08.001.

124. Ashmore, J. A., K. W. Ditterich, C. C. Conley, M. R. Wright, P. S. Howland, K. L. Huggins, J. Cooreman, P. S. Andrews, D. R. Nicholas, L. Roberts, L. Hewitt, J. N. Scales, J. K. Delap, C. A. Gray, L. A. Tyler, C. Collins, C. M. Whiting, B. M. Brothers, M. M. Ryba, and B. L. Andersen. 2019. "Evaluating the effectiveness and implementation of evidence-based treatment: a multisite hybrid design." *Am Psychol* 74 (4):459–73. doi: 10.1037/ amp0000309.

125. Sandén, U. 2016. *... and I Want to Live* (Vulkan, ISBN 9789163913457).

126. Sanden, U., L. Harrysson, H. Thulesius, and F. Nilsson. 2017. "Exploring health navigating design: momentary contentment in a cancer context." *Int J Qual Stud Health Well-Being* 12 (suppl 2):1374809. doi: 10.1080/17482631.2017.1374809.

127. The Lancet. 2019. "Promoting and prescribing the arts for health." *Lancet* 394 (10212):1880. doi: 10.1016 /S0140–6736(19)32796–5.

128. Sanden, U., L. Harrysson, H. Thulesius, and F. Nilsson. 2017. "Exploring health navigating design: momentary contentment in a cancer context." *Int J Qual Stud Health Well-Being* 12 (suppl 2):1374809. doi: 10.1080/17482631.2017.1374809.

129. Sapolsky, R. M. 2017. *Behave: The Biology of Humans at Our Best and Worst* (Penguin Press, ISBN 1594205078). Blakemore, S. J. 2019. "Adolescence and mental health." *Lancet* 393 (10185):2030–31. doi: 10.1016/S0140–6736(19)31013-X.

130. Blakemore, S. J. 2019. "Adolescence and mental health." *Lancet* 393 (10185):2030–31. doi: 10.1016/ S0140–6736(19)31013-X. Nemeroff, C. B. 2016. "Paradise lost: the neurobiological and clinical consequences of child abuse and neglect." *Neuron* 89 (5):892–909. doi: 10.1016/j.neuron.2016.01.019. Nemeroff, C. B., and F. Seligman. 2013. "The pervasive and persistent neurobiological and clinical aftermath of child abuse and neglect." *J Clin Psychiatry* 74 (10):999–1001. doi: 10.4088/JCP.13com08633.

131. The Lancet. 2019. "Lung cancer: some progress, but still a lot more to do." *Lancet* 394 (10212):1880. doi: 10.1016/S0140–6736(19)32795–3. https://www.lung.org/our-initiatives/research/monitoring-trends-in-lung -disease/state-of-lung-cancer/.

132. Alberg, A. J., D. R. Shopland, and K. M. Cummings. 2014. "The 2014 Surgeon General's report: commemorating the 50th Anniversary of the 1964 Report of the Advisory Committee to the US Surgeon General and updating the evidence on the health consequences of cigarette smoking." *Am J Epidemiol* 179 (4):403–12. doi: 10.1093/aje/kwt335. Loeb, L. A. 2016. "Tobacco causes human cancers—a concept founded on epidemiology and an insightful experiment now requires translation worldwide." *Cancer Res* 76 (4):765–66. doi: 10.1158/0008–5472.CAN-16–0149.

133. Lawrence, F. 2019. "Big Tobacco, war and politics." *Nature* 574:172–73. doi: 10.1038/d41586-019- 02991-w. Milov, S. 2019. *The Cigarette: A Political History* (Harvard University Press, ISBN 0674241215).

134. Loeb, L. A. 2016. "Tobacco causes human cancers—a concept founded on epidemiology and an insightful experiment now requires translation worldwide." *Cancer Res* 76 (4):765–66. doi: 10.1158/0008–5472 .CAN-16–0149.

135. The Lancet. 2019. "Lung cancer: some progress, but still a lot more to do." *Lancet* 394 (10212):1880. doi: 10.1016/S0140–6736(19)32795–3. Wild, C. P. 2019. "The global cancer burden: necessity is the mother of prevention." *Nat Rev Cancer* 19 (3):123–24. doi: 10.1038/s41568-019-0110-3. Ganz, P. A. 2019. "Current US cancer statistics: alarming trends in young adults?" *J Natl Cancer Inst* 111 (12):1241–42 . doi: 10.1093/jnci/djz107.

136. Wild, C. P. 2019. "The global cancer burden: necessity is the mother of prevention." *Nat Rev Cancer* 19 (3):123–24. doi: 10.1038/s41568-019-0110-3. Song, M., B. Vogelstein, E. L. Giovannucci, W. C. Willett, and C.

Tomasetti. 2018. "Cancer prevention: molecular and epidemiologic consensus." *Science* 361 (6409):1317–18. doi: 10.1126/science.aau3830.

137. Song, M., B. Vogelstein, E. L. Giovannucci, W. C. Willett, and C. Tomasetti. 2018. "Cancer prevention: molecular and epidemiologic consensus." *Science* 361 (6409):1317–18. doi: 10.1126/science.aau3830.

138. Garcia, P. J. 2019. "Corruption in global health: the open secret." *Lancet* 394 (10214):2119–24. doi: 10.1016/S0140–6736(19)32527–9.

139. Moynihan, R., L. Bero, S. Hill, M. Johansson, J. Lexchin, H. Macdonald, B. Mintzes, C. Pearson, M. A. Rodwin, A. Stavdal, J. Stegenga, B. D. Thombs, H. Thornton, P. O. Vandvik, B. Wieseler, and F. Godlee. 2019. "Pathways to independence: towards producing and using trustworthy evidence." *BMJ* 367:l6576. doi: 10.1136/bmj.l6576.

140. Hanson, R. 1994. "Buy health, not health care." *Cato J* 14 (1). http://mason.gmu.edu/~rhanson/buyhealth.html.

141. https://www.ipcc.ch/srccl/.

142. Chaplin-Kramer, R., R. P. Sharp, C. Weil, E. M. Bennett, U. Pascual, K. K. Arkema, K. A. Brauman, B. P. Bryant, A. D. Guerry, N. M. Haddad, M. Hamann, P. Hamel, J. A. Johnson, L. Mandle, H. M. Pereira, S. Polasky, M. Ruckelshaus, M. R. Shaw, J. M. Silver, A. L. Vogl, and G. C. Daily. 2019. "Global modeling of nature's contributions to people." *Science* 366 (6462):255–58. doi: 10.1126/science.aaw3372. Morecroft, M. D., S. Duffield, M. Harley, J. W. Pearce-Higgins, N. Stevens, O. Watts, and J. Whitaker. 2019. "Measuring the success of climate change adaptation and mitigation in terrestrial ecosystems." *Science* 366 (6471):eaaw9256. doi: 10.1126/science.aaw9256. Diaz, S., J. Settele, E. S. Brondizio, H. T. Ngo, J. Agard, A. Arneth, P. Balvanera, K. A. Brauman, S. H. M. Butchart, K. M. A. Chan, L. A. Garibaldi, K. Ichii, J. Liu, S. M. Subramanian, G. F. Midgley, P. Miloslavich, Z. Molnar, D. Obura, A. Pfaff, S. Polasky, A. Purvis, J. Razzaque, B. Reyers, R. R. Chowdhury, Y. J. Shin, I. Visseren-Hamakers, K. J. Willis, and C. N. Zayas. 2019. "Pervasive human-driven decline of life on Earth points to the need for transformative change." *Science* 366 (6471):eaax3100. doi: 10.1126/science.aax3100.

143. DeGregori, J., and N. Eldredge. 2019. "Parallel causation in oncogenic and anthropogenic degradation and extinction." *Biol Theor* 15:12–24. doi: 10.1007/s13752-019-00331-9.

144. Swinburn, B. A., V. I. Kraak, S. Allender, V. J. Atkins, P. I. Baker, J. R. Bogard, H. Brinsden, A. Calvillo, O. De Schutter, R. Devarajan, M. Ezzati, S. Friel, S. Goenka, R. A. Hammond, G. Hastings, C. Hawkes, M. Herrero, P. S. Hovmand, M. Howden, L. M. Jaacks, A. B. Kapetanaki, M. Kasman, H. V. Kuhnlein, S. K. Kumanyika, B. Larijani, T. Lobstein, M. W. Long, V. K. R. Matsudo, S. D. H. Mills, G. Morgan, A. Morshed, P. M. Nece, A. Pan, D. W. Patterson, G. Sacks, M. Shekar, G. L. Simmons, W. Smit, A. Tootee, S. Vandevijvere, W. E. Waterlander, L. Wolfenden, and W. H. Dietz. 2019. "The global syndemic of obesity, undernutrition, and climate change: The Lancet Commission report." *Lancet* 393 (10173):791–846. doi: 10.1016/S0140–6736(18)32822–8.

145. Quail, D. F., and A. J. Dannenberg. 2019. "The obese adipose tissue microenvironment in cancer development and progression." *Nat Rev Endocrinol* 15 (3):139–54. doi: 10.1038/s41574-018-0126-x.

146. Swinburn, B. A., V. I. Kraak, S. Allender, V. J. Atkins, P. I. Baker, J. R. Bogard, H. Brinsden, A. Calvillo, O. De Schutter, R. Devarajan, M. Ezzati, S. Friel, S. Goenka, R. A. Hammond, G. Hastings, C. Hawkes, M. Herrero, P. S. Hovmand, M. Howden, L. M. Jaacks, A. B. Kapetanaki, M. Kasman, H. V. Kuhnlein, S. K. Kumanyika, B. Larijani, T. Lobstein, M. W. Long, V. K. R. Matsudo, S. D. H. Mills, G. Morgan, A. Morshed, P. M. Nece, A. Pan, D. W. Patterson, G. Sacks, M. Shekar, G. L. Simmons, W. Smit, A. Tootee, S. Vandevijvere, W. E. Waterlander, L. Wolfenden, and W. H. Dietz. 2019. "The global syndemic of obesity, undernutrition, and climate change: The Lancet Commission report." *Lancet* 393 (10173):791–846. doi: 10.1016/S0140–6736(18)32822–8. Nugent, R. 2019. "Rethinking systems to reverse the global syndemic." *Lancet* 393 (10173):726–78. doi: 10.1016/S0140–6736(18)33243–4.

147. Nguyen, K. H., S. A. Glantz, C. N. Palmer, and L. A. Schmidt. 2019. "Tobacco industry involvement in children's sugary drinks market." *BMJ* 364:l736. doi: 10.1136/bmj.l736.

148. https://ec.europa.eu/info/sites/info/files/president-elect-speech-original_en.pdf.

149. Diaz, S., J. Settele, E. S. Brondizio, H. T. Ngo, J. Agard, A. Arneth, P. Balvanera, K. A. Brauman, S. H. M. Butchart, K. M. A. Chan, L. A. Garibaldi, K. Ichii, J. Liu, S. M. Subramanian, G. F. Midgley, P. Miloslavich, Z. Molnar, D. Obura, A. Pfaff, S. Polasky, A. Purvis, J. Razzaque, B. Reyers, R. R. Chowdhury, Y. J. Shin, I. Visseren-Hamakers, K. J. Willis, and C. N. Zayas. 2019. "Pervasive human-driven decline of life on Earth points to the need for transformative change." *Science* 366 (6471):eaax3100. doi: 10.1126/science.aax3100. Swinburn, B. A., V. I. Kraak, S. Allender, V. J. Atkins, P. I. Baker, J. R. Bogard, H. Brinsden, A. Calvillo, O. De Schutter, R. Devarajan, M. Ezzati, S. Friel, S. Goenka, R. A. Hammond, G. Hastings, C. Hawkes, M. Herrero, P. S. Hovmand, M. Howden, L. M. Jaacks, A. B. Kapetanaki, M. Kasman, H. V. Kuhnlein, S. K. Kumanyika, B.

Larijani, T. Lobstein, M. W. Long, V. K. R. Matsudo, S. D. H. Mills, G. Morgan, A. Morshed, P. M. Nece, A. Pan, D. W. Patterson, G. Sacks, M. Shekar, G. L. Simmons, W. Smit, A. Tootee, S. Vandevijvere, W. E. Waterlander, L. Wolfenden, and W. H. Dietz. 2019. "The global syndemic of obesity, undernutrition, and climate change: The Lancet Commission report." *Lancet* 393 (10173):791–846. doi: 10.1016/S0140–6736(18)32822–8.

150. Wilson, B. E., S. Jacob, M. L. Yap, J. Ferlay, F. Bray, and M. B. Barton. 2019. "Estimates of global chemotherapy demands and corresponding physician workforce requirements for 2018 and 2040: a population-based study." *Lancet Oncol* 20 (6):769–80. doi: 10.1016/S1470–2045(19)30163–9. Bray, F., J. Ferlay, I. Soerjomataram, R. L. Siegel, L. A. Torre, and A. Jemal. 2018. "Global cancer statistics 2018: GLOBOCAN estimates of incidence and mortality worldwide for 36 cancers in 185 countries." *CA Cancer J Clin* 68 (6):394–424. doi: 10.3322/caac.21492. The Lancet. 2018. "GLOBOCAN 2018: counting the toll of cancer." *Lancet* 392 (10152):985. doi: 10.1016/S0140 –6736(18)32252–9.

151. Sarto-Jackson, I. 2018. "Time for a change: topical amendments to the medical model of disease." *Biol Theor* 13:29. doi: 10.1007/s13752-017-0289-z. Kauffman, S. I., C.I. Hill, L. Hood, and S. Huang. 2014. "Transforming medicine: a manifesto." *Sci Am worldVIEW*. https://medecine-integree.com/wp-content/uploads/2018 /06/Transforming-Medicine_-A-Manifesto-_-worldVIEW_Kauffman.pdf.

152. Kruk, M. E., A. D. Gage, C. Arsenault, K. Jordan, H. H. Leslie, S. Roder-DeWan, O. Adeyi, P. Barker, B. Daelmans, S. V. Doubova, M. English, E. G. Elorrio, F. Guanais, O. Gureje, L. R. Hirschhorn, L. Jiang, E. Kelley, E. T. Lemango, J. Liljestrand, A. Malata, T. Marchant, M. P. Matsoso, J. G. Meara, M. Mohanan, Y. Ndiaye, O. F. Norheim, K. S. Reddy, A. K. Rowe, J. A. Salomon, G. Thapa, N. A. Y. Twum-Danso, and M. Pate. 2018. "High-quality health systems in the Sustainable Development Goals era: time for a revolution." *Lancet Glob Health* 6 (11):e1196-252. doi: 10.1016/S2214–109X(18)30386–3.

153. Garcia, P. J. 2019. "Corruption in global health: the open secret." *Lancet* 394 (10214):2119–24. doi: 10.1016 /S0140–6736(19)32527–9.

154. Swinburn, B. A., V. I. Kraak, S. Allender, V. J. Atkins, P. I. Baker, J. R. Bogard, H. Brinsden, A. Calvillo, O. De Schutter, R. Devarajan, M. Ezzati, S. Friel, S. Goenka, R. A. Hammond, G. Hastings, C. Hawkes, M. Herrero, P. S. Hovmand, M. Howden, L. M. Jaacks, A. B. Kapetanaki, M. Kasman, H. V. Kuhnlein, S. K. Kumanyika, B. Larijani, T. Lobstein, M. W. Long, V. K. R. Matsudo, S. D. H. Mills, G. Morgan, A. Morshed, P. M. Nece, A. Pan, D. W. Patterson, G. Sacks, M. Shekar, G. L. Simmons, W. Smit, A. Tootee, S. Vandevijvere, W. E. Waterlander, L. Wolfenden, and W. H. Dietz. 2019. "The global syndemic of obesity, undernutrition, and climate change: The Lancet Commission report." *Lancet* 393 (10173):791–846. doi: 10.1016/S0140–6736(18)32822–8.

155. Diaz, S., J. Settele, E. S. Brondizio, H. T. Ngo, J. Agard, A. Arneth, P. Balvanera, K. A. Brauman, S. H. M. Butchart, K. M. A. Chan, L. A. Garibaldi, K. Ichii, J. Liu, S. M. Subramanian, G. F. Midgley, P. Miloslavich, Z. Molnar, D. Obura, A. Pfaff, S. Polasky, A. Purvis, J. Razzaque, B. Reyers, R. R. Chowdhury, Y. J. Shin, I. Visseren-Hamakers, K. J. Willis, and C. N. Zayas. 2019. "Pervasive human-driven decline of life on Earth points to the need for transformative change." *Science* 366 (6471):eaax3100. doi: 10.1126/science.aax3100. DeGregori, J., and N. Eldredge. 2019. "Parallel causation in oncogenic and anthropogenic degradation and extinction." *Biol Theor* 15:1 2–24. doi: 10.1007/s13752-019-00331-9. Stokstad, E. 2019. "Nitrogen crisis threatens Dutch environment-and economy." *Science* 366 (6470):1180–81. doi: 10.1126/science.366.6470.1180. https://ec.europa .eu/info/sites/info/files/president-elect-speech-original_en.pdf.

156. Lancaster, A. K., A. E. Thessen, and A. Virapongse. 2018. "A new paradigm for the scientific enterprise: nurturing the ecosystem." *F1000Res* 7:803. doi: 10.12688/f1000research.15078.1. Alberts, B., M. W. Kirschner, S. Tilghman, and H. Varmus. 2014. "Rescuing US biomedical research from its systemic flaws." *Proc Natl Acad Sci USA* 111 (16):5773–77. doi: 10.1073/pnas.1404402111. Chapman, C. A., J. C. Bicca-Marques, S. Calvignac-Spencer, P. Fan, P. J. Fashing, J. Gogarten, S. Guo, C. A. Hemingway, F. Leendertz, B. Li, I. Matsuda, R. Hou, J. C. Serio-Silva, and N. Chr Stenseth. 2019. "Games academics play and their consequences: how authorship, h-index and journal impact factors are shaping the future of academia." *Proc Biol Sci* 286 (1916):20192047. doi: 10.1098/rspb.2019.2047.

157. Waddington, C. H. 1959. "Behavior and evolution." *Science* 129:203.

158. Dotto, G. P. 2020. "Conjectures, refutations and the search for truths: science, symbolic truths and the devil." *EMBO Rep* 21:e49924. doi: 10.15252/embr.201949924.

159. Tinbergen, N. 1978. *Animals and Behavior* (Time-Life Books, ISBN 0809438917).

160. Goldin, I. 2017. "The second Renaissance." *Nature* 550 (7676):327–29. doi: 10.1038/550327a.

161. Aykut, B., S. Pushalkar, R. Chen, Q. Li, R. Abengozar, J. I. Kim, S. A. Shadaloey, D. Wu, P. Preiss, N. Verma, Y. Guo, A. Saxena, M. Vardhan, B. Diskin, W. Wang, J. Leinwand, E. Kurz, J. A. Kochen Rossi, M.

Hundeyin, C. Zambrinis, X. Li, D. Saxena, and G. Miller. 2019. "The fungal mycobiome promotes pancreatic oncogenesis via activation of MBL." *Nature* 574 (7777):264–67. doi: 10.1038/s41586-019-1608-2.

162. Hunter, P. 2017. "The role of biology in global climate change." *Embo Rep* 18 (5):673–76. doi: 10.15252 /embr.201744260.

163. Crowther, M. D., G. Dolton, M. Legut, M. E. Caillaud, A. Lloyd, M. Attaf, S. A. E. Galloway, C. Rius, C. P. Farrell, B. Szomolay, A. Ager, A. L. Parker, A. Fuller, M. Donia, J. McCluskey, J. Rossjohn, I. M. Svane, J. D. Phillips, and A. K. Sewell. 2020. "Genome-wide CRISPR-Cas9 screening reveals ubiquitous T cell cancer targeting via the monomorphic MHC class I-related protein MR1." *Nat Immunol* 21 (2):178–85. doi: 10.1038 /s41590-019-0578-8.

164. Sontheimer-Phelps, A., B. A. Hassell, and D. E. Ingber. 2019. "Modelling cancer in microfluidic human organs-on-chips." *Nat Rev Cancer* 19 (2):65–81. doi: 10.1038/s41568-018-0104-6.

165. FitzGerald, G., D. Botstein, R. Califf, R. Collins, K. Peters, N. Van Bruggen, and D. Rader. 2018. "The future of humans as model organisms." *Science* 361 (6402):552–53. doi: 10.1126/science.aau7779.

166. Hyun, I., M. Munsie, M. F. Pera, N. C. Rivron, and J. Rossant. 2020. "Toward guidelines for research on human embryo models formed from stem cells." *Stem Cell Reports* 14 (2): 169–74. doi: 10.1016/j.stemcr.2019 .12.008.

167. Duflo, E., and A. Banerjee. 2017. *Handbook of Field Experiments* (Volume 2, Elsevier, ISBN 44640118).

168. Pemovska, T., M. Kontro, B. Yadav, H. Edgren, S. Eldfors, A. Szwajda, H. Almusa, M. M. Bespalov, P. Ellonen, E. Elonen, B. T. Gjertsen, R. Karjalainen, E. Kulesskiy, S. Lagstrom, A. Lehto, M. Lepisto, T. Lundan, M. M. Majumder, J. M. Marti, P. Mattila, A. Murumagi, S. Mustjoki, A. Palva, A. Parsons, T. Pirttinen, M. E. Ramet, M. Suvela, L. Turunen, I. Vastrik, M. Wolf, J. Knowles, T. Aittokallio, C. A. Heckman, K. Porkka, O. Kallioniemi, and K. Wennerberg. 2013. "Individualized systems medicine strategy to tailor treatments for patients with chemorefractory acute myeloid leukemia." *Cancer Discov* 3 (12):1416–29. doi: 10.1158/2159–8290.CD-13–0350. Kodack, D. P., A. F. Farago, A. Dastur, M. A. Held, L. Dardaei, L. Friboulet, F. von Flotow, L. J. Damon, D. Lee, M. Parks, R. Dicecca, M. Greenberg, K. E. Kattermann, A. K. Riley, F. J. Fintelmann, C. Rizzo, Z. Piotrowska, A. T. Shaw, J. F. Gainor, L. V. Sequist, M. J. Niederst, J. A. Engelman, and C. H. Benes. 2017. "Primary patient-derived cancer cells and their potential for personalized cancer patient care." *Cell Rep* 21 (11):3298–309. doi: 10.1016/j.celrep.2017.11.051. Snijder, B., G. I. Vladimer, N. Krall, K. Miura, A. S. Schmolke, C. Kornauth, O. Lopez de la Fuente, H. S. Choi, E. van der Kouwe, S. Gultekin, L. Kazianka, J. W. Bigenzahn, G. Hoermann, N. Prutsch, O. Merkel, A. Ringler, M. Sabler, G. Jeryczynski, M. E. Mayerhoefer, I. Simonitsch-Klupp, K. Ocko, F. Felberbauer, L. Mullauer, G. W. Prager, B. Korkmaz, L. Kenner, W. R. Sperr, R. Kralovics, H. Gisslinger, P. Valent, S. Kubicek, U. Jager, P. B. Staber, and G. Superti-Furga. 2017. "Image-based ex-vivo drug screening for patients with aggressive haematological malignancies: interim results from a single-arm, open-label, pilot study." *Lancet Haematol* 4 (12):e595-e606. doi: 10.1016/S2352–3026(17)30208–9. Lee, J. K., Z. Liu, J. K. Sa, S. Shin, J. Wang, M. Bordyuh, H. J. Cho, O. Elliott, T. Chu, S. W. Choi, D. I. S. Rosenbloom, I. H. Lee, Y. J. Shin, H. J. Kang, D. Kim, S. Y. Kim, M. H. Sim, J. Kim, T. Lee, Y. J. Seo, H. Shin, M. Lee, S. H. Kim, Y. J. Kwon, J. W. Oh, M. Song, M. Kim, D. S. Kong, J. W. Choi, H. J. Seol, J. I. Lee, S. T. Kim, J. O. Park, K. M. Kim, S. Y. Song, J. W. Lee, H. C. Kim, J. E. Lee, M. G. Choi, S. W. Seo, Y. M. Shim, J. I. Zo, B. C. Jeong, Y. Yoon, G. H. Ryu, N. K. D. Kim, J. S. Bae, W. Y. Park, J. Lee, R. G. W. Verhaak, A. Iavarone, J. Lee, R. Rabadan, and D. H. Nam. 2018. "Pharmacogenomic landscape of patient-derived tumor cells informs precision oncology therapy." *Nat Genet* 50 (10):1399–411. doi: 10.1038/s41588-018-0209-6.

169. Maley, C. C., A. Aktipis, T. A. Graham, A. Sottoriva, A. M. Boddy, M. Janiszewska, A. S. Silva, M. Gerlinger, Y. Yuan, K. J. Pienta, K. S. Anderson, R. Gatenby, C. Swanton, D. Posada, C. I. Wu, J. D. Schiffman, E. S. Hwang, K. Polyak, A. R. A. Anderson, J. S. Brown, M. Greaves, and D. Shibata. 2017. "Classifying the evolutionary and ecological features of neoplasms." *Nat Rev Cancer* 17 (10):605–19. doi: 10.1038/nrc.2017.69. West, J., L. You, J. Zhang, R. A. Gatenby, J. S. Brown, P. K. Newton, and A. R. Anderson. 2020. "Towards multi-drug adaptive therapy." *Cancer Res* 80:1578–89. doi: 10.1158/0008–5472.CAN-19–2669.

170. Zhang, J., J. J. Cunningham, J. S. Brown, and R. A. Gatenby. 2017. "Integrating evolutionary dynamics into treatment of metastatic castrate-resistant prostate cancer." *Nat Commun* 8 (1):1816. doi: 10.1038/s41467-017-01968-5. West, J. B., M. N. Dinh, J. S. Brown, J. Zhang, A. R. Anderson, and R. A. Gatenby. 2019. "Multidrug cancer therapy in metastatic castrate-resistant prostate cancer: an evolution-based strategy." *Clin Cancer Res* 25 (14):4413–21. doi: 10.1158/1078–0432.CCR-19–0006.

V WHAT NEXT?

Bernhard Strauss, Marta Bertolaso, Ingemar Ernberg, and Mina J. Bissell

Calling for a revised/extended causal paradigm for cancer research and providing some empirical evidence to support it is one thing; translating it into a novel scientific practice is another. We are well aware that volumes like this with a focus on conceptual change might provoke a "Yes, but ..." response and possibly even "So, what ... ?" Well, then, what should we *do* next?

Each one of the authors has proposed some next steps as envisaged from her or his area of expertise and the content presented in their chapters. From these, and other insights that have emerged over the past two decades, a number of concrete requirements can be identified that need to be addressed in order to move cancer research into new directions. If acted upon, they will prove to be very productive for all of cancer research, if not urgently necessary to improve and save lives of cancer patients. These points can be grouped roughly under the following two overarching themes, and we are aware they will be incomplete:

1) **Education/cross-disciplinary collaboration**

   • *Teaching/language.* As with all scientific innovation, it is important that new ways of thinking about cancer are taught to the next generation of scientists while they are tested in experimental research practice. This also requires a new language that matches the epistemic structure of the new conceptual framework. It also means that "old" language should be discouraged, and it would be important that reviewers of funding bodies and scientific journals and their editors wield their red pens when sentences such as "Cancer is a genetic disease" or "Cancer is caused by mutations in oncogenes" are written as if they were fundamental laws comparable to the laws of gravity. Sentences like these in scientific articles look today more like an attempt to avoid any real discussion in the introduction section, so that one can move on quickly to the technical part of the research. As a scientific community, we are not more immune to the self-hypnotizing effects of language than any other community that spends a lot of time in their respective echo chambers.

   • *Cross-disciplinary collaborations.* However, what the past two decades made abundantly clear is the fact that all of the biosciences as well as medicine have greatly benefited from a broadening of interdisciplinary collaborations. We are entering an era that demands "team science." In particular, input from physicists, mathematicians, engineers, and network scientists has generated a lot of stimulation in basic biology as well as in cancer research (e.g., through initiatives such as the Physical Sciences-Oncology Network of the National Cancer Institute). The idea that solving the cancer problem requires expertise from different epistemic thinking cultures must be further pursued to make sure that cancer scientists understand that working on a complex problem might require the unexpected embrace of nonlinear and probabilistic processes. How to analyze these is usually not routinely taught in a biology or medicine course.

• *Learning from existing debates.* Once embedded in a broader conceptual framework of complex systems, cancer becomes of course just another biological phenomenon that is open to the same old reductionist/antireductionist debates that are very well discussed in fields such as basic biology, evolutionary developmental biology, ecology, or evolutionary theory. Once we accept that, maybe a lot can be learned from such debates, for example, regarding the determination of normal organismic phenotypes through the interplay of physical constraints, gene expression, and biophysical mechanisms of self-organization. However, to train and maintain a cross-disciplinary mindset within a scientific community requires continuous support through dedicated institutions and initiatives that keep that spirit alive and growing.

• At the moment, we can see just the beginnings of such a cross-disciplinary, collaborative culture in the cancer field. And setbacks are common; for instance, funding and support of the abovementioned initiative that fostered the collaboration between cancer scientists and physicists have meanwhile dried up, and former participants are under the impression that its momentum has fizzled out (not a physics term).

2) Translating novel conceptual insights into experimental and clinical research practice

• *Applying more suitable models to explain observational data.* Many empirical observations in the clinic could be better understood by applying principles from complex systems theory, but they would also require experimental validation before translation into therapeutic approaches grounded on these principles. These include the fact that early and late-stage cancers are very different and need to be studied with very different model systems and approaches to treatment. Closely associated with this fact is the finding that metastasis needs to be targeted as early as first detection of a primary tumor in order to save more lives. The current model systems that mostly use very late-stage tumor phenotypes to study cancer and test treatment modalities are not very helpful in understanding the systemic aspects of cancer. Currently, very few resources and therefore innovation are available to study, prevent, and treat very early stages of the disease.

• *Causally relevant factors above the single-cell level.* In basic biology, (tissue) orga-
nizing principles, such as the role of physical forces, self-organizational principles of
cellular life, and higher-order integrative mechanisms, such as bioelectric phenomena
that generate and maintain tissue phenotype have been studied for a long time. Experimentally testing and exploring causal factors relevant for the malignant phenotype
above the single-cell level (representing new "mesoscale constructs") is, however, still
not part of the current cancer research agenda.

• *Validating concepts beyond the single cancer cell.* Although it is very likely that larger
groups of cells, and not just the one mythological, single malignant cancer cell, are
involved in triggering carcinogenesis, we currently have no good experimental framework to study such other units of causal relevance. It has been suggested in the past to
apply, for example, the field concept from physics to organisms in order to define functional units above the single-cell level (as for example in the "field cancerization"
concept). Although conceptually very appealing, and supported by experimental data,
an integrated theory of how these field effects would causally operate, or what their
constituent "forces" are, is still lacking. It is possibly the "cancer niche" concept that
comes close to this idea and has recently gained wider acceptance.

• *Medium- and long-range interactions.* Given that most carcinogenic stimuli affect
initially the whole organism, it is also likely that cancer starts with a systemic damage
that affects many regulatory pathways distant from the organ that will later carry the
first detected primary tumor. We have currently no experimental tools to probe these
long-ranging interactions across levels of organismic organization and over long spans
of organismic lifetime.

• *Patient treatment data for understanding tumor biology.* One neglected resource that
could provide insights into such interactions are the actual tumors in patients and their
response to treatment in real time. We need to make better use of a large body of data
on treatment responders and in particular nonresponders and patients whose condition
worsened as a result of treatment. Many potentially new insights into human tumor
biology could be found right there, when tumors are still part of the organismic system.

• *Investigating the "healthy" tissue state.* We have currently no clear understanding
of the required set of long-range interactions within the body that need to be continuously maintained in order to define the "healthy" state of a tissue. More experimental research into these questions is badly needed, possibly requiring a revival of
"old-fashioned" physiology experimentation that studies the interactions between
larger units of an organism. Possibly data from large cohorts of people taking part
in private-sector P4 medicine services might deliver results on this matter. Perhaps
the rapidly growing field of organoid research and bioengineering attempts to build
organs in a dish might help to uncover some of these fundamentals of healthy tissue
maintenance.

• *Defining cancer susceptibility in tissues.* Due to this lack of knowledge on what the health-defining tissue factors are, we also struggle to understand cancer as a consequence of a breakdown of these tissue/organism intrinsic maintenance and defense mechanisms. We therefore still search for the "active" trigger of malignancy within a single cell. (And, of course, these can be observed in engineered experimental systems, but their suitability as a model for human cancers is increasingly doubted.)

• *Tumors are not "chaotic," random cell masses.* Tumors grow differently in different tissues. More detailed research into how tumors actually grow in three dimensions could help reveal "units" of tumor growth, in a similar fashion as normal tissue growth and regeneration always involve groups of different cell types that need to interact in a coordinated fashion. Once we better understand how groups of tumor cells interact with each other and the tumor microenvironment in order to promote malignancy, disrupting these interactions should be an entry point for treatment approaches.

• *Going beyond single-target treatments.* We know now that throughout carcinogenesis, groups of cancer cells continuously use a number of feedback loops, in both intra- and intercellular interactions. For instance, the specialized local metabolism of cancer cells and the physical properties of the microenvironment and the local as well as long-range inflammatory signaling continuously modify each other. This also made clear that single-shot approaches that aim to treat cancer by "fixing" just one of these aspects of malignancy are mostly failing as would be expected if one considers first principles of complex dynamical systems behavior. Multicomponent combination drugs have long been known to be more effective, and "adaptive therapies" are just being tested very recently. Their success, however, will depend on which causal paradigm they are based on.

In summary, we hope to have made the point throughout this volume that we are by no means "almost there" in our understanding of cancer and in our causal rationalizations of therapeutic approaches. It is more likely that we are "just approaching" a time when we are well enough equipped conceptually and technically to finally get down to business in our attempts to understand and cure cancer. In this endeavor, an open mind and new conceptual frameworks will be essential for success.

Epilogue

Over a hundred years ago, in 1914, Theodor Boveri published his incredibly far-sighted monograph, *On the Origins of Malignant Tumours*, in which he purely by deep deductive reasoning formulated almost all the major conceptual ideas that have emerged over the following century, shaping cancer research to a greater or lesser extent—mostly without ever being referenced. From detailed experimental observations in sea urchin embryos, he proposed that cancer arises due to chromosomal aberrations caused by rare defective cell divisions. Once Boveri had been rediscovered by molecular biologists half a century later, he has ever since been claimed as the father of the idea of the somatic mutation concept that has dominated the past fifty years of cancer research. What a closer reading of his monograph reveals, however, is that Boveri was acutely aware of the importance of the organismic context, as well as the systemic interactions "between the individual cell and the organism as a whole," and that the chromosomal aberrations he proposed as causal agents of carcinogenesis can only play that role when the communication between the cell and its environment is disrupted first—in many ways, perhaps the first systems view of cancer causation explicitly formulated (for relevant quotes from his monograph, see chapter 9, this volume). Boveri was also very aware of the epistemological consequences and challenges of introducing a completely novel paradigm to a science community that had a very concrete practical problem to solve, as is still the case today. He knew that novel concepts are crucial for progress in the experimental sciences, as he writes in the penultimate sentence of the monograph: *"For in this field, as in any other, many important phenomena remain unobserved despite the most assiduous investigation because they are not anticipated by any of our current concepts."*

We are optimistic that it will not take another hundred years until this insight will lead to transformative innovation in cancer research to the greater benefit of patients.

Contributors

David Basanta
H. Lee Moffitt Cancer Centre and
Research Institute

Marta Bertolaso
Institute of Philosophy of Scientific and
Technological Practice

Mina J. Bissell
Lawrence Berkeley National Laboratory

Kimberly J. Bussey
The New College, Arizona State University

Luca Vincenzo Cappelli
Weill Cornell Medicine

Sapienza University of Rome

Peter Csermely
Semmelweis University

Paul C. W. Davies
Arizona State University

Ingemar Ernberg
Karolinska Institutet

Sui Huang
Institute for Systems Biology

Giorgio Inghirami
Weill Cornell Medicine

Christoph A. Klein
University of Regensburg

Courtney König
University of Regensburg

Andriy Marusyk
H. Lee Moffitt Cancer Centre and
Research Institute

Thea Newman
SOLARAVUS

Larry Norton
Memorial Sloan Kettering Cancer Center

Roger Oria
University of California San Francisco,
San Francisco

Laxmi Parida
IBM

Jacques Pouysségur
Université Côte d'Azur

Kahn Rhrissorrakrai
IBM Research

Jacob Scott
Cleveland Clinic

Bernhard Strauss
University of Cambridge

Dhruv Thakar
University of California San Francisco

Emmy W. Verschuren
University of Helsinki

Valerie M. Weaver
University of California San Francisco

Liron Yoffe
Weill Cornell Medicine

Maša Ždralević
University of Montenegro

Index

Figures and tables are indicated by *f* and *t* respectively.